AF847585

Architecture Inspired by Nature

María Rosa Cervera Sardá
Elena-Codina Duşoiu • Tana Nicoleta Lascu
Editors

Architecture Inspired by Nature

Experimenting Bionics

Springer

Editors
María Rosa Cervera Sardá
Universidad de Alcalá
Alcalá de Henares, Spain

Elena-Codina Duşoiu
"Ion Mincu" University of Architecture
and Urban Planning
Bucharest, RO, Romania

Tana Nicoleta Lascu
"Ion Mincu" University of Architecture
and Urban Planning
Bucharest, RO, Romania

ISBN 978-3-031-33146-6 ISBN 978-3-031-33144-2 (eBook)
https://doi.org/10.1007/978-3-031-33144-2

This Springer imprint is published by the registered company Springer Nature Switzerland AG
The registered company address is: Gewerbestrasse 11, 6330 Cham, Switzerland

Paper in this product is recyclable.

Introduction

The application of bionics and detailed study of nature in architecture and design have not been widely approached yet, even if specialists seem to be convinced about its potential and diversity. The book comes to fill this gap, bringing together the opinion of specialists from several countries who are concerned with the field. An international multidisciplinary team was gradually developed around Professor Rosa Cervera, based at the School of Architecture of the University of Alcalá, Spain. This research is probably the most complete and diverse approach in the field. The book can be studied as a source of inspiration by architects, designers, engineers and so on and is also a good tool for students from all fields of visual arts, engineering, physics and many other domains, opening the way to amazing discoveries in the future.

Architecture Inspired by Nature. Experimenting Bionics is the result of some decades of research and experiments, realized by the authors individually or collaboratively. The book is conceived in five sections: the first four sections include the scientific contributions grouped into four different categories; the fifth and last section "Practicing Bionics. Experimental Workshops" contains a short selection of workshop results with the aim of illustrating the translation of bionic thinking into a new approach to design and architecture. The five sections of the book offer an overall view of the object of research from a correct understanding of nature through the tools of bionics, from general to particular and from simple to complex. The book proves how bionic thinking can be employed at various scales: territory and landscape, architecture and construction, interior and object design, construction materials – going into a micro and nano scale in the end.

The first section of the book, Learning from Natural Structures: The Basis for a Bionic Approach, constitutes a basis for the study of nature, mainly referring to the various categories of natural structures and some principles for their correct understanding: specific geometry, configuration, proportion and dimensions, physical properties etc. While it is proved that natural structures are by far more performant than the ones created by humans, it is an inborn impulse to study them and to apply to artificial constructions the criteria that have ensured the success in nature. The section starts with a creative contribution by Rosa Cervera and Iván Curiel, presenting how structural models from nature, both at a macro and micro scale, have been applied to architectural buildings after the industrial revolution and how this type of study can be pushed further and expanded. The next contribution refers to the application of bionics in the conception of tall buildings, by Mark Sarkisian, engineer and researcher with outstanding experience in the field, representing Skidmore, Owings & Merill – SOM. The next work consists in the analysis of large deployment structures, studied by Frei Otto, as a result of his observations of the spider's web and other natural structures, presented by Juan Maria Songel. Next, the section focuses on the more reduced scale of interior design and furniture, defining the study of four categories of natural principles to be considered, by Codina Duşoiu. The first section ends with research on increasing and diversifying possibilities of light perception in architectural space, taking inspiration from the accurate perception of light of various species of organisms present in nature, analyzed by Codina Duşoiu.

The second section, Establishing the Tools: Parametric and Biodigital Architectural Design, makes a further step into the methodology of dealing with bionics in design. The first contribution by Alberto T. Estevez and Yomna K. Abdallah is based on the generative biodigital design concept, which has been studied and practiced by the authors in various projects and experiments, in terms of forms, biomanufacturing, materials, tissues and even biocybernetics. The second one, by Mauro Costa, is an analysis of the cognitive revolution generated by the new media and the extended reality and their effects on human mind, with a complete change of paradigm in the interpretation of the relation with surrounding environment and nature. Ioseph Cabeza-Lainez follows with a personal interpretation of geometric surfaces derived from the conoid and proves the applicability of this family of forms inspired from nature in architecture and design. In the end of the section, Anca Vitcu presents an essay about how new-generation machine learning algorithms are not always a reliable tool, as they can distort a natural, authentic perception of the surrounding environment and its values.

The third section, Studying Scales in Territory: Bionics and Landscape, brings into discussion the concept of scale in design, mainly referring to the study of territory, landscape and sustainable urban planning. Tana Lascu presents an original perspective on the interpretation of landscape as a process, analyzed by using a transdisciplinary vision, including notions of fractal theory, semiotics and other related sciences. Henry Lazarte expounds the disruptive biotope concept, exemplifying it by the case of the Rimac River in Peru and explaining the level of contamination of this river and its neighboring area and its effects on the surrounding environment. Ivan Curiel moves further the concern about the environmental crisis on Earth, wondering whether sustainable design can be a solution for the continuity of life on our planet. Cerasella Crăciun and Codruţ Papina study the origins of the biophilia concept and its potential for future strategies in urban planning. Some modern and contemporary study cases are analyzed and compared (Portland, in the United States, Singapore and the European models of the *Garden City*. Additionally, the authors share conclusions from their own projects, strategically considering landscape at a territorial scale for the cities of Bucharest and Buzău, in Romania. The contribution which concludes the section, belonging to Ana Mohonea, brings an original and useful approach, to be applied both to architecture and urban planning, referring to a proposal of bionic criteria for sustainable labelling of buildings.

The fourth section, Researching on Materials: from Macro to Nano, focuses on the analysis of innovative materials applicable in architectural bionics and sustainable design, considering a variety of scales. All the contributions in this section are based on original laboratory substantiations, with direct and immediate applicability. Maria Segura, representing the company AlgaEnergy, Spain, explains the potential of the use of micro-algae as an industry for the future, as a source of energy in the fields of agriculture, food and feed, aquaculture, energy, pharmaceutical and cosmetics. The contribution explains the conditions necessary to the cultivation of micro-algae as well. Rosa Cervera and Maria Rosa Villalba share the experience of translating the process of cultivation of micro-algae to architectural design and its benefits, obtaining a zero-carbon print. The next three contributions propose innovative material inspired by bionic laws. The first proposal refers to nanocomponents for new constructions and technologies, by Hugo Varela, Gonzalo Barluenga, and Arnaud Perrot. The next one deals with PET residual recycled concrete and post-fire self-healing concrete, obtained through a bacteria encapsulation method by Ajitanshu Vedrtnam. The last contribution of the section remains in the same investigation area, referring to cement-based materials with self-modulating and self-sensing properties, by Javier Puentes, Irene Palomar, Gonzalo Barluenga, and Cynthia Guardia Martin.

The fifth section of the book, the last and the concluding one, brings a different perspective, being in fact a visual and experimental second part. The section is based on a selection of projects that were produced during the workshops dedicated to research of architectural bionics, organized by the University of Alcalá, in 2011, 2012, 2017, 2018, 2019, 2020, as well as at the "Ion Mincu" University of Architecture and Urban Planning in 2009, 2010, 2014, 2017, 2019, 2022, with the contribution of participating lecturers and students. All these events were generated by María Rosa Cervera Sardá, some of them with the contribution of Javier Pioz, Cristina Ochinciuc, Elena-Codina Duşoiu and Tana-Nicoleta Lascu. The projects are classified in eight categories, offering eight possible methods of investigation offered by the deep study of natural laws and principles: vertical structures, bioinspired forms, fluid topological compositions, design with fractal geometry, building with modules, recycling and self-healing structures, perception of light, lightweight structures.

The stimulation of continuous transversal mental representations leads the student to connect with what the school institutionally separates. In this case, knowledge is enriched through analogical metaphors that can return a more organic and complementary perception of the complex epistemological universe we are living in.

Using bionics as an interface between the natural and man-made world, the workshops have been a model of systemic and empathetic education about the real world, leading to new cycles of conceptual thinking that offer new challenges in the noosphere. Truely creative laboratories that discovered geometric patterns and bio-morphological laws, have generated a transversal perception able to understand complex potentialities and going beyond the limits of disciplinary didactics. It has also activated a type of evolutionary sensibility based on a holistic vision that allows us to understand the dynamism of the organic world, and to perceive human creation as an integral part of natural creation.

We believe this book will represent a useful tool for academics and professionals belonging to a variety of fields: architects, designers, engineers, biologists, physicians, etc., as well as for students from all fields of architecture and visual arts – design, media, etc. and for students in engineering, physics, biology, to name just a few areas. At the same time, we hope the book will provide a pleasant reading quality time for everybody who is interested in a deeper understanding of nature and in encountering solutions for respectful adaptation to life on our planet.

Alcalá de Henares, Spain — María Rosa Cervera Sardá
Bucharest, Romania — Elena-Codina Duşoiu
Bucharest, Romania — Tana Nicoleta Lascu

December 22, 2022

Foreword: Uncertainty. Physics and Structures of Life

After 400 years, physics has reached the limit of the study of simple and linear systems. Now, we are entering the new adventure of trying to understand complex structures and situations. Moreover, here we can learn from living systems.

In physics, we have many unsolved problems dealing with subjects like fluid turbulence, nonlinear waves, weather and climate, the structure of stars, and particularly the sun, and many others. All these systems are characterized by having many different parts, sizes and time scales in nonlinear interactions, and this is a characteristic of living structures.

The nonlinear interactions produce uncertainty in the time evolution of the physical systems and that is also a property of living structures that are anything but deterministic. Can we learn from living beings in these aspects?

When a fluid moves in a spatial domain with big scales, or moves with a high velocity, the fluid organizes itself in three-dimensional vortexes that decompose into smaller and smaller ones and combine into bigger systems, filling up the available space with individual structures of all available spatial scales. We are, until now, unable to explain this phenomenon. The motion is uncertain. As the atmosphere is a fluid moving in big space and time scales, and at times with very high velocities, its motion is turbulent, forming big vortexes that represent storms and in certain cases, hurricanes and typhoons, that move in an uncertain way.

The climate is the analysis of the weather on long time scales. In scales of centuries, the weather patterns change in a notable way. Around 800 AD, the average temperature in Europe increased by around 1.5°C, promoting the surge of population in the Scandinavian region, and massive migrations towards the south (the viking cruises). The reverse happened around 1650, when the average temperature decreased around 1.5°C all around the world, causing famines, revolutions and abandonment of residence sites, for instance in New Mexico. Climate change, and its evolution, is uncertain.

Uncertainty is the rule in the universe, although the human beings try hard to eliminate it. However, it is impossible, and it is much better to assume that it is, and will be, always present, and design our structures accepting this reality.

This assumption should force designs that provide alternative ways and solutions, and planning for many different possibilities.

For instance, urbanism. Cities are made new, or are reformed when existing, with the idea of a regular way of living, when the reality is that living in them has important chaotic elements. As this chaos is often not taken into account, there are no departments in municipal administrations that deal with alternative plans.

For instance, in January 2021, a big episode of snowing took place in Madrid. No plan was extant to deal with this possible occurrence, with the result that the city was paralyzed for ten days, with huge economic losses and a considerable disruption of social life.

Now, life acknowledges uncertainty. Bacteria mutate constantly. Insects produce hundreds of eggs, fishes, millions of them, because it is uncertain if anyone will fructify.

Inside trees and the bodies of big animals there are no traffic jams, although fluid motion is crucial for the life to keep on. Traffic inside the bodies is carried smoothly due to multiple branching. Walls surround human cities as a relic of warlike times, when attacks were at the level of the soil. Now the attacks are done by air, so walls are obsolete, but we have surrounded cities by the walls of circular highways with an extremely reduced number of entrances. Traffic jams are the norm in all cities of the world. There is neither branching nor capillarity.

To solve this problem, the only solution that city planners develop is to concentrate the people in saturated buses and trains, which are the contrary to the smooth circulation. No living system, be it vegetal or animal, save the human beings, concentrate its cells with a daily periodicity to produce the resources needed to live.

If we learn from the universe of plants, hierarchical cities with "a center" must be abandoned. Instead, we should distribute people and institutions in many highly connected nuclei, where the connections should be smooth instead of concentrated. This can be obtained by a system of moving platforms with increasing speeds, in a way that any walking person could jump from the street into the first platform and reach the central one in 4 or 5 jumps if she must travel long distances between centers of population.

Life is smooth and distributed. People inhabit today monster agglomerations that present no evident advantages and clear unsolved problems.

We must learn from living systems and their solutions to confront inescapable uncertainty.

University of Alcalá
Alcalá de Henares, Spain

Antonio Ruiz de Elvira

Acknowledgements

This book was made possible due to the research and experimental workshops on bionics and its connection with architecture carried out at the University of Alcalá de Henares (Spain), and at the "Ion Mincu" University of Architecture and Urban Planning (Bucharest, Romania) between 2009 and 2022. The results owe to the collaboration with all the professors and professionals involved, as well as with the institutions and companies to which they belong. We mention here especially Javier Pioz, who was part of the team in the first editions developed in Romania and Spain.

We would like to express our gratitude for the collaboration of Prof. em. Ph.D. Arch. Cristina Victoria Ochinciuc, former Director of the Doctoral School of IMUAU, one of the first promoters of academic cooperation between the Schools of Architecture of the Universities from Alcalá and Bucharest. We would also like to express our special thanks to the experts who were members of the scientific committee of the book. We mention here Andreea Anamaria Anghel, Lecturer, Ph.D. Arch. (Polytechnic University Timişoara, Romania). Subhes Bhattacharyya, Prof. Ph.D. MEng. (University of Surrey, UK); Angelica Stan, Prof. Ph.D. Arch., Director of the Council of Doctoral Studies (IMUAU, Bucharest, Romania). Pablo Millán Millán, Prof. Ph.D. Arch.; Santiago Quesada García, Prof. Ph.D. Arch.; and Inmaculada Rodríguez Cunill, Prof. Ph. D. Arch., from the University of Seville, Spain. Our appreciation also goes to the reviewers and translators who helped with editing of the final materials, mainly to Dana Lupu ("Ion N. Socolescu" College of Architecture and Public Works, Bucharest), Victoria-Elena Vlad, Ana Cervera Benito and Maria Mohonea. Furthermore, we would like to thank the people who helped with the layout of the last part of the book, namely to Iván Curiel and to all the students who participated in the different editions of the workshops and events we have organized together since 2009.

Contents

About the Editors

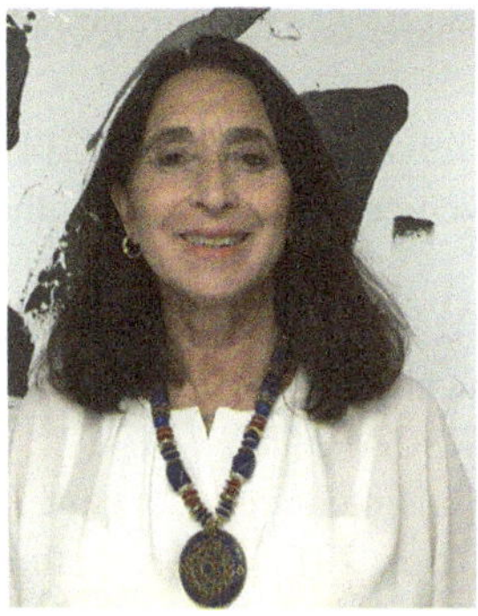

María Rosa Cervera Sardá Ph.D. Architect, Professor and former Dean of the School of Architecture at the University of Alcalá. Current Director of the Master's degree in Advanced Architecture and City Projects. She is the author of writings and books on architecture, among which we highlight: *Bionics, Biomimetics and Architecture* (2019); *Space and Time in Architectural Composition* (2018); *Recycling Mumbai. Re-envisioning the Slum* (2012); *Madrid, Recycled City* (2011); *Iron in 19th Century Madrid Architecture* (2006). A regular speaker in professional and academic circles, she has given lectures in Spain, China, India, USA, Bolivia, Peru, El Salvador, Venezuela, Romania, Italy, etc.

As an architect, Rosa Cervera has received several awards: the "Antonio Maura" award, the COAM award, the "Transfer of Knowledge" award, and has won several international architecture competitions.

Rosa Cervera is a pioneer in the research of Bionics, Biomimetics and the application of biological structures to innovative and efficient architecture and urban design. A direct result of these studies is the Self-Sustainable Vertical Garden City, Bionic Tower.

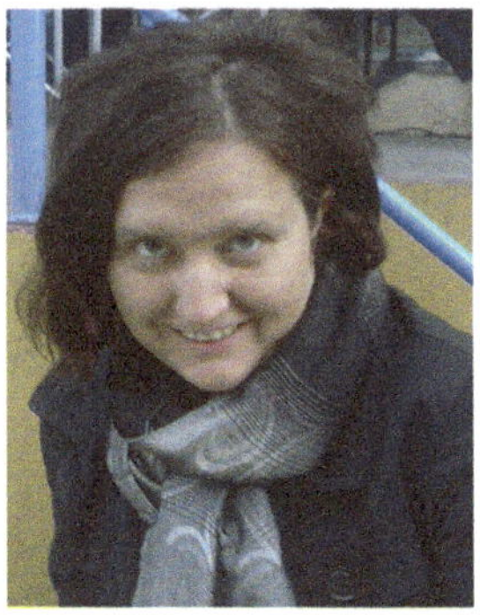

Elena-Codina Duşoiu is an Architect (graduated in 2000, IMUAU) and Professor within the "Ion Mincu" University of Architecture and Urban Planning from Bucharest, Romania. She also accomplished study periods within the Polytechnic University of Catalonia, Barcelona (Master in Restoration of Monuments, 2001) and Venice International University (*Dottorato di Eccelenza. Storia della Citta, dell'Architettura e del Restauro,* 2004). Her main fields of research are: rehabilitation and conversion of buildings (published books: *The Dynamics of the Sacred Space. The Influence of Function, 3 Breweries and Their Destiny*, "Ion Mincu" Publishing House, 2009), vernacular architecture and ecology (organizer of 12 editions of Spanish-Romanian workshops on vernacular architecture 2006-2018), design and bionics (research studies realized within the IMUAU and the Alcalá University, 2021–2022). She received various research grants and published about 100 scientific articles and 8 authored and co-authored books in the mentioned research fields. Outstanding activity as visiting professor in Spain, Italy, Greece, Czech Republic, Liechtenstein, Argentina etc. Several national and international prizes and nominations in architectural research and architectural design, owner of a personal architecture studio since 2003.

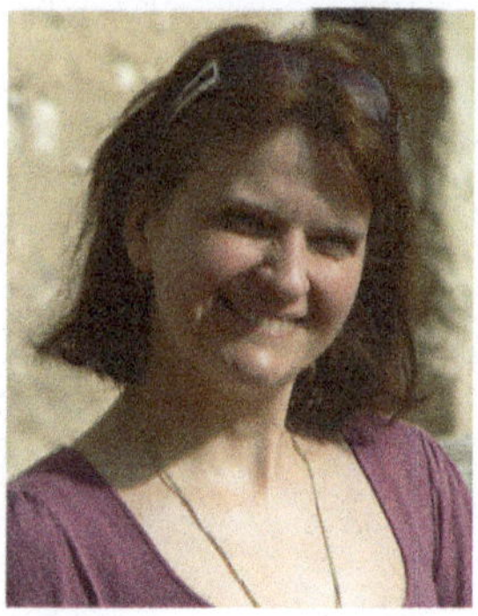

Tana Nicoleta Lascu Ph.D. Architect and Urban Planner, Lecturer at the Basics of Architectural Design Department, Faculty of Architecture of the "Ion Mincu" University of Architecture and Urban Planning from Bucharest.

Graduated at IMUAU in 1992 and having a postgraduate specialization in restoration and conservation of monuments and historic sites, she has been involved in several restoration projects in Romania, France, and Italy, thereafter developing more than 100 projects during the 7-year activity at Cornelis de Jong Architektenburo bna, Middenbeemster, Noord-Holland, the most significant in Urk, De Rijp, Marken and Beemster, elaborating the project for the listing of Beemster polder on UNESCO World Heritage List.

She was a Visiting Professor at the University of Liège (2014), the University of Architecture in Venice – IUAV, and the University of Alcalá (2021).

Since 2012, she has represented IMUAU within The Network of Universities for Studies and Education according to the European Landscape Convention – UNISCAPE.

Her doctoral thesis "Landscape as an Integrated Concept in the Sustainable Development", finalized in 2011, and her studies published in four books and over 20 international conference Proceedings, as well as the three research grants and more than 50 international curated exhibitions, seminars and workshops, emerged as a result of her constant interest in connecting architecture with landscape, within a transdisciplinary integrated approach.

Part I

Learning from Natural Structures: The Basis for a Bionic Approach

Bio-inspired Lightweight Structural Systems: Nature-Inspired Architecture

María Rosa Cervera Sardá and Iván Curiel Martínez

Abstract

The growing worldwide urbanisation trend, together with an equally global energy crisis, is forcing a new stance in building processes. The studies made to date conclude that the global contribution of energy consumption of the building sector has surpassed that of other industries. Nowadays, great care is taken regarding the reduction of operational energy consumption during the useful life of the building. However, looking into minimising the embedded or incorporated energy consumption is fairly recent. This relates to the energy that the making of the building has consumed right up to minute zero, that is, the very beginning of its useful life. The revolution 4.0 offers opportunities that allow for new investigations in the reduction of material during the building and structural processes in construction. And taking a closer look into the examples nature has to offer is the starting point. In the process of creating living forms, nature takes into account the relationship between mass and energy. Evolution in living organisms has been progressing towards a lightening up of mass to the point of making vacuum a constant in every step of the scale. And this is the very reason why air is a great ally for both the stability and the resistance of structures. Its great advantage is that air does not carry embodied energy. Some examples of naturally forming lightweight structures are revised in this chapter. The composition of a fragment of a tree trunk is taken here as a case study by delving into it through the computer modelling of its composition and geometry. Thanks to digital computing, it has been found that in a given tree trunk bulk, the presence of vacuum is quantified as 58% of the total. And with the tree being one of the slenderest natural shapes in the planet, of mighty resistance and longevity, this chapter opens up a line of investigation that establishes bridges with nature towards creating the architecture of building "as light as air".

Keywords

Vacuum/void · Structural micro-vacuum · Nature · Lightweight structure · Bio-inspired architecture · Tree trunk

Introduction: Towards a Reduction in the Embedded Energy

There is a growing global narrative of rampant urbanisation and an unstoppable expansion of constructions of all and any range and class, which, together with the fossil fuel energy crisis, forces a new positioning in the constructive processes. Since the first space voyages that took place in the 60s presented us with the view of a fragile blue planet, there has been a waking up of consciousness. The first green movements came into being aware of the necessity of preserving life on Earth, our shared home. With the sum of pressing environmental, energy, demographics, etc., issues, any and all efforts put into reducing the use of natural resources is a must. In this chapter, we will analyse how nature's morphogenesis tends to lighten up the weight of solid matter to a maximum.

And it manages so by utilising the air as the lead element, hence optimising the energy consumed. The objective is to build bridges with nature and learn from it so that it might lead us to find novel construction systems that are truly efficient.

An awareness has evolved over the last few decades of the twentieth century of the need to avoid any further squandering, and a solid theory has been developed over the operational energy or consumed energy of a building during the period of time comprising its useful life. Practice has followed theory first by the use of certificates of energy-saving systems that promote energy-efficient architecture, something that has been undertaken by firms on a voluntary basis

M. R. Cervera Sardá (✉) · I. C. Martínez
Universidad de Alcalá, Alcalá de Henares, Spain
e-mail: rosa.cervera@uah.es; ivan.curiel@edu.uah.es

M. R. Cervera Sarda et al. (eds.), *Architecture Inspired by Nature*, https://doi.org/10.1007/978-3-031-33144-2_1

and later on with the relatively recent Building Technical Codes dictated by governments and public administrations, which mandate that the totality of any new construction scheme optimises energy use.

But the interest in reducing embedded or embodied energy is nevertheless a rather recent affair. Embedded energy comprises the amount of energy a building has already consumed only just during construction up to the very start of its useful life: at minute zero. There is a requirement to revise materials, how these are produced, how they are transported and introduced in the building site, as well as a revision of all construction criteria. The studies made to date tell us that the global construction contribution to energy consumption has surpassed that of other sectors such as industry and transport (Pérez-Lombard et al., 2008). They also agree that embedded or embodied energy in a building of an average useful life of approximately 50 years can comprise up to 60% of the total, which is actually larger than the operational energy of the building depending on construction systems and typologies (Quispe Gamboa, 2016).

The birth of industry in the eighteenth century radically changed construction methods and the pre-industrial materials used. From building with little processed materials locally sourced such as soil, clay, straw, stone and wood, the industry shifted to using materials that require complex extraction and elaboration processes. Among those most used in contemporary construction and that consume the most embodied energy are iron, aluminium and plastics. In fact, each ton of steel consumes 35 MJ/kg (Cepeda Gutiérrez & Mardaras Larrañaga, 2004). If we add to that the knowledge of its high density, 7800 kg/m^3, it turns out that one steel cubic metre needs 273,000 MJ, or to put it another way, 75.8394 kWh. Even if steel is primarily used in linear rather than cubic-shaped elements, its density and therefore weight implies a large amount of embedded energy. Ranking directly opposite among those that consume less energy is wood. Midway on the scale, we find ceramics, plaster, cement and concrete, to name but a few of the most universally used. In the case of concrete, it is one of the lead materials in contemporary construction, and significantly more efficient than steel; production requires 1.1 MJ/kg or 2.775 MJ/m^3 of energy use (Cepeda Gutiérrez & Mardaras Larrañaga, 2004), though it is used in very large quantities particularly on the foundations and buried parts, so it has a high impact of the embodied energy of a building. Revising the literature on this subject, several studies agree that the structural part of a building, particularly when it includes underground storeys, might reach up to 60% of the total embedded energy (Cepeda Gutiérrez & Mardaras Larrañaga, 2004).

The opportunities offered by the revolution 4.0 and current advances in the investigations into biomaterials allow to further expand new research that looks to minimise the amount of material involved in the construction and structural processes by observing the examples nature has to offer as a starting point. The natural world takes into account the relationship between matter and energy in the process of creating living structures. The diverse materials responsible for the completion of natural textures require not only different amounts of energy spent per unit of mass but also the different proportions of such mass as determined by structural and functional requirements (Cervera, 2019). In the process of effort optimisation in the evolution of living organisms, the lessening of the weight of matter has been progressing in such a way that one of the lead materials utilised in forms and structures is actually air. The air is a great ally of structures for both stability and resistance, and it does not imply embodied energy. The benefit is obvious and indisputable. Throughout this chapter, we will revise a diversity of vegetal organisms and animal structures, focusing on analysing the tree trunk and delving into its process of systematically lightening up its structure. In the balance of matter and air and the geometry utilised to reach the vital requirements of stability and resilience lay the clues to a novel approach of constructive processes.

Bio-inspired Lightweight Structures: Precedents

The nature-inspired search for lightening up structures presents remarkable examples throughout history. Such is the case of Antoni Gaudí, the utmost visionary representative in the art of learning from nature and merging this knowledge with his architecture. Recently though, we might also refer to other relevant professionals who have studied alternatives to reduce the amount of construction material. Nature has been the starting point for Le Ricolais, Nervi, Torroja, Candela, Fuller and Frei Otto, among others.

The French engineer Robert le Ricolais (1894–1977) showed an interest in the internal structure of bones and their three-dimensional arrangement and was also interested in the complex structural geometry of radiolarians. These studies provided him with a knowledge that evolved into the experimental designs of lightweight structures with a reduction in the amount of solid matter. He went through a series of prototypes to explore the limits of optimisation of a structure with maximum light and minimum weight by utilising hollow forms and pentagonal and hexagonal geometries. He also looked into the functioning of the textures produced by soap, as well as into the tension generated by spiderwebs when stresses were put upon them, and applied this to tensegrity models (Juárez Chicote, 1996) (Figs. 1, 2, and 3).

Engineers such as Pier Luigi Nervi (1891–1979), Eduardo Torroja (1889–1961) and the architect Félix Candela (1910–1997) worked on optimising structures using containers and reinforced concrete ribbed shells inspired on folded struc-

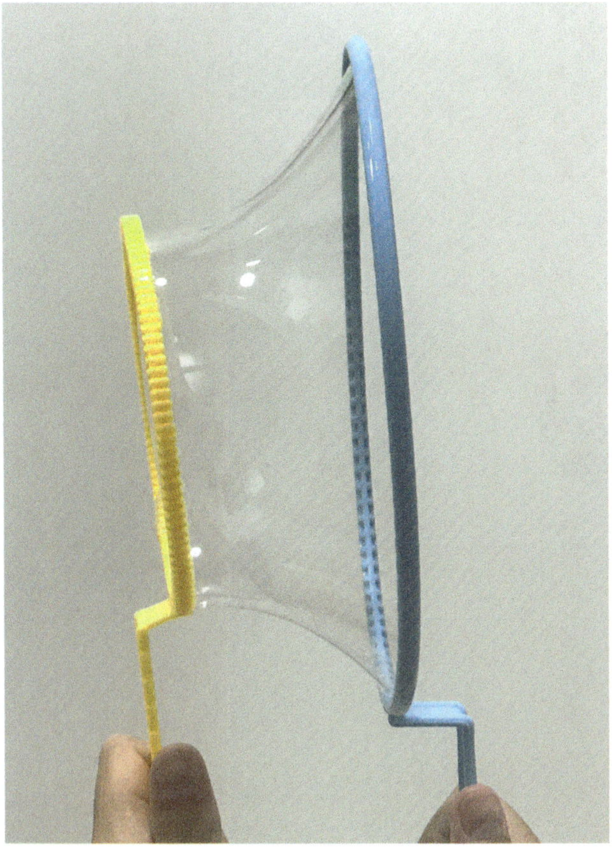

Fig. 1 Structural formations with soap and water from two skeletons. (Photo I. Curiel)

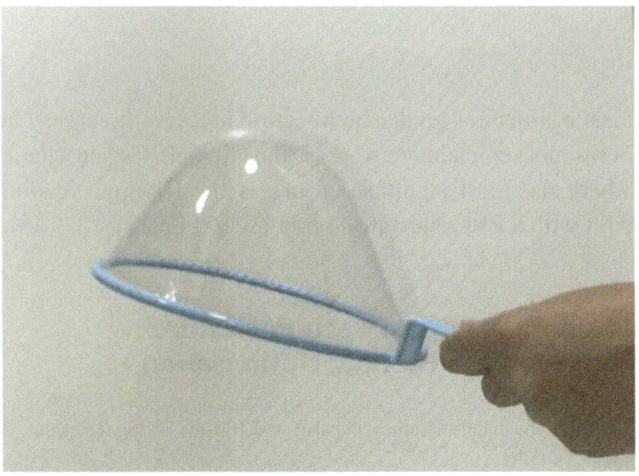

Fig. 2 Structural formations with soap and water from skeleton and movement. (Photo I. Curiel)

tures. The objective was to save on construction material in the zero load-bearing sections to achieve very thin structures, thanks to the ribs directed loads and the geometry used. Nervi's ribbed designs have great geometric strength and large span structures while minimising material on those areas where there is no structural activity going on. Some of his more widely known works are the Hangar of Orvieto,

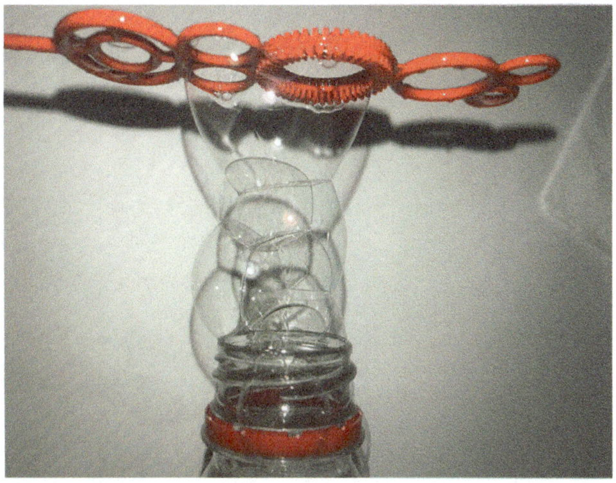

Fig. 3 Complex structural formations with soap and water from skeletons. (Photo I. Curiel)

which succeeds in covering a space of 45 m by 111.5 m with 1.1 m by 15 cm ribs, and another work, realised together with Vitellozzi, the Sports Arena in Rome that has the shape of a spheroid cap of 69 m diameter that is 12 cm thick (Fig. 4). One of the major reference sources he studied was the geometrical and structural organisation in the leaves of the water lily Victoria Amazónica (Fig. 5), as they served as inspiration for his reinforced concrete ribbed shapes (Granados, 2018).

Eduardo Torroja looked to cover large areas with the minimum possible thickness (Fig. 6) and concentrated his search in the exoskeletons in molluscs. Thanks to their curvature, it is not only the now optimised material that offers resistance but its geometry. One of his most distinguished works is the Hipódromo de la Zarzuela (Zarzuela's Racecourse) where he manages 13 m high cantilevered slab roofs with curved sheets of variable thickness that reach a mere 5 cm in the extremes (Fig. 7).

Meanwhile, Félix Candela also applied the double curvature structure. In this case, he used the paraboloid hyperbolic as the prevailing geometry in his works. An early example is the Capilla de Palmira (Palmira Chapel). It is a space exclusively covered by a type of geometry known as saddle surface or saddle roof in this case. It reaches 21 m in height and it is just 4 cm thick per structural sheet. There are many tree leaves that adopt this most characteristic shape to carry loads such as water or snow while they themselves being extremely lightweight (Figs. 8, 9, and 10).

Buckminster Fuller (1895–1983) was a visionary architect that studied the properties of materials and developed the so-called geodesic dome. He based it on the principles of tensegrity, using structures configured by forces of traction and compression, where the elements work as though they were a skeleton. So the geodesic dome is about a light and stable structure that can span large dimensions with the least

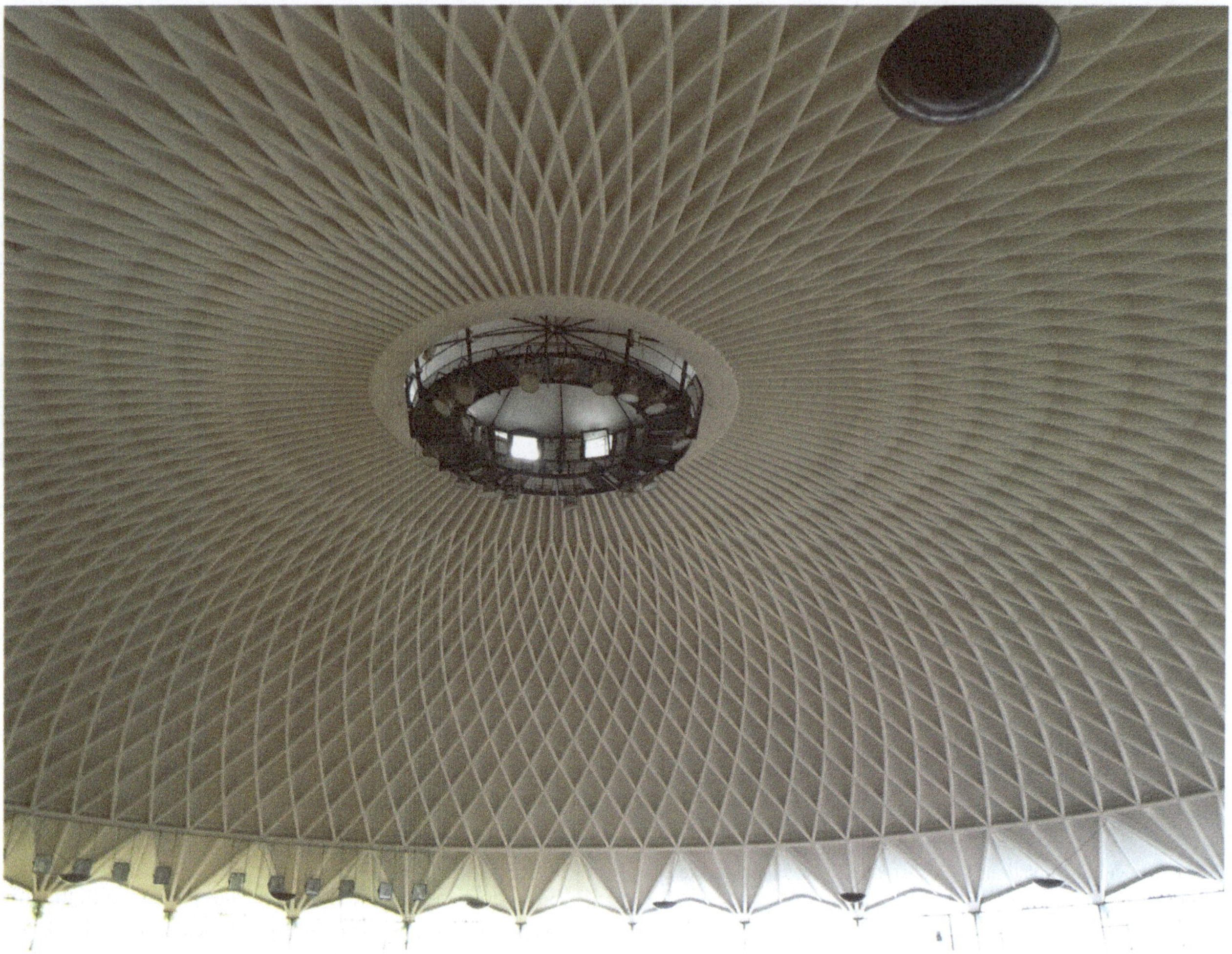

Fig. 4 Palazzetto dello Sport, Rome, Italy. (Photo R. Cervera)

amount of material involved. Even if Fuller's work was not inspired by nature, it so happened that later on, in 1985, a type of carbon molecule was found that had the same configuration as his dome does. This molecule was named "Fullereno" in his honor (Fig. 11). Equal to the structure produced by Buckminster Fuller, these are greatly resilient molecules, thanks to the configuration of their hexagonal and pentagonal unions between atoms as well as for being covalent bonds (Martínez, 2020).

Frei Otto (1925–2015) was an architect and engineer who contributed to the innovation of lightweight and extensible structures inspired by natural phenomena and biological forms such as spiderwebs and birds' skulls. He created constructive environmentally conscious solutions basing it all on the self-generation processes of form such as the structures in a net of cables, pneumatic structures, etc. Just as in the case of Le Ricolais, he was inspired by soap bubbles to generate large structures. He developed all kinds of models, looking to find methods of using the least possible amount of material and energy in order to create large covered spaces. Thanks to technology, prefabrication and mass production, tensile structures guided architecture in a new direction. One of his most emblematic works is the Olympic Park in Munich where the most significant work is the Olympic Stadium built with a PVC sheet just 4 mm thick (Fig. 12).

Vacuum and Micro-vacuum: Nature's Strategy for Structural Optimisation

> In nature materials are expensive and form is cheap. (Vincent, 1997, p. 9)

In the process of morphogenesis in the near infinite variety of existing living species, nature tends to constantly reduce the amount of material in use. The reason behind is ever so simple: Producing materials requires energy and nature tends towards minimising the use of energy, given its marked low efficiency. This is why forms and shapes in nature end up organising themselves with the minimum amount of embodied or embedded energy possible. And as such, they can turn out to be a training book for artificial

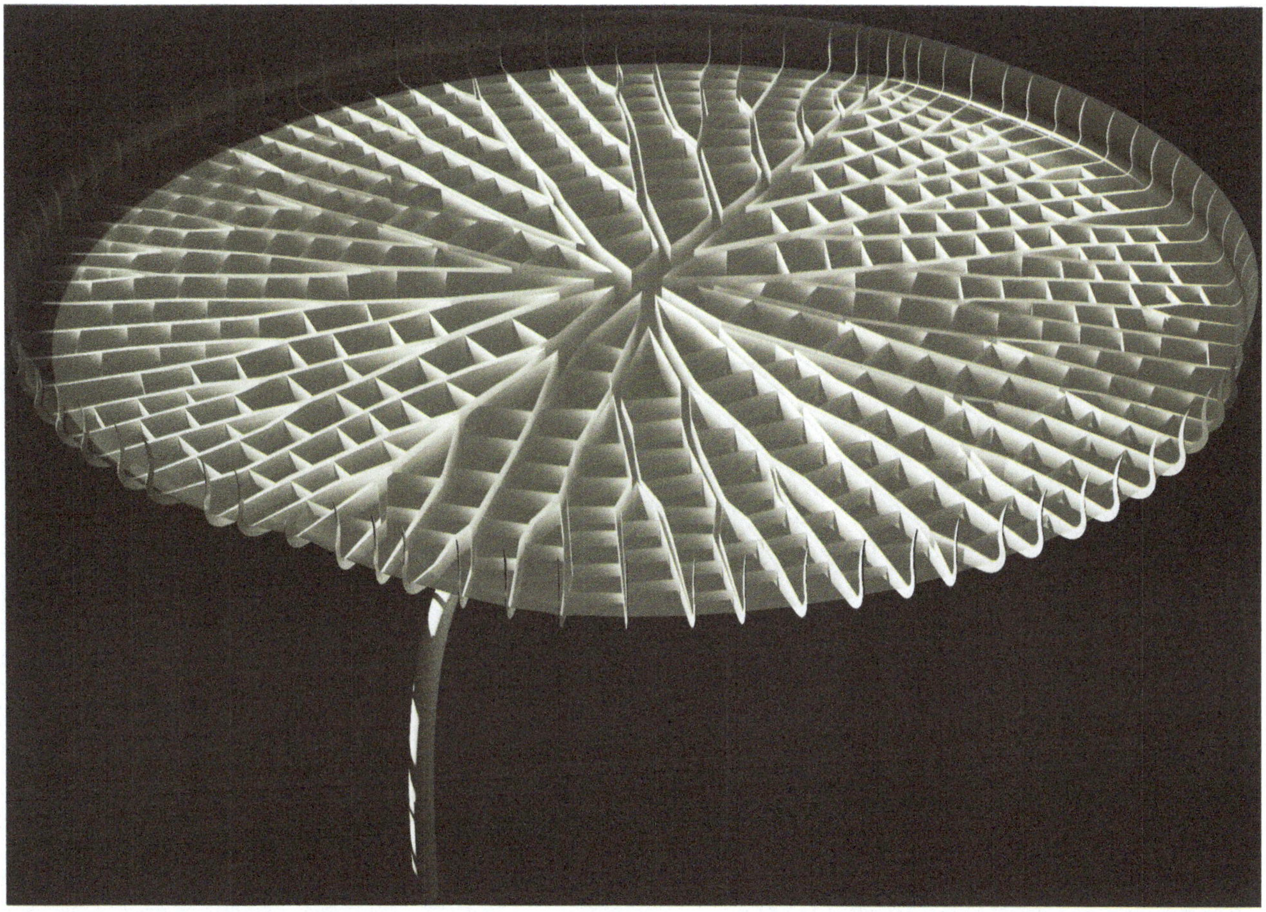

Fig. 5 The geometry of the reverse side of the water lily leaf. (Cervera & Pioz)

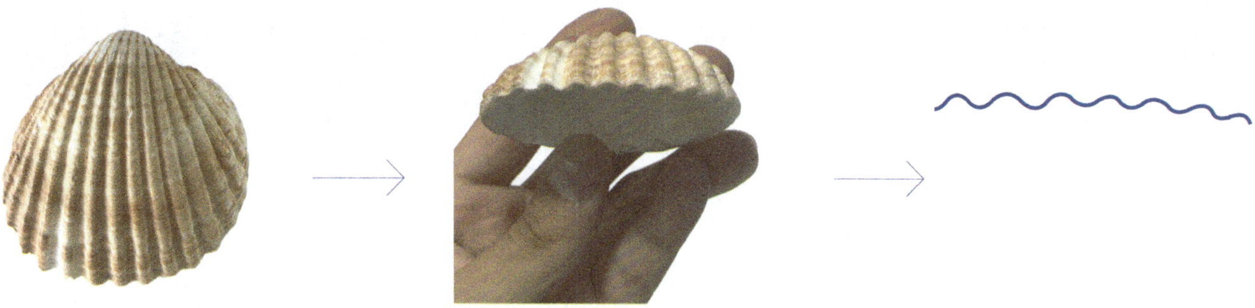

Fig. 6 Geometry of the bivalve shell, Playa de Gavá, Barcelona. (I. Curiel)

man-made constructions. In the process, natural forms follow to optimise energy, and one of the most interesting strategies we are able to observe is their choice of air as a primary element. Envisioning the world at a macroscopic scale, it seems to be brimming with solid forms, massive and robust, when they are in fact lightweight forms following a relentless process of lessening the load. And this is iterated at every scale, from macro to micro to nano and even atomic level, where vacuum percentage is 99%. This is the reason the configuration of forms in nature is spongy and puffy like, balancing matter and the air that complements it in the empty cavities that make up the whole. The air becomes a material that cooperates to achieve strength and stability against the impacts of gravity, winds, aggressions and earthquakes, their own weight plus added ones, and all in all, the many forces that reign planet Earth. Le Ricolais knowingly interpreted this in his well-known reflection: "If instead of working with solid elements one thinks of the empty spaces between them, truth reveals itself" (Juárez Chicote, 1996, p. 8).

In this chapter, we shall focus on the macro and micro scales as they are as affected by the forces of gravity and wind aggressions as man-made constructions are. We could

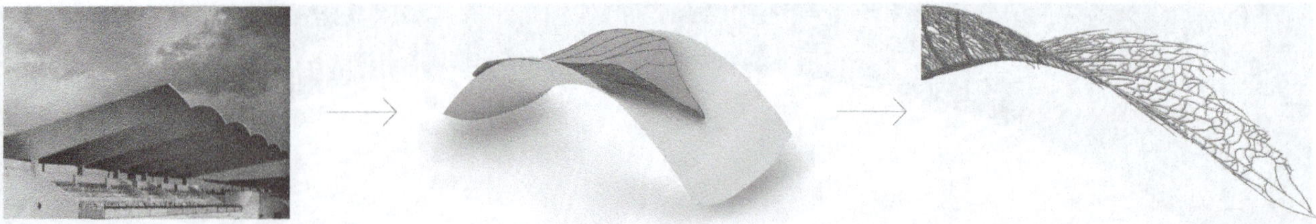

Fig. 7 Construction and nature. Geometry of the Zarzuela racecourse and study of the curvature of a leaf. (I. Curiel)

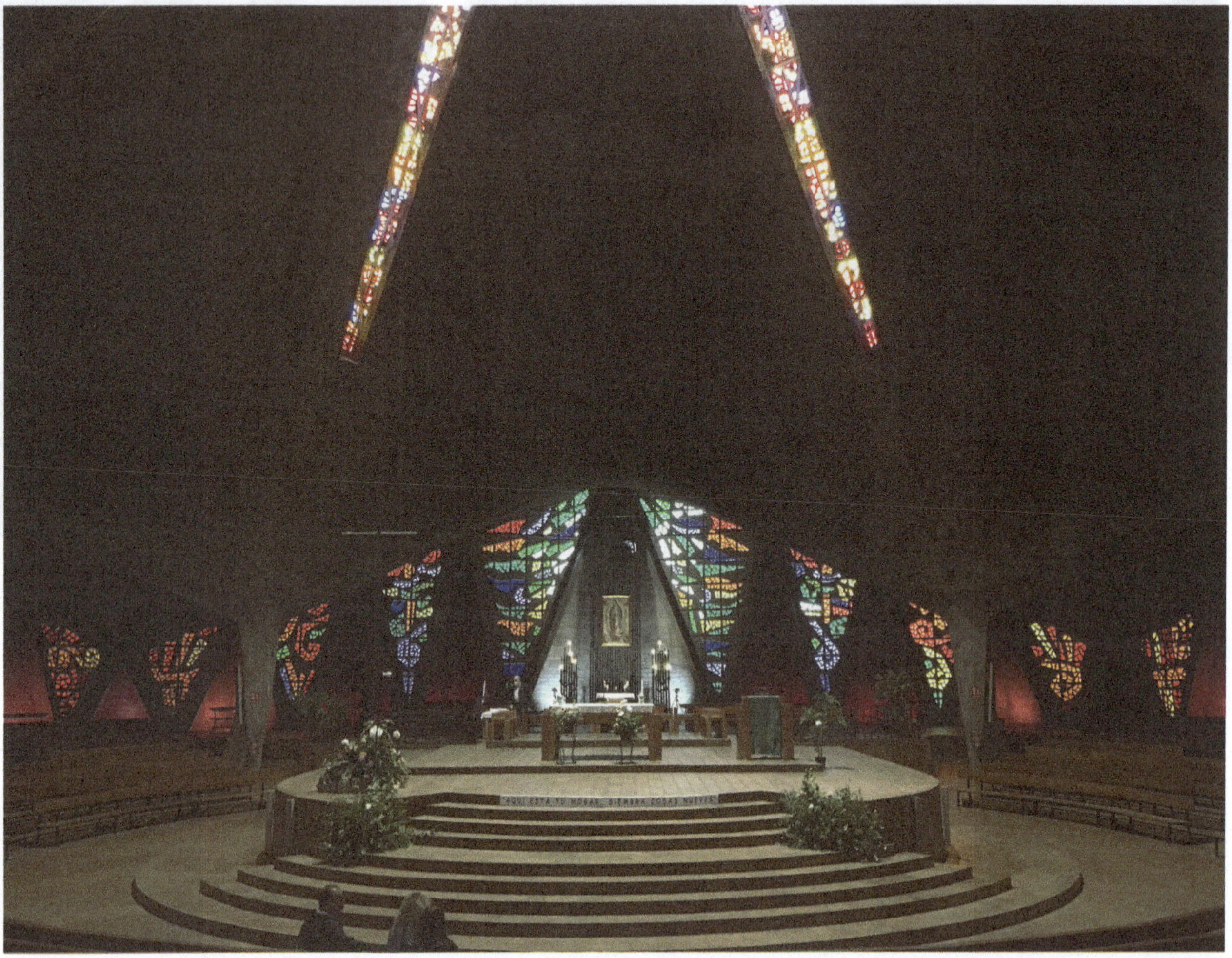

Fig. 8 Church Nuestra Señora de Guadalupe, Madrid. (Photo R. Cervera)

present here an infinite number of examples of living forms with a porous constitution. By way of an example, we shall revise some of the most notorious cases before delving into the final objective of this chapter: the tree trunk.

One of the flora species that best represents the vacuum as the lead element of its structure is bamboo. It is the slenderest vertical structure that exists. Some varieties can reach a diameter to height ratio of 1/200, while others can grow up to 40 m high having a mere 20 cm thick diameter. Structurally speaking, bamboo is made up of a hollow tube surrounded in its perimeter with a thick wall. At a macro scale, this hollow tube presents a ratio between matter and vacuum of approximately 50%, a bit less further down the base and up to 60% in the higher smaller parts at the top. Descending now to the microscopic scale, we find the wall in the perimeter presents a porous structure of heterogeneous density that is accentuated in the external layers of the bark. So the bark is dense the further out from the centre the layers are. The configuration of this wall can reach an average ratio of matter to vacuum of 25–75%, respectively (Table 1). Finally, should we

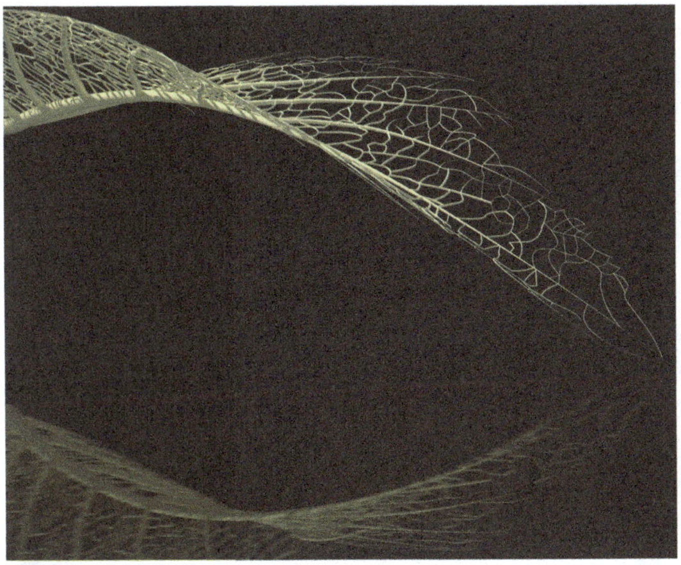
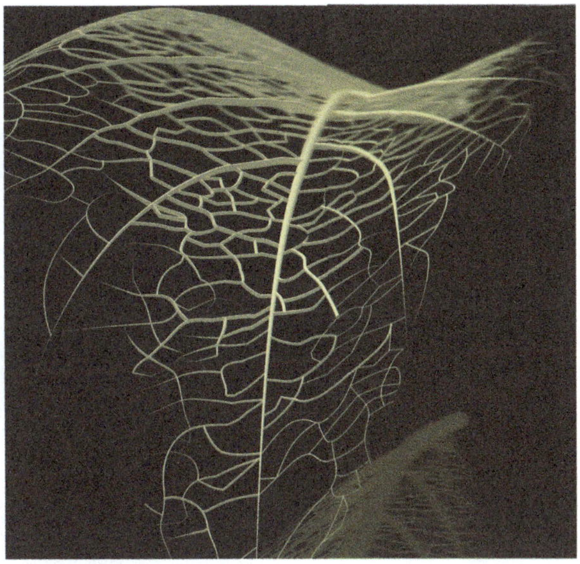

Figs. 9 and 10 Hyperbolic paraboloid in a leaf. (Cervera & Pioz)

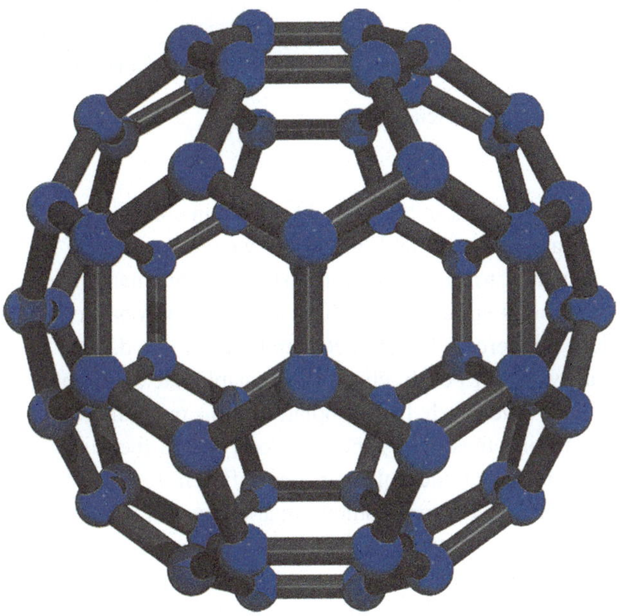

Fig. 11 Fullerene molecule. (I. Curiel)

add up the vacuum perceived simply by looking to the vacuum inside the most inner structure, we find that of all the possible total volume of bamboo, only approximately 15% of it is actually made up of matter.

The pneumatic structure of animal bones reaches its utmost expression in birds. The pneumatic structure has the goal of reducing weight to improve flight. They adopt the form of hollow tubes with criss-crossing struts for structural strength. These trabeculae or diagonal projections of osseous tissue work as braces of a structure and are arranged in all directions, creating an interesting three-dimensional network able to withstand the stresses of flying. If we take a feather and cut through its pen, we will find the same bone formation mentioned before: a structural hollow tube with an inner three-dimensional network where void has an important role. If we repeat the process on a barb with a 3 mm section, on a barbule with a 1.5 mm section or on the very tip of a barbule (0.5 mm), we'll find the same formation. All in all, it is a fractal process that seeks to lower weight at all scales. With this mechanism of sponginess, material and non-material parts cooperate to withstand the forces when moving through the air. It is relevant the weight of the skeleton of birds in comparison with their total weight (Cervera, 2019). For example, the Bald Eagle weights about 4 kg, and its skeleton is only 300 g due to the air inside the bones and amounts to less than 1/13th of the total weight, and its plumage reaches up to 600 g (Fig. 13). But even the skeleton of large land mammals maintains the concept of porosity. It has a crown of compact exterior bone with a larger density and an interior spongy one inside with a variable density depending on its resistance to stresses (Fig. 14).

The water lily leaf is a floating sheet of astounding structural efficiency that has inspired artists, engineers and architects alike (Fig. 15). Varieties such as the giant Amazonian water lily, Victoria Amazonica, can reach up to 2 m diameter and can bear 40 kg of weight. The geometrical organisation of radio concentric nervation presents a fractal system that makes them subdivide themselves the further out from the centre they are. It is thanks to this subdivision that a series of polygons between the radial and concentric nervation are generated that never exceed the optimum dimension so as to not increase the thickness of the fine layer of the sheet. The water lily leaf is highly porous in both the nervation and in the parenchyma, which reduces its weight and makes it float.

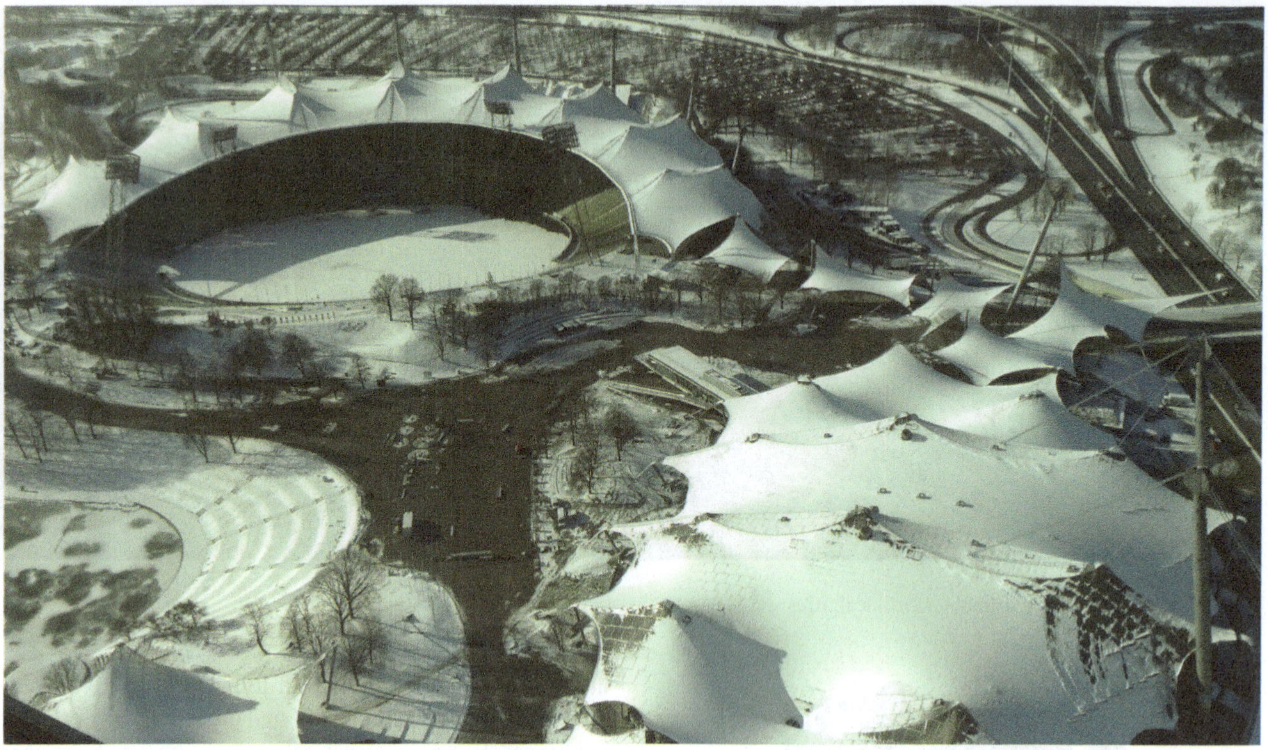

Fig. 12 Tensile structure, Olympiastadion and Olympiapark, Munich, F. Otto. (Photo R. Cervera)

Despite this characteristic, and thanks to its ribs, it is highly resistant (Cervera, 2019, p. 106). It has been studied that the Amazonian water lily leaf is an economy of matter that could have evolved in response to the cost-benefit associated with leaf size (i.e., benefit accrued by photosynthesis balanced by the costs of construction and maintenance) (Box et al., 2022).

Analysing the Void Inside a Tree Trunk

Trees are structures vertically developed and embedded at one of its ends. They could simulate a cantilever with a peak load produced by the wind beared at the top (Cervera, 2019). Varieties such as sequoias can shoot up to 100 m high and display a slender ratio that exceeds 1/30, something that is unthinkable in buildings. This structure's efficiency exists thanks to the large percentage of void at the micro level. The apparently solid wood can manage a proportion of void of up to 65% of its volume as we are about to see. Trees are the slenderest forms on the planet, and they manage to lessen their load by having an internal puffy configuration of fibres and veins. The fractal increase of the wall surfaces of the numerous empty tubes that make up the wood does not condense the volume, but it does reduce the embodied energy. It also carries other functions such as allowing water and its nutrients in from the roots to any and all the elements of the tree by a process of capillarity. The vast number of nanotubes organised in all three directions in space cooperate to increase resilience and lighten the weight.

In order to study the relationship between matter and void in arboreal structures, we have found it appropriate to conduct a differential study between types of wood, such as the hard wood of beech, cherry, oak, maple, etc., and soft ones such as pine, birch, black poplar, etc. The main difference for classifying these two types of wood lay on the microstructure growth of each of them, the different tissues that conform them and their geometrical arrangement. The basic cell types in wood are tracheids, vessels, fibres and parenchyma. Depending on these elements' specialisation, their function leans more towards carrying liquids and nutrients or towards being a skeleton-like structural support. In other cases, they help in hormonal communication. Tracheids and parenchyma prevail in softer woods, whereas vessels, fibres and parenchyma do so in hard ones; however, differences might appear depending on assorted species.

These basic components are all tubular like and have perforated walls, as if they were perforated layers that allow water and nutrients to flow. These are larger in softer woods than in harder ones. They are arranged in both radial and axial manners. The conduits in axial disposition are the ones with vertical dominance that runs parallel to the trunk's axis and plays a leading role in its constitution. And the radial disposition conduits are perpendicular to the bark to ensure that water and nutrients flow laterally (Spicer, 2014) and

Table 1 Study of macro-vacuum and micro-vacuum in the structure of bamboo. (R. Cervera; I. Curiel)

MACRO SCALE	CALCULATING	RATIO
Small internode 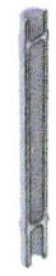	Total Volume = 505752.48 (+/-0.0052) mm3 Vacuum Volume = 202379.60 (+/-0.0052) mm3 Material Volume = 303372.87 (+/-0.0052) mm3	40,02% Vacuum 59,98% Material
Medium internode 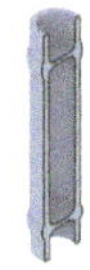	Total Volume = 3261776.73 (+/-0.021) mm3 Vacuum Volume = 1501990.5(+/-0.0071) mm3 Material Volume = 1759786.23(+/-0.021) mm3	46,05% Vacuum 53,95% Material
Large internode 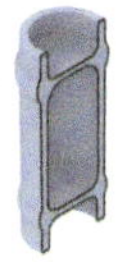	Total Volume = 3019735.3 (+/-0.032) mm3 Vacuum Volume = 2884894.4 (+/-0.21) mm3 Material Volume = 3134840.8 (+/-0.21) mm3	47,93% Vacuum 52,07% Material
MICRO SCALE		
Excerpt section 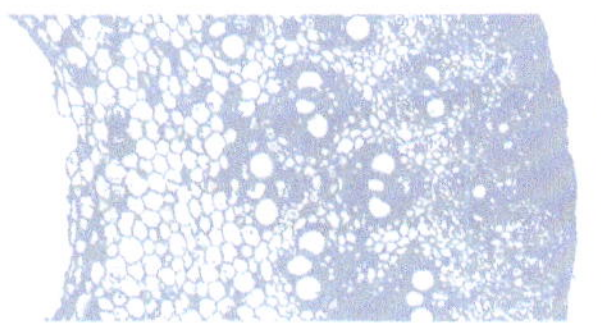	Total Area = 6776.87 (+/- 1e-01) u2 Vacuum Area = 5094.44 (+/- 1e-01) u2 Material Area = 1682.43 (+/- 1e-01) u2	75,17% Vacuum 24,82% Material

contribute towards the three-dimensional bond of the structure (Table 2). Also to take into account is the helicoidal growth of the whole that adds a double alternating direction helix that, although this varies as per species, presents a structural solution to the absorption of the lateral stresses produced by wind thrusts.

This chapter has followed a process of digitally replicating, as faithfully as possible, the makeup of the structure of tree trunk fragments in two types of wood—hard and soft. Thanks to digital modelling, the percentages of void and solid matter have been subsequently quantified in units within the total volume of said fragments.

The geometry of a tree trunk fragment has been replicated by analysing each of the elements one by one, be them fibres, vessels, axial parenchyma, ray or radial parenchyma. The proportion of void in the total volume of the fragment with respect to the proportion of solid matter has been calculated, and results show that hardwood presents up to 58% (Table 3) of vacuum and softwood even up to 69% (Table 4). Even if these amounts cannot be considered 100% accurate in a scientific sense, they do however offer an approximation into how the system works towards lightening up its structure. The trunk is made up of multiple small cooperating hollow pieces and perforated walls, none of which depreciate structural capacity, and it also avoids grand gestures in structural shape that would add to the total embedded energy. The continuity in the presence of vacuum explains the low density in wood compared to that of other materials, as it oscillates between 400 kg/m^3 and 550 kg/m^3 depending on tree species. Interestingly, it has been evidenced that the elemental wood component that is the cellulose cell replicates at a micro scale the same hollow tube system that exists at the macro scale and also alternates layers and helicoidal movement, reinforcing in this manner and at a cellular scale the same structural answer to the twists and turns that wind incurs on the tree (Figs. 16 and 17). And low density does not by any means equate to lower resilience. It is quite the opposite in fact. The tension force of cellulose is 10,000 kg/cm^2, which is greater than that of steel in spite of its density being 15 times lower. Lignin has

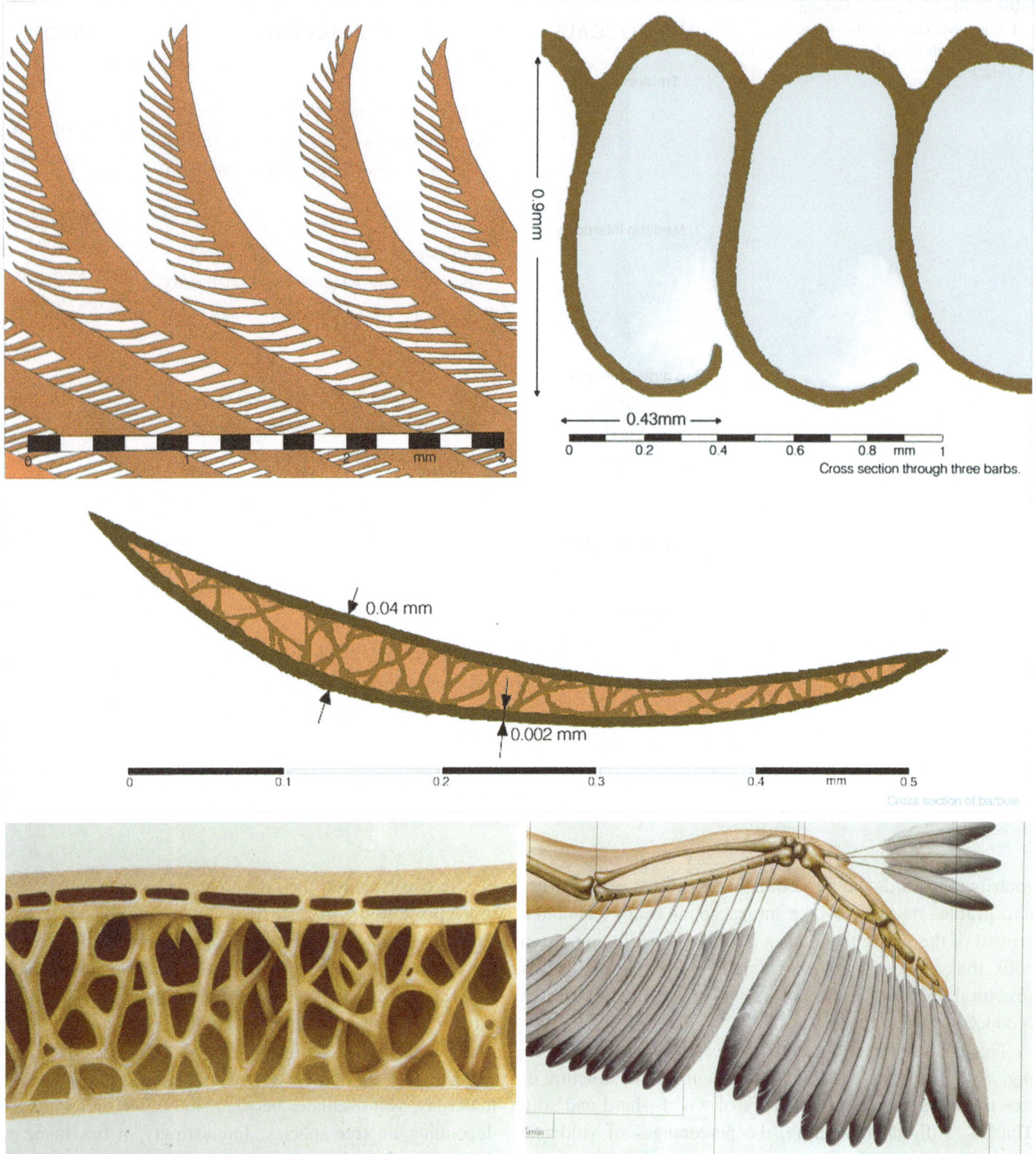

Fig. 13 Pneumatic structure in birds. (Cervera & Pioz)

a compression strength of 2400 kg/cm^2, which is greater than that of concrete, and that makes it lightweight in relation to its strength. We cannot compete with natural materials at the moment but we can learn from their display and geometrical disposition and from their mechanism of lightening up matter (Cervera, 2019, p. 247).

The parallel views on the morphological organisation in living beings and the contemporary artificial constructions in architecture could pave the way towards getting past the model that has been generically in place since it was introduced by the industrial production system. The possibilities of new construction processes anticipated by the digital revo-

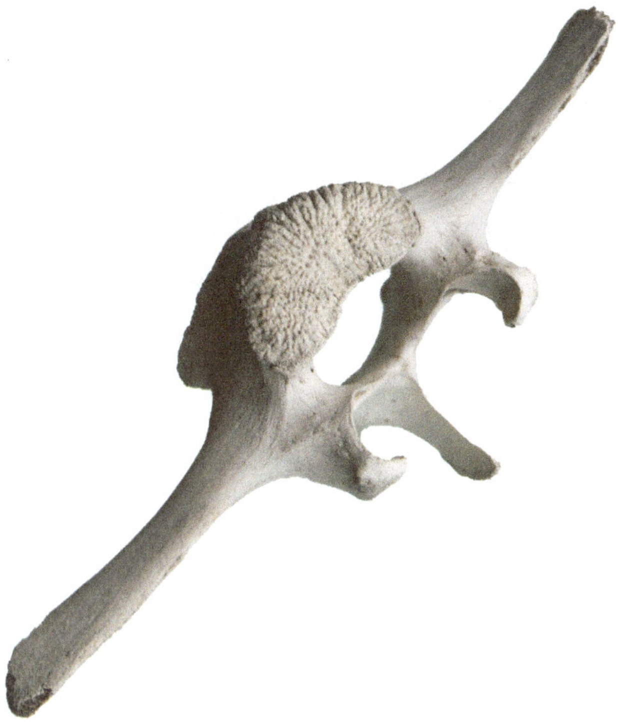

Fig. 14 Cow bone. (Photo R. Cervera)

lution and recent scientific advances in bio- and nano materials foresee a change of cycle that is in turn imperative in the aim to reduce the consumption of natural resources.

Conclusion

Reducing the embedded or embodied energy in architectural constructions is a clear need, as we are by now well aware of the fact that it has a high impact in the overall life cycle of the building. Investigations have taken place that looked into different living forms in nature in order to understand how they are structurally and morphologically organised. It has been observed in every case that they consistently lighten up the amount of solid matter utilised and consequently use the surrounding void as part of its structure. This particular quality has already been investigated by relevant architects and engineers, and it has been further examined in detail in this chapter by looking into a tree trunk fragment and digitally replicating its composition and geometry. Thanks to digital modelling, the presence of vacuum in a given tree trunk has been quantified at more than 58% and even up to 69% depending on species. With the tree being one of the slender-

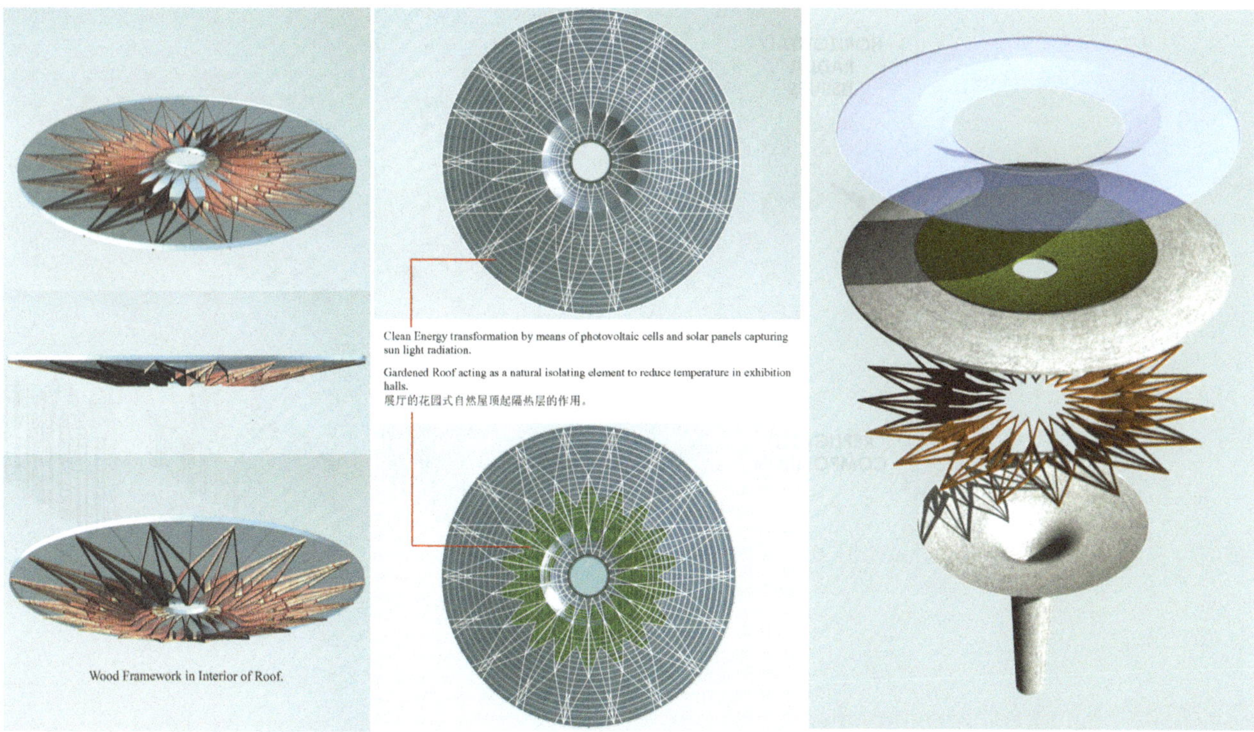

Fig. 15 Xixi national water museum in Hangzhou. Structural design inspired by the water lily leaf. (Cervera & Pioz)

Table 2 Summary of the study of the micro-vacuum composition of hardwoods and softwoods. (R. Cervera, I. Curiel)

	CELLULAR COMPOSITION OF WOOD	
	HARDWOOD	**SOFTWOOD**
MICROSCOPIC IMAGES	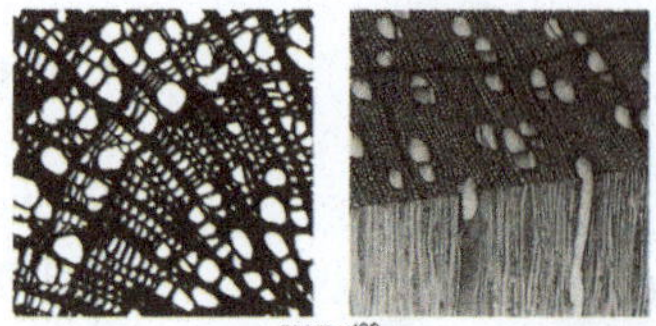 zoom x400	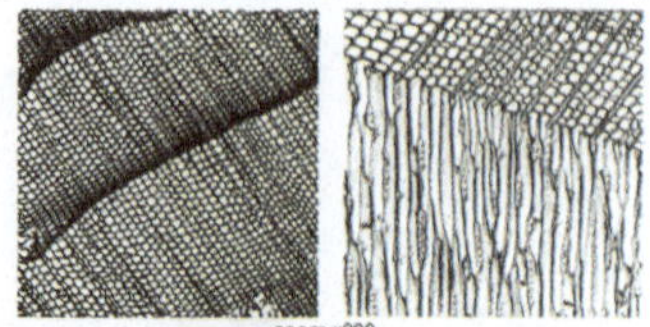 zoom x200
VERTICAL TISSUES 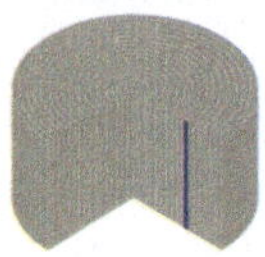	Hardwood is distinguished by the presence of pores or vessels in a vertical direction. It is the fibres that provide the structural support that occupies most of the volume. **Fibres** **Axial parenchyma** **Vessels** 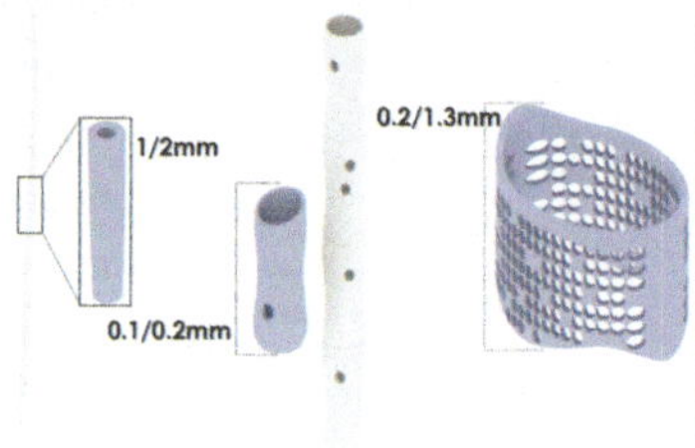 1/2mm 0.1/0.2mm 0.2/1.3mm	Softwood is composed of 90-95% longitudinal tracheids for structural support and water conduction. **Axial parenchyma** **Axial tracheids** 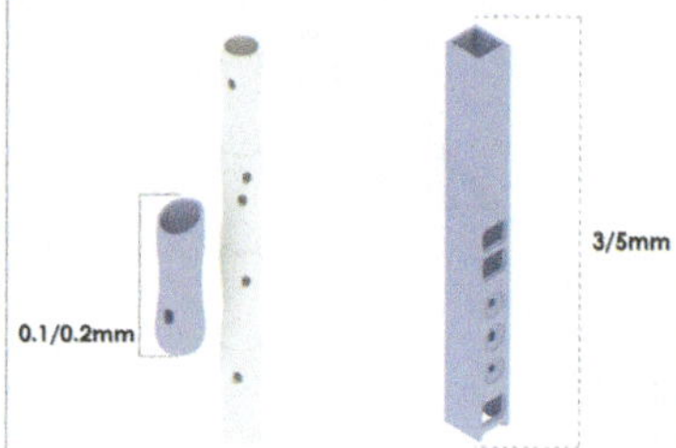 0.1/0.2mm 3/5mm
HORIZONTAL/ RADIAL TISSUES 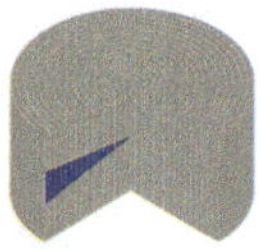	The radial parenchyma mainly plays a storage, water and nutrient role. 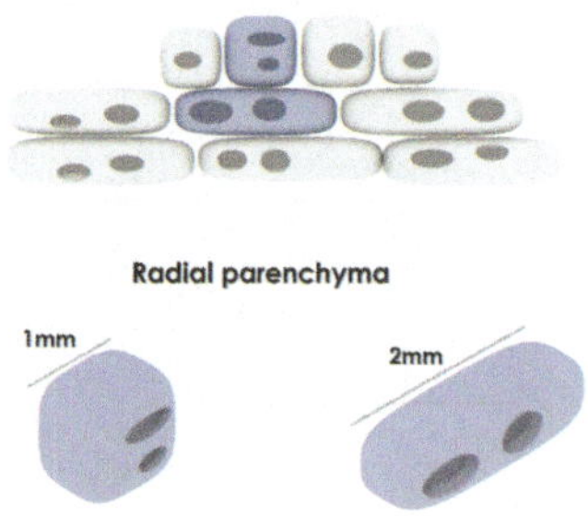 **Radial parenchyma** 1mm 2mm	Softwood is also composed of radial tracheids similar to parenchyma but usually more irregular in shape. 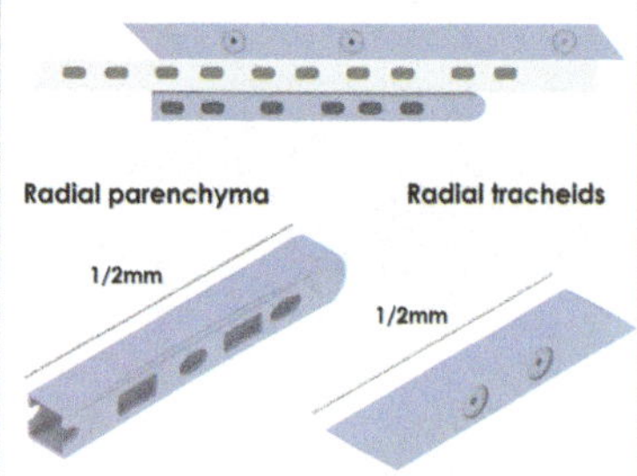 **Radial parenchyma** **Radial tracheids** 1/2mm 1/2mm
REPLICATED COMPOSITION	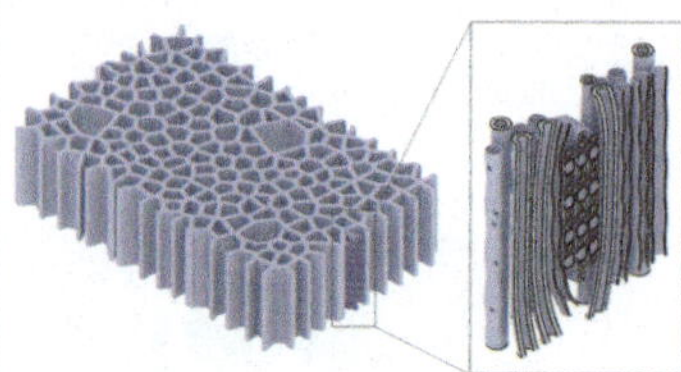	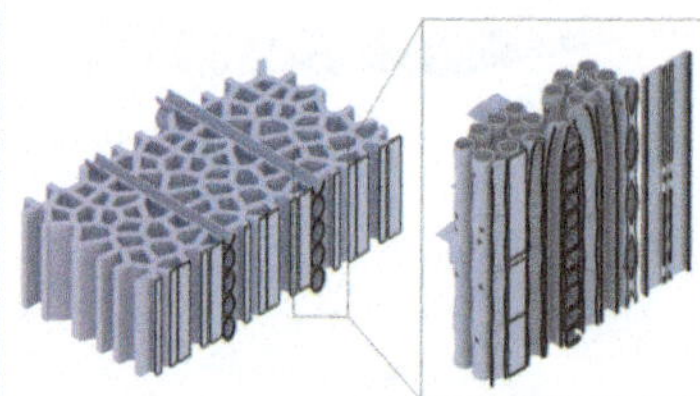

Table 3 Hardwood case study. (R. Cervera; I. Curiel)

HARDWOOD CASE STUDY	CALCULATING	RATIO
Excerpt section 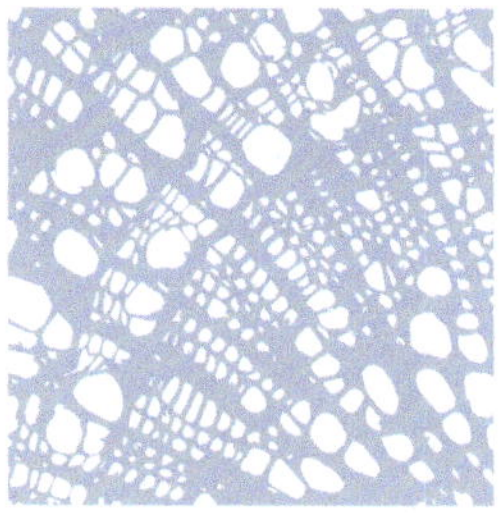	Total Area= 9228,68 (+/- 1e-01) u2 Vacuum Area = 5108,25 (+/- 1e-01) u2 Material Area = 4120,43 (+/- 1e-01) u2	55,35% Vacuum 44,64% Material
Fibres 	Total Volume = 10.110985 (+/- 3e-06) u3 Vacuum Volume = 3.46807165 (+/- 3.3e-08) u3 Material Volume = 6,64291335 (+/- 3.3e-08) u3	34,29% Vacuum 65,71% Material
Axial Parenchyma 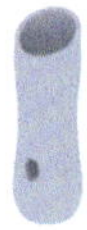	Total Volume = 90.5458825 (+/- 3.6e-06) u3 Vacuum Volume = 65.2487978 (+/- 3.6e-06) u3 Material Volume = 25.2971147 (+/- 1.6e-06) u3	72,06% Vacuum 27,93% Material
Vessels 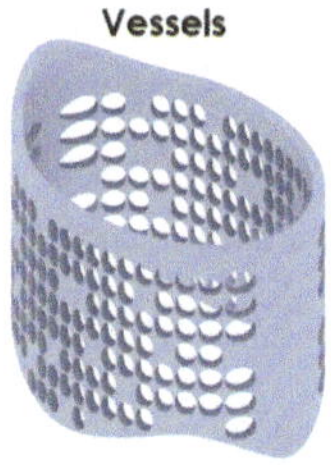	Total Volume = 39,942674 (+/- 1e-07) u3 Vacuum Volume = 26,443399 (+/- 1e-07) u3 Material Volume = 13,499275 (+/- 1e-07) u3	66,20% Vacuum 33,79% Material
Radial parenchyma 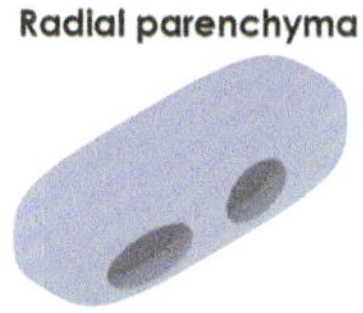	Total Volume = 2.65936271 (+/- 3.8e-08) u3 Vacuum Volume = 1.7519302 (+/- 3.7e-08) u3 Material Volume = 0,907432451(+/- 3.7e-08) u3	65,87% Vacuum 34,13% Material
Replicated Composition 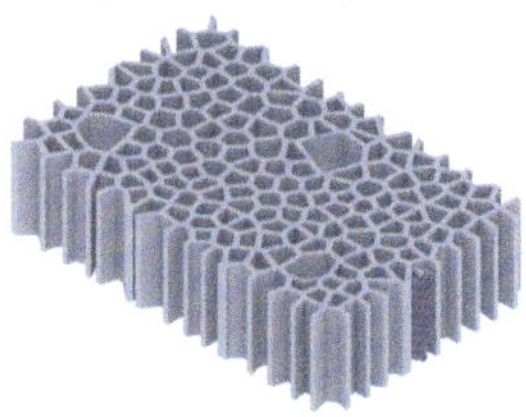	Once we have studied the vacuum-material ratio of each of the tissues that make up the tree, we find a ratio of **58.75%** vacuum to **41.25%** material.	

Table 4 Softwood case study. (R. Cervera; I. Curiel)

SOFTWOOD CASE STUDY	CALCULATING	RATIO
Excerpt section 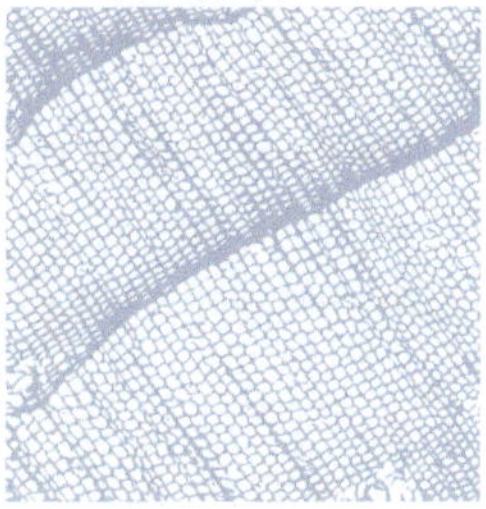	Total Area = 12011,33 (+/- 1e-01) u2 Vacuum Area = 9159,57 (+/- 1e-01) u2 Material Area = 2851, 76 (+/- 1e-01) u2	76,25% Vacuum 23,74% Material
Axial Parenchyma 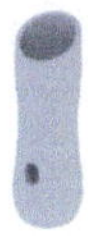	Total Volume = 90.5458825 (+/- 3.6e-06) u3 Vacuum Volume = 65.2487978 (+/- 3.6e-06) u3 Material Volume = 25.2971147 (+/- 1.6e-06) u3	72,06% Vacuum 27,93% Material
Axial Tracheids 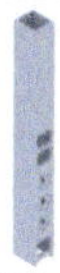	Total Volume =20,4559370 (+/- 1e-08) u3 Vacuum Volume = 14.4233929 (+/- 1e-08) u3 Material Volume = 6.0325441 (+/- 3.2e-08) u3	70,50% Vacuum 29,49% Material
Radial parenchyma 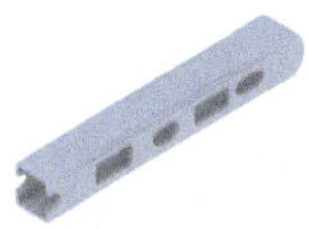	Total Volume = 353.8818 (+/-1e07)u3 Vacuum Volume = 230.958291 (+/- 1e-07) u3 Material Volume = 122.92354 (+/- 3.9e-07) u3	65,26% Vacuum 34,73% Material
Radial Tracheids 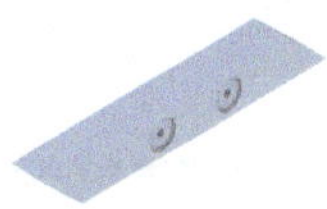	Total Volume = 5,59011037 (+/- 1e-09) u3 Vacuum Volume = 3.50402925 (+/- 1e-09) u3 Material Volume = 2.08608112 (+/- 1e-09) u3	62,68% Vacuum 37,31% Material
Replicated Composition 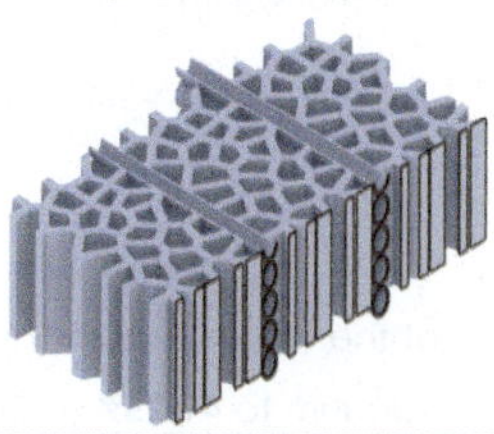	Once we have studied the vacuum-material ratio of each of the tissues that make up the tree, we find a ratio of **69.35%** vacuum to **30.65%** material.	

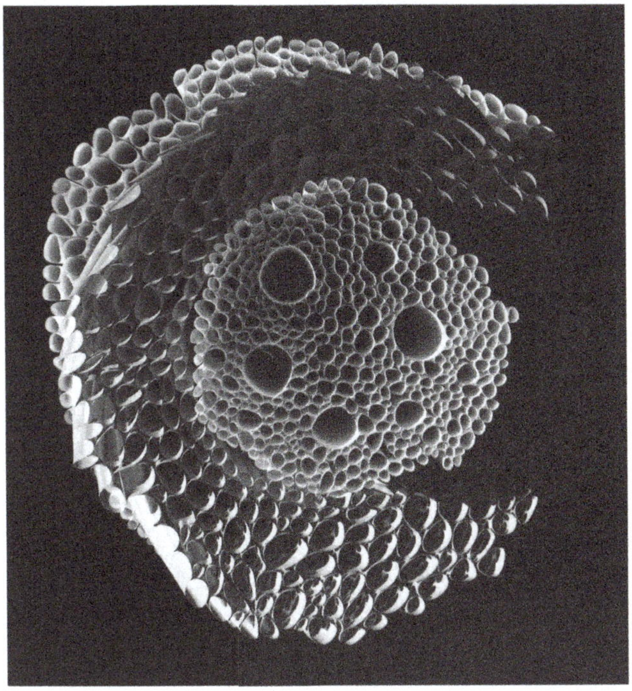

Fig. 16 Cellulose cell structure. (Cervera & Pioz)

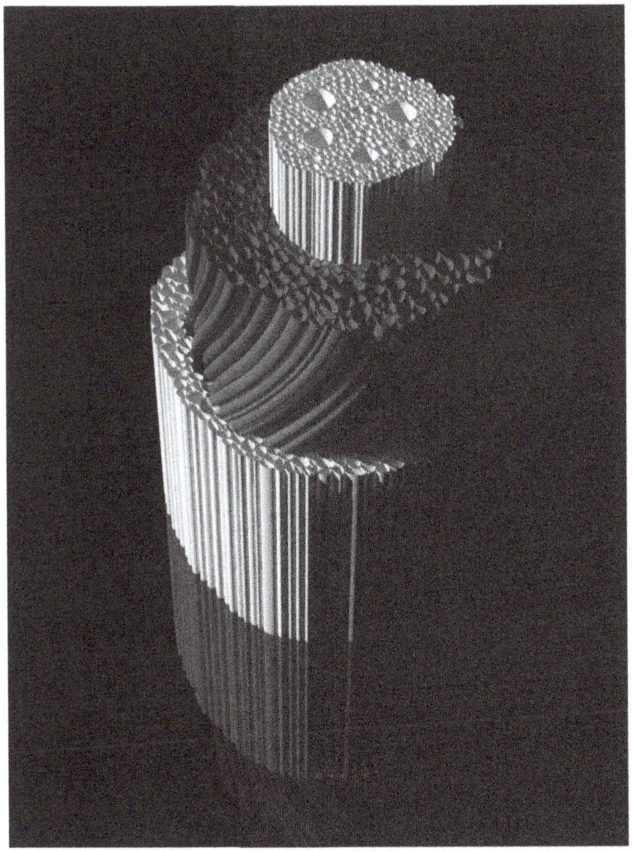

Fig. 17 Torsion growth of the cellulose cell. (Cervera & Pioz)

est natural shapes in the planet, of mighty resistance and longevity, this chapter opens up a line of investigation that establishes bridges with nature towards creating the architecture of building "as light as air".

References

Box, F., Erlich, A., Guan, J. H., & Thorogood, C. (2022). Gigantic floating leaves occupy a large surface area at an economical material cost. *Science Advances, 8*(6), eabg3790. https://doi.org/10.1126/sciadv.abg3790

Cepeda Gutiérrez, M., & Mardaras Larrañaga, I. (2004). Cuantificación energética de la construcción de edificios y el proceso de urbanización. *Conarquitectura, 12*, 65–80. Retrieved from http://www.conarquitectura.com/articulos%20tecnicos%20pdf/12.pdf

Cervera, M. R. (2019). Biónica, biomimética y arquitectura: Aprendiendo de la naturaleza = Bionics, biomimicry and architecture: Learning from nature Barcelona: By Architect publications.

Granados, J. (2018). El ingeniero que eclipsó a los arquitectos: Pier Luigi Nervi. Retrieved from https://arquitecturayempresa.es/noticia/el-ingeniero-que-eclipso-los-arquitectos-pier-luigi-nervi

Juárez Chicote, A. (1996). El arte de construir con agujeros. reflexiones en torno a Robert Le Ricolais. Madrid: E.T.S. Arquitectura (UPM).

Martínez, L. (2020). Creador de las cúpulas geodésicas, Buckminster Fuller acuñó el término tensegrity. Retrieved from https://www.revistaad.es/diseno/iconos/articulos/creador-cupulas-geodesicas-buckminster-fuller-acuno-termino-tensegrity/25482

Pérez-Lombard, L., Ortiz, J., & Pout, C. (2008). A review on buildings energy consumption information. *Energy and Buildings, 40*(3), 394–398. https://doi.org/10.1016/j.enbuild.2007.03.007

Quispe Gamboa, C. N. (2016). Análisis de la energía incorporada y emisiones de Co2 aplicado a viviendas unifamiliares de eficiencia energética. Universidad Politécnica de Cataluña, Escuela Técnica Superior de Arquitectura de Barcelona.

Spicer, R. (2014). Symplasmic networks in secondary vascular tissues: Parenchyma distribution and activity supporting long-distance transport. *Journal of Experimental Botany, 65*(7), 1829–1848. https://doi.org/10.1093/jxb/ert459

Vincent, J. (1997). Stealing ideas from nature. *RSA Journal, 145*(5482), 36–43. Retrieved from https://www.jstor.org/stable/45402294

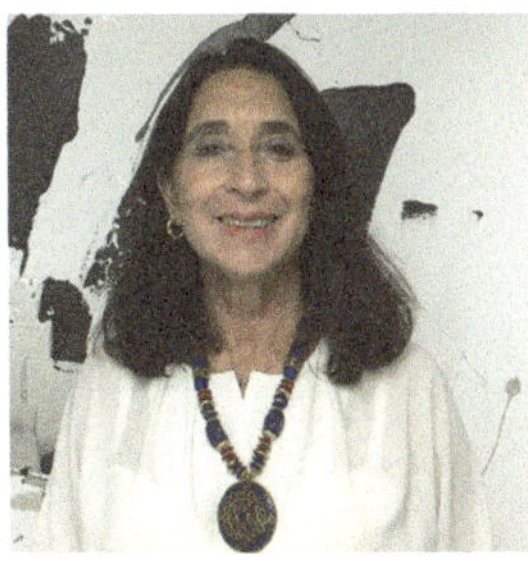

María Rosa Cervera Sardá Ph.D. Architect, Professor and former Dean of the School of Architecture at the University of Alcalá. Current director of the Master's degree in Advanced Architecture and City Projects. She is the author of writings and books on architecture, among which we highlight: Bionics, Biomimetics and Architecture (2019); Space and Time in Architectural Composition (2018); Recycling Mumbai. Re-envisioning the Slum (2012); Madrid, Recycled City (2011); Iron in 19th Century Madrid Architecture (2006). A regular speaker in professional and academic circles, she has given lectures in Spain, China, India, USA, Bolivia, Peru, El Salvador, Venezuela, Romania, Italy, etc.

As an architect, Rosa Cervera has received several awards – the "Antonio Maura" award, the "COAM" award, the "Golden Global award", the "Transfer of Knowledge" award – and has won several international architecture competitions.

Rosa Cervera is a pioneer in the research of Bionics, Biomimetics and the application of biological structures to innovative and efficient architecture and urban design. A direct result of these studies is the Self-Sustainable Vertical Garden City, Bionic Tower.

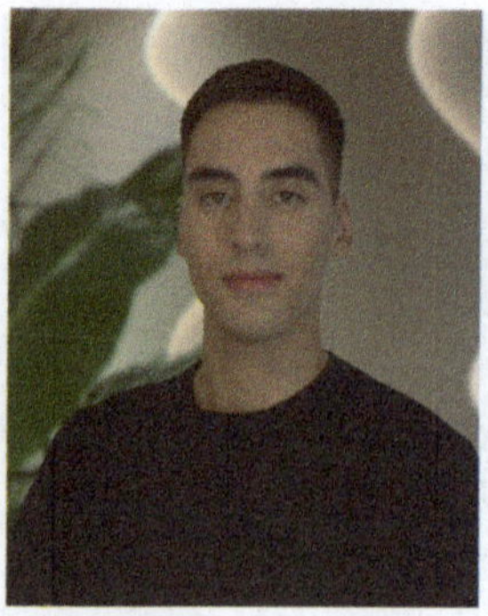

Iván Curiel Martínez is an Architect and Urban Planner from the School of Architecture of the University of Alcalá and graduated in the Master's degree in Architecture in 2022. He is currently studying the Master in BioDigital Architecture at the International University of Catalonia, Barcelona. Furthermore, he has participated as support in a couple of projects for the General Foundation of the University of Alcalá de Henares.

Bionics and the Future of Tall Building Design

Mark P. Sarkisian

Abstract

Research is essential for developing new ideas, creating resilient structures that have extended and predicted life cycles. Through creative use of materials, construction techniques, and building performance, this research can address broad issues related to the life of structures starting with urban planning and ending with health monitoring. Awareness is key to solving complex issues and the collaboration between academic study, research, and practice is critical.

This chapter will focus on how research has manifested new approaches to design. Examples include the use of mechanized approaches to design, optimization theory, and machine learning. Ideas range from components in buildings to entire cities, and the work includes projects from various places around the world.

Keywords

Carbon · Systems · Nature · Research · Resilience · Optimization · Seismic · Friction · Structural system · Tower · Concrete · Steel · Building · Frame · Post-tensioning · Elastic · Ductile · Carbon · Pin-Fuse · Machine learning

Introduction

Recent research inspired by natural form and support by advanced analysis tolls that utilize structural optimization techniques has led to investigations of resilient structural systems that are both efficient and influential in creating new architecture (Fig. 1). Historically, natural growth patterns have influenced art and useful products that use principles of assembly to create long-life durable goods. The concept of interweaving structural elements is an appropriate approach to ideas of all scales whether in constructing a basket or an ultra-tall building (Figs. 2 and 3).

Although optimization analysis techniques used to date have been largely used to understand behavior rather than provide results that can be used for final designs, the process has been used to help define new structural system ideas for major tower structures. These structures tend to be highly customized with little repetition or mass application. Efforts recently have been made to develop ideas with greater application and use on structures of all scales. In addition, the process has led to even greater understanding of position structural elements that not only response to force flow but specific programmatic building requirements.

Efficient structural design leads to reduction in embodied carbon. In addition to materials, construction processes and time, and probabilistic damage of structural components or systems, all contribute to full-life carbon. Enhanced structural engineering systems reduce construction time and improve performance during adverse conditions such as strong seismicity all leading to better solutions for the environment.

Machine learning offers great potential for structures considering quality control and perhaps future processes for construction. This chapter illustrates inventions that use machine learning to not only assess damage to structures but compare in-place construction to design drawings (Fig. 4).

Rational Structural Response to Force Flow

The China Merchant competition submission was influenced by natural force flow patterns and specific programmatic uses (Fig. 5). A reinforced concrete frame was designed to transform over the building height and respond directly to internal uses. Sketches on the structural system were refined during the

M. P. Sarkisian (✉)
Structural Engineering Partner, Skidmore, Owings & Merrill, Chicago, IL, USA
e-mail: Mark.Sarkisian@som.com

M. R. Cervera Sarda et al. (eds.), *Architecture Inspired by Nature*, https://doi.org/10.1007/978-3-031-33144-2_2

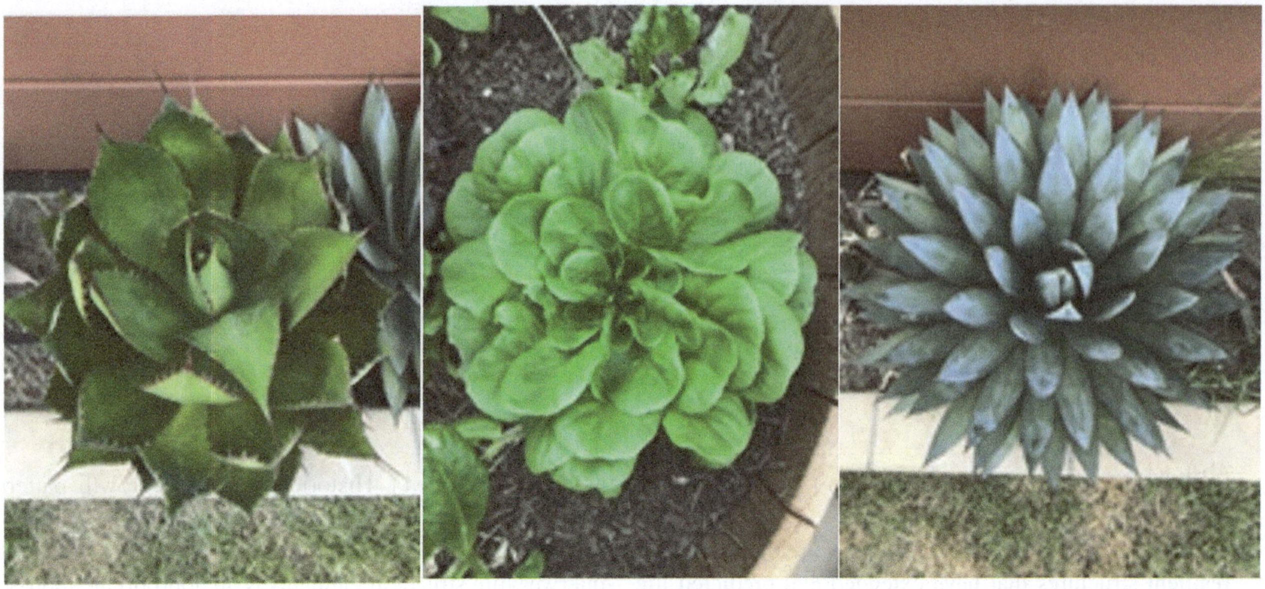

Fig. 1 Natural plant formation

Fig. 2 Woven basket

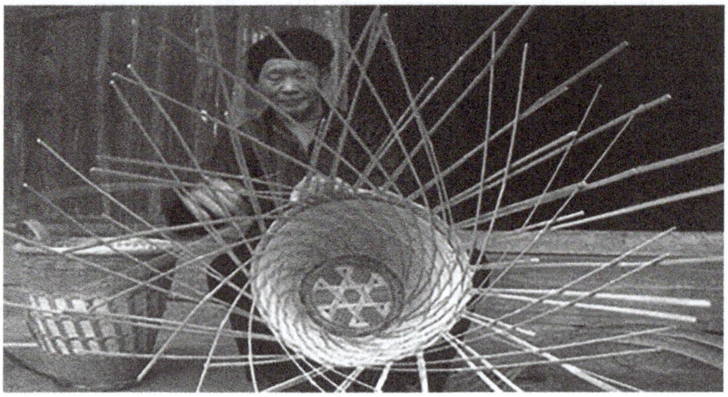

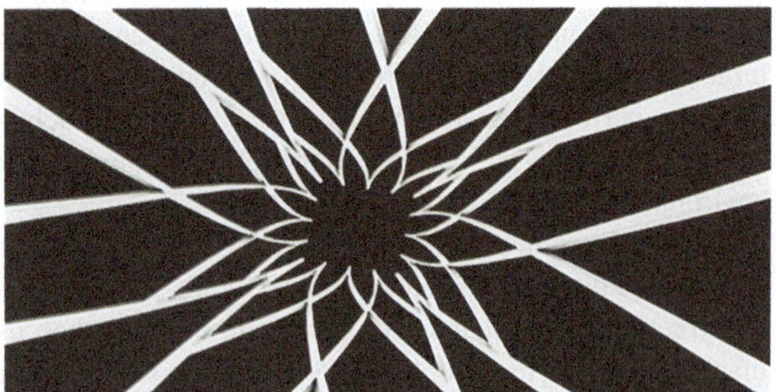

Fig. 3 Organic frame concept

process, finally resolving the structural solution with a frame that could be used in combination with a central reinforced concrete core interconnected with steel outrigger trusses or concrete walls to resist lateral loads imposed from wind or seismic conditions. These early sketches reflect a programmatic change approximately 3/4 of the height of the tower.

Further refinement of the initial concept resulted in a structural system that:

1. Included a wider perimeter, reinforced concrete column spacing at the top of the tower where executive offices would include greater unobstructed views and flexible office spaces within
2. Incorporated concrete frame transitions and primary joint public spaces approximately located at the 1/4 and 3/4 height elevations in the tower also corresponding to outrigger truss or wall interconnections with a central reinforced concrete core well
3. Introduced closer perimeter, reinforced concrete column spacing and the tower's midsection where space would be leased and any structural premium for perimeter span eliminated
4. Included a wider perimeter, reinforced concrete frame and the lower portion of the tower to accommodate large open tradition floor spaces

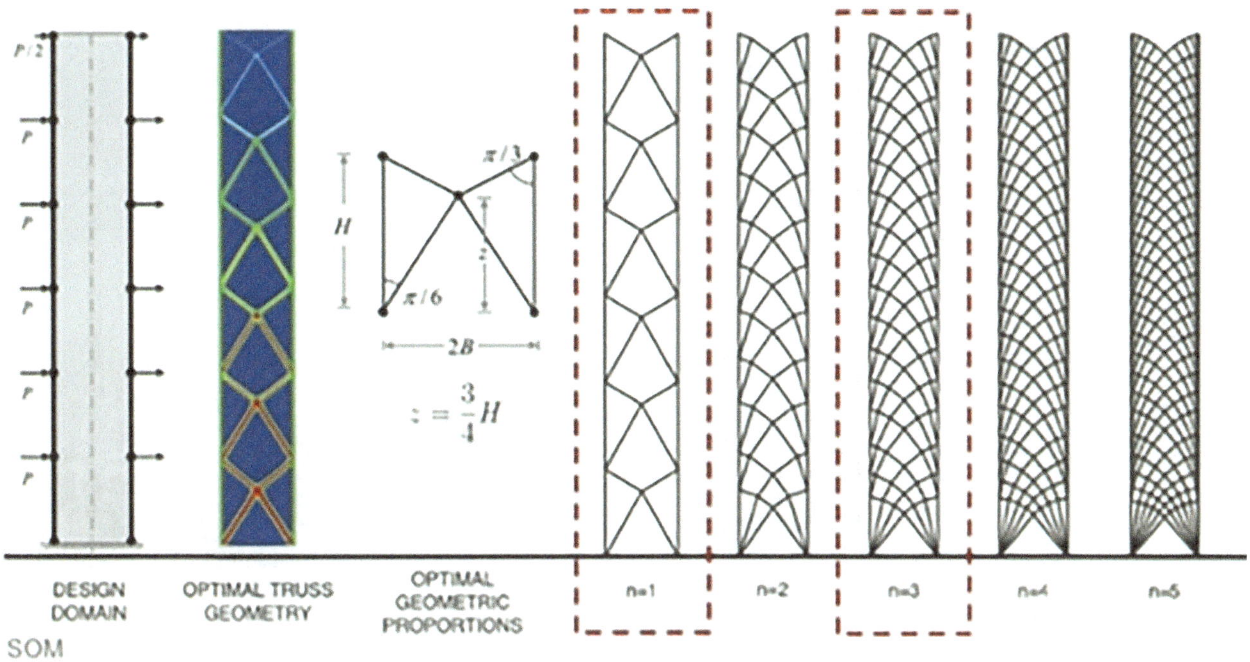

Fig. 4 Optimization analysis and potential lateral load-resisting frames

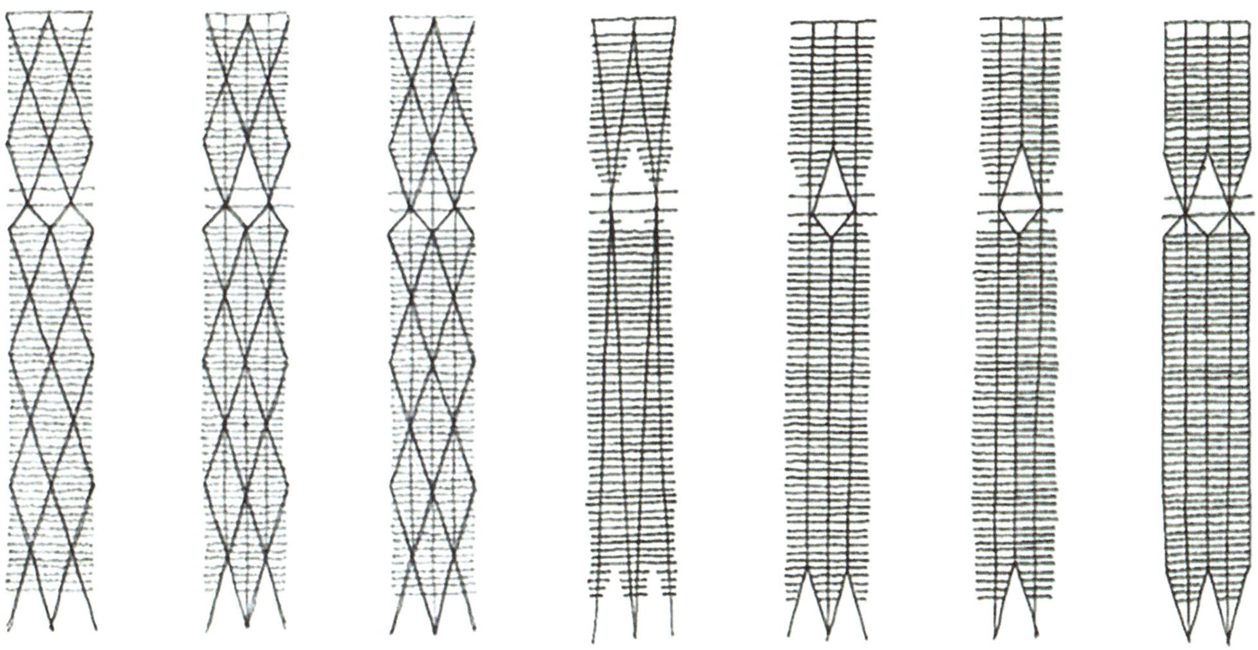

Fig. 5 Conceptual drawings of structural system for the China Merchant Bank Tower, Shenzhen, China

Even though transitions of the frame were incorporated over the height of the tower, the system allows for a continuous force flow without transfer of load (Fig. 6). Gravity load in the perimeter frame offsets any potential uplift due to typhoon wind or moderate seismic demands.

Optimized Slab Systems with Other Potential Applications

Post-tensioned slabs are commonly used worldwide in concrete structures of all scales. The slabs generally provide an efficient solution to long-span conditions with relative thin

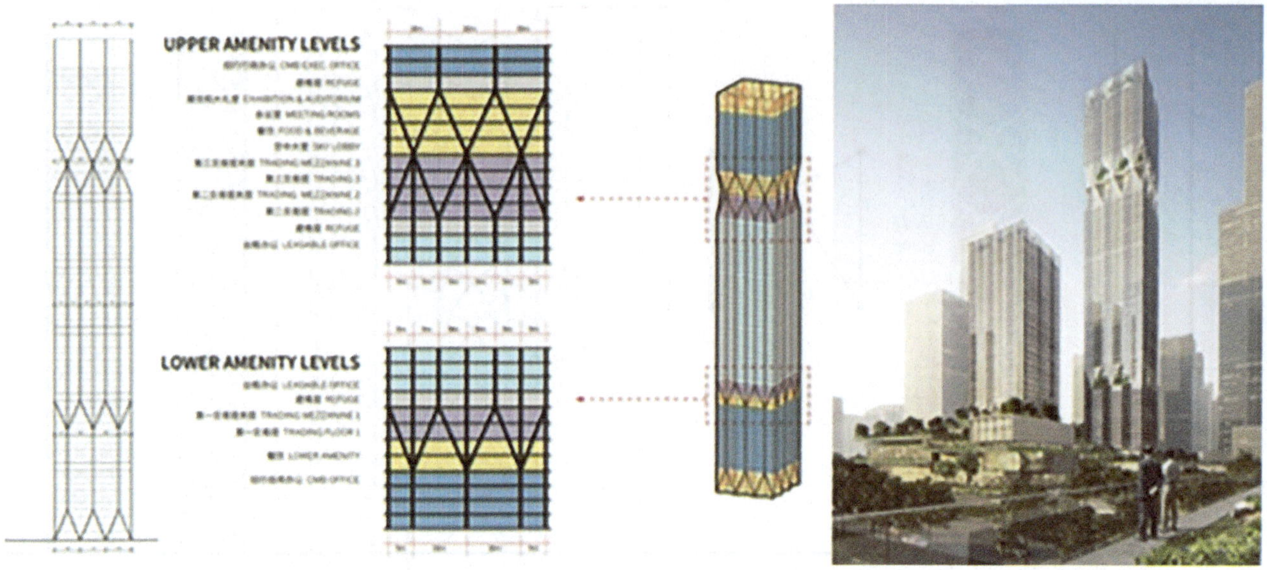

Fig. 6 Concrete frame elevation, program description, final rendering of China Merchant Tower

slab thicknesses. Historically, the quantity of post-tensioning required to achieve usable spans in commercial and residential construction has been 0.9 psf to 1.0 psf. Attempts to reduce the amount of post-tensioning has not been successful even when considering banding and carefully placement of material. However, research has shown that placing tendons on optimized load paths have resulted in reductions of post-tensioning of 25–30%. Therefore, the quantity of post-tensioning historically required is now on the order of 0.65 psf to 0.75 psf. This reduction not only represents a reduction of raw material required but a reduction in construction time and carbon emitted into the atmosphere due to both the production and placement.

Even though the 1111 Sunset Boulevard Project in Los Angeles incorporates an unusual core only lateral system, the post-tensioned floor system is regular, utilizing an optimized post-tensioning layout for tendons (Figs.7 and 8).

The tendons are mapped based on analyses considering the vertical wall elements in plan. Post-tensioned slabs are commonly used worldwide in concrete structures of all scales. The slabs generally provide an efficient solution to long-span conditions with relative thin slab thicknesses. Historically, the quantity of post-tensioning required to achieve usable spans in commercial and residential construction has been 0.9 psf to 1.0 psf. Attempts to reduce the amount of post-tensioning has not been successful even when considering banding and careful placement of material. However, research has shown that placing tendons on optimized load paths have resulted in reductions of post-tensioning of 25–30%. Therefore, the quantity of post-tensioning historically required is now on the order of 0.65 psf to 0.75 psf. This reduction not only represents a reduction of raw material required, but a reduction in construction time and carbon emitted into the atmosphere due to both the production and placement.

Repetitive, equally spaced columns result in a particularly good application of optimized post-tensioning for the slab framing. It should be noted that repetitive circular plan geometries emerge from this concept that allows for easy markings on formwork for the layout of the tendons during construction (Figs. 9, 10, and 11).

Dynamic Responses of Structures: Creative Use of Post-Tensioning

All structures are in motion. Natural dynamic responses of structures are sources of inspiration for new structural systems. Free vibration of systems illustrates modal behavior, and in seismic events, structural demands are directly related to site and superstructure characteristics (Fig. 12).

The use of post-tensioning in vertical structural elements will lead to better performance when subjected to seismic loadings. Reflecting on research performed for slab systems results in a geometric nonlinear placement of post-tensioning. When placed in this manner for vertical column elements, the post-tensioning creates both net compression on the members but also provides resistance to lateral loads applied in any direction. Post-tensioning could be placed three-dimensionally in columns subjected to biaxial bending.

Similarly, geometric nonlinear post-tensioning could be placed in frame beam members to resist lateral cyclic loads

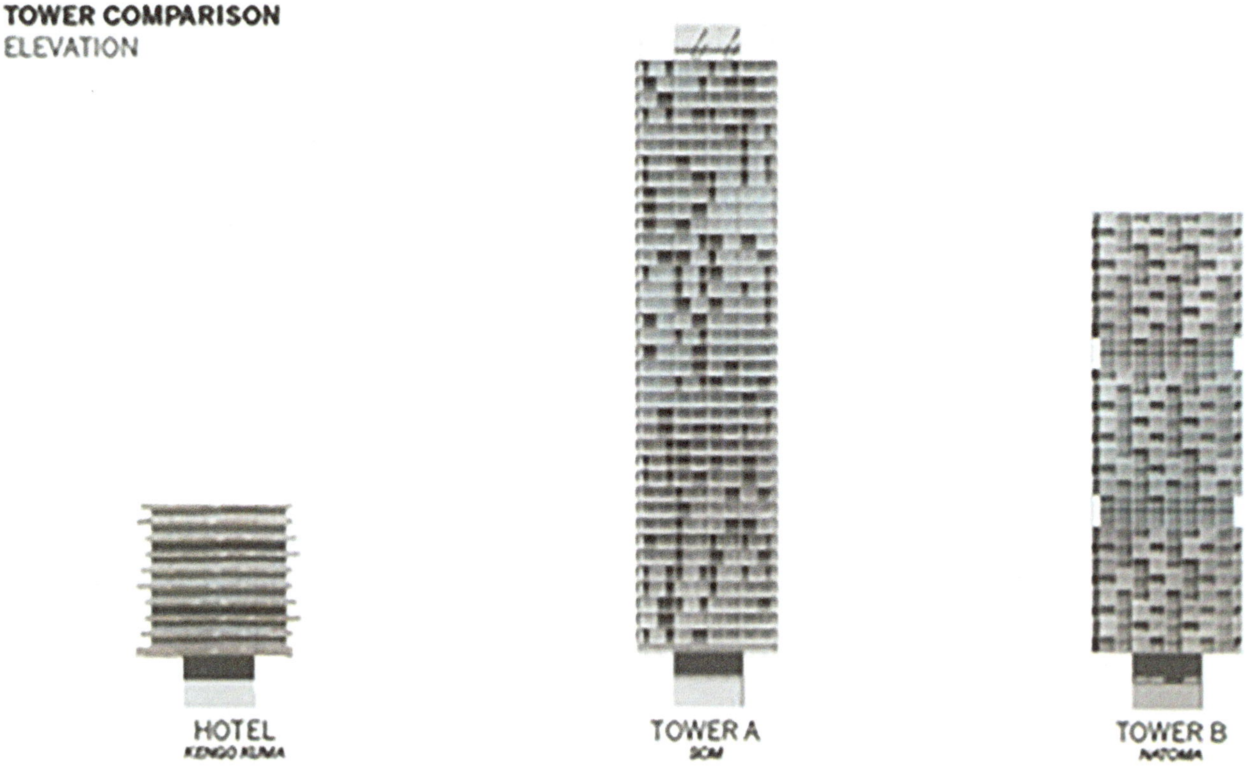

Fig. 7 1111 Sunset Boulevard Towers, Los Angeles, CA

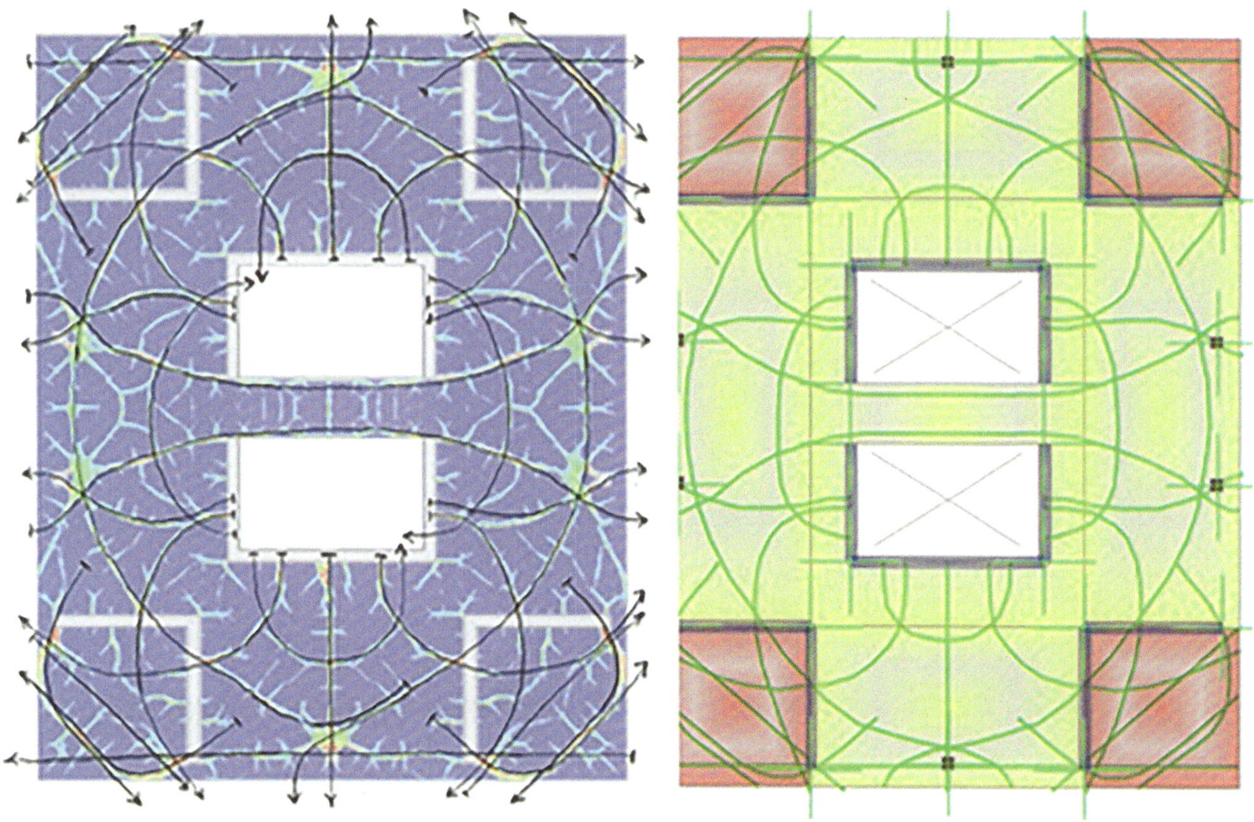

Fig. 8 Optimized post-tensioning layout in plan – 1111 Sunset Boulevard Towers, Los Angeles, CA

applied in any direction. This placement of post-tensioning in both columns and beams acts to achieve essentially elastic behavior in a seismic event and creates elastic strain energy in the overall system that helps to establish recentering after a significant seismic event (Figs. 13 and 14).

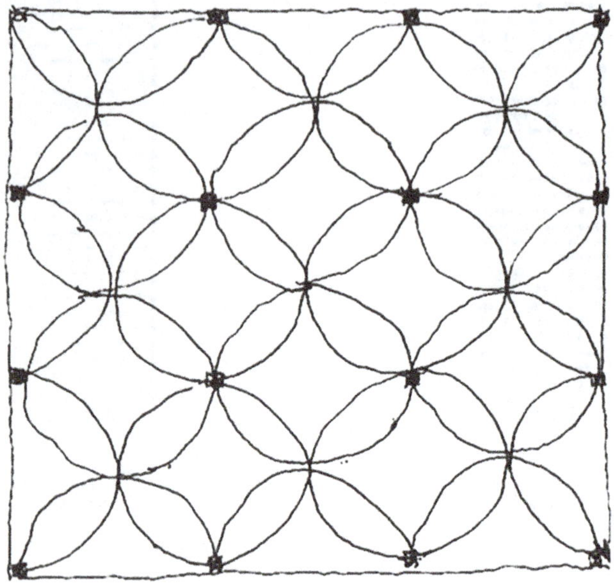

Fig. 9 Theoretical layout of completely symmetrical optimized post-tensioning layout in plan – 1111 Sunset Boulevard Towers, Los Angeles, CA

Fusing Tall Building Frames

The challenge of braced framed tall building structures is that stiffness most be combined with softness. Stiffness is typically required for controlling displacements during service conditions (wind loadings and moderate seismicity). However, these frames typically result in low ductility during extreme seismic events. Therefore, the key to solving both issues is to introduce fuses into the frames that allow for the dissipation of energy only when needed.

The design of the structure for the super tall Shenzhen Citic Financial Tower was optimized to achieve the maximum stiffness of the building with least material, by considering tip displacement of the building when subjected to load. The resulting design is the stiffest structure that can be obtained for the volume fraction of material considered and satisfies the performance requirements under gravity and lateral loads (wind and seismic) (Figs. 15 and 16).

The optimal geometry was derived for elastic material behavior, which was accurate for wind loads and earthquakes with a relatively low return period for the site. However, when earthquakes with a longer return period were considered, the structure exceeded the elastic limit and ductile elements needed to be incorporated in the stiff lateral system to ensure energy dissipation with consequent reduction of the inertial loads on the structure.

The seismic fuses could consider use of a conventional ductile steel link similar to those used with eccentric braced

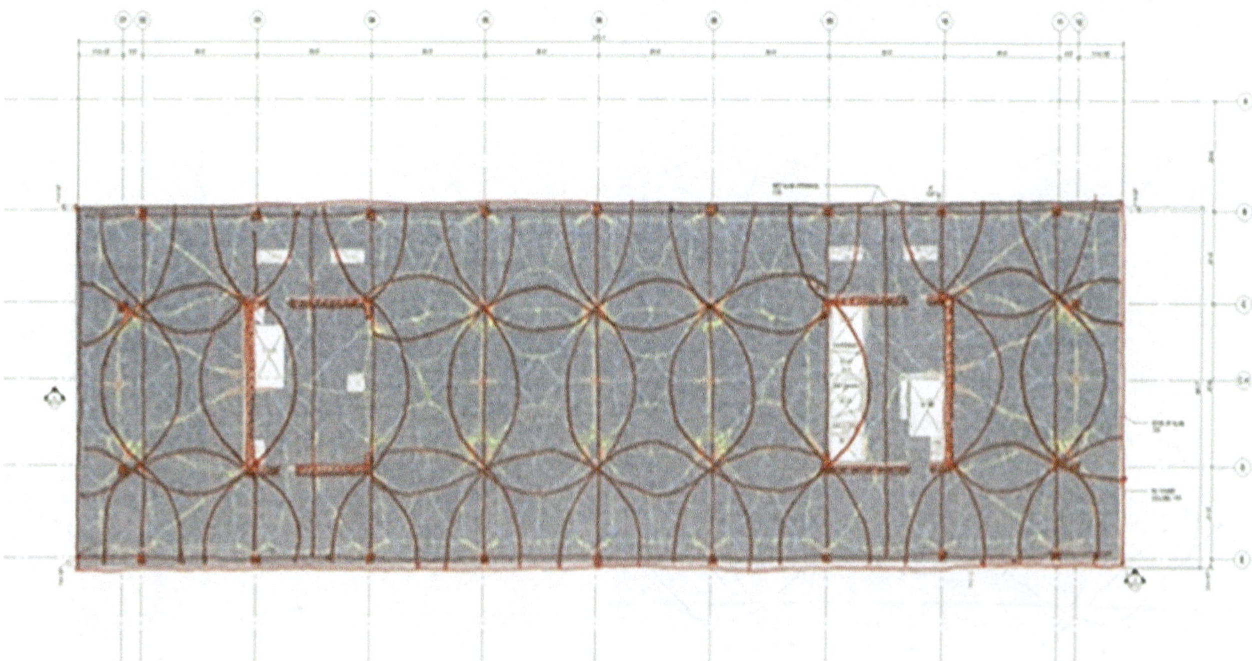

Fig. 10 Overlay of potential lines of post-tensioning over optimized slab analysis

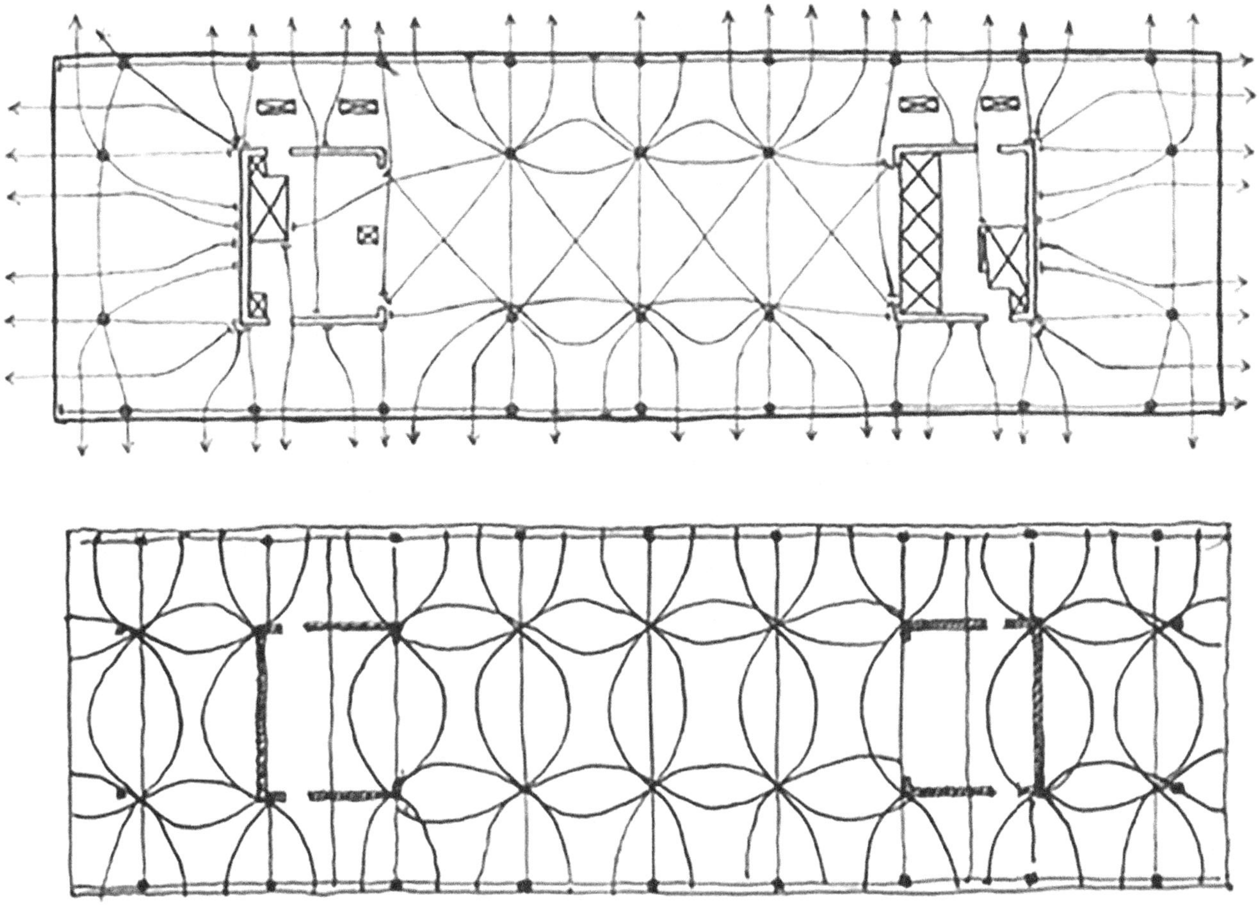

Fig. 11 Initial pass at post-tensioning layout (above) and the final interpretation of the layout (below)

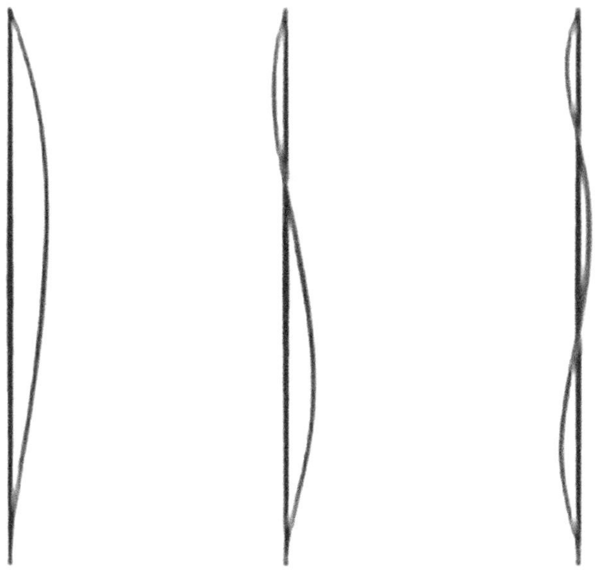

Fig. 12 Structural dynamics – mode shapes

frames or could use a modified link-fuse joint that is designed to protect the base building structure from permanent damage. The conventional ductile steel link is designed to yield during a significant seismic event where the link-fuse uses a clamped pin connection, dissipating energy through sliding of the joint (Figs. 17 and 18).

Conceptually, the systems mimic plant behavior where trunks carry water vertically to branches and leaves, with active movements from the ground to the top. For this structure, the principles are the same; only forces replace water and active flow from the ground is replaced with resistive forces originating in the superstructure, the seismic mass of floor framing systems, or winds applied to each floor diagram passing into the braced frame down to the foundations. Shear and bending moments increase as these forces near the ground and so does the density of the structure. The structure could have multiple purposes and would be resilient, capable of acting elastically with minimum damage in strong earthquakes with optimal performance for wind and frequent seismic events (Fig. 19). This frame system could be considered

Fig. 13 Vertical post-tensioning in columns

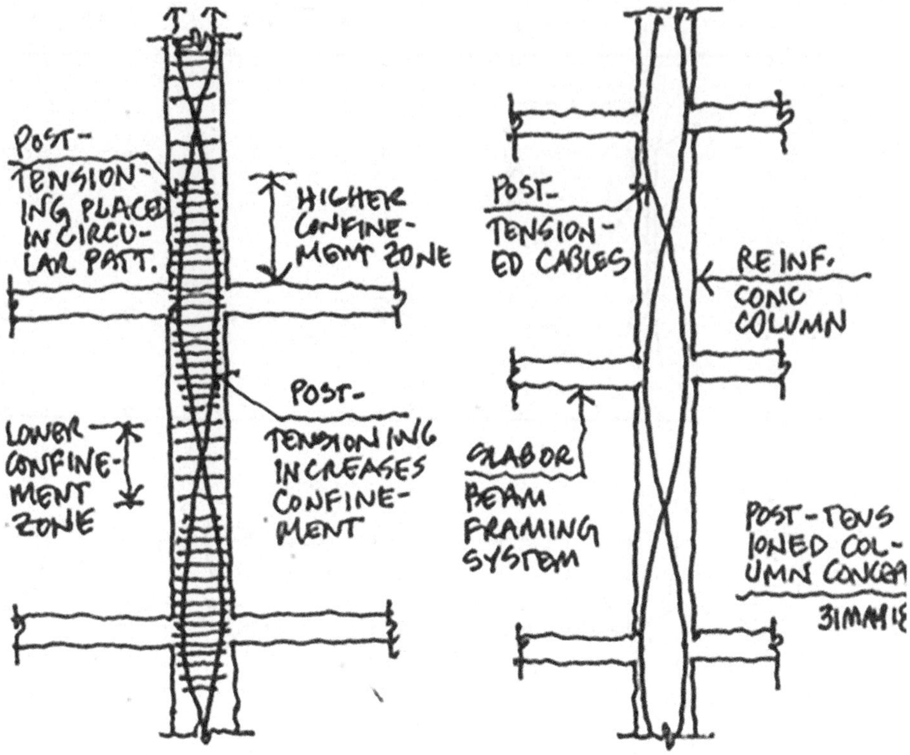

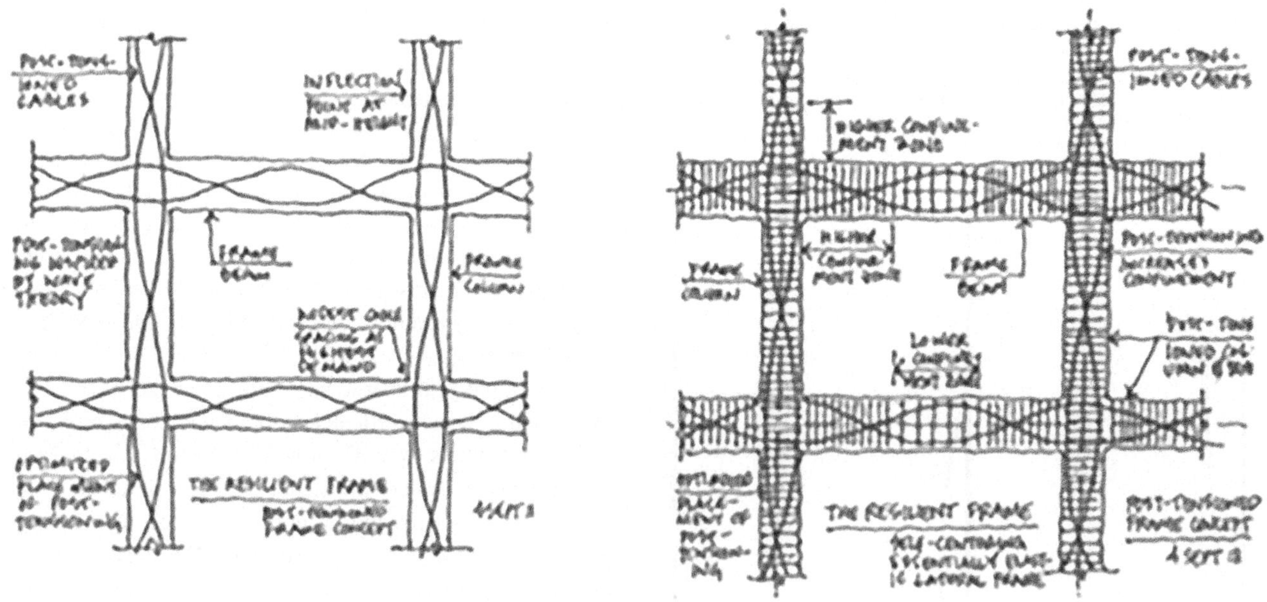

Fig. 14 Post-tensioning in moment-resisting frames subjected to seismic loading

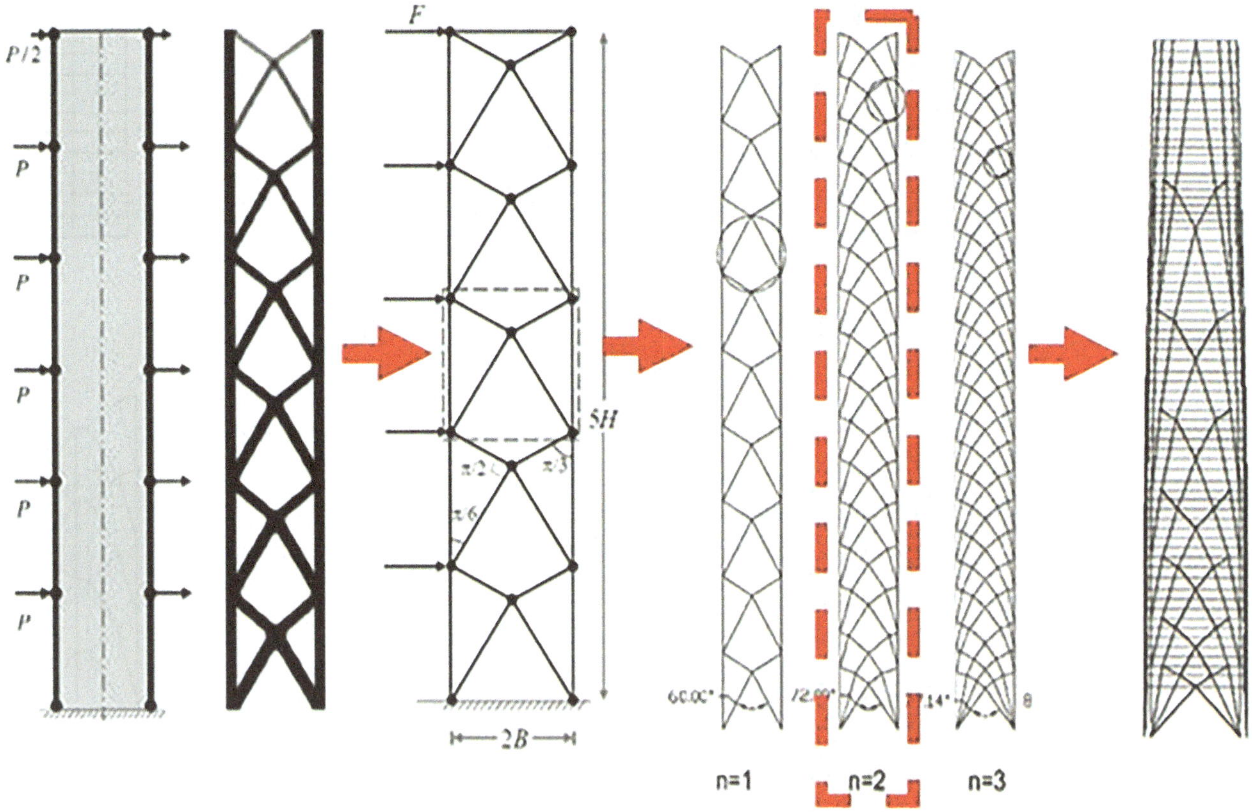

Fig. 15 Citic Financial Tower

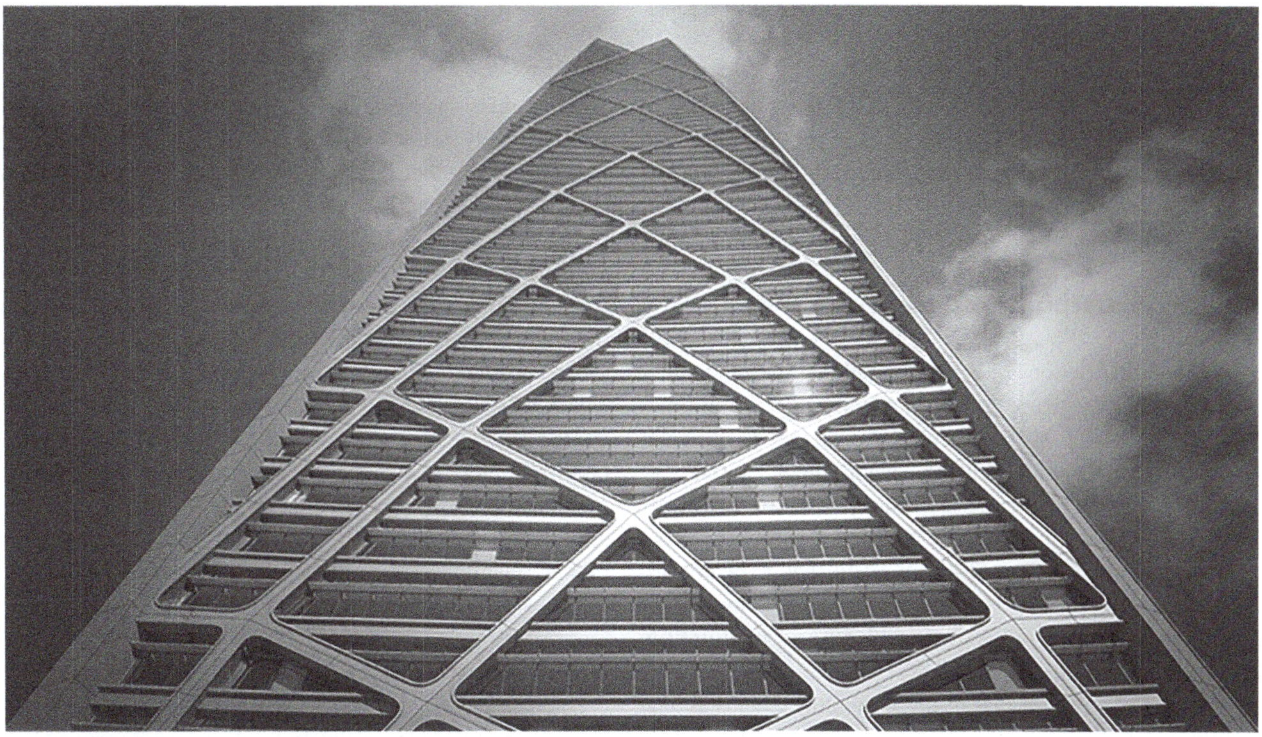

Fig. 16 Evolution of theory to concept for structural system

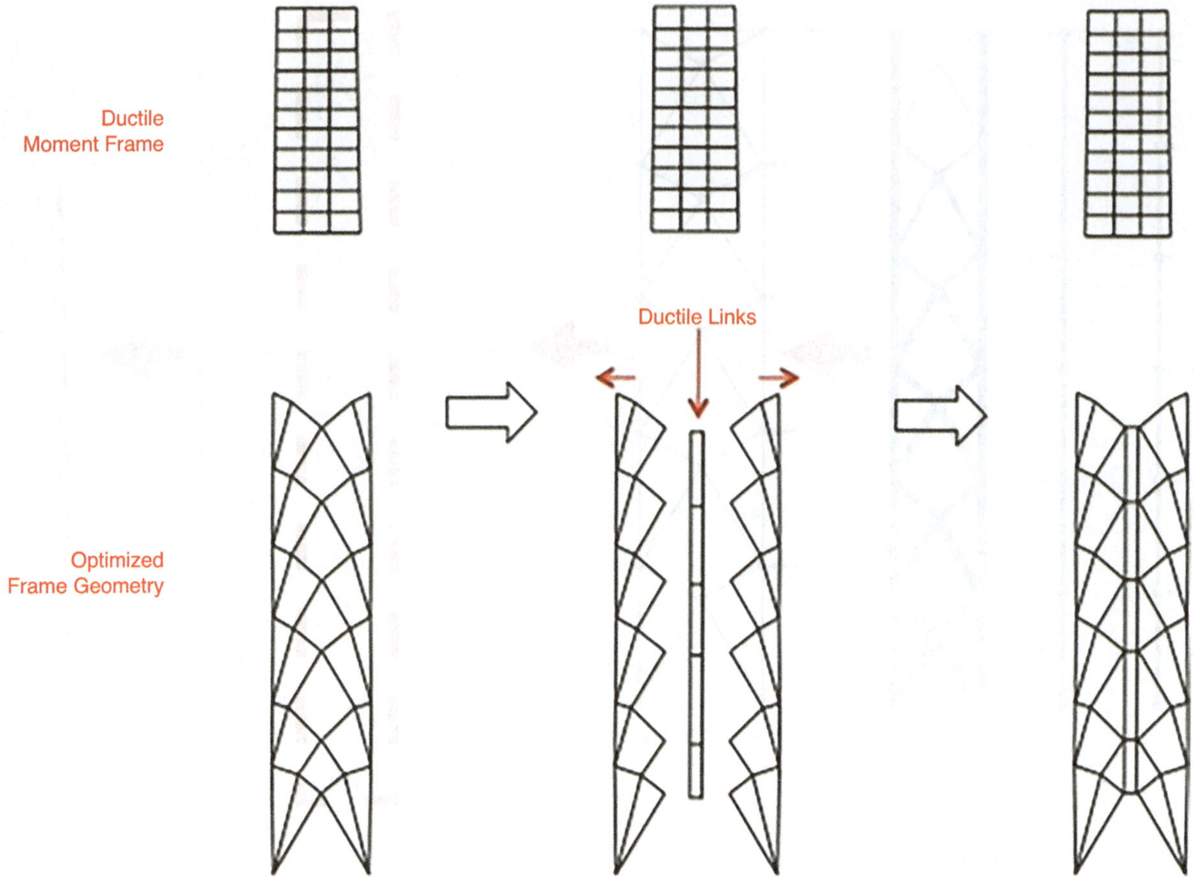

Fig. 17 Braced frames interconnected with seismic fuses

as steel, concrete, or composite. For the Citic Tower, concrete-filled tubes will be used with a central shear wall core of reinforced concrete and lightweight structural steel framing for the floor system (Fig. 20).

Embodied Carbon: Proposal for Change

Millions of metric tons of carbon are emitted into the atmosphere during the extraction, refinement, and installation of structural materials such as steel, concrete, wood, and cold-formed metal framing. With the environment at risk, the structural engineering profession must carefully reconsider design approaches to structures. Embodied carbon of structural systems in buildings has been established to be considerable in the overall environmental impact of buildings. By some estimates, it could comprise 15–50% of the total life cycle of carbon emission buildings (Fig. 21).

One could envision that in the near future, international building codes will not only address safety but also life cycle performance, imposing limits on carbon emissions attributable to building construction and lifetime use. As a profession, we need to create greater awareness of damaging carbon emissions resulting in global warming and embrace the most advanced systems available, creating a holistic awareness of their beneficial impact on performance, finances, damage, and environmental impact.

The majority of sustainability-related efforts in the building industry have focused on carbon emissions resulting from building operation. To assist designers in quantifying the embodied carbon of structural systems, the Environmental Analysis Tool™ (EA Tool) has been released for public use. The EA Tool can assess probabilistic seismic damage using the HAZUS methodology for conventional structures and an empirical method for enhanced structural systems such as base isolation. Other efforts have considered different approaches to assessing the environmental impacts of probabilistic seismic damage such as EnvISA and ATC 86.

Data and Embodied Carbon

Over several decades, a database of over 200 designed and constructed buildings has been complied, and embodied carbon levels have been assessed for each project using the EA Tool™. Below, data subsets are graphically represented con-

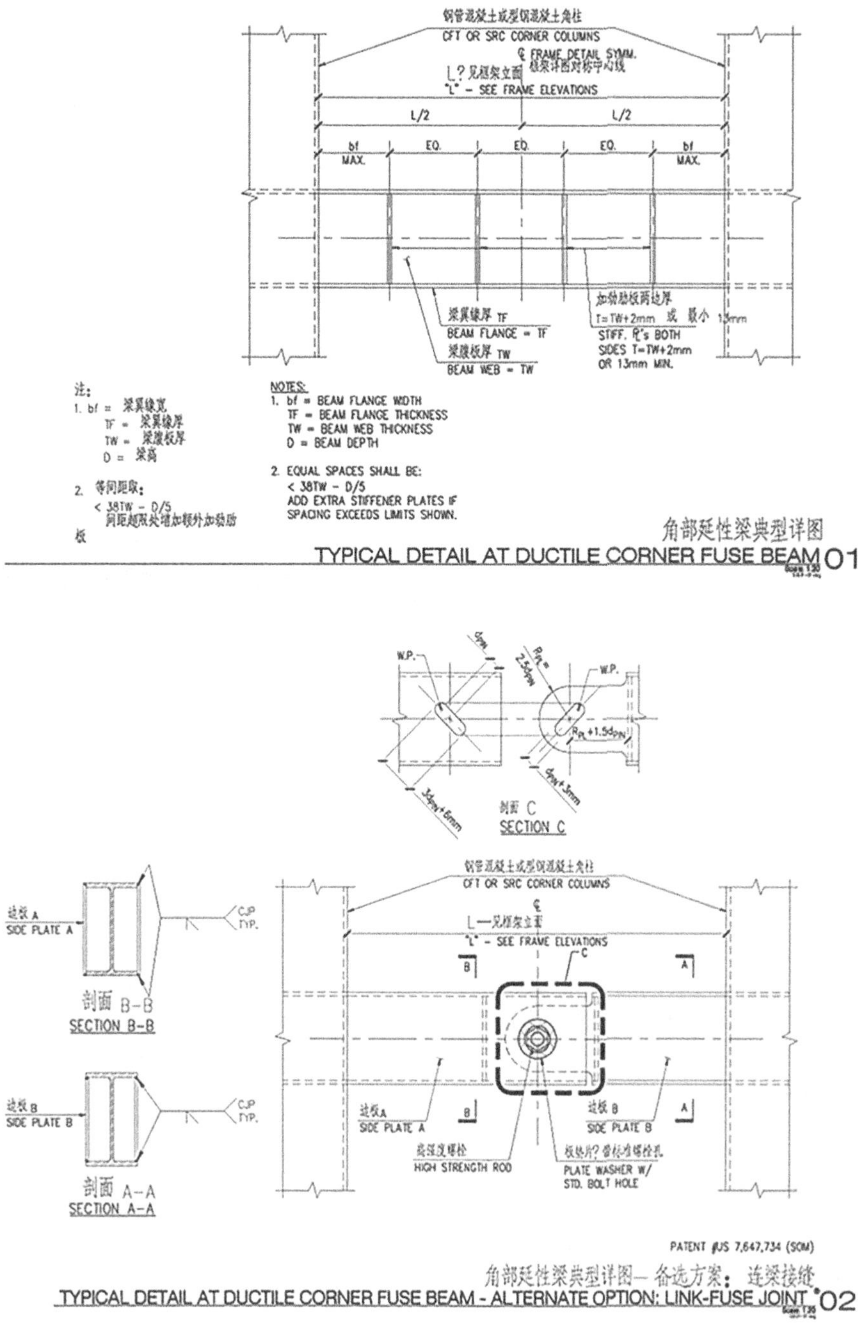

Fig. 18 Ductile steel seismic fuse, futuristic link-fuse seismic joint

sidering key parameters such as building use and primary structural material. These embodied carbon values include a variety of structural materials, construction, and probable seismic damage associated with foundations, substructure, and superstructure. In the graphs below, "composite" refers to a reinforced concrete core with steel gravity framing that may or may not be combined with steel moment frames and/or braces. "Mixed" refers to a building with multiple occupancy types such as office and residential. All emissions data is provided in CO_2eq. The average $C0_2$eq/m2 value of 586 kg/m^2 is provided in Fig. 22 alongside individual data points.

Analytical Investigation

As observed in Fig. 22, buildings above approximately 60 stories typically produce higher CO_2eq intensities than those below 60 stories. To better understand these and other per-

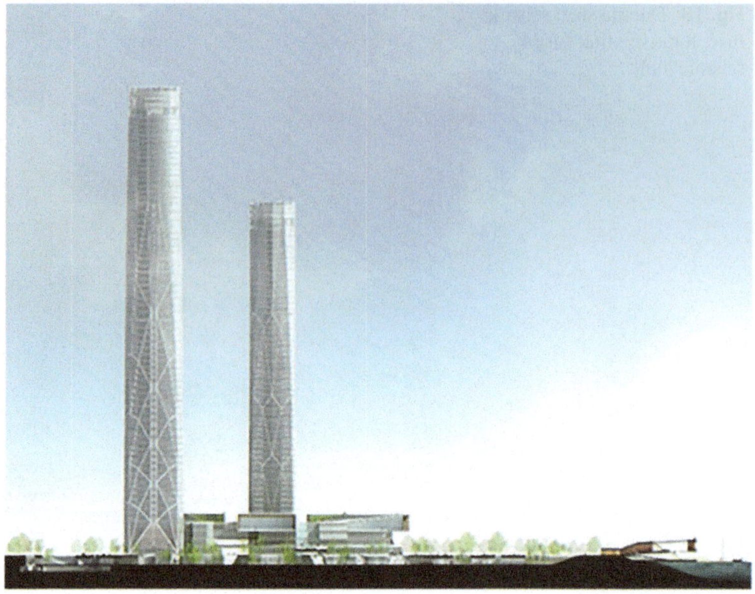

Fig. 19 Rendering of Citic Financial Towers – office, hotel, and residential uses

ceived trends, Fig. 23 provides CO_2eq averages across height and material categories. Potential relationships between parameters such as material and height with embodied carbon are provided in the following correlation study.

Correlation Study

Some inferences can be made from the correlation matrices. It is apparent that for buildings over 60 stories, structural material choice has a significant impact on embodied carbon as opposed to those under 60 stories where many factors contribute relatively equally.

Developing a Basis of Sustainable Structural Design

What determines if a project is "sustainable"? What is an embodied carbon limit that results in an appreciable difference in the environment? Should embodied carbon limits be absolute or relative to a baseline design? Should different structural systems have different embodied carbon limits? What should be included in the embodied carbon life cycle?

A basis of carbon-conscious design needs to be proposed and developed. It should guide clients and designers to develop more sustainable structures while providing flexibility for project-specific conditions, encouraging synergy between building systems. Requirements should influence design decisions but not be over burdensome on project budgets, design teams, or contractors. Balancing these considerations is important in successful systems implementation and should be developed with the total environmental impact of the built environment in mind. An identification of the salient parameters that influence embodied carbon levels is key; those presented in this study and in other literature should be considered.

Embodied and operational carbon are correlated in many ways. Although it has been shown that taller buildings have higher levels of embodied carbon, high-rise mixed use buildings in dense urban settings greatly reduce transportation-related carbon emissions of occupants. These types of relationships need to be quantified to understand their respective and correlated impacts. Their synergetic relationships should be encouraged and recognized by building codes and incentive programs.

Previously discussed case studies substantiated that significant reductions in embodied carbon ranging from 20 to 50% can be realized, especially in regions of high seismicity using currently available tools. As an initial goal, a 15% reduction, shown as a red line in Fig. 24, from the dataset average of embodied carbon is proposed. This reduction can be realized in the construction industry immediately given the resources and expertise available. Potentially, a more significant required reduction of 30% could be implemented in as little as 10 years.

As a means of beginning this process, the following CO_2eq limits are proposed for a 15% embodied carbon reduction in Fig. 24. Here parameters are aggregated by structural height category to illustrate a metric that can be used to facilitate tailored implementation. Furthermore, provisions should be made for low and high seismic regions that have been determined to exhibit −10% and +10% CO_2eq levels, respectively.

The Transbay District of San Francisco is currently under redevelopment. Several height limitations have been increased to encourage replacement of older, less efficient buildings, and a large mass transit center is being replaced.

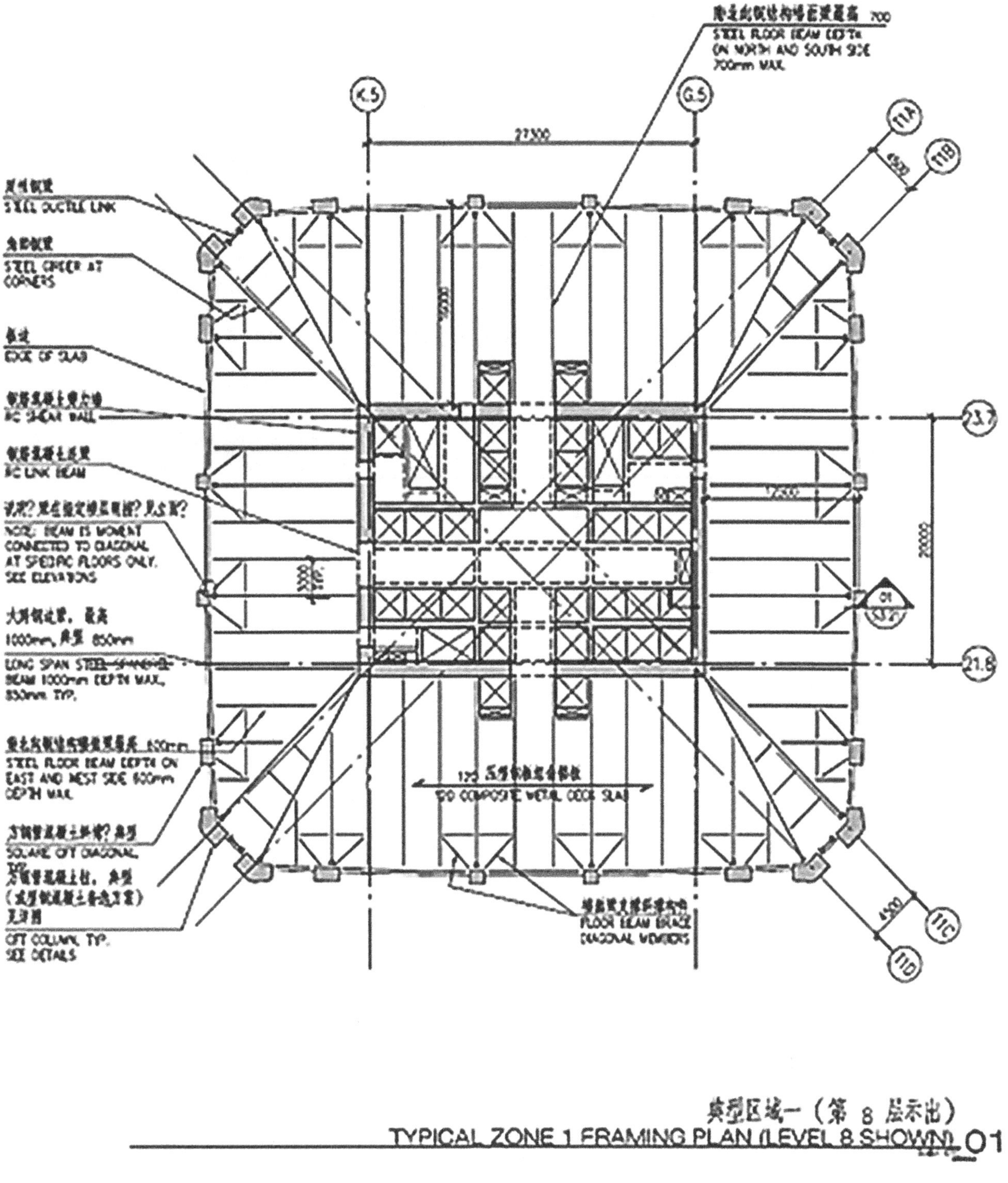

Fig. 20 Citic Financial Tower – floor framing system

Some parcels' limits have been increased to over 220 meters; currently, a single parcel with a 300-meter limit is under construction and will become the tallest building in San Francisco.

As a case study investigation, PCM will be employed to evaluate NFA and embodied carbon of potential building forms. As opposed to evaluating current building forms, the investigation will focus on the potential of each given parcel extents and height limitations. A uniform offset of 15% from parcel edge to building form is assumed.

By knowing the parcel limits and height limitations, an initial form extent can be realized. Viewing this form as a

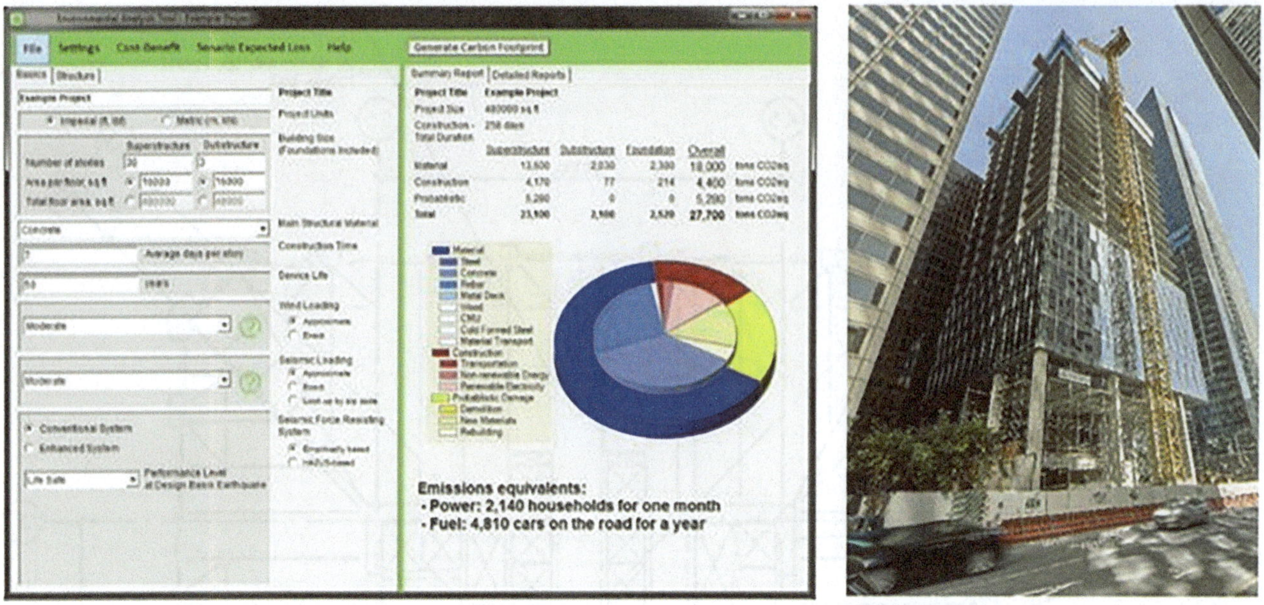

Fig. 21 Tool software user interface and EA tool components (left) and SF office tower (right), a project on which EA Tool™ was used to reduce embodied carbon

Fig. 22 Environmental Analysis Tool™ methodology in assessing structural embodied carbon EA

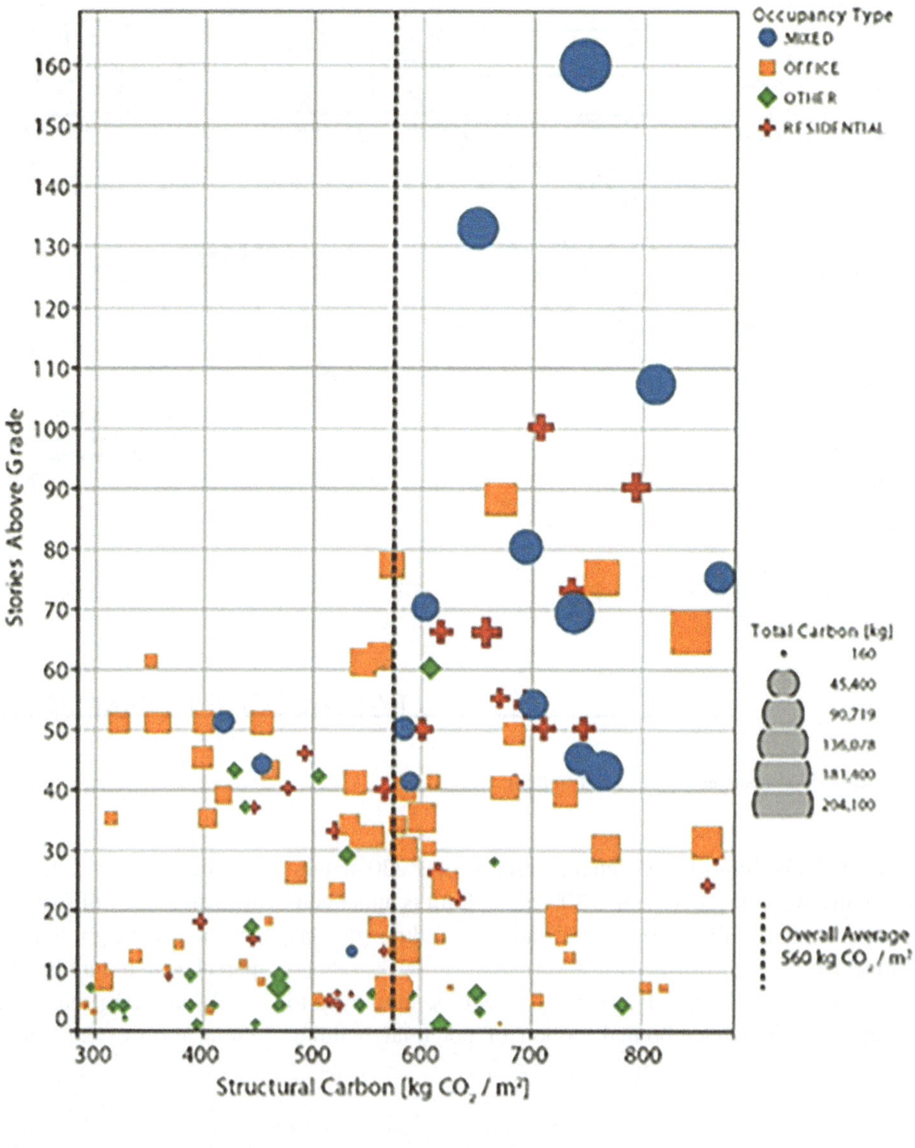

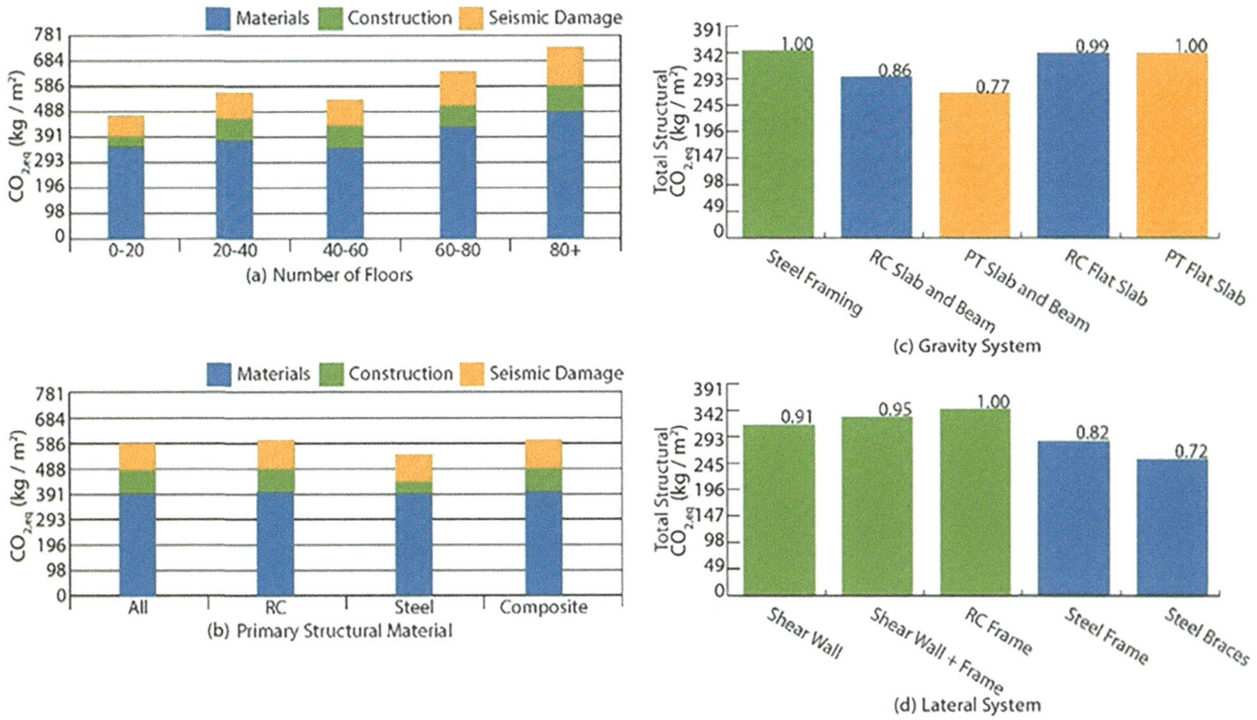

Fig. 23 CO_2eq correlations for structural materials and systems

GFA limit of the parcel, a study of the potential performance of the parcel can be conducted with PCM. The methodology can identify parcels that are well suited for their designated height limits as well as parcels not likely to develop their desired potential.

To investigate the potential impact planning and design decisions have at a district scale using PCM, a series of analyses are conducted considering building height, material, and resiliency (Fig. 25). In Fig. 25, red indicates buildings with low or poor NFA values and green indicates parcels with NFA values exceeding 75%. First, the current Transbay District plan is evaluated using PCM. As can be observed, a large number of small parcels that have been zoned with tall height limits cannot reach their desired potential due to poor NFA values and corresponding financial performance. Next, each parcel's height limit is adjusted to produce an NFA value of 75%. The resulting urban form is relatively uniform but can potentially facilitate nearly 50% more GFA – an important consideration with the increasing densities of our future cities. The effect of structural material selection is also considered; even taller buildings can be facilitated with steel construction and yield a nearly 70% GFA increase.

Structural embodied carbon considerations are quantified for the three abovementioned cases. As presumed, tall buildings often have higher embodied carbon values. Furthermore, for the optimization height limits with concrete as the structural material, the overall environmental performance is very good while the opposite is true for the steel scenario. This could be due to the material and taller height limits.

This environmental issue can be mitigated through enhanced seismic performance. For low-rise buildings, this could be achieved with technologies now in use such as base isolation and in taller building with novel energy dissipating elements such as the Pin-Fuse seismic systems.

Fused Timber Systems

The building community is challenged to improve the global built environment, both reducing greenhouse gas emissions and addressing community resilience to earthquakes. Structural material selection is critical in limiting embodied carbon (the equivalent global warming potential created to manufacture, fabricate, deliver, and construct structural materials). For this reason, mass timber has emerged as a viable alternative to concrete and steel. However, the performance of these timber systems in high seismic zones must be improved to provide reliable lateral systems that guarantee ductility at large magnitude seismic events. Alone, timber does not possess the ductility of steel or reinforced concrete and can experience more brittle failures.

Traditional wood-framed buildings are categorized as either light-framed buildings using stud walls and joists or heavy timber buildings using large sawn timber elements.

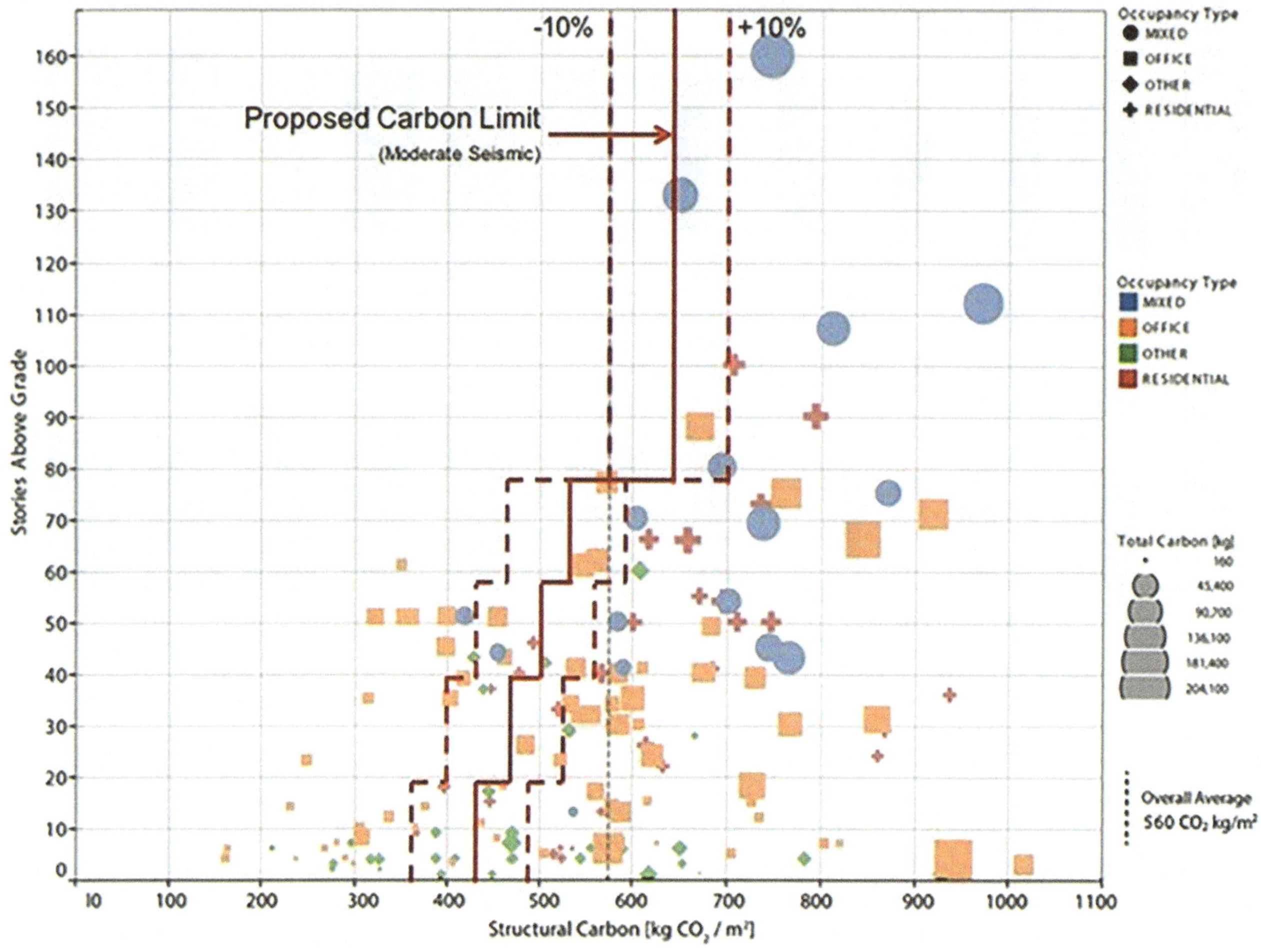

Fig. 24 Proposed embodied carbon limits (CO_2eq)

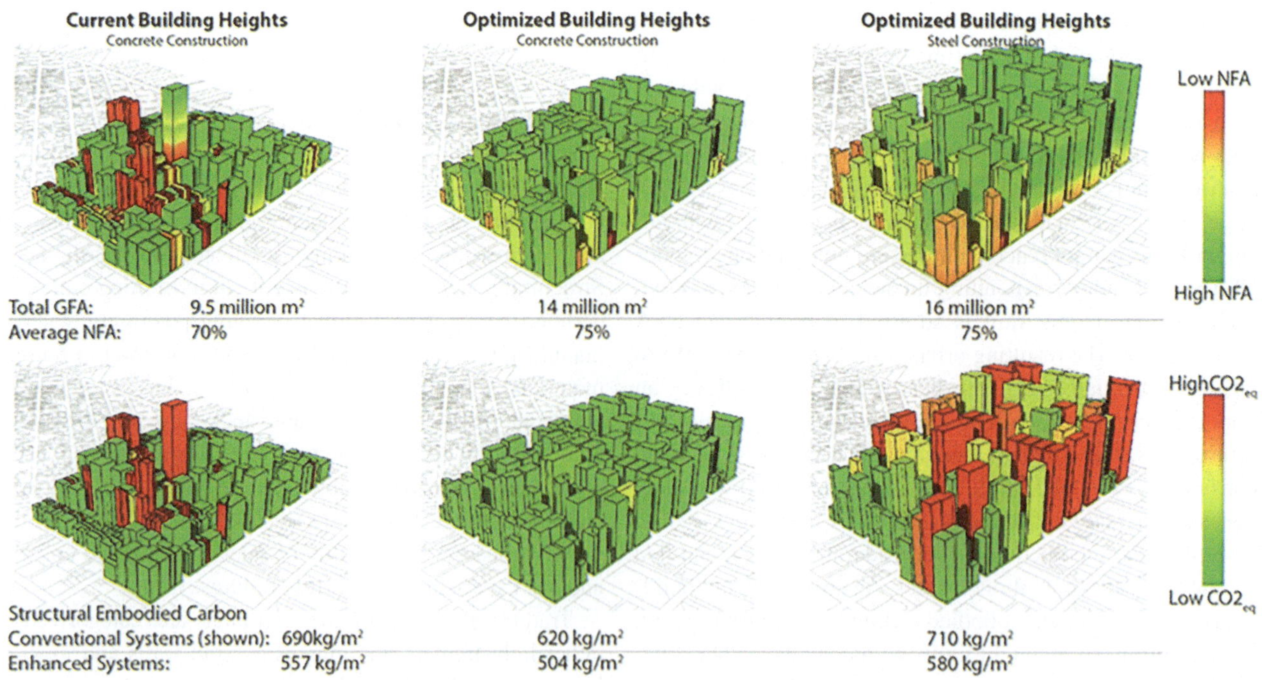

Fig. 25 Results of PCM analysis at the district scale

Heavy timber has a higher fire rating than light-framed because the mass of timber elements forms protective char layers during a fire. Contemporary timber buildings rarely use heavy timbers for support and instead use "mass timber" – framing elements incorporating small dimensional lumber and adhesives to build up massive elements behaving like heavy timber in fire. Mass timber elements include glued-laminated (glulam) timber columns and beams, cross-laminated timber (CLT) floor and wall panels, laminated veneer lumber (LVL) floor and wall panels, and nail-laminated timber (NLT) floor panels. The wood materials in mass timber have several key properties. The density of wood is typically 482–561 kg/M3 that results in a lightweight framing system. The weight and seismic mass of a mass timber building is typically 50 to 60% that of a concrete building, reducing foundation demands.

Mass timber framing systems have several benefits in multistory buildings. Elements are prefabricated and can be erected quickly on site. These construction and cost benefits make mass timber buildings competitive, but their sustainability benefits have arguably had the largest impact on their increasing share of the market. Wood is a renewable resource that promotes healthy ecosystems when sourced from sustainably managed forests. Wood is a natural carbon sink, consisting of approximately 50% carbon.

Typical framing systems for residential mass timber buildings consist of CLT bearing/shear walls located at demising walls between units. CLT floor panels span between the bearing walls that creates a flat soffit condition like a concrete flat plate. The floor panels bear directly on the wall panels in a "platform" framing style where bearing walls are discontinuous. The lowest floor of mass timber buildings is often framed in concrete for ground floor flexibility and durability (Fig. 26).

Wood elements have low to moderate ductility and typically rely on yielding of metal connectors for ductility. The most common connections for CLT shear walls are metal "L" brackets are attached to walls and floors. The ductility of this system is relatively low as most energy dissipation comes from the yielding of the fasteners and of the metal brackets in through-thickness bending. Steel connectors are relatively small compared to the size and strength of the walls, resulting in a low amount of dissipated energy compared to the elastic energy stored in the CLT. The low ductility of CLT shear walls systems offsets the seismic mass benefit of the material.

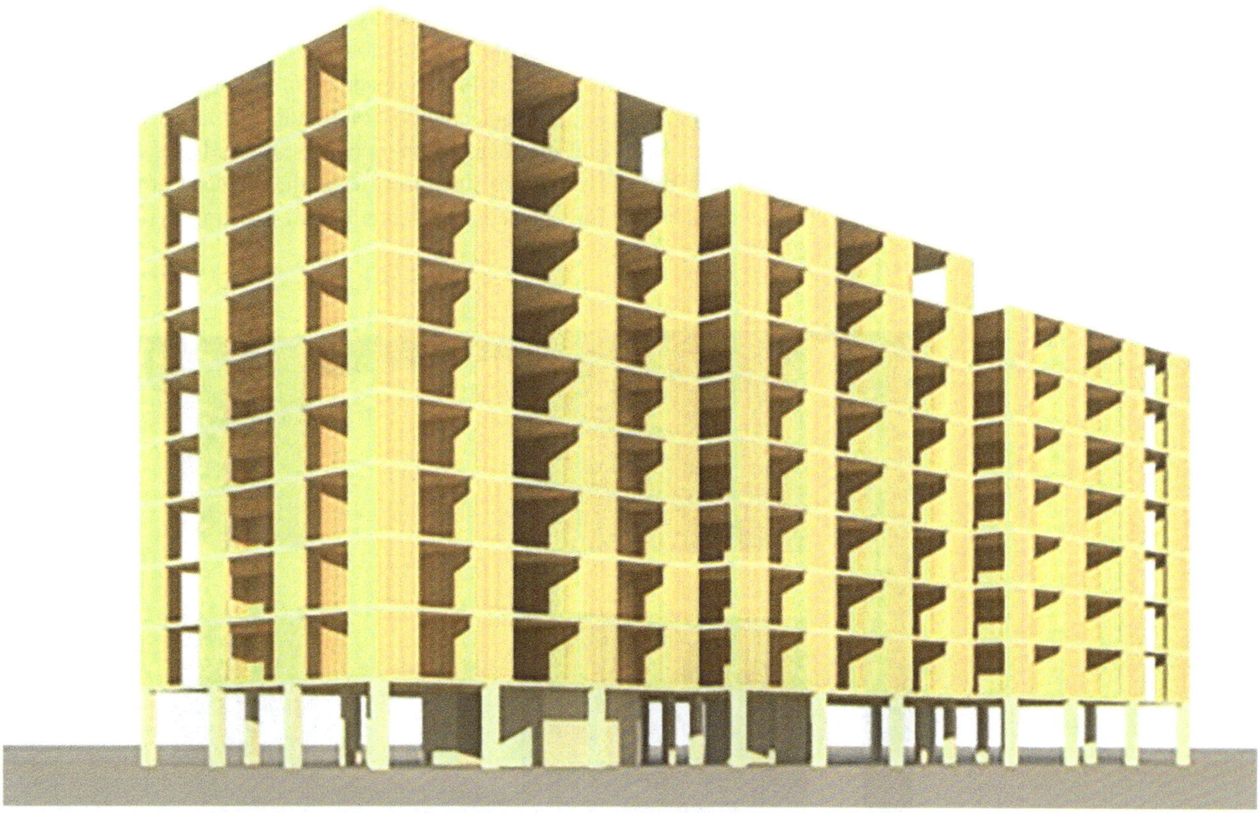

Fig. 26 Typical mass timber building with CLT walls and floors above a concrete podium

Pin-Fuse Seismic Systems for Timber

Even with these developments, barriers still exist to designing timber tall buildings (~15 stories) with enhanced seismic systems comparable in performance and construction to traditional high-rise materials:

- Feasibility of rectangular structural cores for increased overturning stiffness
- Modular and rapid construction of prequalified moment frame systems

For these reasons, ongoing analytical research is being completed to incorporate the Pin-Fuse® collection of seismic systems, originally developed for steel and concrete systems, into mass timber structures.

Timber Pin-Fuse® systems consist of mass timber structural components combined with steel Pin-Fuse® Joint or Link-Fuse Joint™ energy dissipating devices. Systems are designed such that the timber elements remain elastic during a seismic event. The steel Pin-Fuse® elements are expected to be prefabricated, shipped to the timber manufacturer, and preassembled before delivery to site.

The Pin-Fuse® Joint connects glulam beams and columns (Fig. 27). One implementation of the connection utilizes a notch in the timber beam at the steel plates to limit the total width of the beam sandwich and provides opportunity for architectural expression of the ball and socket joint inspiration. For the timber Link-Fuse Joint™, the large single shear pin of the original Link-Fuse Joint™ is replaced with multiple bolts (Fig. 27). This design allows for a use of off-the-shelf products and is feasible with the relatively lower link beam shear demands in timber structures. The details depict dowel-type connectors through the wood. These connections could use self-tapping screws or timber rivets for low to moderate seismic load applications or require epoxy-connected steel plates embedded in the timber elements for highly loaded joints.

Prototype Mass Timber Building

The TMBR™ Minneapolis building is a 10-story mass timber condominium that is expected to start construction in early 2020. The project is being developed by Todd Simning of TMBR Development LLC and designed by Dwyer/Oglesbay Architects, with timber structural engineering by SOM. The project consists of nine stories of mass timber-framed condos over a one-story post-tensioned concrete podium and three concrete-framed basements. The project is 125 ft-tall and has approximately 160,000 square feet of above ground area.

TMBRT™ is designed to be a luxury product for the local market and priced to compete with conventional concrete construction. The mass timber systems selected for the project must satisfy the stringent performance demands of a condominium project while also meeting the budget. The proposed structure consists of a timber-concrete composite floor system, glue-laminated timber beams and columns, and two cross-laminated timber shear wall cores. The cross-laminated timber cores are located at the central elevator and service core and a remote stair at the northern exterior wall. CLT link beams are made up of two separate segments, each a fully connected portion of adjacent larger panel segments, and span across core openings. Lapped screwed connections provide for shear transfer at the midspan of the link beams. This placement of lateral load resisting systems allows for

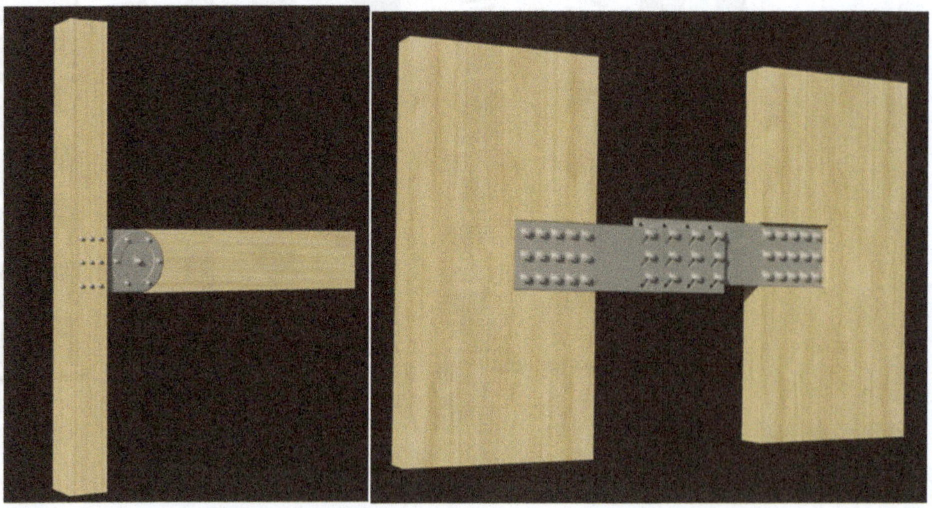

Fig. 27 Timber Pin-Fuse® joints at frame and shear wall

variable unit configurations over the height of the building similar to concrete-framed buildings. This approach is expected to be a competitive structural system in the residential market.

The success of this model will depend on the ability for CLT shear walls to satisfy seismic demands required elsewhere in the country. The ductility of the CLT shear wall cores is highly dependent on the types of connections and fasteners provided. Conventional CLT wall connections and fasteners have limited ductility, offsetting the lightweight benefits of timber construction in seismically active regions. The seismic performance of the mass timber system could be improved with the application of friction energy dissipating devices in the CLT cores, as well as the perimeter glulam framing.

Machine Learning and Futuristic Approaches to Resilient Structures

The concept of artificial intelligence (AI) has existed for decades, but the world is only now on the verge of an AI revolution due to the following factors: (1) access to large amounts of digital data and (2) access to powerful and affordable cloud computing power.

In parallel, the need for a technology revolution in the architecture, engineering, and construction (AEC) industry is imminent. The current world population of 7.6 billion is expected to reach 8.6 billion in 2030. Infrastructure demands will increase drastically, and projects will need to be delivered more quickly to meet this rapid population growth. Design and construction processes need to be made exponentially more efficient and sustainable in order to meet these demands in a responsible manner.

There are many opportunities to enhance the built environment through artificial intelligence. Incorporating AI into construction verification and infrastructure health monitoring could improve quality control and decrease construction time. AI-enabled infrastructure health monitoring tools could be used to improve maintenance and safety practices by evaluating the deterioration of both nonstructural components, such as exterior walls and structural components. Photo recognition-based systems can be applied to continuously evaluate and monitor bridges and highlight potential maintenance and safety issues in advance. AI damage assessment tools could make recovery and rebuilding efforts more efficient by rapidly and automatically evaluating structural integrity following major natural or man-made events. In addition to practical applications, AI can also be incorporated into current design processes to optimize for multiple parameters and develop new design options that may not have been thought of otherwise.

Post-Earthquake Building Damage Identification

One of the greatest economical losses associated with earthquake events is due to downtime from repair and rebuilding. Home and business owners of damaged buildings can wait months for an inspection to be completed by an expert. In some cases, they cannot occupy their home or operate their business until this inspection is complete. Photo recognition technology can be employed to expedite this evaluation process and mitigate this lost time. A comprehensive and vetted machine learning-based application could provide users with immediate feedback regarding the level of damage of their building, whether or not it can be occupied and whether or not it requires major repairs (Diaz et al., 2017).

In addition, in the aftermath of a major earthquake, local and international reconnaissance teams upload hundreds of images to various databases, not all of which can be rapidly evaluated by experts. Integrating machine learning into these databases could make the photo cataloguing process not only more uniform and efficient but also more comprehensive. Following the 7.1 magnitude Central Mexico earthquake on Tuesday, September 19, 2017, SOM sent a team of engineers to Mexico City to contribute to post-disaster recovery efforts by documenting building damage and providing technical support to local structural reconnaissance efforts. In an effort to enhance post-earthquake reconnaissance capabilities, SOM performed tests using machine learning to identify and classify building damage from the photos taken by the SOM reconnaissance team. Some results from these tests are shown in Fig. 28 below.

An object detection-based machine learning model was trained using approximately 40 annotated images in the training set. The model was trained to differentiate between the following eight classes: (1) nonstructural light damage, (2) nonstructural medium damage, (3) nonstructural heavy damage, (4) nonstructural severe damage, (5) structural light damage, (6) structural medium damage, (7) structural heavy damage, and (8) structural severe damage. The model was based on a Faster RCNN (Ren et al., 2015) approach using Resnet-101 (He et al., 2016). It was initialized with the pre-trained weights from the MS COCO dataset. One model was trained using 3000 iterations, while the other was trained using 10,000 iterations. The model trained to 10,000 iterations most likely overfit the data. This means that it very accurately represents the training data, at the expense of

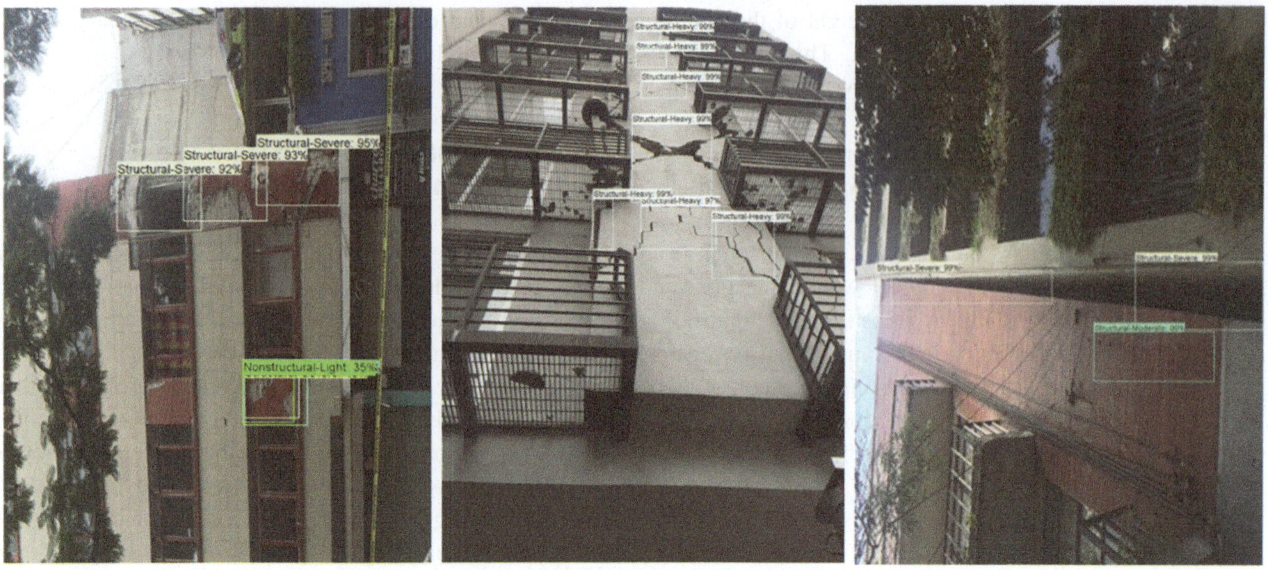

Fig. 28 Examples of building damage identification and classification using machine learning

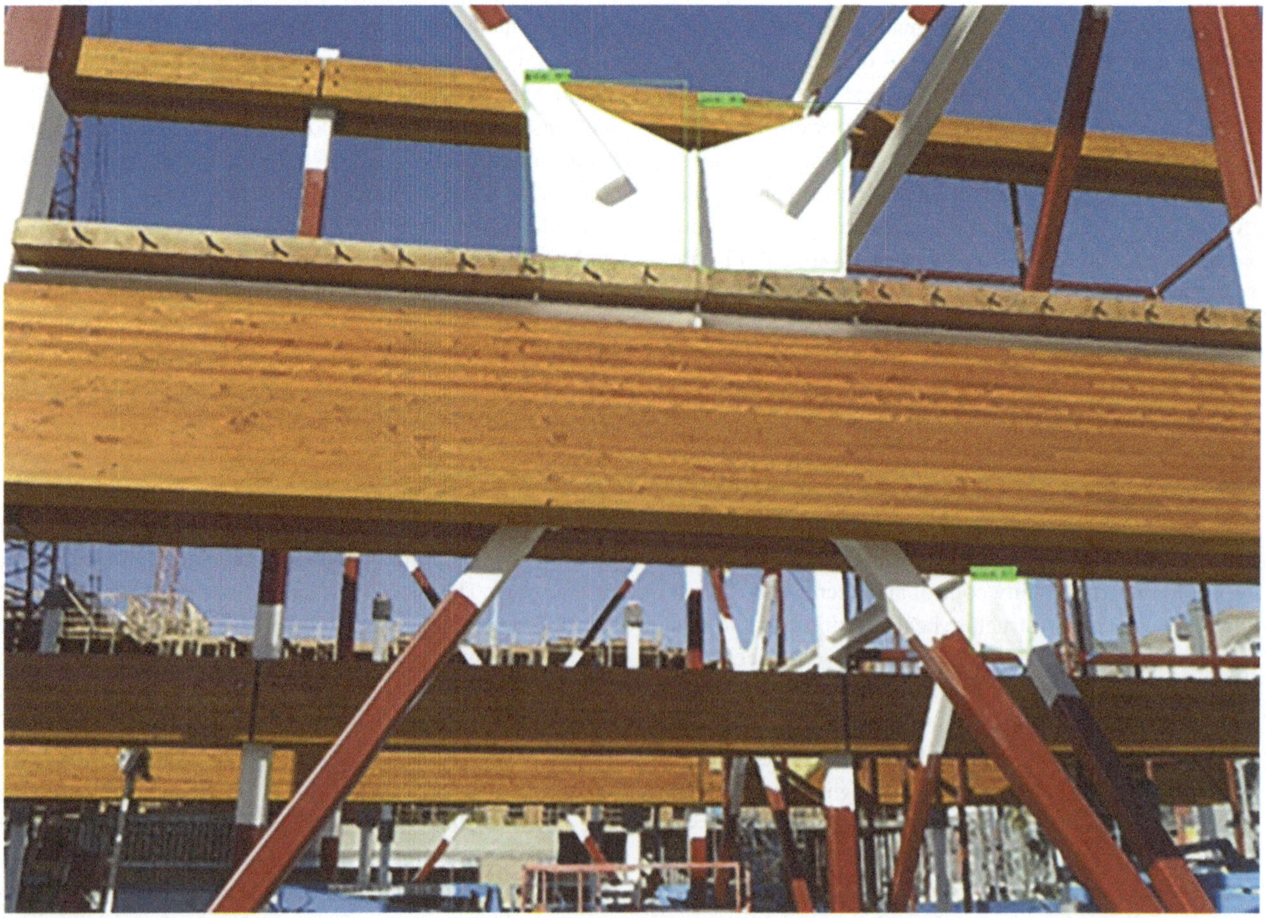

Fig. 29 Identification of gusset plates from site photos using object detection

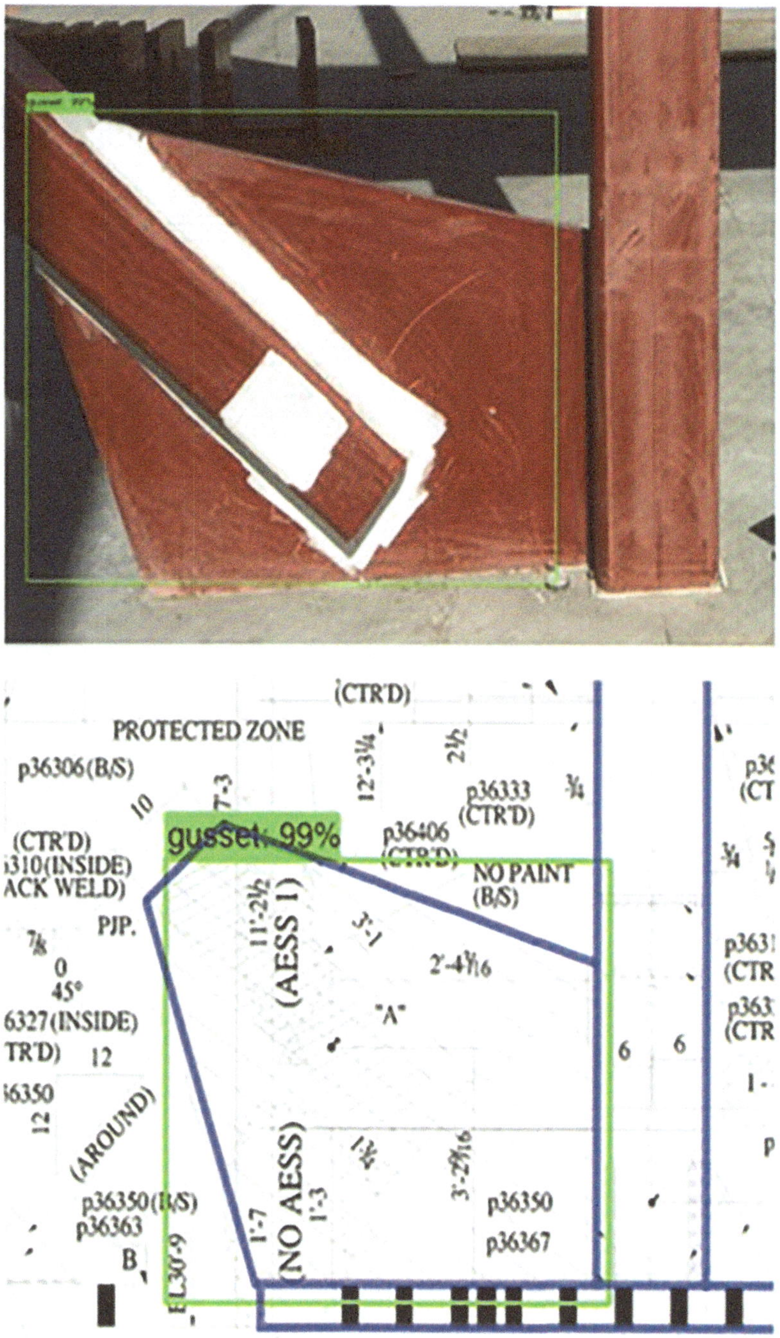

Fig. 30 Identification of a gusset plate from both a site photo and shop drawings (blue outline added manually to highlight the steel shapes)

being as useful to novel data. Using a similar method with a training set of at least 1000 images may yield a model that generalizes well to new data.

Construction Verification

Machine learning models can be trained to identify gusset plates from photographs using object detection. The model was able to identify gusset plates with a high degree of confidence from photographs taken at various angles. For example, in Fig. 29, the model detected not only the two gusset plates in the foreground facing the camera but also the gusset plate in the background at an angle to the camera. The success of this model may be due to the unique shape of gusset plates, which makes them clearly distinguishable from other elements on a construction site.

In order to compare objects identified from site photos to the construction documents, a machine learning model was also trained to identify gusset plates from shop drawings.

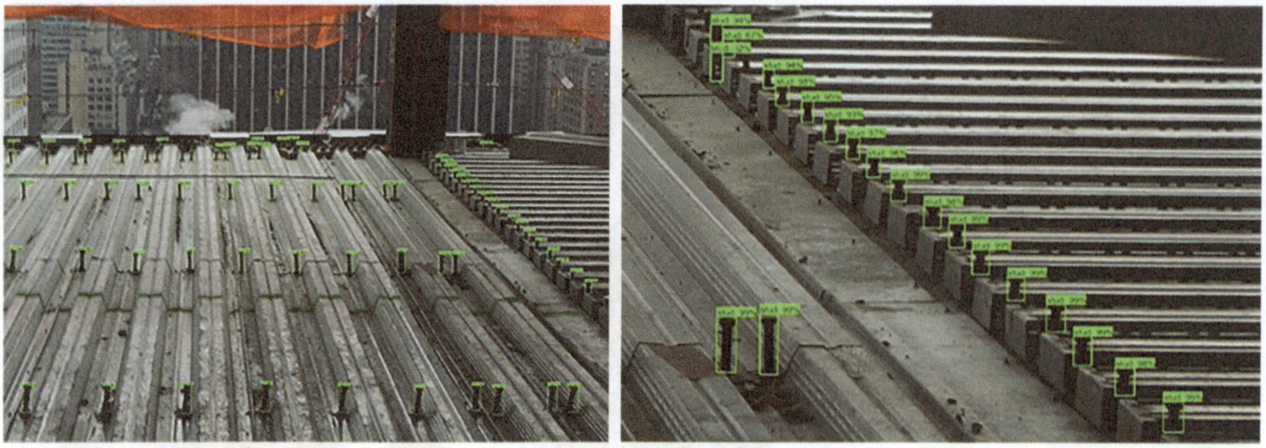

Fig. 31 Shear studs identified with object detection

Fig. 32 Bolts identified from a site photo using object detection

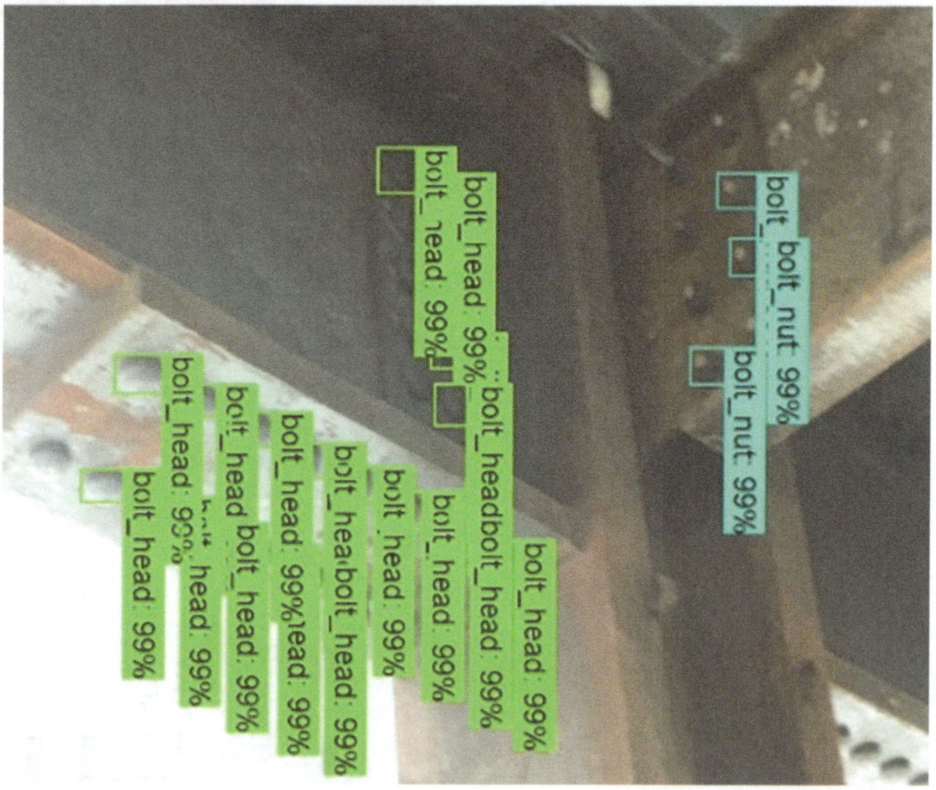

Figure 30 shows a gusset plate being detected both from a site photo and a shop drawing using object detection. The machine learning model was able to identify gusset plates in shop drawings to a high degree of confidence. This performance is likely due to both the distinguishable shape of gusset plates and the consistency and precision of steel shop drawings.

Using object detection, another machine learning model was trained to recognize shear studs from site photos. Because shear studs are small and are typically counted over a large area, a digital single lens reflex (DSLR) camera was used to take high-resolution photographs of large areas containing shear studs. However, machine learning models are very limited in terms of resolution. It is computationally intractable to feed a full DSLR image into a typical machine learning model on today's hardware. To get around this limitation, the images were divided into nine sub-images and then each sub-image was processed independently. Afterward, the photos were recombined. This allowed the onsite user to take a single photograph while enabling the

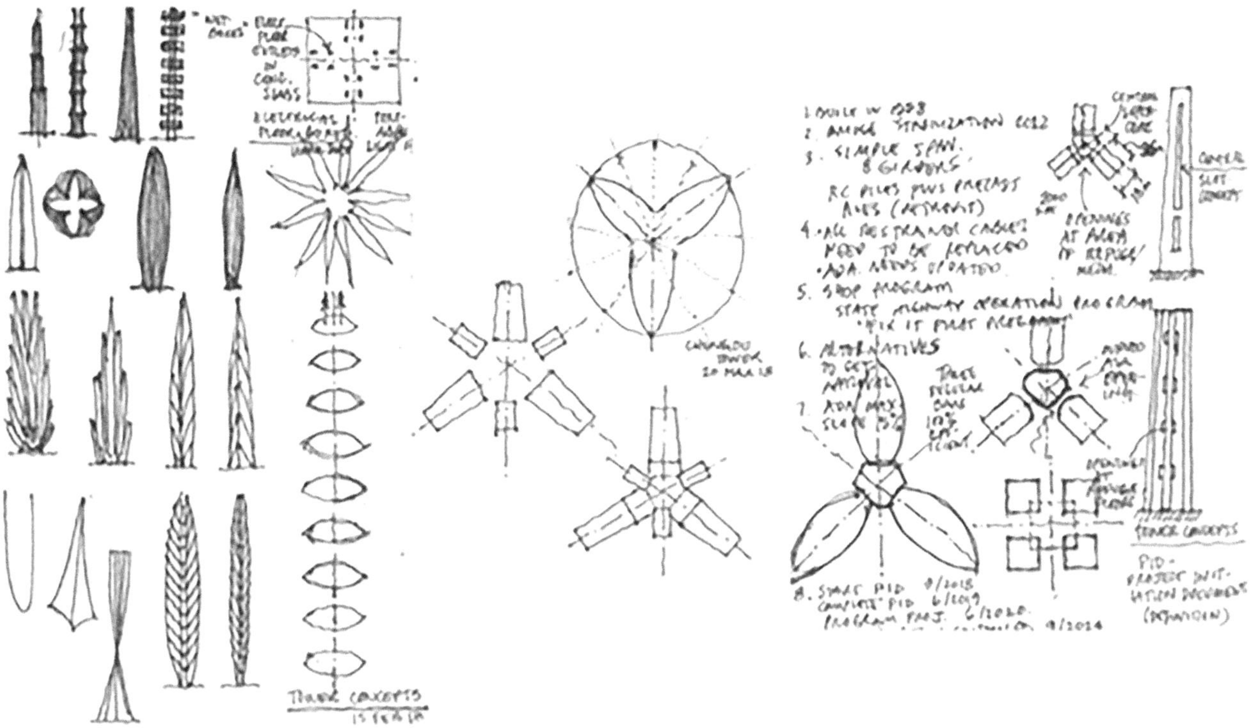

Fig. 33 Initial structural system concepts – Chengdu Tower, Chengdu, China

system to process the images at an effective resolution. A sample of a processed sub-image is shown in Fig. 31 below.

A sample of a recombined photo is shown in Fig. 31. Although some studs in far corners of the photo were not identified, the lack of false positives in these results indicates that a larger training set could produce a more robust model.

In another application, a machine learning model was trained to recognize bolts from site photos. Figure 32 illustrates that not all the bolts were successfully identified, potentially due to the angle and lighting conditions of the photo. However, no false positives were identified in the photo. Therefore, a larger and more varied training set could lead to improved results. Issues with identifying bolts in perspective could also be controlled by limiting the machine learning model to head-on photos of bold groups.

The Chengdu Tower

A competition for a 600-meter-tall tower in Chengdu, China, created the opportunity to explore the use of natural growth patterns into a structure that combined structural steel and reinforced concrete in a creative solution. Initial concepts were developed around a stable tripod form where a reinforced concrete central core would resist torsional effects of wind and seismic loads with the legs of the tripod resisting the combined effects of gravity and overturning forces (Fig. 33).

Further development of the structural system led to design where segmented diagonal steel braces were introduced into the building perimeter and linked together at perimeter, segmented linear reinforced concrete walls. These links would also provide ductility during a seismic event, design to fuse under significant seismic demand (Fig. 34).

The primary service elements including express elevators, stairs, and mechanical spaces were placed in the central core with local elevator located at the outside of the tripod-shaped floor plan. The floors were typically column-free with large reinforced concrete elements used at the perimeter to accept the bracing system (Fig. 35).

Air Foil Concept

It is possible that the limitations on forming techniques are the single greatest obstacle in creating concrete structures that result in the most fluid and efficient response to load. Forming systems to date are largely limited by the material type used that focuses on wood. Alternate systems including those of synthetics such as plastics or fabrics will lead to more fluid structures. Reinforcement limitations must also be addressed, and the prefabrication of systems with com-

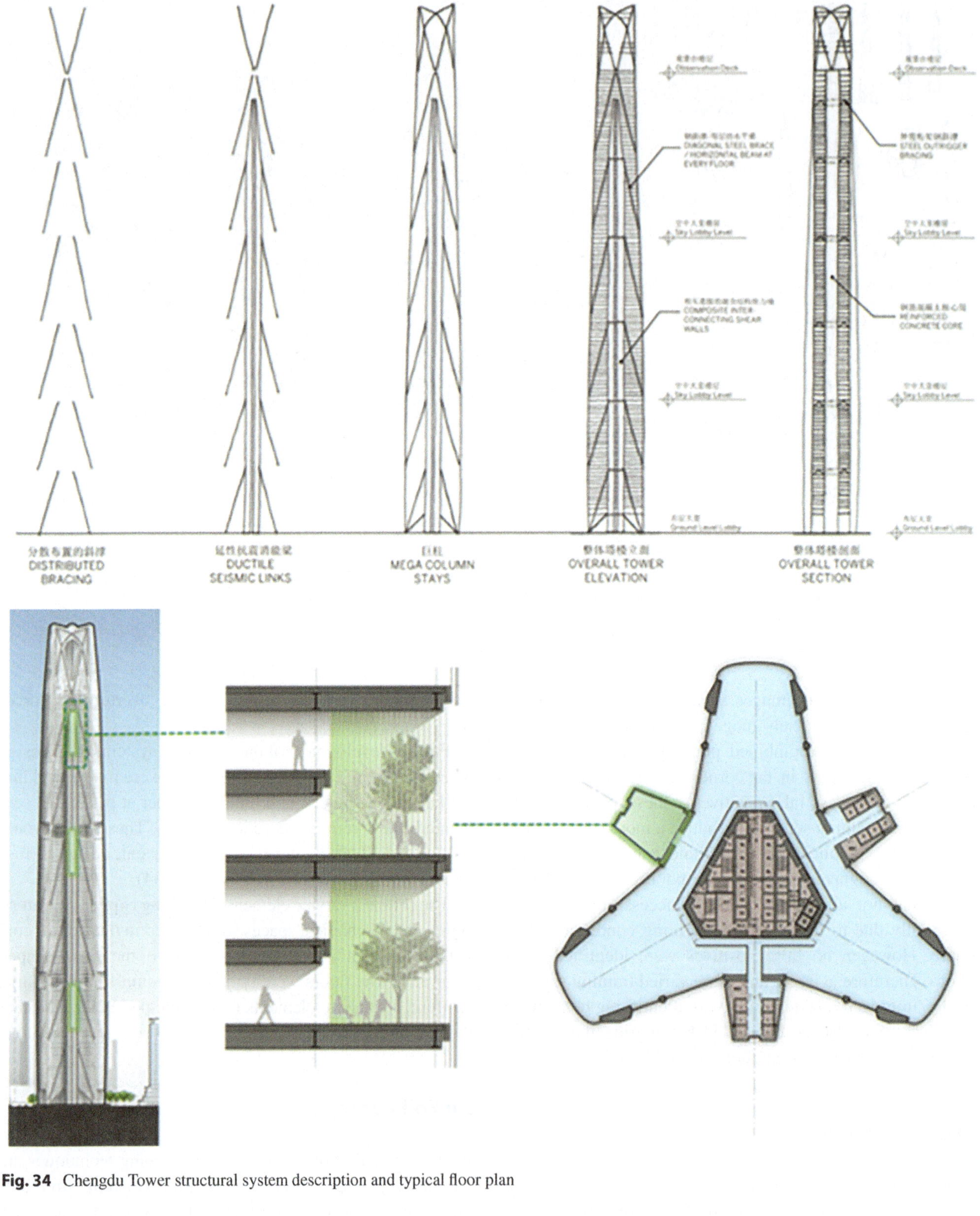

Fig. 34 Chengdu Tower structural system description and typical floor plan

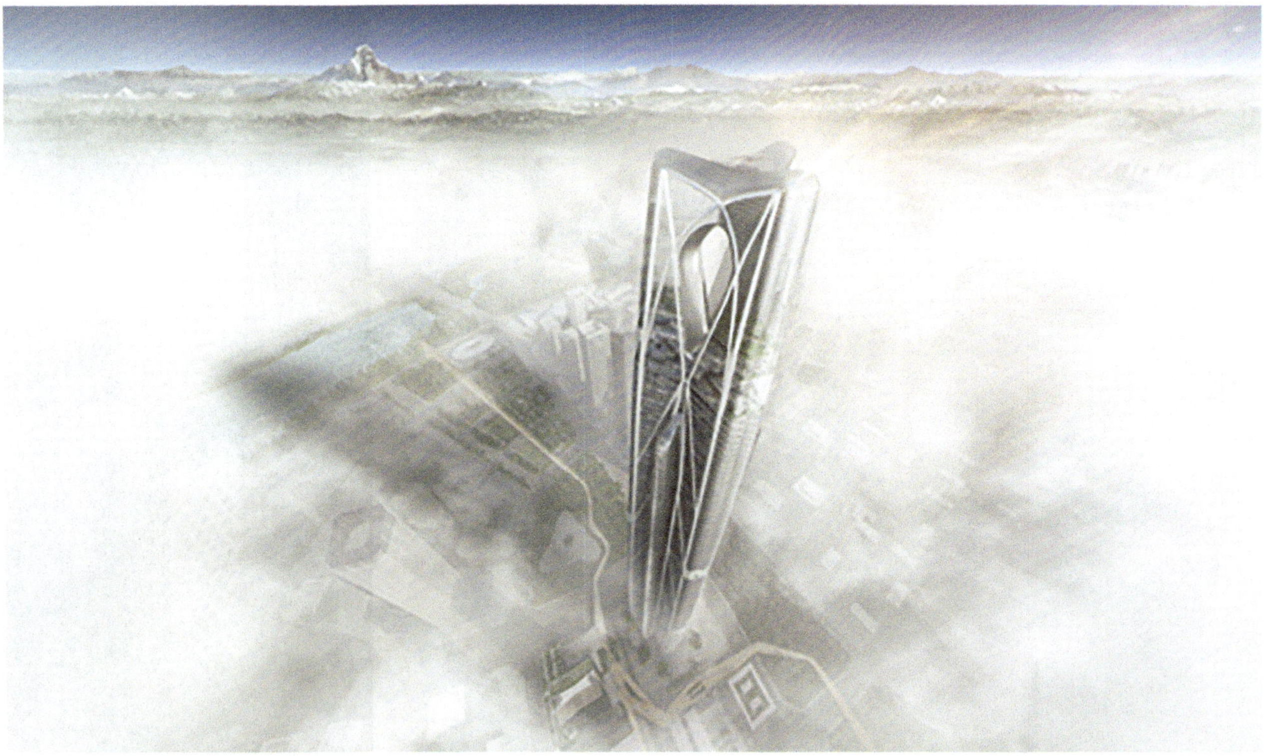

Fig. 35 Chengdu Tower rendering

plex geometries are also needed. Advancements in pumping systems will lead to greater flexibility on the placement of concrete and perhaps most importantly the use of pressurized systems where concrete properties could be enhanced, concrete matrices densified, and advanced highly efficient structures created.

Structures designed and constructed with concrete systems provide opportunities for combined technology represent the future. Perforated concrete structural wall systems will not only create beautiful interior spaces but could provide spaces where power generation or even combined systems control structural behavior through aerodynamics and air foils (Figs. 36, 37, and 38).

The Urban Sequoia Tower

Achieving net-zero carbon for building projects falls short of the need to heal the earth. If buildings could absorb carbon over time, significant advancements could be made to reduce carbon that has been emitted over time. Beyond the building, if it was possible to integrate these ideas at a district of city scale, the impact would be far more significant.

The concept of the Urban Sequoia is one that combines optimized structural design with materials that use carbon capture technologies. The Sequoia includes ideas that can be applied to buildings at all scales. For example, the floor framing system could be used in a building with only a few floors or one that has 100 floors. Nature-based solutions emit far less carbon than conventional system and absorb carbon over time. Materials could include Bio-Brick, hempcrete, timber, and Biocrete that could reduce the carbon impact by 50% compared to concrete and structural steel. The complete system could reduce carbon by as much as 95% compared to conventional buildings (Figs. 39, 40, and 41).

Conclusions

Self-sufficiency and full resiliency are the goal buildings with the best life cycle. Great advancements have been in design and construction of structures, and China has been important in these advancements. Ideas have fed on previous successes. Material fluidity, whether with the use of steel or concrete, is important to exploring the next generation of structures.

To achieve full resiliency and the best life cycle design, structural components and systems must perform elasticity for all environmental conditions of the site. In regions of high seismicity, particular consideration must be given to components and systems where typical code compliant design would result in plastic deformation of key elements

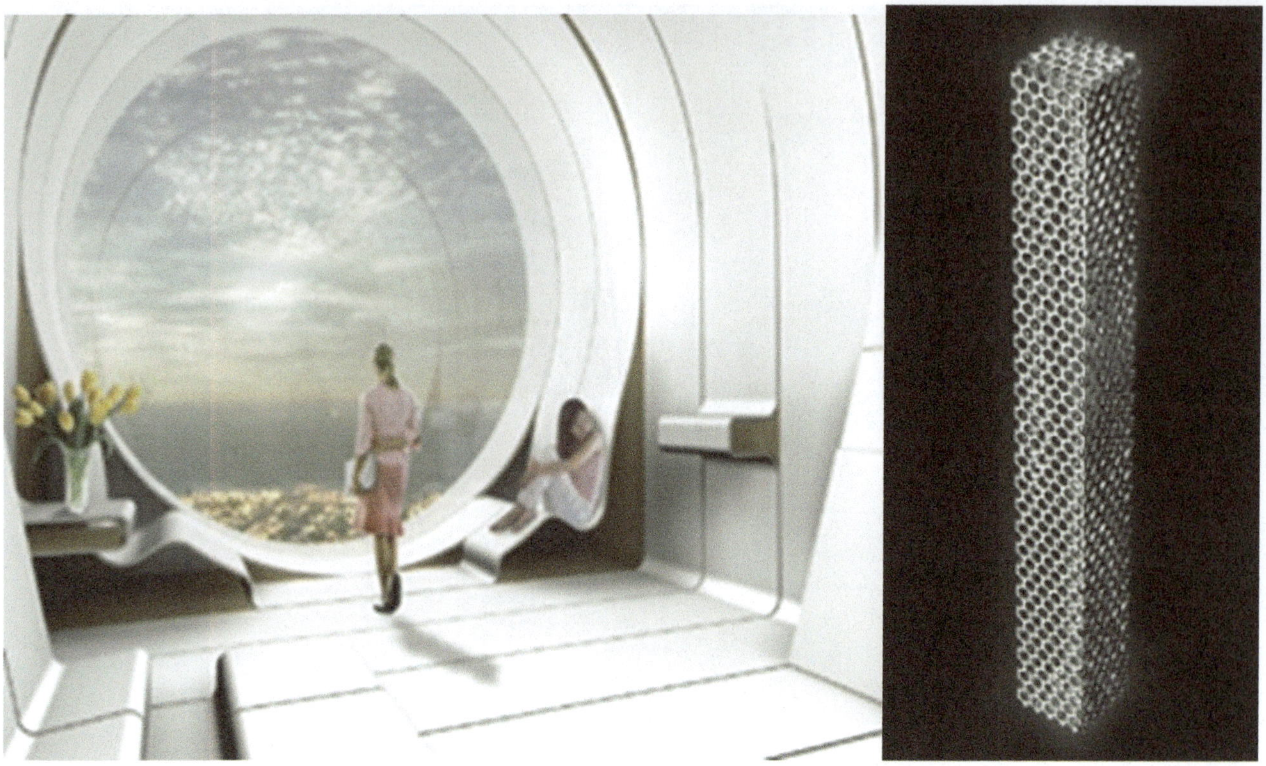

Fig. 36 Concept for concrete tall buildings – circular holes in perimeter wall system

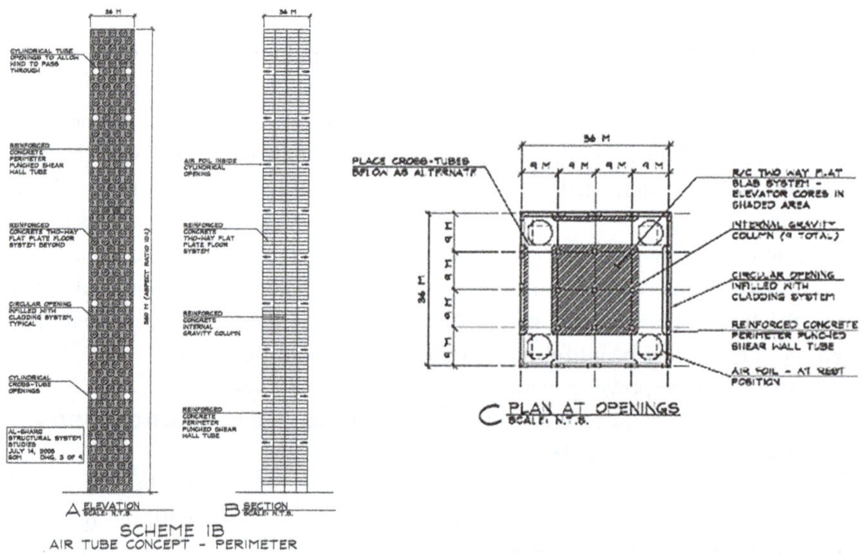

Fig. 37 Structural system combined with air foil concept

within the structure. This plastic deformation requires repair or replacement of these elements. A more resilient or perhaps fully resilient system could be one that introduces friction connections into these conditions with energy dissipated through the generation of friction/heat without plastic deformation. All building material will remain elastic. Carbon dioxide that would be emitted for the repair or even the reconstruction of a plastically damaged building after a significant event would not occur.

Braced-framed structural systems in tall buildings typically lead to lower material quantities because lateral loads are resisted in tension and compression with minimal bending. This system is ideal for controlling building displacements due to wind and moderate seismicity but could perform poorly in major seismic event. The solution would be to create a combination of stiffness and ductility. Seismic "fuses" introduced into key locations of the braced frames allow for the dissipations of energy during strong ground motion. If

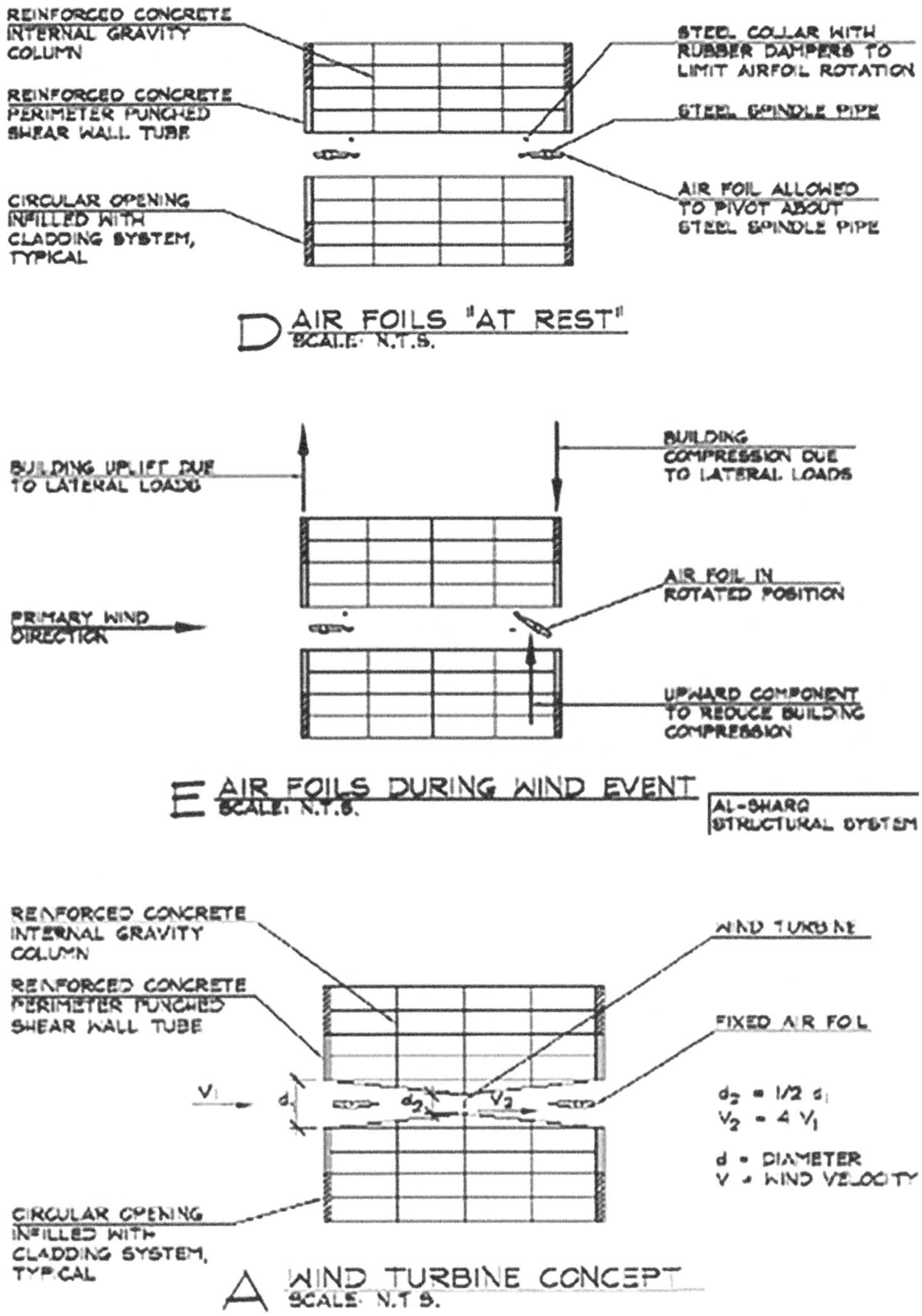

Fig. 38 Details of air foil concept

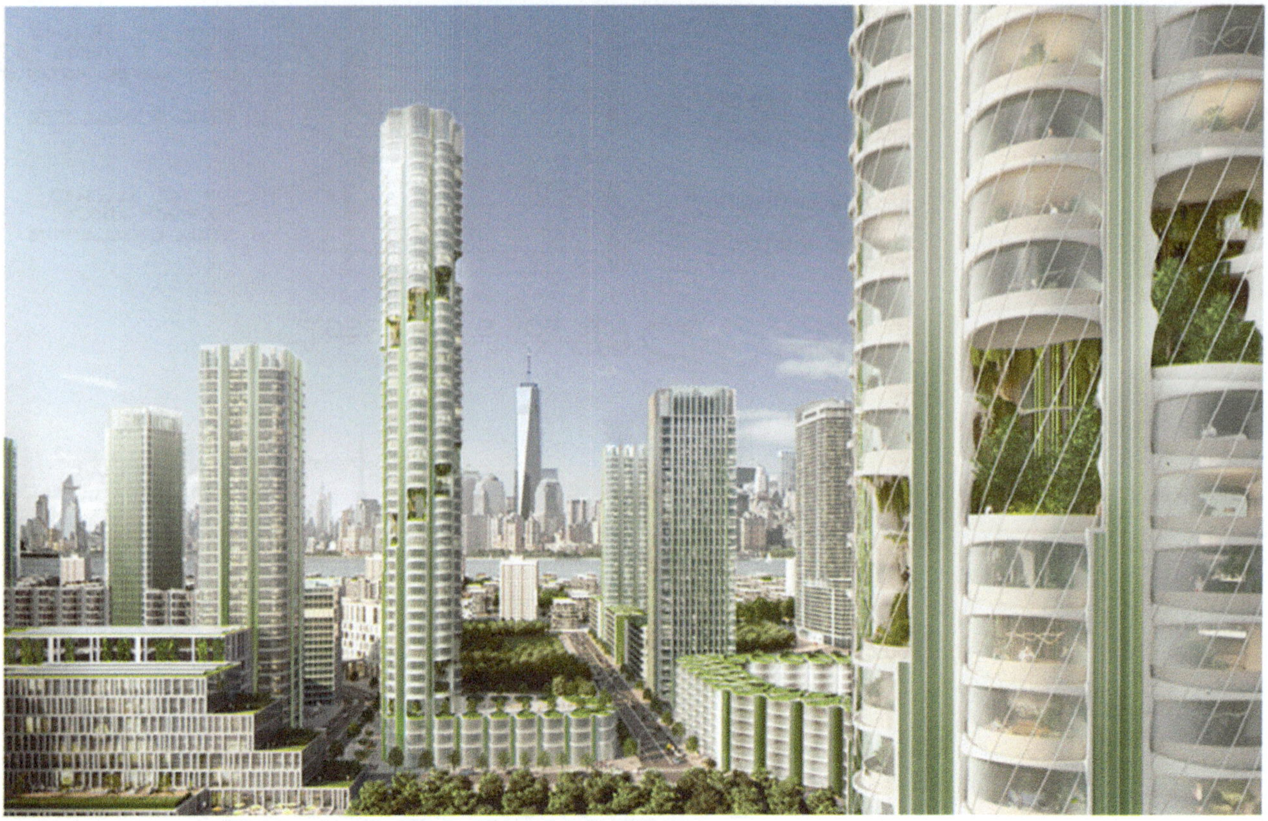

Fig. 39 Urban Sequoia in city context

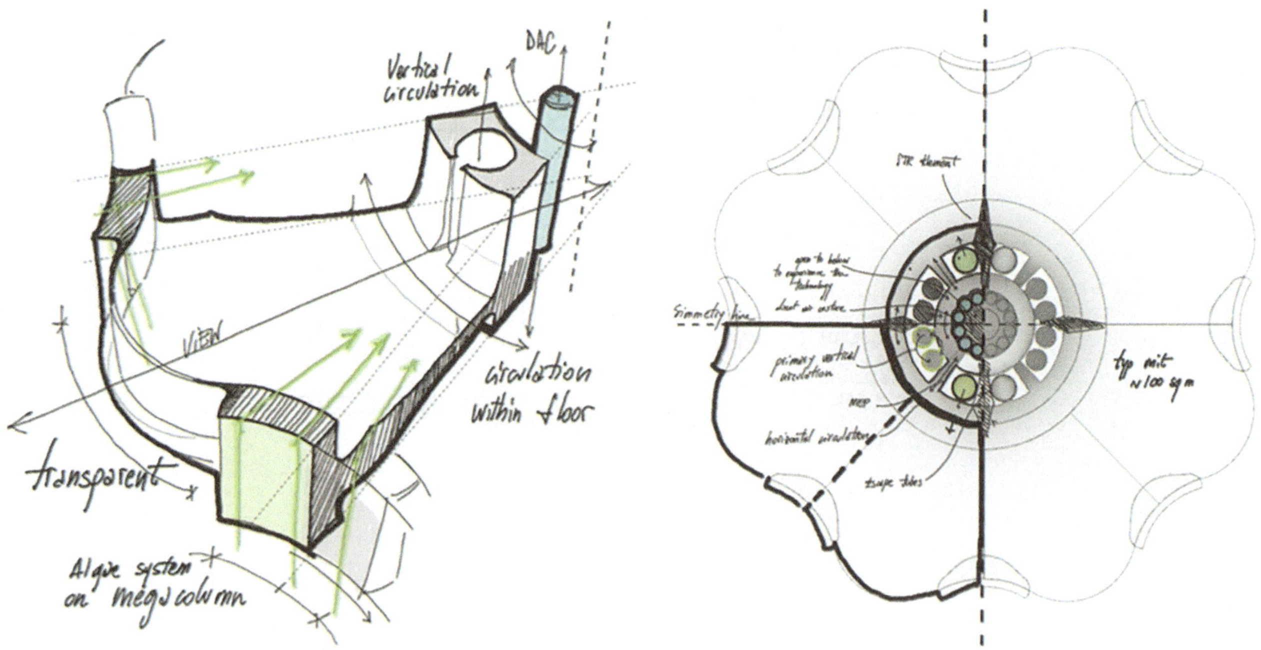

Fig. 40 Overall system concept (axonometric and plan)

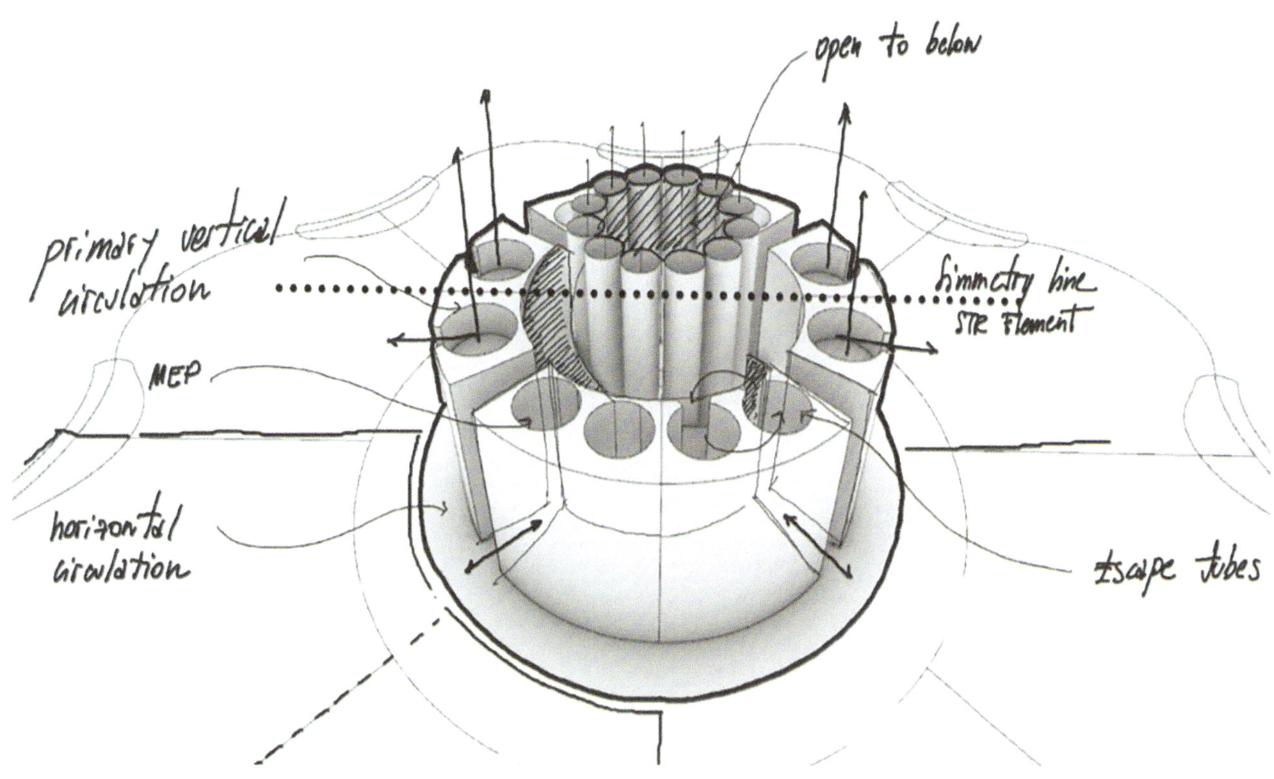

Fig. 41 Enlarged core axonometric/key operational characteristics

these fuses used friction-type connections, then these elements would remain elastic.

With the continued and perhaps increased use of timber in major building structures especially in major seismic regions, inherent ductility is essential. To optimize resiliency, this ductility could be achieved with friction-type connectors integrating structural steel at key joints within the timber system.

Ultimately, machine learning will be an integral part of designing, building, and verifying structural systems. Using this process to compare as designed structures to built structures will minimize errors. In the future, machine learning could be used for construction, directly translating design to building.

References

Diaz, A., Murren, P., & Walker, S. (2017). *Preliminary observations in the aftermath of the September 19, 2017 Puebla-Morelos earthquake.* Skidmore, Owings and Merrill (SOM).

Freddi, F., Tubaldi, E., Zowa, A., & Dall'asta, A. (2020). Seismic performance of dual systems coupling moment-resisting and buckling-restrained braced frames. *Earthquake Engineering and Structural Dynamics, 50*, 329–353.

He, K., Zhang, X., Ren, S., & Sun, J. (2016). *Deep residual learning for image recognition.* Proceedings of the Institute of Electrical and Electronic Engineers (IEEE) conference on computer vision and pattern recognition.

Ren, S., Kaiming, H., Ross, G., & Jian, S. (2015). Faster R-CNN: Towards real-time object detection with region proposal networks. *Advances in Neural Information Processing Systems*, 91–99.

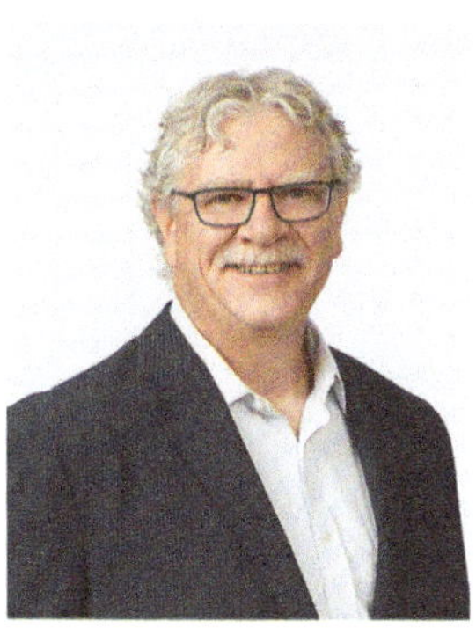

Mark P. Sarkisian PE, SE, NAE, LEED BD+C, Partner of Seismic and Structural Engineering in the San Francisco office of Skidmore, Owings & Merrill LLP, has developed innovative engineering solutions for over 100 major building projects around the world including some of the tallest and most complex. Mark holds ten US Patents and five International Patents for high-performance seismic structural mechanisms designed to protect buildings in areas of high seismicity and for seismic and environmentally responsible structural systems. In 2021, Mark was elected to the prestigious United States' National Academy of Engineering. He is the author of Designing Tall Buildings – Structure as Architecture, and teaches at UC Berkeley, California College of the Arts, Stanford University, Cal Poly, Northeastern University, NC State University, the Pratt Institute and Alcala University in Madrid. He has a BS-CE Degree from the University of Connecticut and is a Fellow of the Academy of Distinguished Engineers, an MS-SE Degree from Lehigh University, an Honorary Doctorate Degree from Clarkson University and an Honorary Master's degree from the Politecnico di Milano.

[illegible]

Currently [illegible] designing buildings [illegible] this process [illegible] could be used [illegible] directly [illegible] design [illegible] building.

[illegible] 2021. [illegible] Academy [illegible] College [illegible] University [illegible] Politecnico di Milano.

References

[illegible] Structure [illegible]
[illegible] Influence of [illegible]
He, X., Zhang, X., Ren, S., & Sun, J. (2016). [illegible] Proceedings of the [illegible]

Frei Otto: Light Structures Inspired by Nature. Experiments with Physical Self-Forming Processes: Soap Solutions, Viscous Fluids, Branching Structures and Funicular Forms

Juan María Songel

Abstract

Frei Otto is well known for his pioneering contributions to light structures. Observing natural structures and physical phenomena through experiments with different everyday materials was a key tool in his work to pursue and understand lightweight construction. His intention with these experiments was not so much to find forms in nature to be imitated, as to understand the natural processes of form generation. His experiments focused therefore on materials and objects whose form is driven by physical self-forming processes, which involve maximum efficiency with minimum amount of material.

The purpose of this chapter is to present a systematic overview of some of the experiments carried out by Frei Otto and his collaborators, showing a close link between form and its generating process: soap solutions, viscous fluids, branching structures, and funicular forms. Accordingly, the ordering or classifying criterion used to present this overview has been based on materials, generating forces and resulting forms.

An important conclusion concerns the validity of physical models nowadays in relation to computers and parametric design software. When confronted with this issue, Frei Otto argued that experiments with physical models provide the chance to invent, the possibility of finding the unsearched, as they contain the laws of nature, whereas computers would not have that ability, as the laws governing them have been created or prescribed by man. Form generation through physical self-forming processes and careful observation of natural phenomena were indeed pivotal in Frei Otto's experience for the invention of new light structures.

J. M. Songel (✉)
Centro de Investigación en Arquitectura, Patrimonio y Gestión para el Desarrollo Sostenible (PEGASO) Universitat Politècnica de València, Valencia, Spain
e-mail: jsongel@cpa.upv.es

Keywords
Frei Otto · Experiments · Models · Soap solutions · Viscous fluids · Branching structures · Funicular forms

Introduction

Frei Otto is well known in the history of architecture and structural engineering of the second half of the twentieth century for his pioneering contributions to light structures (Nerdinger et al., 2005). The search for lightweight construction and its governing principles has been a constant feature throughout his career. A key focus of his research was the relationship between form, force, and mass of physical objects, whether natural or artificial, analyzing their efficiency in the transmission of maximum load to the maximum distance with the least amount of mass or material.

His work can be understood within the tradition of leading structural or civil engineers in their endeavor to develop strength of materials with efficient structural forms. Robert Le Ricolais was a French civil engineer who worked and taught in the USA, and was also a pioneer of light structures. He expressed that unattainable goal or ideal towards lightness in a very eloquent way: "zero weight, infinite span." He also described the art of structural design in very significant terms: "The art of structure is how and where to put the holes, to build with holes, to use things which are hollow, things which have no weight" (Le Ricolais, 1997).

Frei Otto defined lightweight construction as "a way of building with a minimum consumption of material, energetic, and economic means" (Brinkmann, 1990, p. 273). Observing natural structures and physical phenomena through experiments with different everyday materials was a key tool in Frei Otto's work to pursue and understand lightweight construction. They reflect a principle of economy that can be perceived in nature and the universe. The mathematician Matila C. Ghyka formulated it as the principle of least action for the inorganic world and the principle of economy of material for the organic world (Ghyka, 1977, p. 88).

M. R. Cervera Sarda et al. (eds.), *Architecture Inspired by Nature*, https://doi.org/10.1007/978-3-031-33144-2_3

Frei Otto's intention with his experiments was not so much to find forms in nature to be imitated, as to understand the natural processes of form generation, or self-forming processes governed by the laws of nature. He speaks of "natural structures," also in engineering, "if their forms are not manipulated by man, but simply develop, organize, form and optimize themselves in accordance with requirements" (Bach et al., 1988, p. 7). His experimental methodology and his ability to explore physical phenomena involves therefore a conception of form not as a precondition, but as a result of a searching process, in which strength and lightness must be formally unveiled. Natural structures accordingly should not derive from planned processes but should result from self-forming or self-organizing processes.

Experiments with Physical Self-Forming Processes

The purpose of this chapter is to present an overview of some of the experiments carried out by Frei Otto and his collaborators, always linking form to its generating process. His intention was to check whether objects of inanimate nature, of living and dead nature, as well as those coming from technical production processes have a characteristic form that is an expression of its generating process. His experiments focused therefore on objects whose form is driven by physical self-forming processes (Otto et al., 2017). Accordingly, the ordering or classifying criterion used to present this overview of experiments has been based on materials, generating forces, and resulting forms. These are the experiments that will be addressed in this overview: soap solutions, viscous fluids, branching structures, and funicular forms.

Soap Solutions

One of the best known physical self-forming processes are the experiments carried out with soap films, bubbles, and foam. Very thin membranes can be obtained by inflating air or by dipping a frame in a solution of water with detergent or Pustefix fluid. These membranes develop themselves a minimal surface within the boundary conditions provided, as they tend to contract to the smallest surface possible, due to the surface tensioning. This minimal surface is self-formed when all the points in the membrane are subject to the same tension in all directions, and a state of equilibrium is reached, resulting from the lowest energy level (Roland, 1972).

Soap Films

Soap membranes generated when immersing a frame or a plaque in a soap solution develop minimal surfaces or a network of minimal ways.

Minimal Surfaces

Many different forms of minimal surfaces were obtained in Frei Otto's experiments with soap films, playing with the rigidity, flexibility, shape, and positioning of boundaries and supports.

Within Rigid Boundaries

A frame made up of flat or spatial, rectilinear or curved, rigid bending-resistant steel rods immersed in a soap solution generates, when lifted, a minimal soap film surface, which usually adopts doubly curved anticlastic, also called "saddle," forms (Fig. 1).

Within Flexible Boundaries

If the edge is a thin flexible wire or a thread, the minimal surface develops from the equilibrium of forces between the edge and the soap film. The edge curves are drawn in and concave, and have a constant curvature, due to the even distribution of stresses within the membrane (Gass, 1990, p. 2.54).

Anticlastic Doubly Curved Four-Point Minimal Surfaces

They are saddle-shaped minimal surfaces developed between two diagonally opposed high and low points (Fig. 2).

Wave Minimal Surfaces with Ridges and Valleys

The main formal feature of this type is the alternating sequence of high and low points along the membrane edges, generating a pattern of alternating ridges and valleys. The surfaces between them are anticlastic. This sequence of ridges and valleys creates an undulating longitudinal cross section, and depending on their layout, the waves can be parallel (Fig. 3) or radial (Fig. 4).

Pointed, Humped, and High-and-Low-Point Minimal Surfaces

Prestressed anticlastic surfaces can be generated by inserting internal high points, or low points, or a combination of both. A key problem here is the stress concentration at those internal points of the membrane. Frei Otto designed different intermediate elements between these internal points and the membrane to gradually reduce and distribute evenly these high stresses within the membrane (Fig. 5).

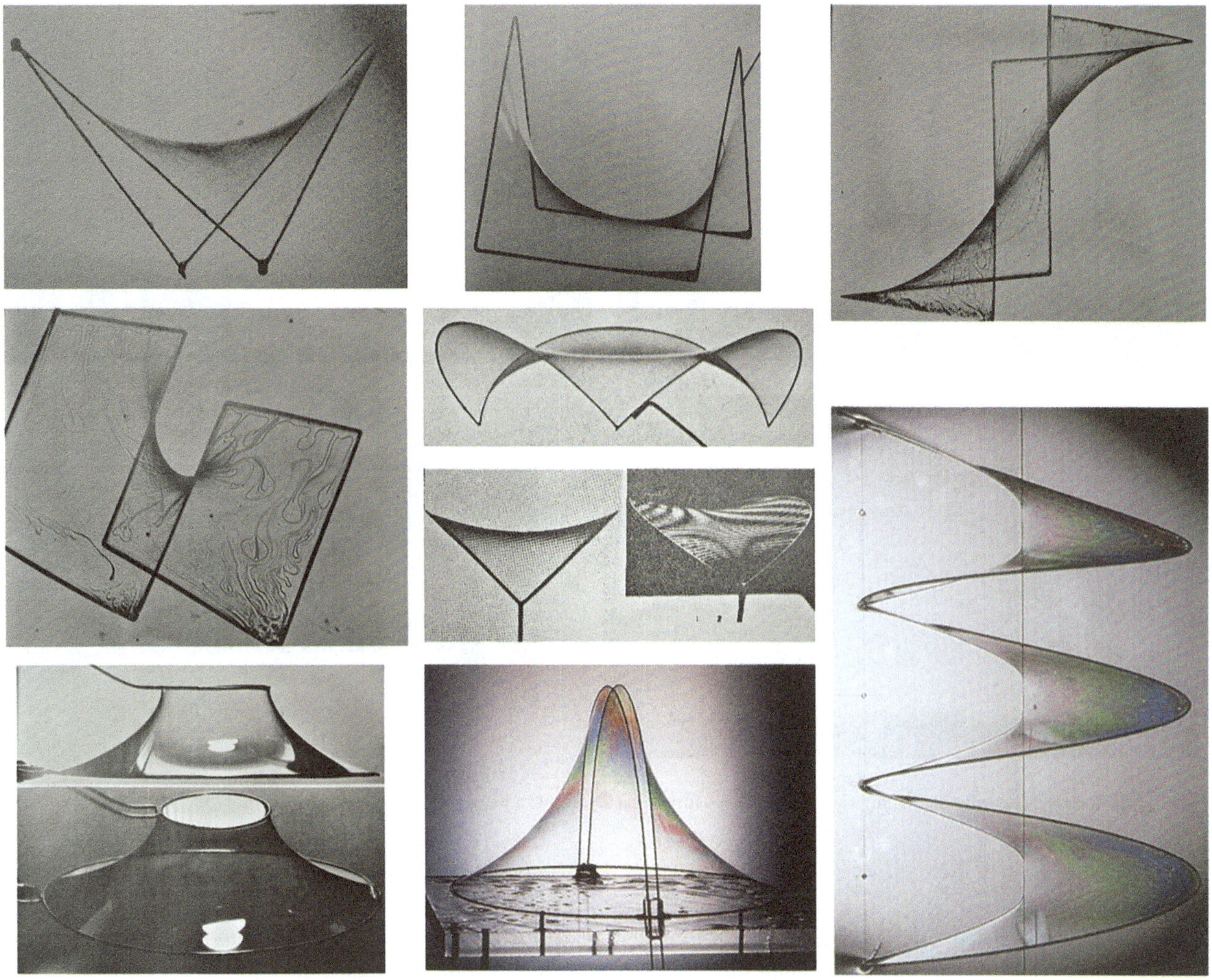

Fig. 1 Minimal surfaces developed with soap films within rigid boundaries. (Source: ©Atelier Frei Otto + Partner)

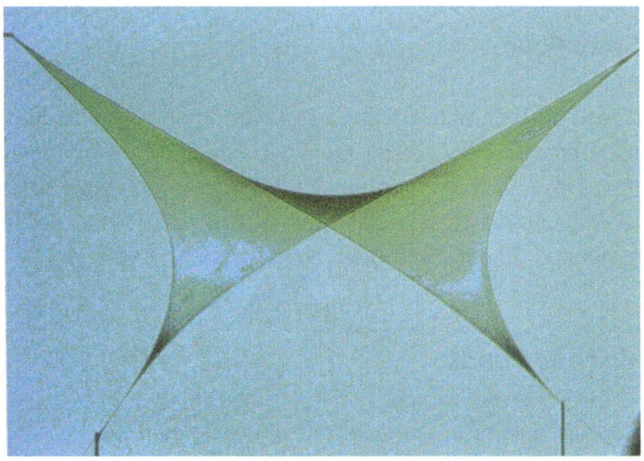

Fig. 2 Minimal surface developed between two diagonally opposed high and low points. (Source: ©Atelier Frei Otto + Partner)

Minimal Ways

There is an issue in Mathematics, called the "Steiner" problem, which is the search for a minimal way system connecting a group of randomly distributed points in a plane. This minimal network can be obtained experimentally by vertical soap film strips stretched between a horizontal glass plate and the surface of a soap solution, which can be lowered slowly. The points are arranged by needles, pressed against the glass plate from below, which are fixed at the end of fingers that can be positioned from the edge of the basin. Vertical soap film strips are then self-produced between the needles and the space between the plate and the water surface. Due to the surface tensioning of the soap film strips, the overall length of the network of vertical film strips connecting the needles is minimal (Fig. 6).

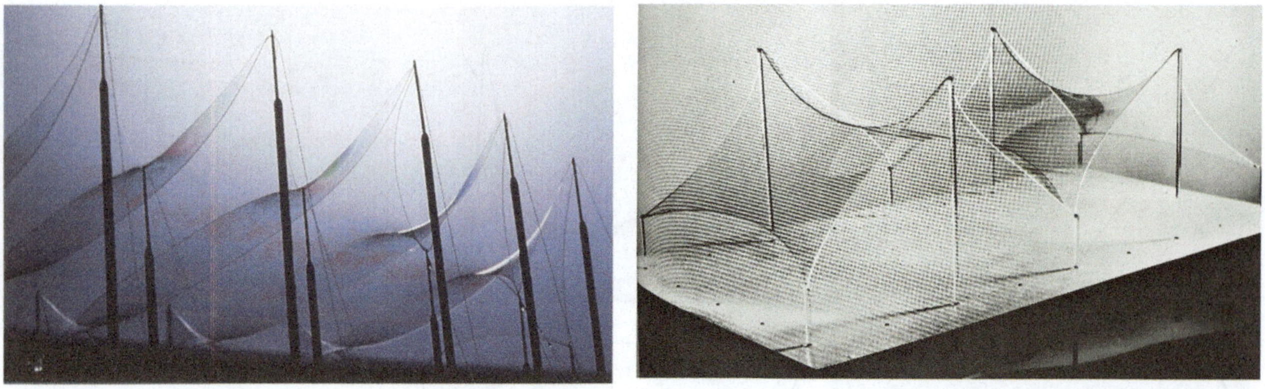

Fig. 3 Minimal surfaces with alternating ridges and valleys forming a layout of parallel waves. (Source: ©Atelier Frei Otto + Partner)

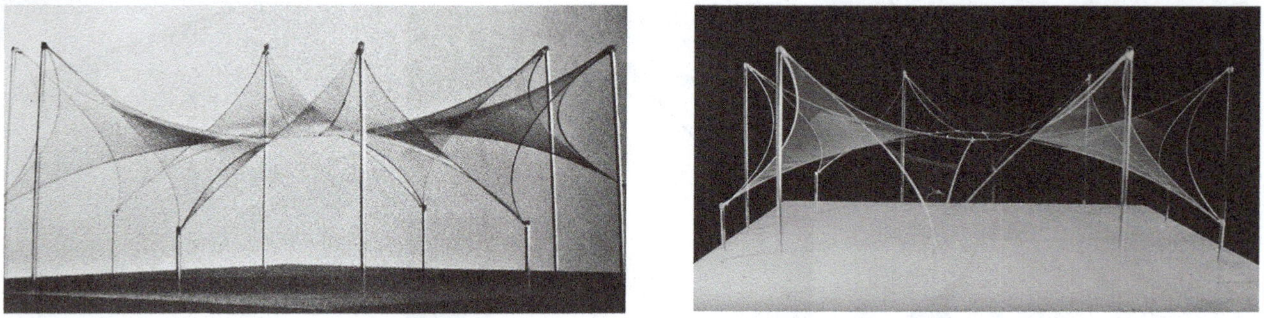

Fig. 4 Minimal surfaces with alternating ridges and valleys forming a layout of radial waves. (Source: ©Atelier Frei Otto + Partner)

Soap Bubbles

Soap membranes inflated with air generate bubbles, which enclose a maximum volume with a minimum surface. Their size and shape depend on the difference between the inner and outer air pressures as well as on the surface tension of the soap membrane (Bach et al., 1988, p. 222).

Bubbles on Rigid Boundaries

It is possible to form bubbles with dome shapes above circular, triangular, quadrangular, polygonal, or free plan shapes (Fig. 7).

Bubbles on Rigid Circular Rings. Semi-torus

Different bubble forms can be obtained with an outer rigid circular boundary and inner circular low points of various sizes and positions. If a smaller inner ring is concentrically placed in relation to an outer circular boundary, a semi-torus develops (Fig. 8).

Bubbles with Free Boundaries. Hemisphere

When bubbles float on liquids or stand on a wet surface, hemispherical shapes with unconstrained round boundary edges develop, always keeping the rule of minimum surface enclosing maximum volume (Fig. 9).

Linear Addition of Spherical Bubbles

When bubbles come into contact, partitions develop, these being flat if the bubbles and their internal pressure are equal, or bulge from the smaller into the larger bubbles. If they stand on a horizontal plane and have different sizes, partitions incline toward the smallest bubbles (Fig. 10).

Soap Foam

Three-dimensional bubble clusters form polyhedral foams. If spherical bubbles cluster from all sides, polyhedral foams develop, and their partition walls behave as braces of the overall cluster structure. Different structural forms can be generated, like flat and steep cones or arches (Fig. 11).

Viscous Fluids

It is interesting to compare the self-generated forms of viscous fluids with those obtained with soap solutions, as they behave, when subject to tensile stress, in a similar way. Viscous fluids, such as honey or glue, also tend to develop minimal ways and surfaces, but when viscosity increases, the internal friction prevents an even distribution of the material over the entire membrane surface (Gass, 1990, p. 2.60).

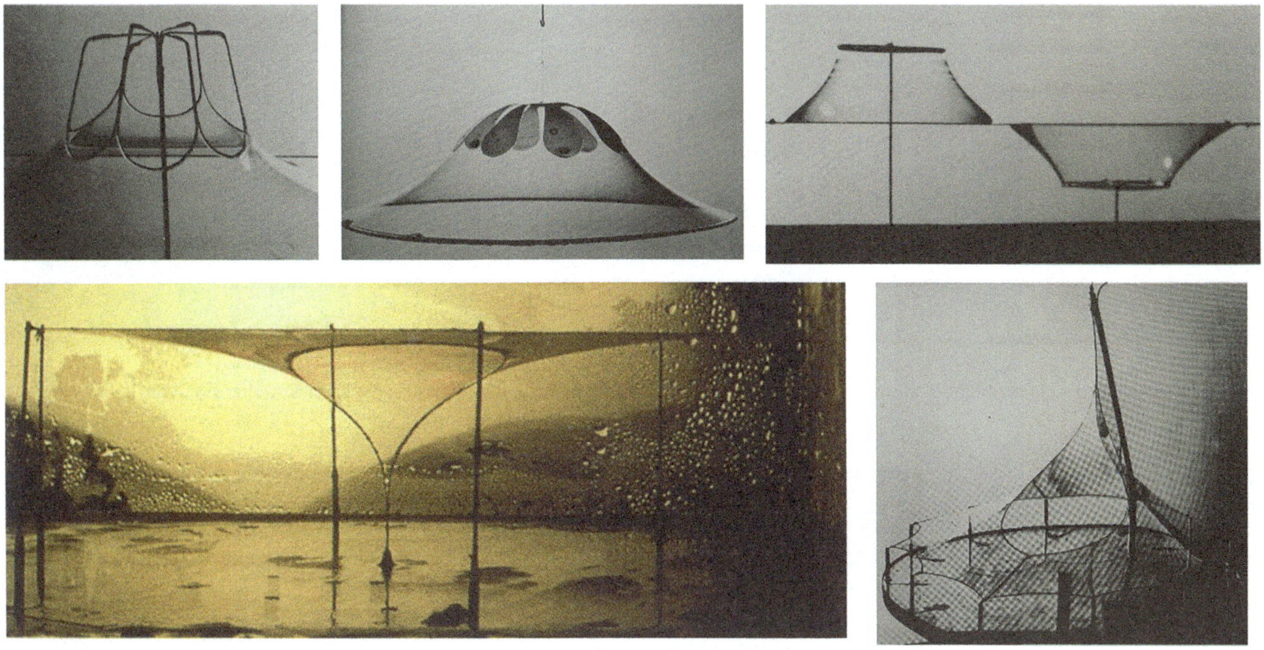

Fig. 5 Different intermediate elements between membranes and internal support points. (Source: ©Atelier Frei Otto + Partner)

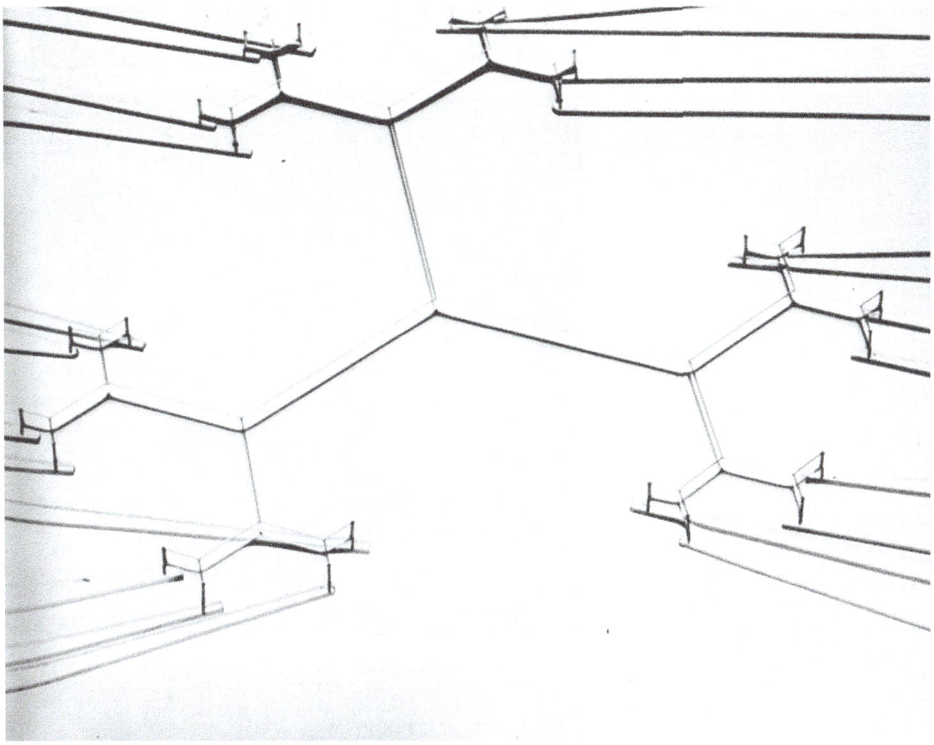

Fig. 6 Minimal network of ways connecting a group of points in a plane, developed by vertical soap film strips. (Source: ©Atelier Frei Otto + Partner)

Threads

Viscous fluids, in contrast with soap solutions, can generate both planar and space networks of branched threads, usually with three-armed nodes. When two plates smeared with a viscous fluid between them are separated, a thready spine structure develops (Fig. 12).

Membranes

The uneven distribution of material in membranes of viscous fluids leads to a concentration of material along ridges and edges, needing no extra flexible or rigid elements for the membrane linear boundaries. Holes in the membranes of viscous fluids are stabilized with a rearrangement of material,

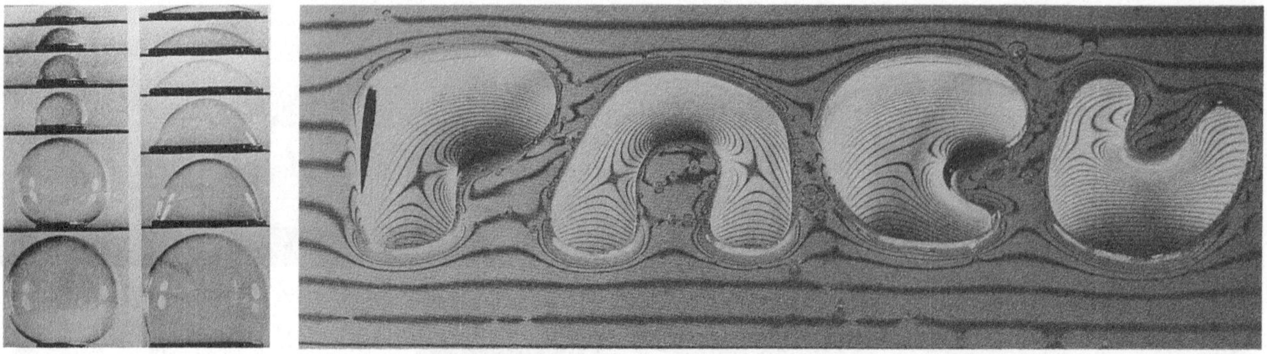

Fig. 7 Dome-shaped soap membranes above circular, quadrangular, and free plan rigid boundaries. (Source: ©Atelier Frei Otto + Partner)

Fig. 8 Soap membranes above rigid circular rings. (Source: ©Atelier Frei Otto + Partner)

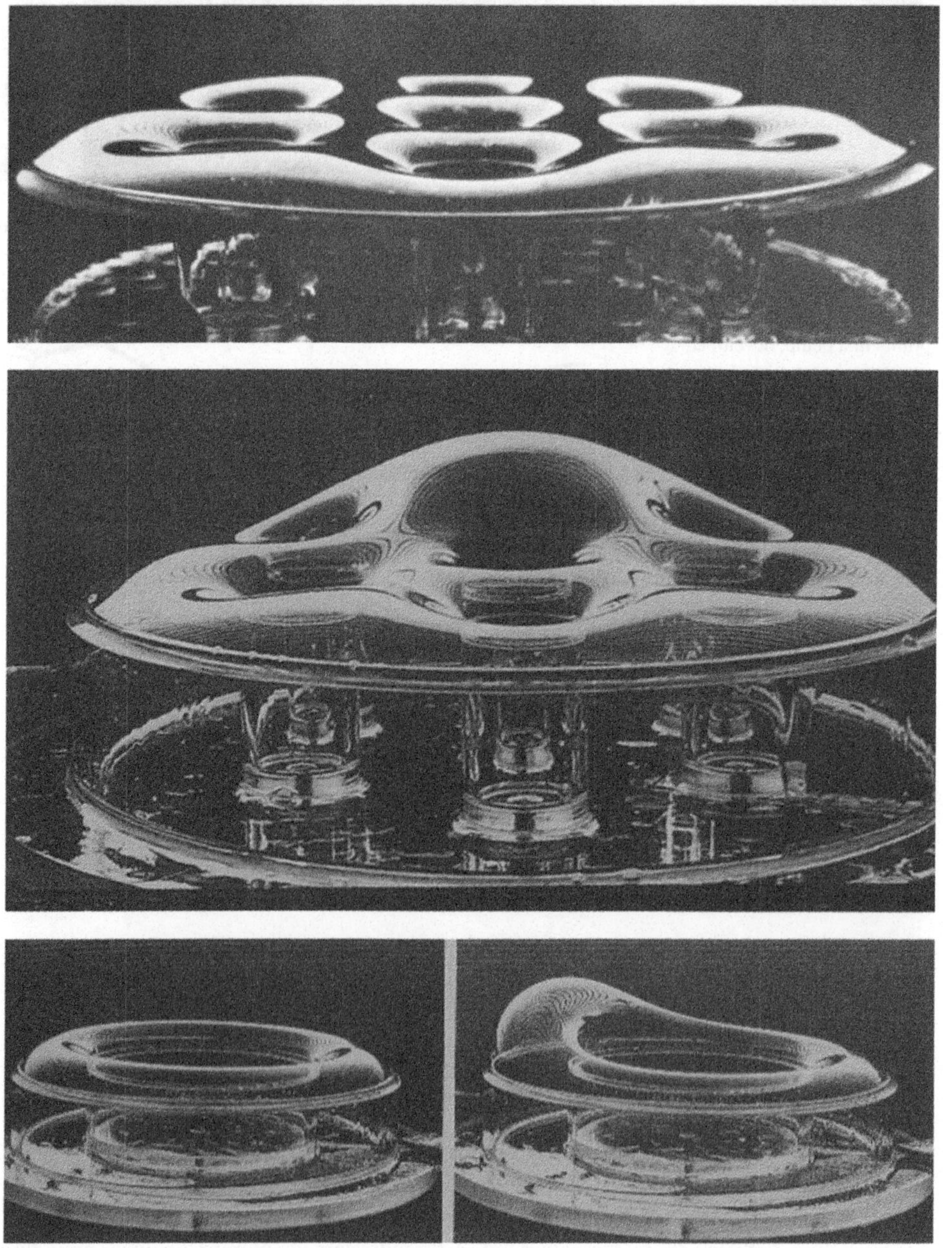

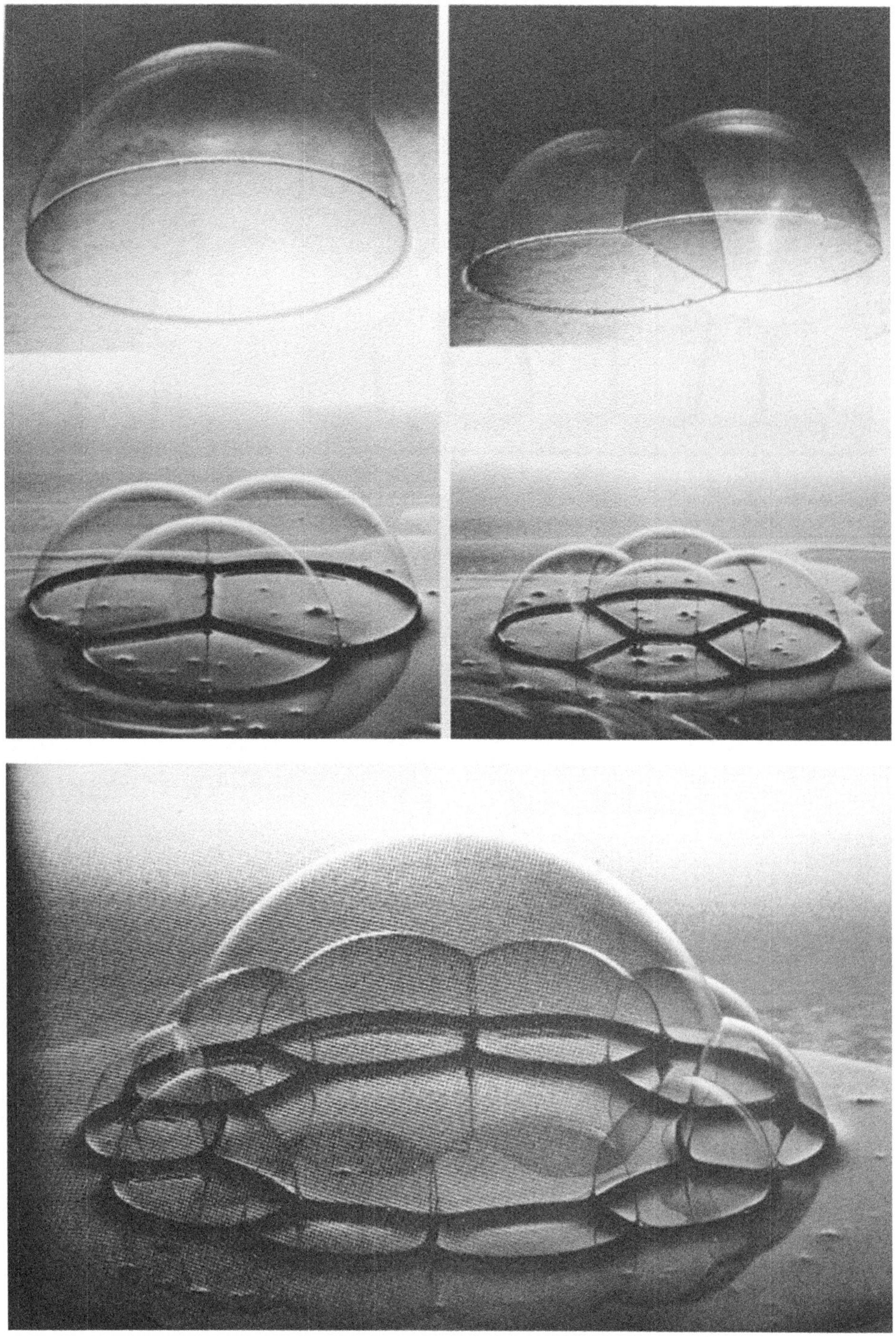

Fig. 9 Soap bubbles with free boundaries and hemispherical shapes. (Source: ©Atelier Frei Otto + Partner)

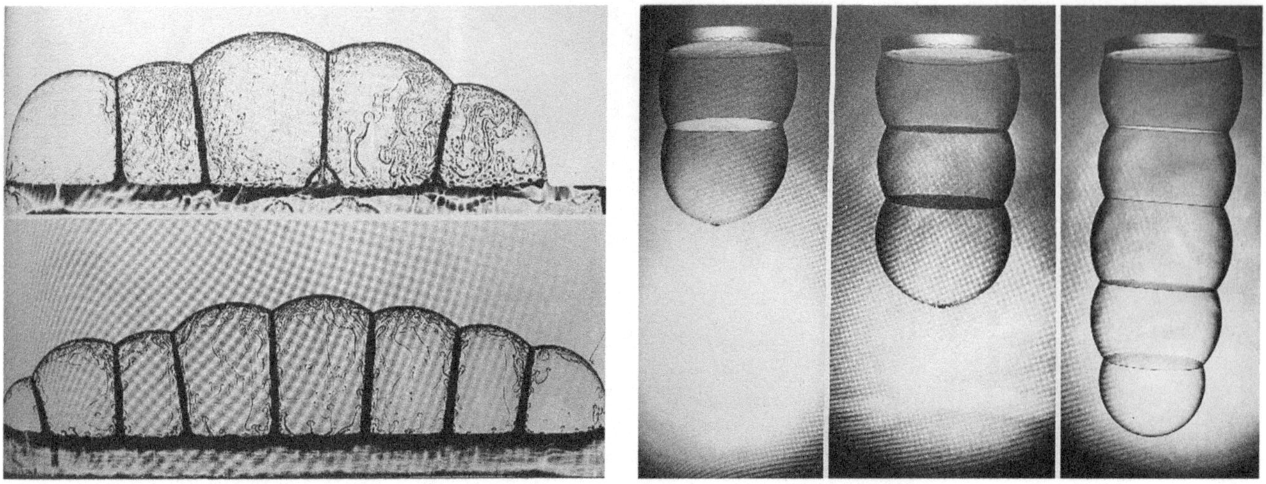

Fig. 10 Horizontal and vertical linear additions of spherical bubbles. (Source: ©Atelier Frei Otto + Partner)

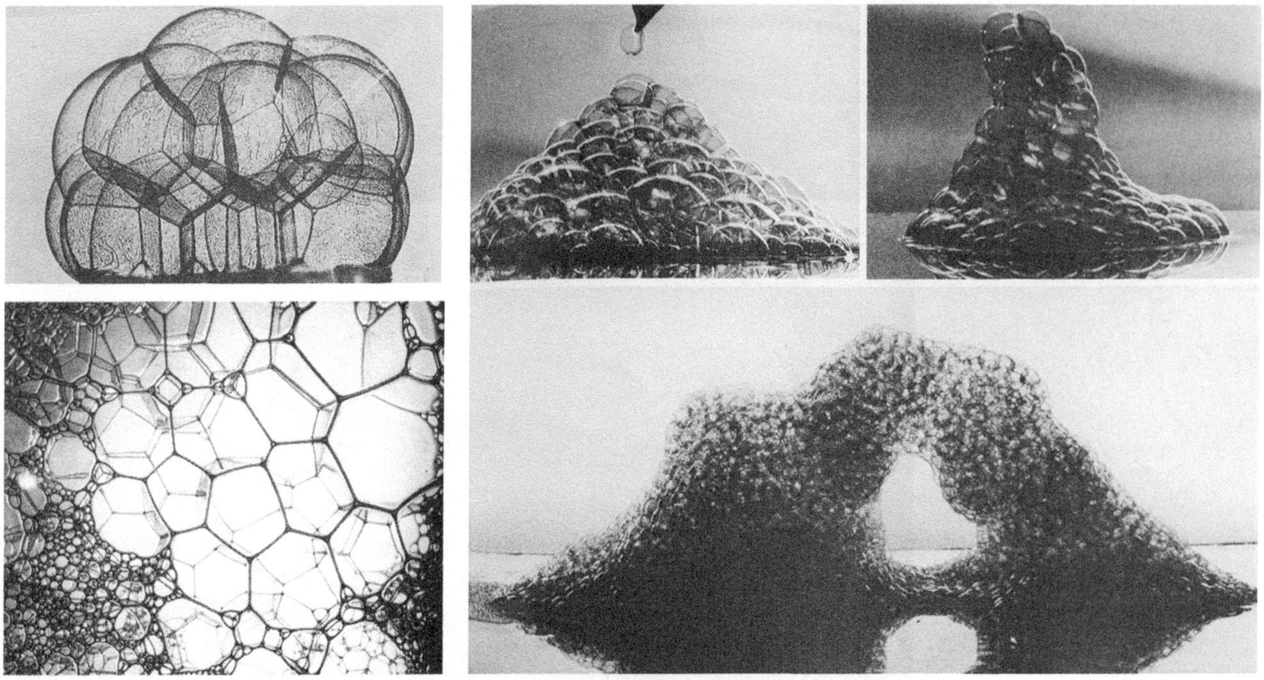

Fig. 11 Bubble clusters forming polyhedral foams. (Source: ©Atelier Frei Otto + Partner)

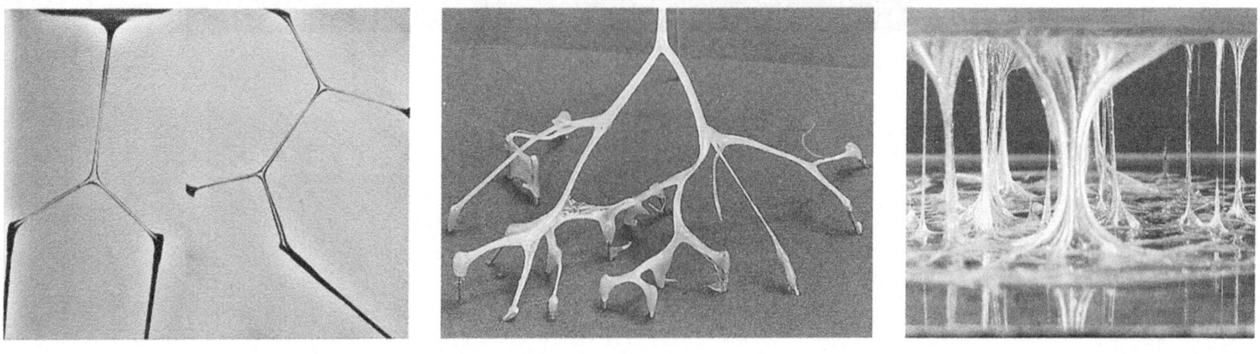

Fig. 12 Branched threads and spine structure developed with viscous fluids. (Source: ©Atelier Frei Otto+Partner)

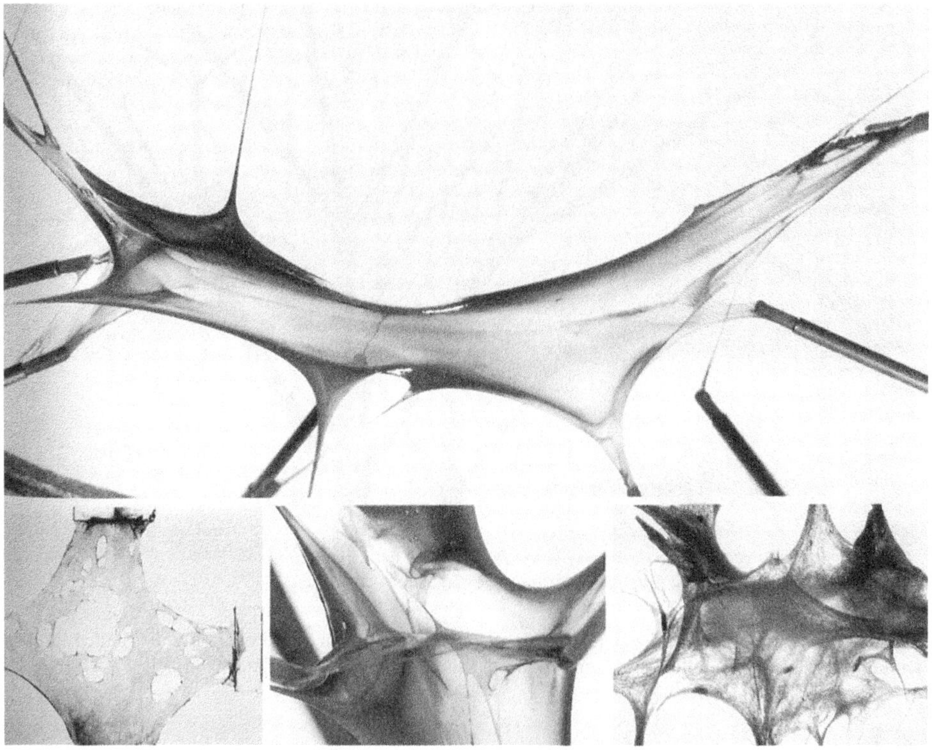

Fig. 13 Membranes developed with viscous fluids. (Source: ©Atelier Frei Otto + Partner)

which concentrates along the hole edges, forming a cord. This would never happen in a soap film, where any hole would lead to the immediate withdrawal of the membrane to the edge elements (Fig. 13).

Branching Structures

Many natural and artificial objects resulting from physical self-forming processes show branchings (Gass, 1990, p. 2.38). Minimal way networks developed with soap films are in fact branching patterns with three-armed nodes. Frei Otto's experiments in the field of branching structures are mostly made with threads. They are related to either hanging models to produce an efficient load-bearing funicular layout made up of linear elements subject to compression, or optimized path systems connecting a number of points with minimum detours. A basic distinction can be made between open and closed-mesh branching structures.

Open Branchings

The challenge of supporting large span roofs avoiding structural elements subject to bending, in favor of compressed or tensioned elements, led Frei Otto to experiment with different types of hanging funicular models, developing an open branching pattern. Its rectilinear members should be working in compression only, and therefore the danger of buckling made it necessary to develop a system of straight elements with a balance between their length, section, and slenderness.

In the first suspension model, a horizontal flat thin plaque was hung from a grid of evenly distributed threads converging into four points. The threads were moistened with liquid polyester, which bundled them through its surface tension, and developed a branching layout. Once dry and stiff, the model was inverted, producing an open branching system of straight elements with reduced buckling lengths and thin sections in the upper part, and longer buckling lengths with thicker sections, due to thread bundles, in the lower part (Fig. 14).

Another branching hanging model was made with chains and springs, to determine the optimum branching angles and rod lengths of a three-layered branching structure with four-armed nodes (Fig. 15).

Following the forms, angles, and rod lengths obtained in the two previously mentioned suspension models, a model of a tree structure was made with thin steel tubes (Bach et al., 1988, p. 66) to transmit gravity loads from an upper horizontal plaque into four lower supporting points through a three-layered branching layout (Fig. 16).

Other open branching models were made with free-hanging threads, which bundled after dipping in water by the capillary tension of the water film (Fig. 17).

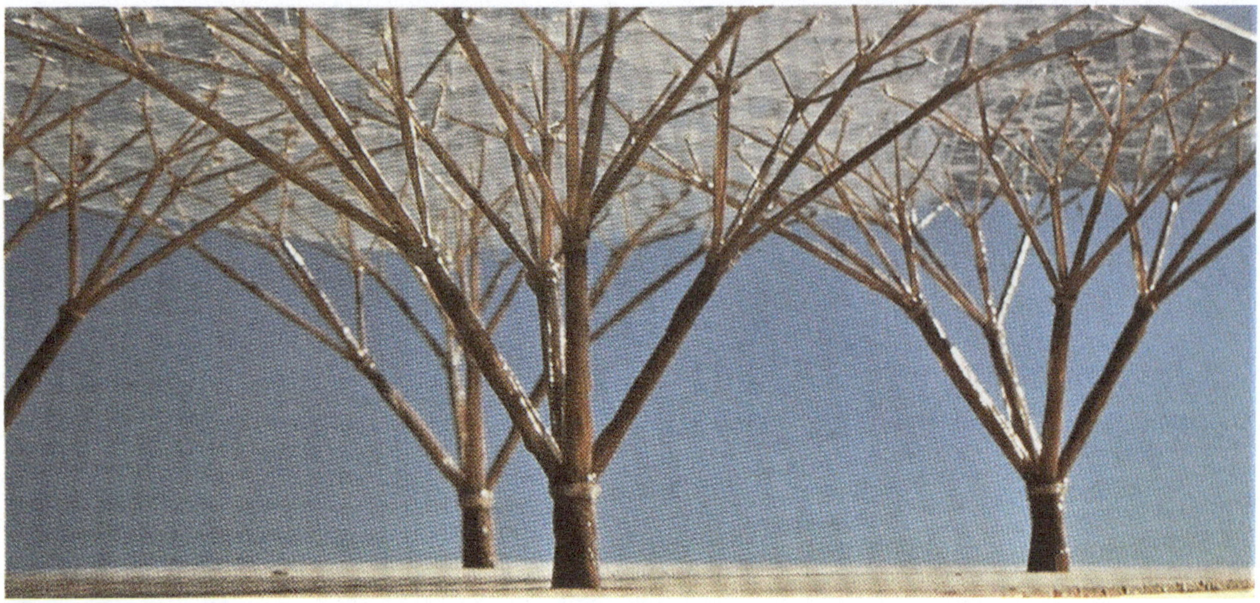

Fig. 14 Branching structure developed with a plaque hanging from a grid of evenly distributed threads converging into four points, moistened with liquid polyester, bundled, dried, and inverted. (Source: ©Atelier Frei Otto + Partner)

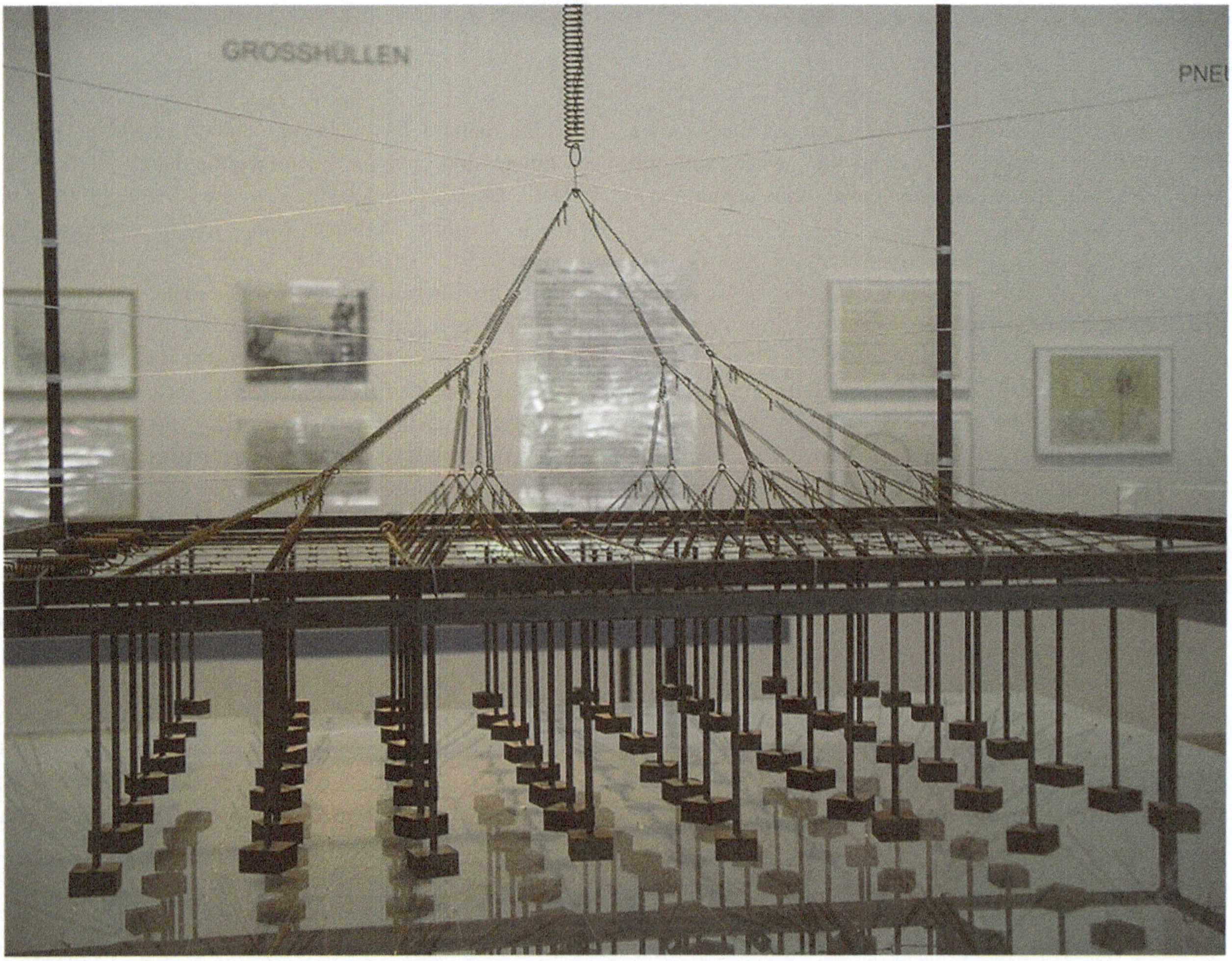

Fig. 15 Branching hanging model made with chains and springs. (Source: author's photograph)

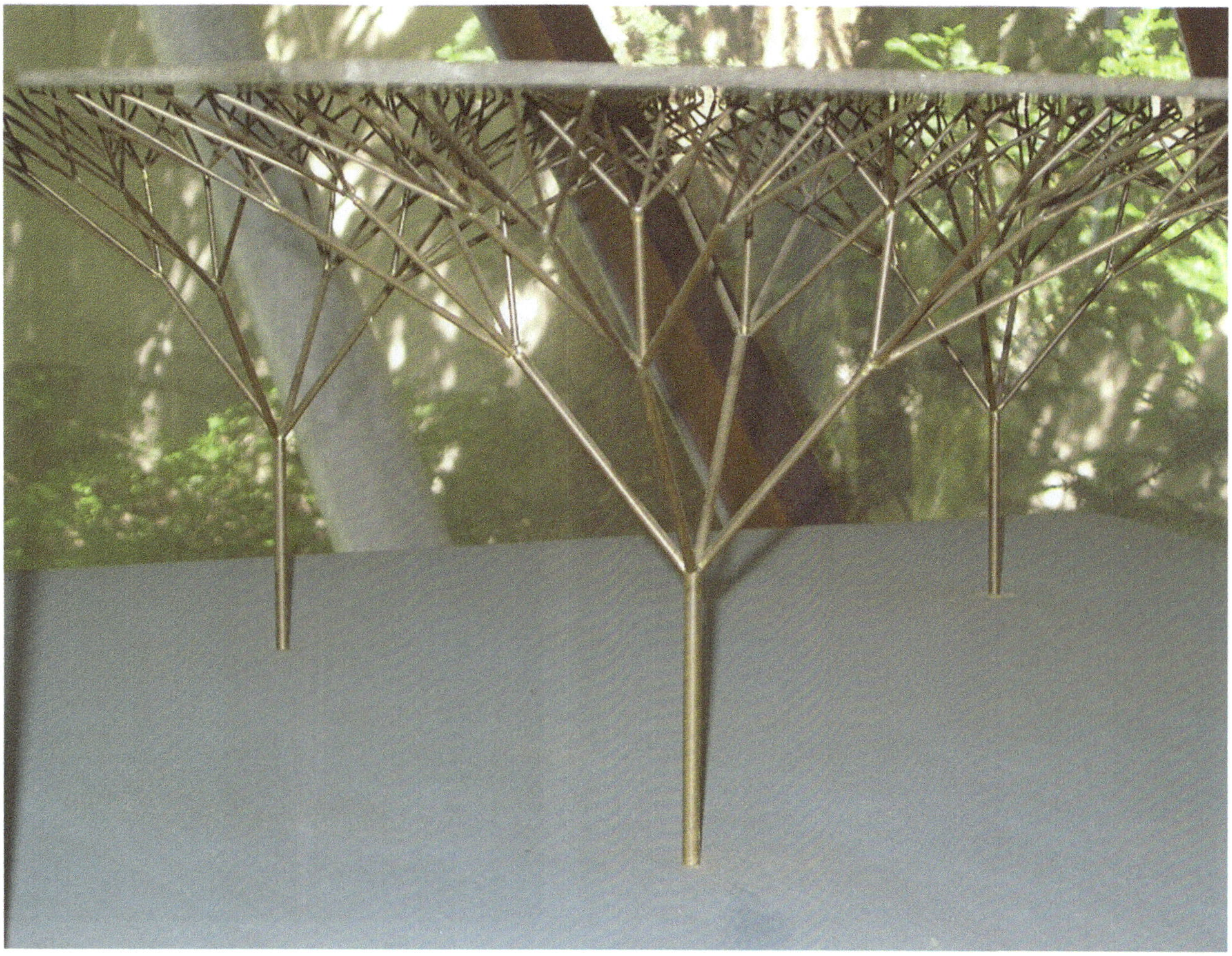

Fig. 16 Model of a tree structure made with thin steel tubes. (Source: author's photograph)

Closed-Mesh Branchings

This type of branching comes up with an experiment Frei Otto developed to investigate what he called "optimized path systems." A number of points to be connected are marked around a circular ring. Each point is connected to every other point by a wool thread in a straight line, making a direct path system (Fig. 18a). This layout is ideal, because all points are connected to each other with the shortest distance and no detours. But the overall path length and the area covered are very large. If the threads are loosened and given an extra length of 8%, for example, and the whole model is dipped in water, then the wet threads will tend to stick together and make bundles due to the capillary tension of water (Fig. 18b). The paths connecting two points will be 8% longer, but the area covered and the overall length is often only 30–50% of the direct path system (Otto et al., 1995, p. 68). The result of this self-forming process is a closed-mesh branching layout of a minimal detours system, also called an "optimized path system."

Funicular Forms

Another important group of forms generated by self-forming processes are funicular forms. They were widely investigated with Frei Otto's experiments. Whether with hanging chains, nets, or cloths, he sought forms to avoid bending forces (therefore thick sections and heavy structures) and to develop axial forces only: either compression or tension along the longitudinal axis of the structural element. Axial forces involve minimum amount of material and are therefore a good way to achieve lightweight construction. Funicular comes from the Latin *funiculus*, which means "string." Funicular forms experience only axial forces. If the form of a structure is not funicular, it will experience bending forces. Chains, cables, nets, cloths, ropes, and strings adopt funicular forms because of their flexibility, which prevents them from resisting bending forces. Given a new set of loads, they always reshape themselves in such a way that they experience only axial tension (Allen et al., 2010, p. 22).

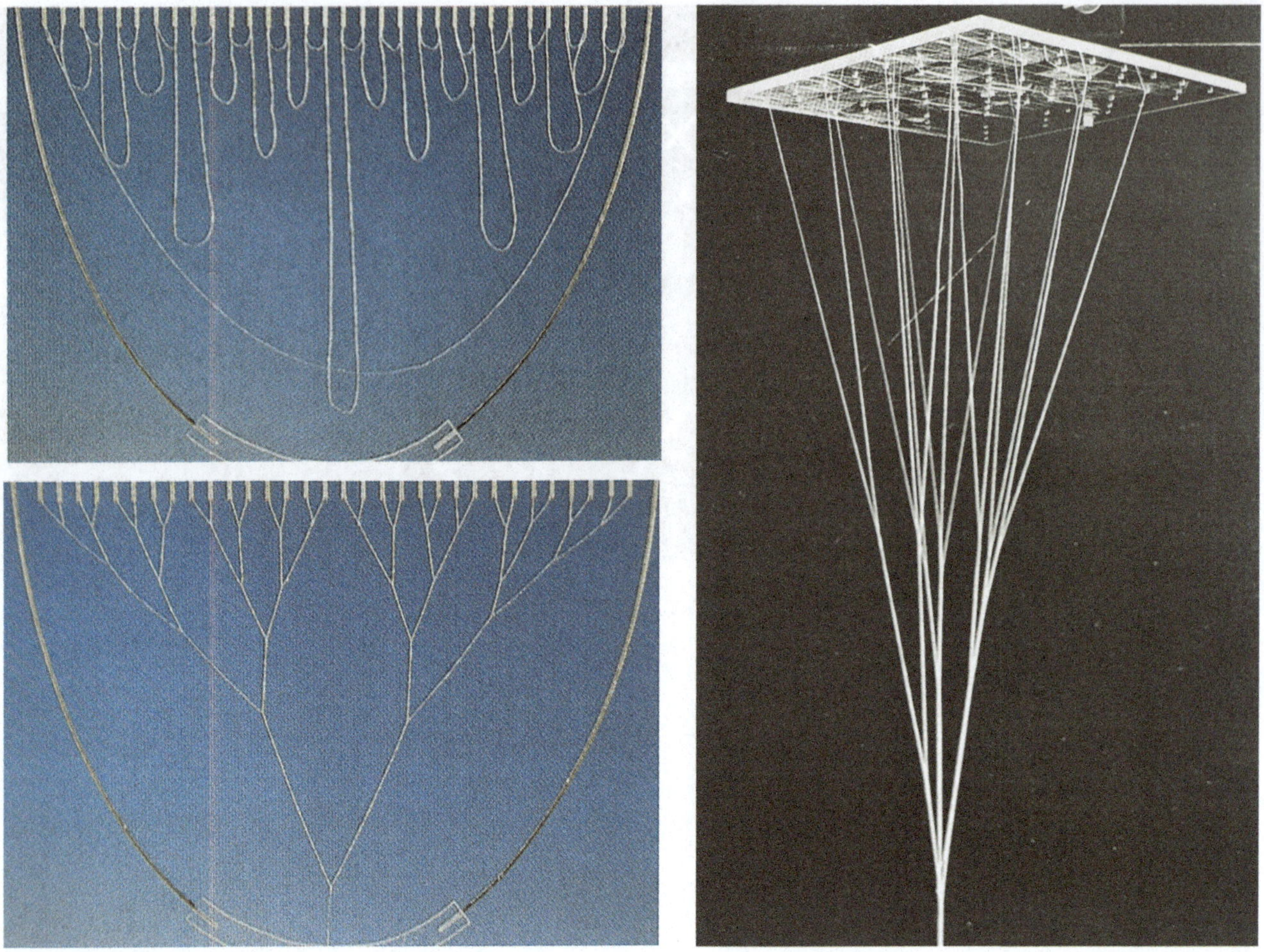

Fig. 17 Open branching models developed with hanging threads moistened in water and bundled by capillary tension. (Source: ©Atelier Frei Otto + Partner)

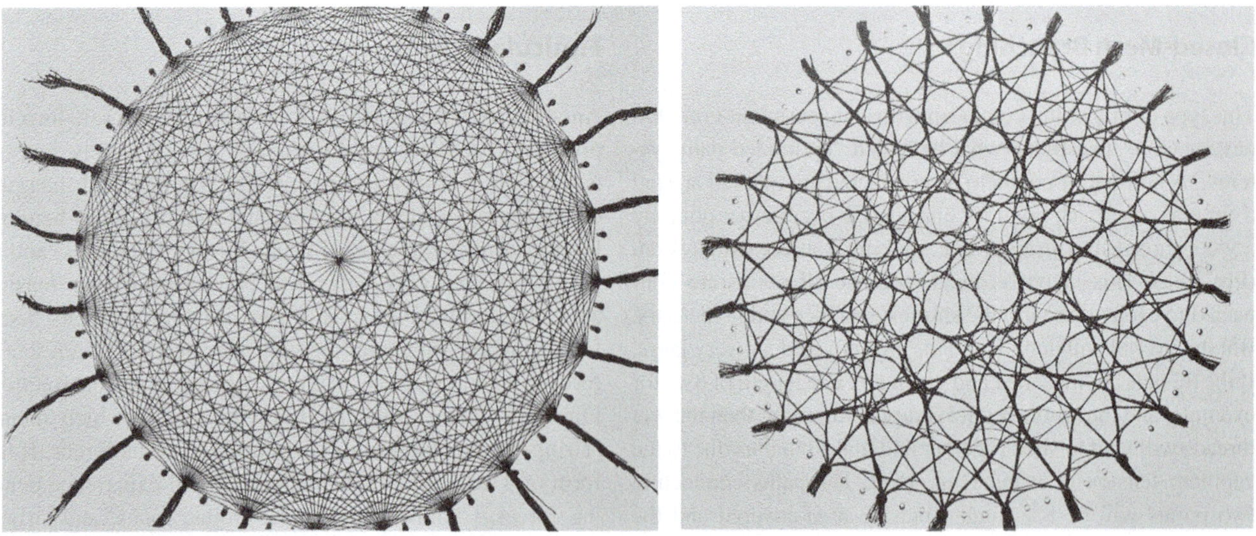

Fig. 18 (**a**, left) direct path system connecting all points to each other. (**b**, right) optimized path system with a reduced overall length and a minimal detour layout. (Source: ©Atelier Frei Otto + Partner

Hanging Chains

A chain suspended between two fixed points with an evenly distributed load along its length and subject to gravity adopts the most suitable form to resist tension forces only: the catenary curve. This is a typical and classical self-forming process extensively explored by Frei Otto with his experiments (Otto, 1973, p. 18–23).

Simple Chains

A chain of constant length suspended between two points initially coincident and gradually separating horizontally at equal stretches (Fig. 19a). Pattern developed by chains suspended with a constant span, gradually increasing length to expand sagging at equal stretches (Fig. 19b).

Freely Suspended Chains That Form Surfaces

Different types of surfaces can be obtained with a series of suspended chains at equal distances. Depending on the lines of suspension points (rigid or flexible, straight or curved, concave or convex), and on the sagging of suspended chains (constant or variable, with largest sag inside or outside), the surfaces developed can be singly or doubly curved, synclastic, or anticlastic. Series of chains can also be arranged in a parallel or radial layout (Fig. 20).

Compressive Inverted Arches

The famous English scientist Robert Hooke put forward in the seventeenth century the so-called principle of inversion. He found that a hanging chain is shaped by gravity to take the form of a catenary. If this funicular form is stiffened and inverted, it will turn out to be the best form for an arch to support the same set of loads: "As hangs the flexible line, so but inverted will stand the rigid arch" (Fig. 21). He highlighted the relationship between the self-formed catenary curve of a hanging chain working in tension only, and the ideal shape of an arch with the same span, rise, and set of loads, to work in compression only, avoiding in both cases bending.

Frei Otto built in 1952 a "standing chain" or "multi-hinged arch," an arch model able to take different lines of thrust or arch outlines with varying patterns of loads. Its voussoirs had convex contacting faces, and could roll over each other, and were also secured against slipping by threads.

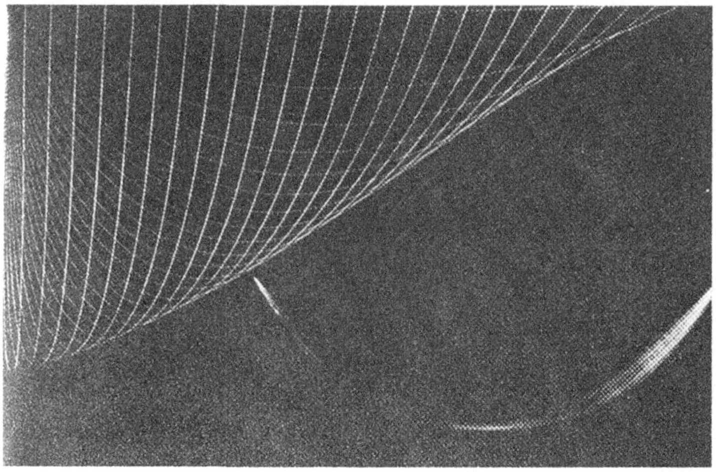

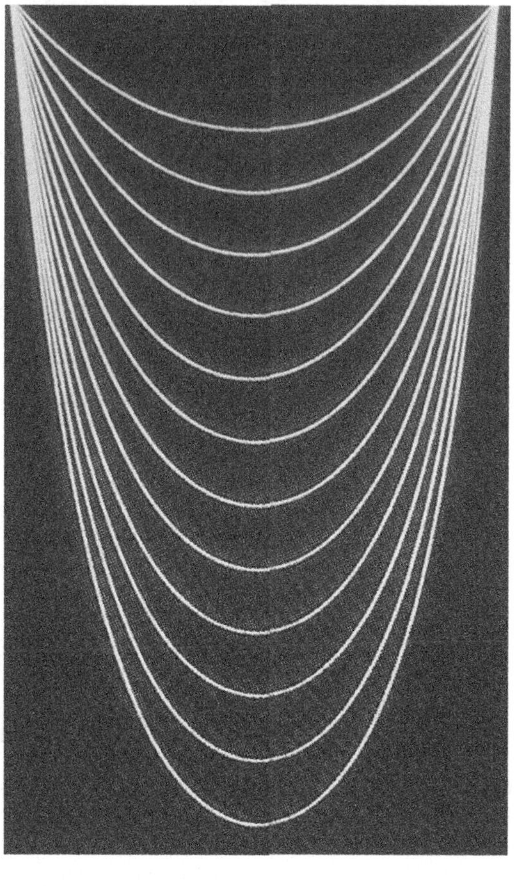

Fig. 19 (**a**, left) Hanging chain of constant length with supports separating gradually. (**b**, right) Hanging chains with a constant span gradually increasing length. (Source: ©Atelier Frei Otto + Partner)

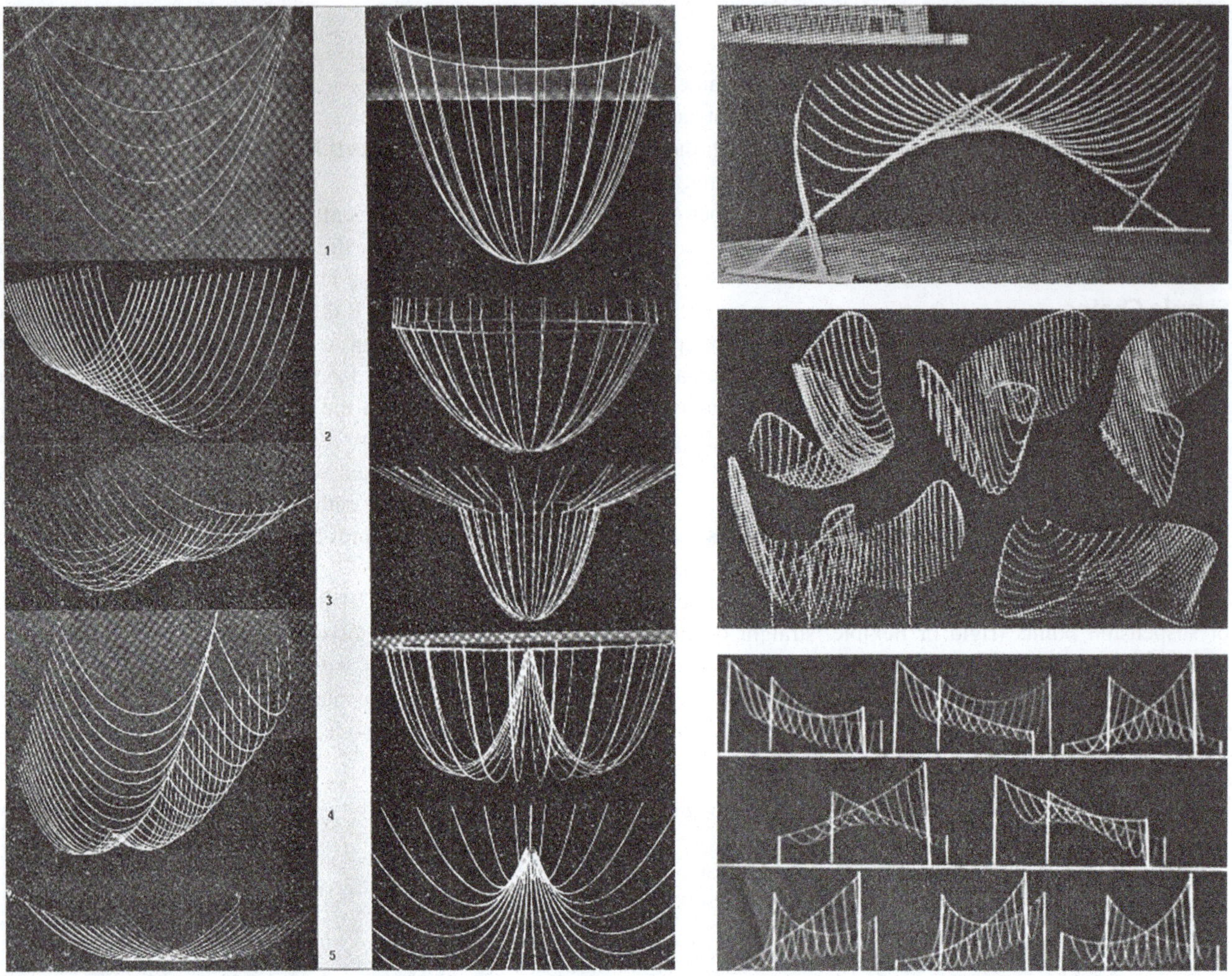

Fig. 20 Parallel and radial series of hanging chains developing different types of surfaces. (Source: ©Atelier Frei Otto + Partner)

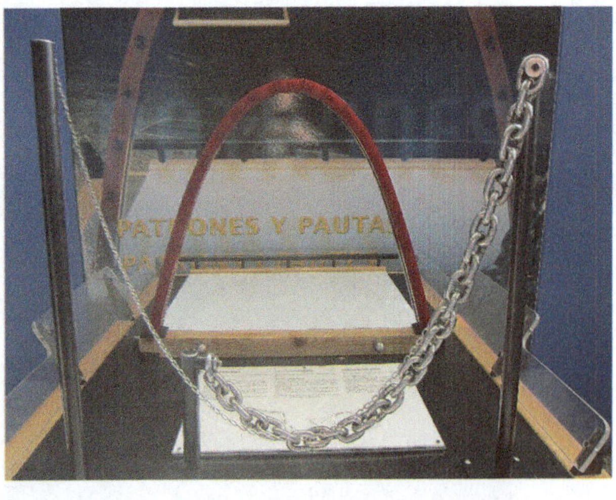

Fig. 21 Principle of inversion (Robert Hooke): "As hangs the flexible line, so but inverted will stand the rigid arch." (Source: author's photograph)

The contact points between the voussoirs mark the line of thrust (or the inverted catenary). Their position changes in accordance with the acting loads, and the overall line of thrust is reshaped, developing a self-forming arch similar to a freely suspended chain (Graefe, 2021, p. 96) (Fig. 22).

Hanging Chain Nets

When chains are interlaced to form a net, they behave in a different way than when grouped in a series of independent parallel or radial elements. Hanging chain nets develop funicular models of surfaces made up of linear elements working in tension only. When inverted, they provide the form of gridshells, made up of a grid of continuous linear elements working mostly in compression. The most suitable chain nets to produce funicular models for gridshells are those with a flat horizontal pattern of square meshes. When

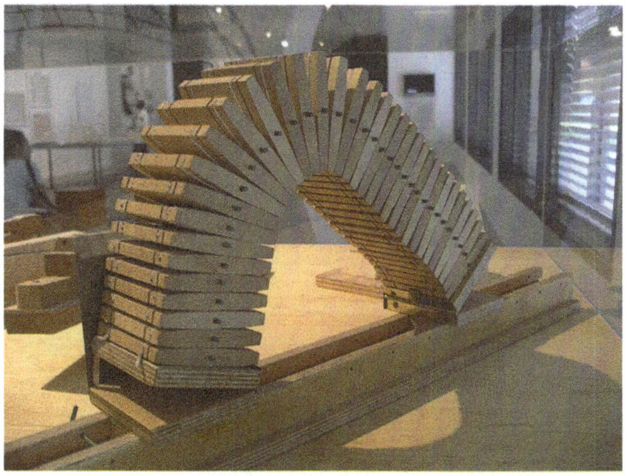

Fig. 22 Standing chain or multi-hinged arch designed by Frei Otto in 1952. (Source: author's photograph)

these chain nets are hung, a self-forming process generates, whereby the squares transform into rhombi with gradually varying sharp angles, to develop a curved surface. Its form depends on the type of boundary meshes and the type and layout of suspension. Two basic categories can be identified (Hennicke, 1974):

Complete Mesh Chain Nets

All the meshes of the net are closed and complete, and the net boundaries are net lines. The outline of the flat net is a square or a rectangle. Supports are punctual and coincide with net nodes. The net edges are flexible and coincide with net lines between two supporting points. They develop self-forming funicular arches (Fig. 23).

Incomplete Mesh Chain Nets

The edge of the net can take any form and does not coincide with any net line. The peripheral meshes of the net are cut, open, and incomplete. Supports are linear and rigid, and can be external to the net, like a wire or a board trimmed with an inner hollow. The final overall pattern of this type of chain net must be defined on the basis of the suspended form, by stretching or shortening gradually the loose ends of the incomplete edge meshes until a uniform and continuous curvature of the whole net is reached, avoiding any local sagging (Fig. 24).

Gridshells

Frei Otto developed a new type of light thin shells, which he named *Gitterschalen*, usually translated in English as "gridshells." They are vaults whose surface is openwork, as a skin pierced with large holes, since it is generated by linear elements, instead of continuous and opaque surface elements. Gridshells are doubly curved surface thin vaults made up of a lattice of thin wooden slats, intended to work mainly in compression. Their form is therefore antifunicular and must be defined by hanging chain net models.

Another important feature is their building process (Songel, 2020, p. 238). A flat planar orthogonal square-mesh slat grid lying on the floor, with loose bolts at every slat crossing, is hoisted up at some points in the central area and is gradually deformed into a doubly curved surface. During this process, the thin slats or laths are slightly bent, and rotate at each grid crossing or node, changing the original right angles into oblique ones, and the initial mesh squares into rhombi. Once the flexible slat grid has reached the funicular shape obtained through the hanging chain net model, it is stiffened by fixing the angles and tightening the bolts at each grid crossing (Fig. 25).

As gridshells and hanging chain nets are so closely related, there are also two similar different types of gridshells, in accordance with the already mentioned types of hanging chain nets.

Open Mesh Arch Boundary Gridshells

The edges of the gridshell coincide with the peripheral grid lines. All the meshes of the grid are closed and complete. The outline of the flat grid is a square or a rectangle. Supports are punctual (Fig. 26).

Closed Flat Boundary Gridshells

The edges of the gridshell do not coincide with any grid line. The boundary edges can have any form, and the ends of the laths are fastened to them. The peripheral meshes of the grid are cut, open, and incomplete. Supports are linear (Fig. 27).

Hanging Cloths

Frei Otto knew inverted forms since he was a child, when he used to play in his father's sculptor workshop with hanging cloths soaked in plaster, dried and inverted, to produce different funicular forms (Otto & Songel, 2010, p. 30). He also experimented with hanging rubber membranes loaded with evenly distributed nails (Fig. 28a) and with models made from medical plaster bandages dipped in water or sprayed with it (Fig. 28b). Folding patterns developed by hanging cloths were also explored in his experiments (Fig. 28c). Heinz Isler, a Swiss engineer and one of the last shell builders of the twentieth century, also used hanging cloth models to design his shell structures (Fig. 28d).

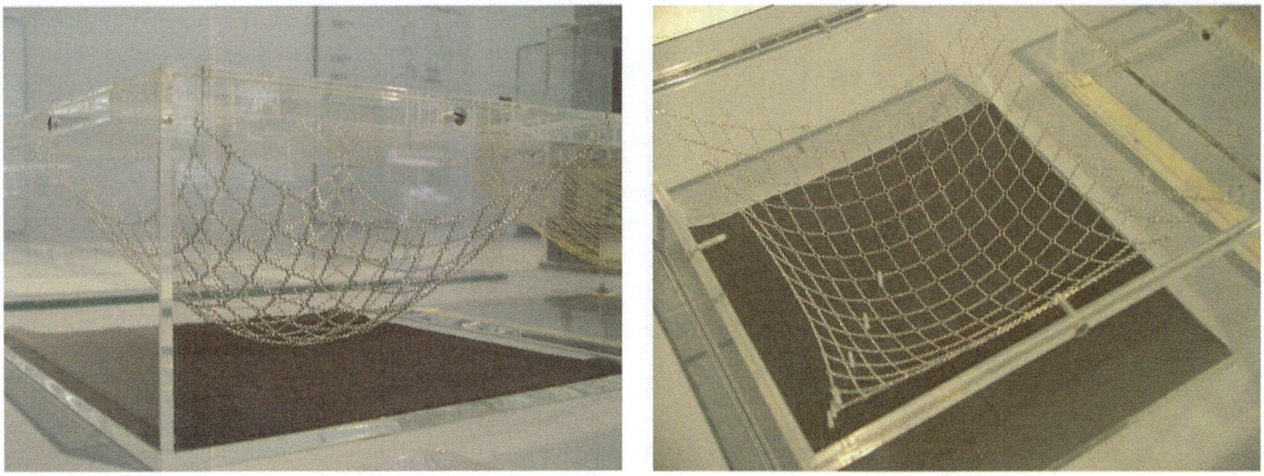

Fig. 23 Complete mesh hanging chain net. (Source: author's photographs)

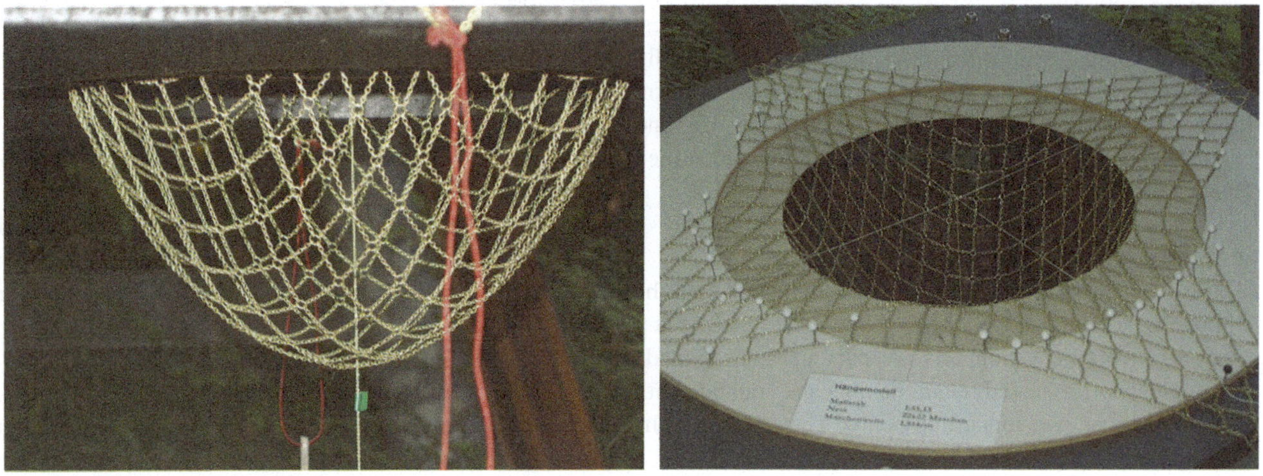

Fig. 24 Incomplete mesh hanging chain net. (Source: author's photographs)

Compressive Inverted Shells

Experience gained by the author in workshops to build funicular models of shells shows the importance of cutting the cloth pattern correctly, in order to ensure that flat cloth surfaces can unfold into three-dimensional forms, in accordance with the intended design, generating an even and seamless surface, where no wrinkles should come up (Songel, 2015, p. 623). It is also worth noting the difference between the form of the hanging cloth before and after soaking in plaster. It is only when the cloth is loaded with plaster that gravity springs into action and develops a smooth and continuous funicular surface (Fig. 29). Many different funicular compressive shell forms can be obtained: either concave or convex, synclastic, or anticlastic. Folds can be generated by inserting strings in the piece of cloth, or by splitting the cloth pattern in different pieces, which will later be sewn together (Fig. 30).

Concluding Remark

It could be argued that computers and parametric design software have superseded physical models, as tools to simulate material behavior to generate light structures or optimal and minimal tensile surfaces. When confronted with this issue in a conversation with the author in 2004, Frei Otto said:

> The computer can only calculate what is already conceptually inside of it; you can only find what you look for in computers. Nevertheless, you can find what you haven't searched for with free experimentation." [...] "The truly important things [arose] largely from fortuitous or casual observations made during experiments, some of which were planned in a completely systematic style. I have always combined systematic experimentation with the fortuitous or casual, where chance plays a role; if something is accidentally discovered, it would be stupid to reject it simply because it doesn't fit within the systematization. (Otto & Songel, 2010, p. 38 and 32)

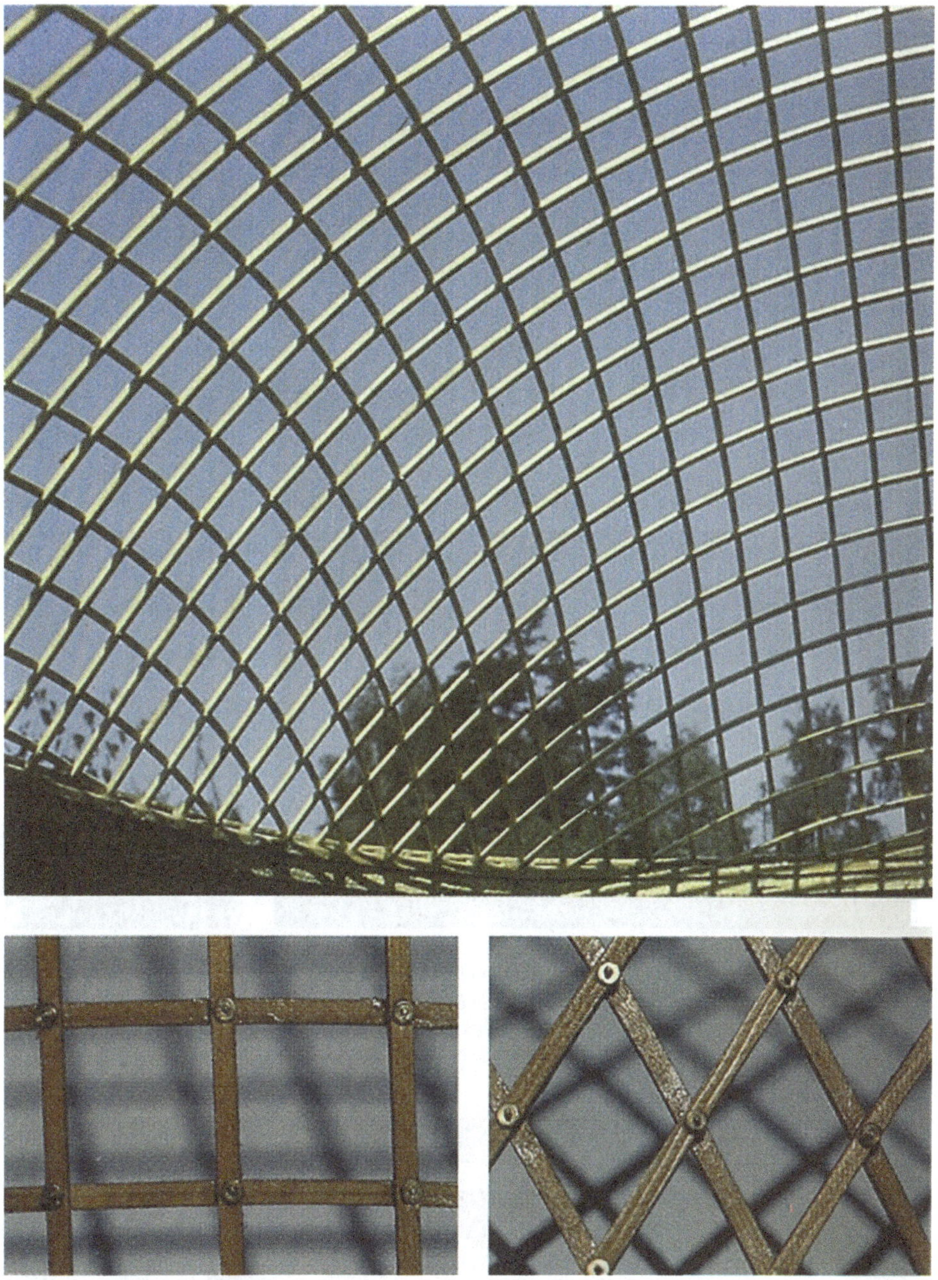

Fig. 25 Deformation of an orthogonal flat slat grid, hoisted to generate a gridshell. (Source: ©Atelier Frei Otto + Partner)

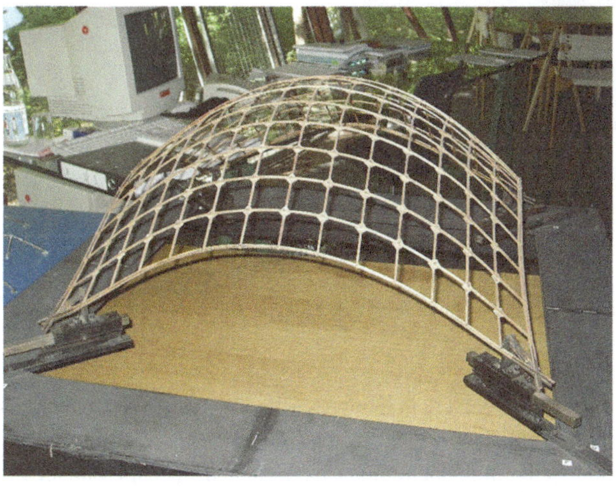

Fig. 26 Open mesh arch boundary gridshell. (Source: author's photograph)

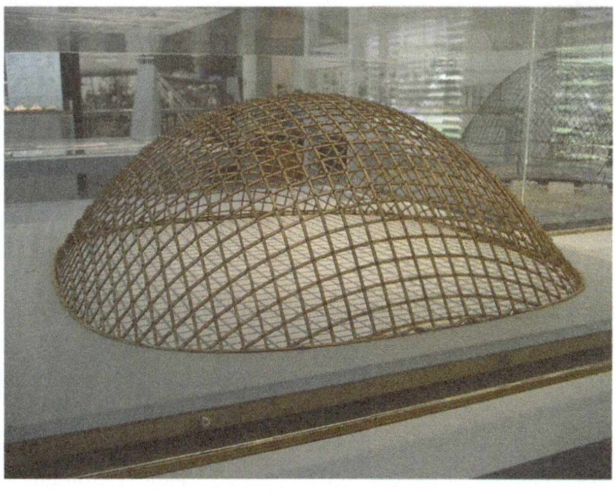

Fig. 27 Closed flat boundary gridshell. (Source: author's photograph)

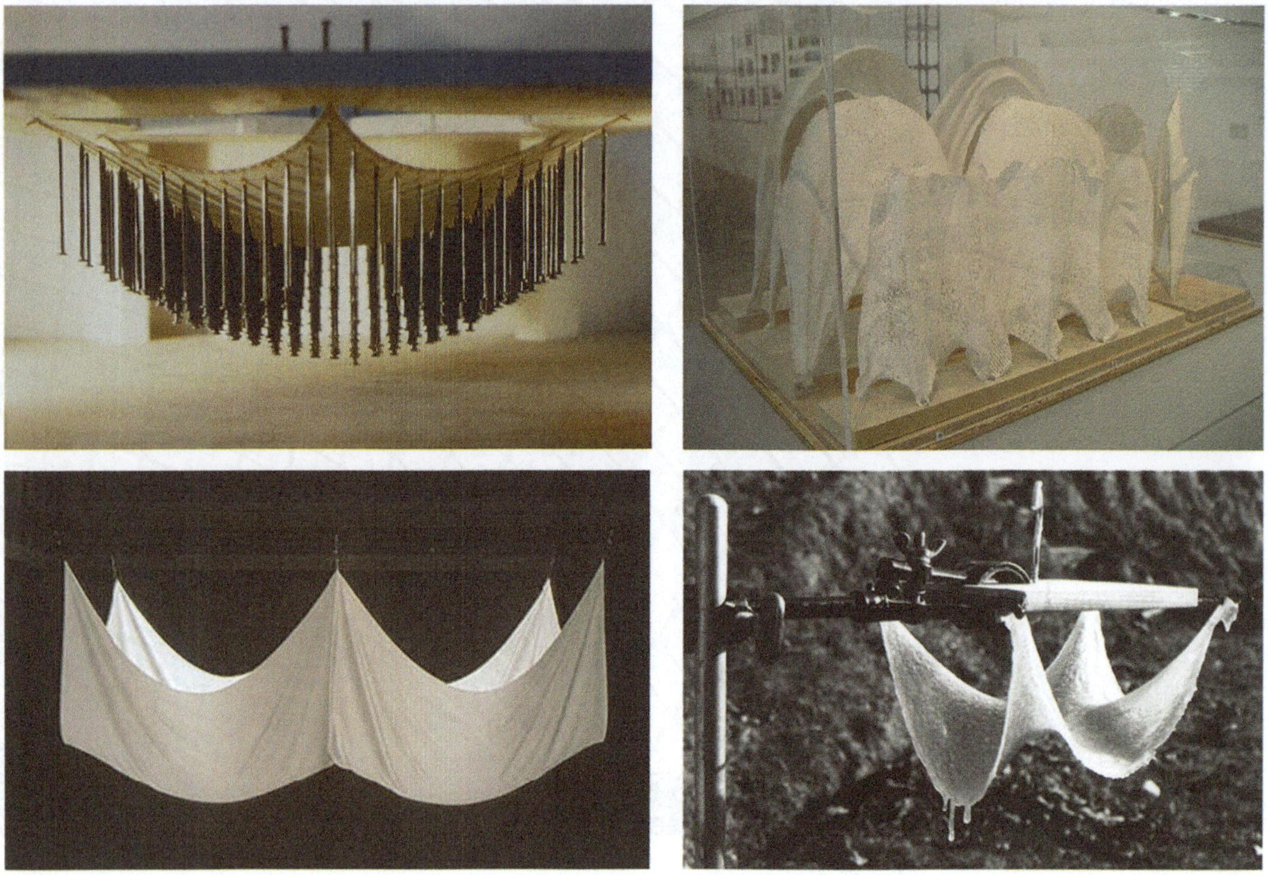

Fig. 28 (**a**, Above left) Hanging rubber membrane loaded with nails. (**b**, Above right) Hanging medical plaster bandages moistened in water. (**c**, Below left) Folding patterns developed by hanging cloths. (**d**, Below right) Heinz Isler's hanging cloth model of a shell. (Source: ©Atelier Frei Otto + Partner (a,c), author's photograph (b), Heinz Isler gta Archiv ETH Zürich (d))

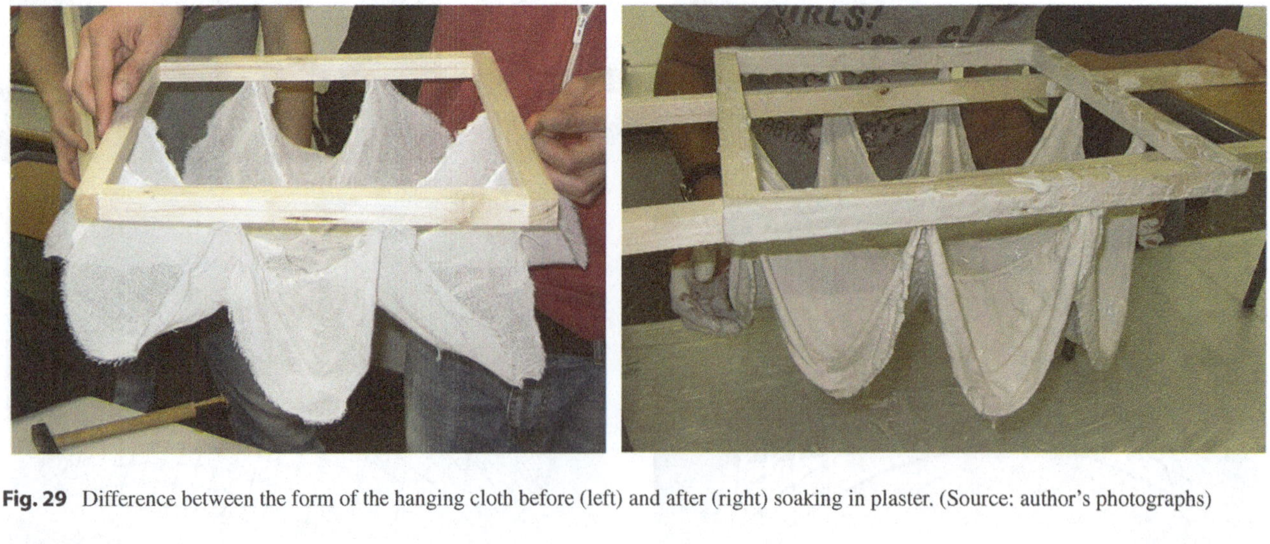

Fig. 29 Difference between the form of the hanging cloth before (left) and after (right) soaking in plaster. (Source: author's photographs)

Experiments with physical models, according to Frei Otto, provide the chance to invent, the possibility of finding the unsearched, as they contain the laws of nature. Computers, however, would not have that ability, as the laws governing them have been created or prescribed by man. Form generation through physical self-forming processes and careful observation of natural phenomena have indeed been pivotal in Frei Otto's experience for the invention of new light structures.

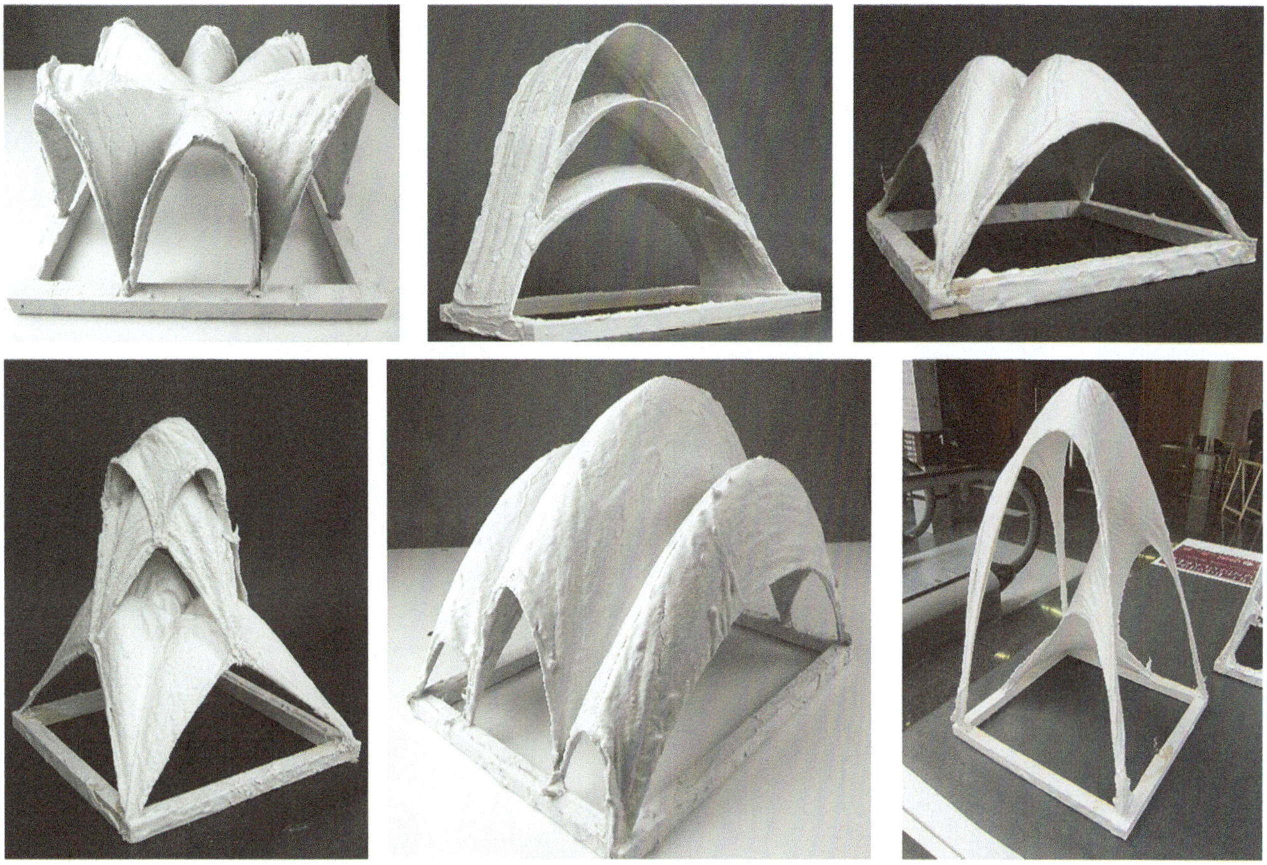

Fig. 30 Different funicular compressive shell forms. (Source: author's photographs)

References

Allen, E., Zalewski, W., & Michel, N. (2010). *Form and forces: Designing efficient, expressive structures*. John Wiley & Sons.

Bach, K., Otto, F., & Burkhardt, B. (1988). *IL 18. Seifenblasen. Forming bubbles*. Krämer.

Brinkmann, G. (Ed.). (1990). *Leicht und Weit*. VCH.

Gass, S. (1990). *IL 25. Experimente. Experiments*. Institut für Leichte Flächentragwerke.

Ghyka, M. C. (1977). *The geometry of art and life*. Dover.

Graefe, R. (2021). The catenary and the line of thrust as a means for shaping arches and vaults. In B. Addis (Ed.), *Physical models. Their historical and current use in civil and building engineering design*. Ernst & Sohn.

Hennicke, J. (Ed.). (1974). *IL 10. Gitterschalen. Grid shells*. Institut für Leichte Flächentragwerke.

Le Ricolais, R. (1997). *Visiones y paradojas. Visions and paradox*. Fundación Cultural COAM.

Nerdinger, W., Meissner, I., Möller, E., & Grdanjski, M. (2005). *Frei Otto: Complete works: Lightweight construction natural design*. Birkhäuser.

Otto, F. (1973). *Tensile structures*. MIT Press.

Otto, F., & Songel, J. M. (2010). *A conversation with Frei Otto*. Princeton Architectural Press.

Otto, F., Rasch, B., & Schanz, S. (1995). *Frei Otto, Bodo Rasch: Finding form, towards an architecture of the minimal*. Axel Menges.

Otto, F., Vrachliotis, G., Kleinmanns, J., Kunz, M., & Kurz, P. (2017). *Frei Otto. Thinking by modeling*. Spector books.

Roland, C. (1972). *Frei Otto: Structures*. Longman.

Songel, J. M. (2015). Form follows forces. Building funicular models to show how gravity shapes form. In *EDULEARN15 proceedings* (pp. 621–626). IATED.

Songel, J. M. (2020). Sustainability lessons from vernacular architecture in Frei Otto's work: Tents and gridshells. *The International Archives of the Photogrammetry, Remote Sensing and Spatial Information Sciences, XLIV-M-1-2020*, 233–240. https://doi.org/10.5194/isprs-archives-XLIV-M-1-2020-233-2020

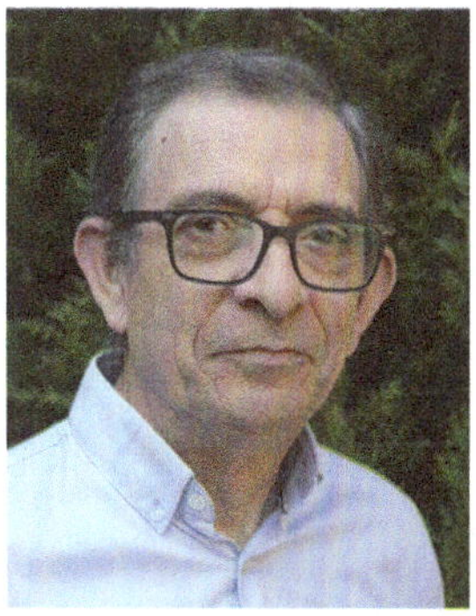

Juan María Songel is a Ph.D. Architect and an Associate Professor at the Technical University of Valencia (UPV), where he has taught courses on Architectural Composition, History of Architecture, and Art and Aesthetics in Civil Engineering. His research areas include different topics such as Theory and History of Architecture, Aesthetics in Civil Engineering, and Frei Otto and Light Structures. He has been a member of the research projects "Living & virtual visiting European World Heritage" and "Roman theatres in Hispania. Conservation, restoration criteria and use enhancement", among others. He held a position as an assistant director for Research at the School of Architecture at UPV, and was also the editor-in-chief of *VLC Arquitectura* Research Journal, published by the School of Architecture. He has also taken part as a jury member in research assessment panels for the Foundation for Science and Technology (FCT) of the Portuguese Government and for the State Research Agency (AEI) of the Spanish Ministry of Science and Innovation.

References

Categories of Natural Principles and Their Adaptation to Bionic Design

Elena-Codina Duşoiu

Abstract

The relation between nature and design has a millennial history. While art found aesthetic inspiration and proportion in nature, technique used natural principles for inventions, adaptable to everyday life or industrial production. Design pushed further the research in terms of ergonomics, dynamics, functionality and efficacy. Therefore, the aim to define and order the categories of inspiration sources taken from nature may appear as useful. Form, geometry, and proportion may appear as the easiest way to study and transmit the principles of nature (an approach used in biomimicry). However, the dynamic study of natural mechanisms may change the perspective. A different approach is brought by understanding natural substance as composed of interconnected small elements, creating perfectly functional networks. And last but not least, let us not forget about the beauty of nature, about its inner proportion and mathematical composition that bring absolute harmony.

The present ecologic crisis obliges us to be smart and resilient and bionics appeared to be one of the solutions to this crisis. The strategic concept of sustainable development, consecrated by the Summit from Rio in 1992, produced the vision of sustainable design. Later on, it was refined and turned into the notion of *ecologically sustainable development* (and design), such as formulated by R. Harding in 2006. The present discourse produces a concrete onset for this concept and illustrates each defined category with a specific design project inspired by nature, presented as a study case. The final purpose is to approach the study of nature in a conscient way, using a series of effective tools. This may lead to a more intelligent and efficient design conception at any scale (from urbanism and landscape design to conception of buildings, interior architecture, or object design), in terms of general thinking, form, elected materials, and efficient functioning. The easiest way to be ecologically sustainable is by correctly understanding and applying the lesson of nature.

Keywords

Nature · Morphology · Natural mechanism · Bionic design · Ecologically sustainable design · Network-inspired design

Introduction

We are accustomed to see technique and art as two completely distinct fields – one aimed by the power of rational thinking, logic, and calculation and the other one based on inspiration, harmony, and aesthetics. There are disciplines (such as architecture) that demonstrate the inaccuracy of this reasonment. There are spirits that have perfectly melted the two branches usually dedicated to science/technique and art/artistic creation. Maybe the best example is Leonardo da Vinci, usually considered to be an artist but recognized as well as engineer, architect, and scientist. We can wonder which element produced the nucleus of such an amazing career. There is an easy answer: The study of nature is the deepest school that has formed Leonardo as an artist and a scientist. In nature, he has discovered in the same time the laws of harmony and the scientific principles for his inventions: parachute, flying machine, submerging costume, a type of robot, different types of weapons, etc. (Laurenza & Taddei, 2006). Considering the field of architecture, we may observe that nature, represented both by plants and animals, has always been a constant inspiration source, both in vernacular architecture and in creations of some famous architects: Antoni Gaudí, Renzo Piano, Frank Gehry, Zaha Hadid, etc. However, these names are just a reference to the modern

E.-C. Duşoiu (✉)
"Ion Mincu" University of Architecture and Urban Planning, Bucharest, Romania

M. R. Cervera Sarda et al. (eds.), *Architecture Inspired by Nature*, https://doi.org/10.1007/978-3-031-33144-2_4

and contemporary era. In fact, I wonder if there is any period in architecture history when architecture had *no* relation with nature (in terms of proportion, details, constructive logic, etc.). We can observe the long-lasting relation between architecture and nature in the huge papyriform Egyptian columns or in the perfect proportion of the Greek orders, reproducing the human body. Nature is present as well in the logic of the Gothic structure, reproducing the fractal growth of trees, in the precision and serenity of the Renaissance space, or in the richness of the modern Art Nouveau decoration, endowed with a multitude of vegetal elements.

Nowadays, one century after the abstract and minimalist Vanguard, design is walking through a deep and conscious oncoming on nature, which seeks inspiration and scientific justification in bionics and its components (biomimicry, biophilia, etc.). Today, the *ecologically sustainable design* is an aim, in search of intelligent and technically controlled models for the life frame of tomorrow (Harding, 2006). Nevertheless, seldom we do not observe that solutions for our problems are closer than we expect. Janine Benyus was the first author to formulate the difference between "learning about the natural world" and "learning from the natural world," which is the process applied in *bionic design* (Benyus, 2002).

In this sense, a sequence of directions to approach the study of nature are presented, materialized in four main categories. Though bionic architecture and design inspired by nature have been treated and largely explained in various occasions (Benyus, 2002; Tanov, 2018; Yeang, 2006; Keller, 2018), the idea of defining which part of natural laws is studied exactly for a certain design and how this study works has not been defined clearly yet. The four proposed categories are illustrated with personal design studies dedicated to furniture objects and with projects realized by students in Product Design ("Ion Mincu" University) who participated in the course "Ecology," taught in the academic year 2018–2019. These studies constitute concrete examples illustrating the four categories proposed and described below. In this moment, the study is formulated in a general frame, however it can be adapted to mathematic models of calculation and behavior.

Some Key Concepts in Design and Their Correspondence in Nature

From the beginning, design has constituted a combination between aesthetics and efficacy, in terms of form, processed materials, and employed technology. Design has consecrated concepts such as ergonomics, comfort, efficacy, and later on adaptability and flexibility, in order to adapt to spaces that are in a constant functional evolution. Conceived as the science of conciliating the human body with the surrounding ambience, ergonomics has defined itself through the use of geometric dimensions and shapes. However, natural geometry is by far more adaptive to the human body, since it uses the same system of proportions and measures, based on the golden ratio and not on the artificial metric system. Any kid may discover that the branch of his favorite tree can be more ergonomic than any sofa or bench. Design is as well associated with comfort – meaning soft, dense surfaces, with a smooth touch and a perfect ambience for the environment – in terms of temperature, humidity, and air ventilation. Let us count how many aesthetic and technologic efforts are necessary in order to complete all these requests and compare everything with the chance of lying on the grass in a flowered meadow, on a sunny spring day. Can we invent something more adaptable than the skin of chameleon or more flexible than the body of a caterpillar? Efficacy in living organisms has inspired a lot of technological inventions, such as hulls of boats imitating the thick skin of dolphins, medical ultrasound technology imitating animal echolocation, prototypes for a dispositive for carrying water inspired by the Namib desert beetle, etc. Moving the study to an architectural scale, we may observe that the most efficient building structures created by architects and engineers are (consciously or not) copies of natural structures: shells, trees, mushrooms, honeycombs, organic membranes, etc.

Speaking about aesthetics, nature gives us nothing less than the geometric formula of beauty, contained in the golden spiral, the golden ratio (the famous number *Phi* 1.6180339887…), and the 137.5° angle, which generates the movement of the golden spiral. So far, the importance of studying nature in order to produce quality design is evident. The only remaining problem is how to approach the huge diversity of natural patterns in order to find specific solutions to design problems.

Four Categories for Approaching Nature in Design Research

Form/Morphology

Maybe the most obvious contact with nature is the study of form. The observation of volumes, including their composing elements and their envelopes functioning as skins and membranes, is the first type of study that comes into mind. We can make a deep study on the geometric form of the natural object, its generative law, its texture and material properties, and its color and shape. We should study the micro-tissue of the envelope, as well as the inside surface. There is no need to translate all the discovered properties to the designed object, just the ones considered as more significant. The final object should be a clear reference of the chosen model without being a direct imitation.

Study Cases (Form/Morphology)

The sketches below constitute the idea for a project for a furniture piece based on the deep study of the shape of the egg (Fig. 1). The challenge is to realize a continuous shell in one material (laminated plastic could be an option). The geometry of the egg form is important to be studied and understood (the sculptor Constantin Brâncuşi called the egg "the beginning of the world"). The inside material of the shell should be different and comfortable (as the membrane that strengthens the egg shell). The sitting part should be a soft cushion with a round form. The armchair is proposed to be realized imitating the natural colors of the yolk and albumen. A functional purpose leads to the proposal of setting the object both on the ground and hanging it. The egg form will lead to similitudes with consecrated design objects (for instance, with the Ovalia Egg Chair created by Thor Larsen or the Egg Chair realized by Arne Jacobsen). The Italian designer Simone Micheli has an interesting interpretation as well. In this case, the purpose is to push further the study of natural form and tissue by using a multitude of senses (sight, touch, acoustics, smell, etc.)

A project realized in 2018–2019 by a student in Product Design (Sara Vahdani) is presented as well, in order to illustrate the mentioned category. The project proposes the design of a lamp inspired by the form of the nest of a weaver, observing the details and imitating the texture of the tissue created by the bird (Fig. 2). The resulted surface is flexible and resistant, due to the multitude of irregular nodes and joints, allowing the lamp to change its form according to the design conception of the space and providing a protection envelope against a too big quantity of light. The material proposed for the realization of the lamp is recycled metallic wire that allows the morphologic change of the lamp according to the concept of atmosphere generated in the conceived space.

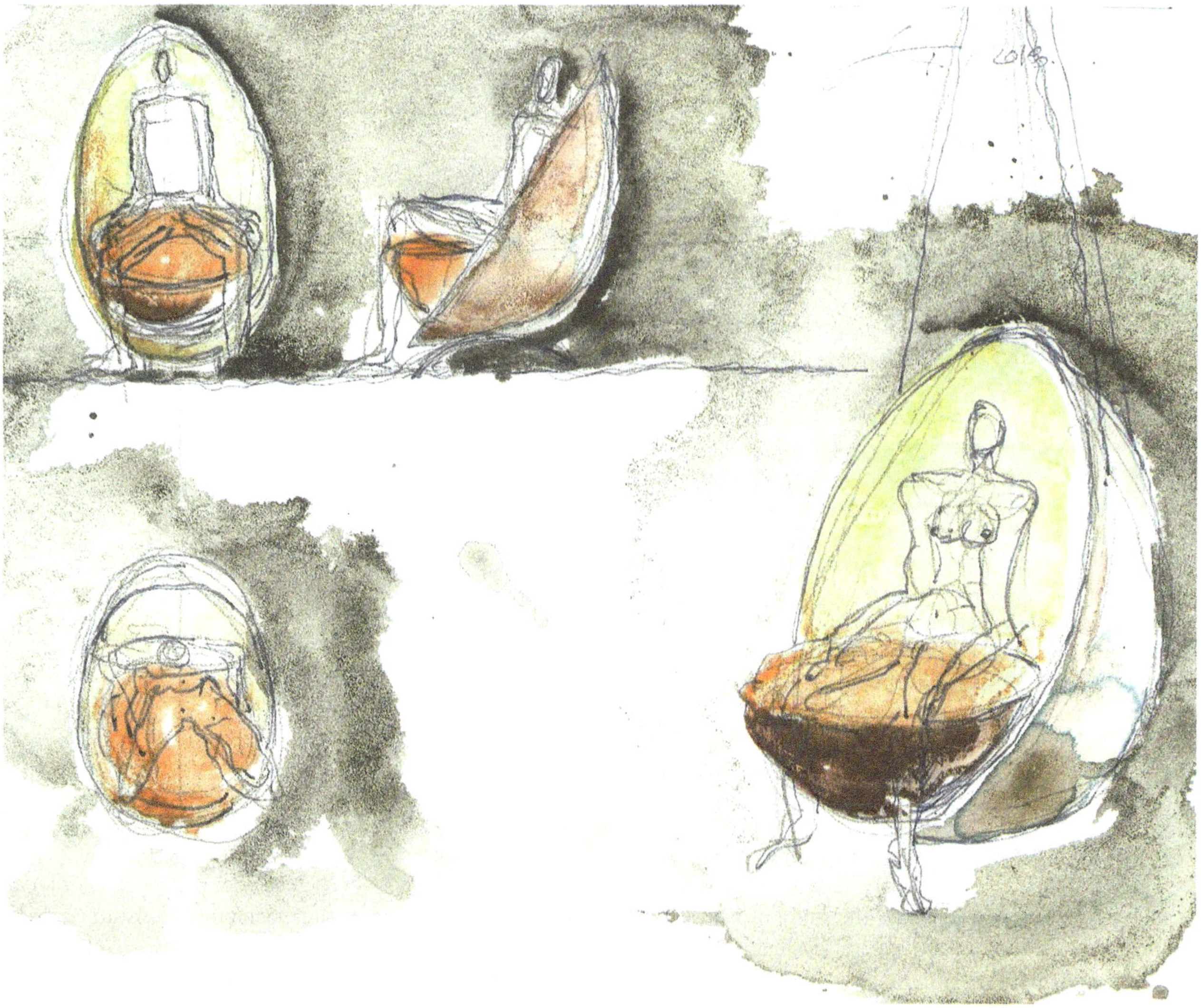

Fig. 1 Sketches of a chair inspired by the form and properties of an egg, by the author

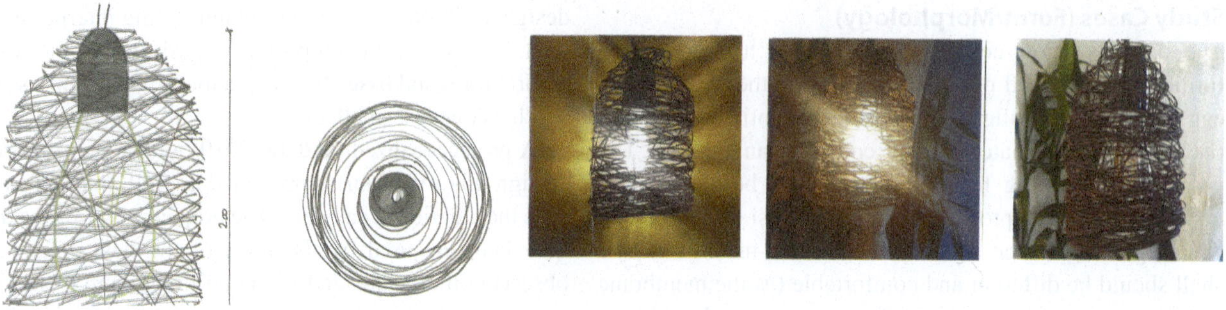

Fig. 2 The project of a lamp inspired by a weaver nest, by Sara Vahdani, Faculty of Product Design, IMUAU (University project, 2018–2019, illustration of the category Form/Morphology)

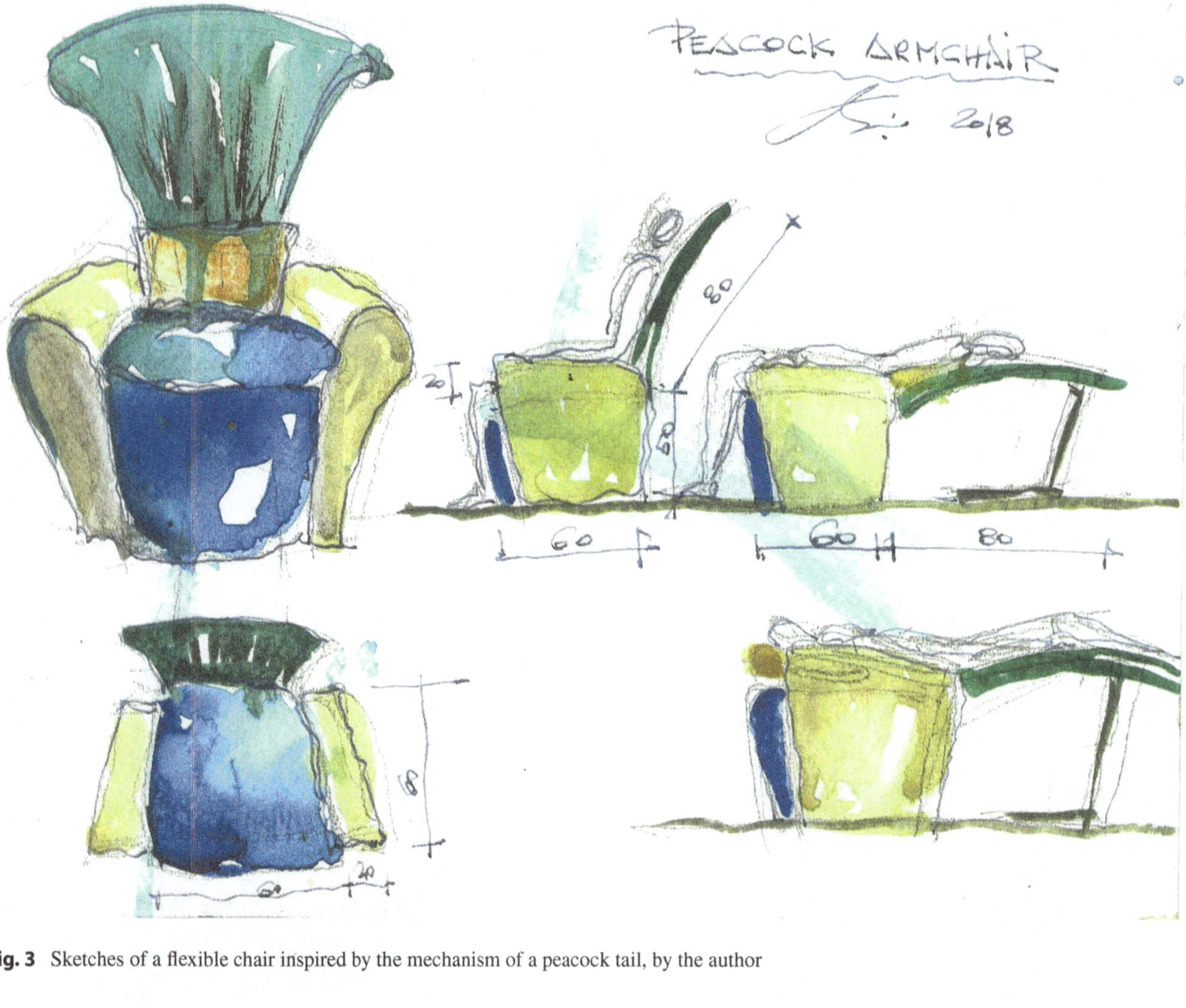

Fig. 3 Sketches of a flexible chair inspired by the mechanism of a peacock tail, by the author

Dynamics/Function

A more dynamic perspective on the study of nature can be achieved by focusing on the natural mechanisms and their laws of functioning. Life itself means movement and change, properties that are inherent to living substance. We can both observe the mechanical and thermodynamic functioning of living organisms (thermal insulation from the environment, ventilation through respiration, humidity through the permissivity of their skin). The envelope of a vegetal or an animal organism (the covering membrane constituted by the skin) should be studied as a dynamic system, in relation with local climate conditions – wind, temperature, and humidity. Mechanical functioning of living organisms can be a rich source of inspiration as well, easy to translate to structural problems that we need to solve in engineering and design.

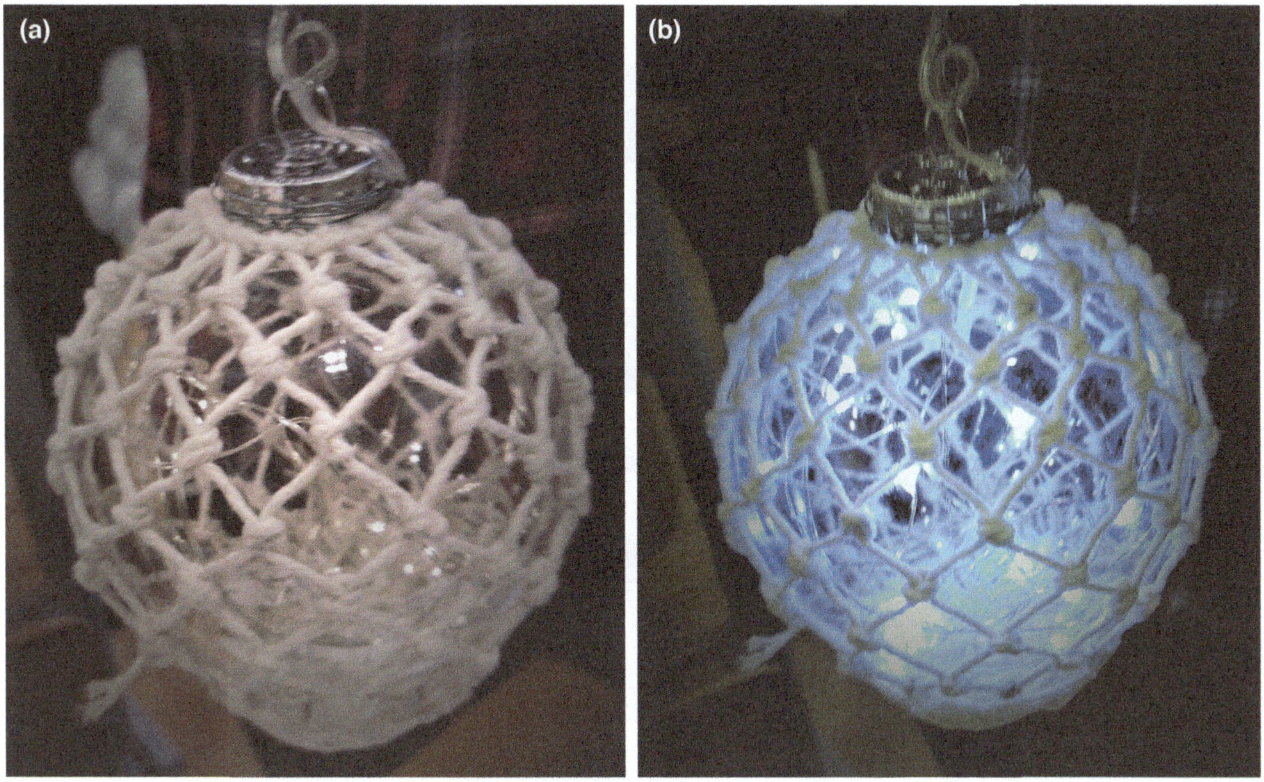

Fig. 4 (**a, b**) The project of a lamp inspired by the way of producing light of fireflies, university project (Faculty of Product Design IMUAU, illustration of the category Dynamics/Function)

Special attention should be paid to mechanic details, easy to transpose in a technical thinking, such as articulations, joints, and embedments.

Study Cases (Dynamics/Function)

Here is a proposal of a flexible transformable armchair, inspired by the functioning of a peacock tail (Fig. 3). The essence of the object is the opening and closing of the backrest, which also allows its transformation into a *chaise longue*. In this sense, the study of its articulation with the sitting volume is important to be realized. The project revealed that an extra element (metal stem) is needed for the support of the backrest in horizontal position. Special attention should be paid to textures and colors that imitate the feathers of the peacock. The concept is to be as close as possible to the inspiration source, using biomimicry, because the genuine colors of the peacock have a specific role in nature (ensuring communication between the fellows of this species, reproduction, and, of course, a magnificent appearance of the male in the surrounding environment). A survey through the examples of armchairs inspired by the peacock tail revealed that they all have a formal inspiration in the shape of the bird, with no deep study on the opening mechanism of the tail, so the creation of such a piece of furniture would be quite an invention.

Another project for this category is the one of a lamp, entitled "Firefly lamp," possible to be used equally in domestic and public spaces (Fig. 4a, b). The model belongs to the series of studies created by the students of the Faculty of Product Design at the "Ion Mincu" University (2018–2019). In this case, the focus is on dynamics and functioning. The lamp imitates the relation with light of a firefly, who actually controls the beginning and end of the light emission (due to a chemical reaction based on the addition of oxygen in its case, property called bioluminescence). The project proposes a lamp conceived to control the quantity of light according to the specific needs of a certain space and the weather conditions outside.

Aesthetics/Proportion

Some natural organisms or details of their body have the property to be just beautiful, above all. If we can find the geometric key of beauty in proportion and the golden ratio, it is hard to define what makes a color or texture beautiful. By studying nature, one may discover the laws of harmony are quite simple and the same golden spiral generated by the Fibonacci series, the golden ratio 1.618…, as well as the 137.5° angle, appears everywhere. However, sometimes we cannot explain why things are beautiful; we just feel like it. In this situation, to imitate nature by drawing and understanding it may be the correct solution, as the great Leonardo demonstrated.

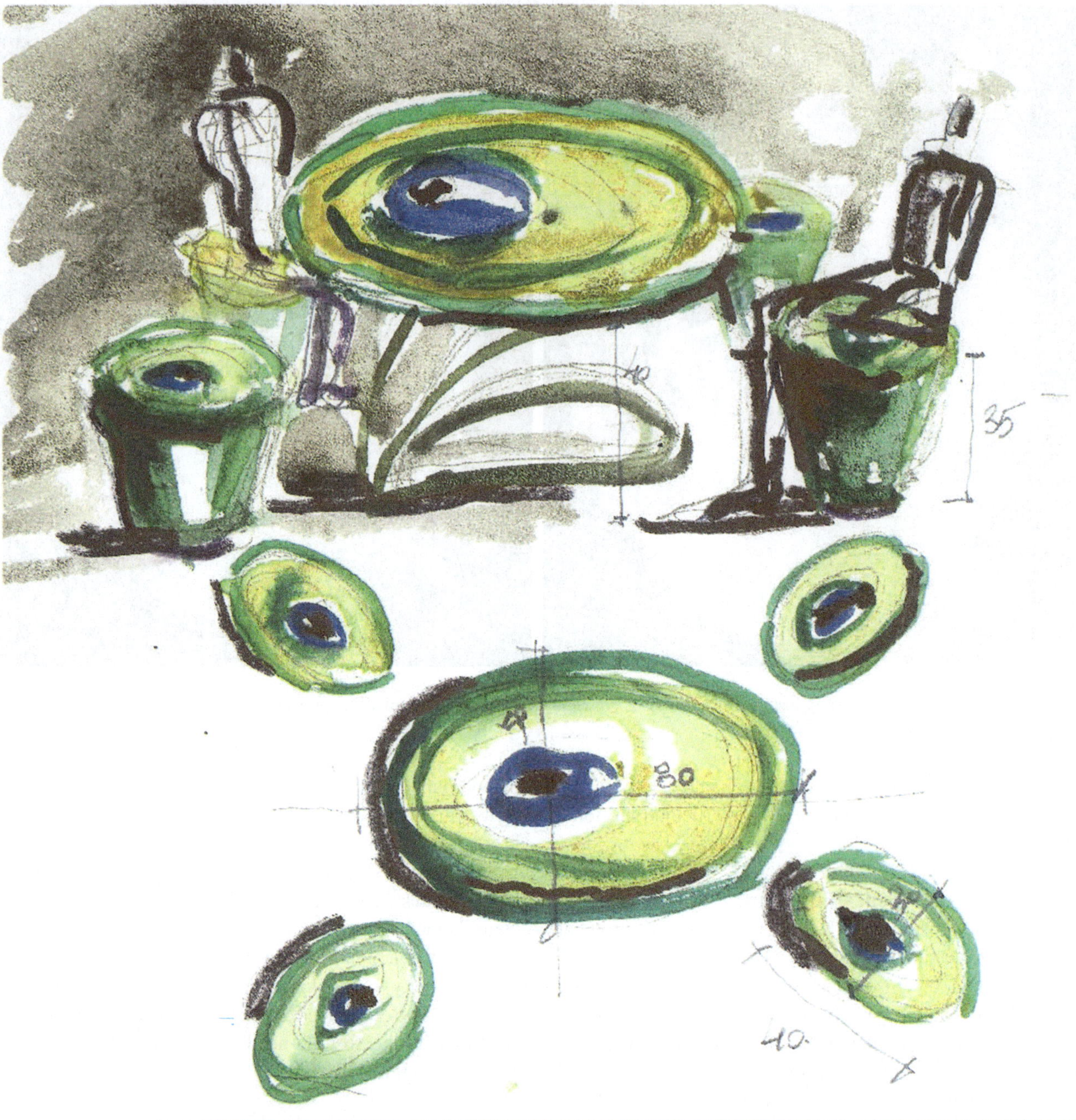

Fig. 5 The project of a coffee table inspired by the form and textures of the details of the peacock tail, by the author

Study Cases (Aesthetics/Proportion)

A proposal of inspiration source is to study "eyes" and other details of the peacock tail. The project of a series of furniture pieces focuses on material properties (translucence, color) and form, which is inspired by the one of the "eyes" that decorate the peacock tail (ellipsis). Some sketches for a decorative table conceived for the living room or for the ambience of a cafeteria are presented below (Fig. 5). The height of the table is low (maximum 55 centimeters), being proper for serving coffee and discussions between friends. A contrast should be perceived between the translucence of the table and the opaque soft chairs (velvet could be a good option for them).

The next project based on aesthetics, realized by Ilinca Jitaru, student in Product Design at the Ion Mincu University (2018–2019), is illustrated by another proposal of lamp design, based just on the wood texture of a tree trunk (acacia). In this case, the intention of the project is to put into value the aesthetics of the growing rings of the tree, which mark the age of the trunk, in a metaphoric and artistic way (Fig. 6).

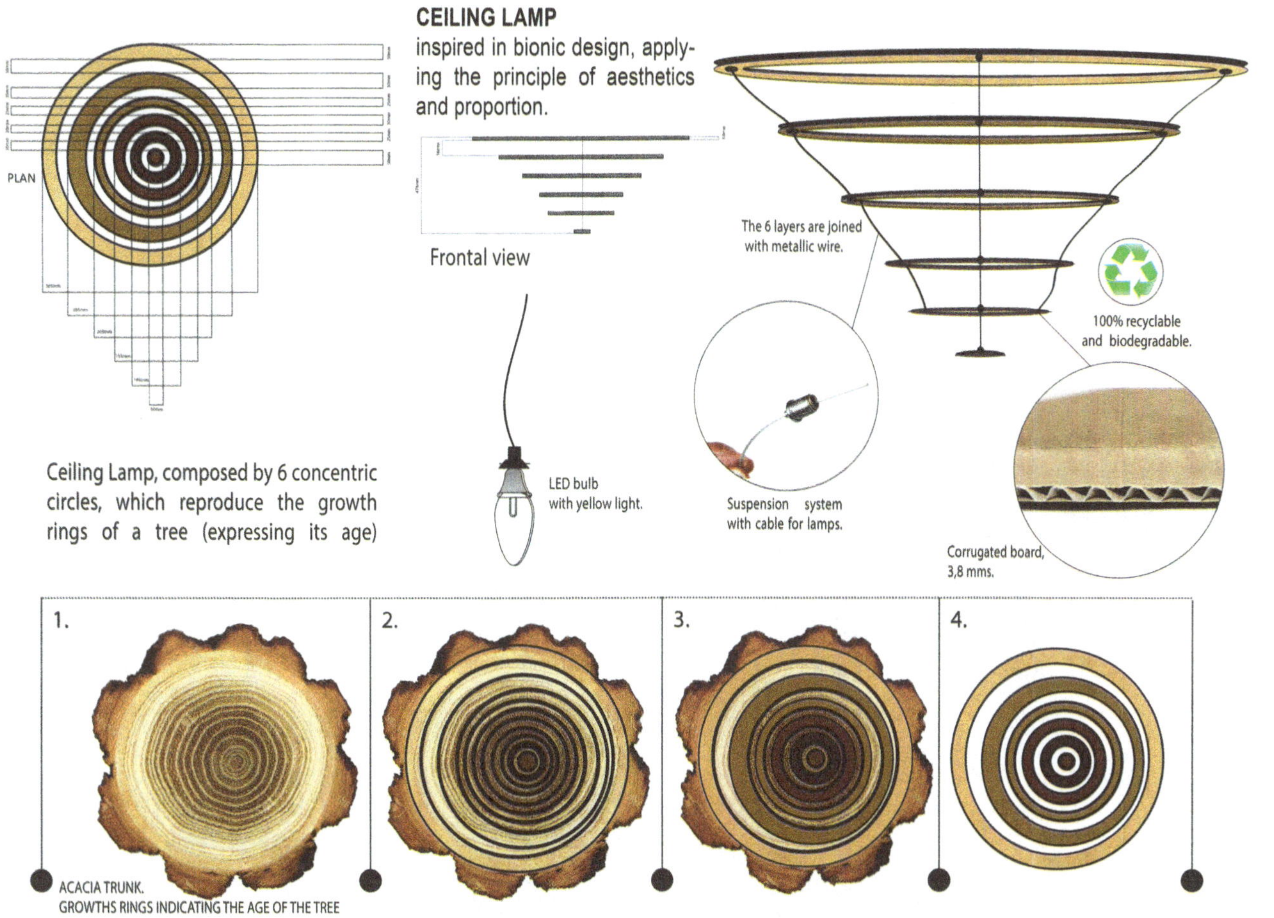

Fig. 6 The project of a lamp inspired by the growing rings of an acacia trunk, by Ilinca Jitariu, university project (Faculty of Product Design IMUAU, illustration of the category Aesthetics/Proportion)

Interrelation/Network

Some natural organisms or habitats are mainly defined by being a net, consisting in a multitude of similar elements in continuous interrelation, interdependence, and cohesion and constituting a complex system based on simple elements (modules). In such cases, the necessary approach is to understand the morphology of the basic modular element and the categories of relationships between basic elements. The scale of the study is also an approach to be retraced, considering networks of micro-elements or macro-elements, ensembled in an efficient way.

Study Cases (Interrelation/Network)

The honeycomb is one of the most common networks present in nature and largely applied in architecture and design. The first example proposes modular sitting spaces interconnected between each other, inspired by a honeycomb. The interior of the hexagons is furnished with sofas, allowing intimate discussions in small groups and ensuring soft surfaces (Fig. 7). A 3D interpretation of the concept is also possible, considering the ceiling, hanging elements, etc. Another use of similar modules could be as playground for children, since it is proved that children are fond of exploring and discovering nature. Natural warm materials are proposed (wood, soft tissues, etc.).

The second design proposal is an uncommon shelf with a free form. The modules could be light elements in some adaptive plastic material, with a round organic form, fixed on several central metallic axes with circular section (Fig. 8). Stability on the ground should be studied and solved, as well as the transformability of the ensemble. The proposed color respects the natural tone of the coral.

The next project illustrates this fourth and last category, imagining a separating panel, or mobile wall (Fig. 9). The structure of the panel is inspired by the micro-texture of a

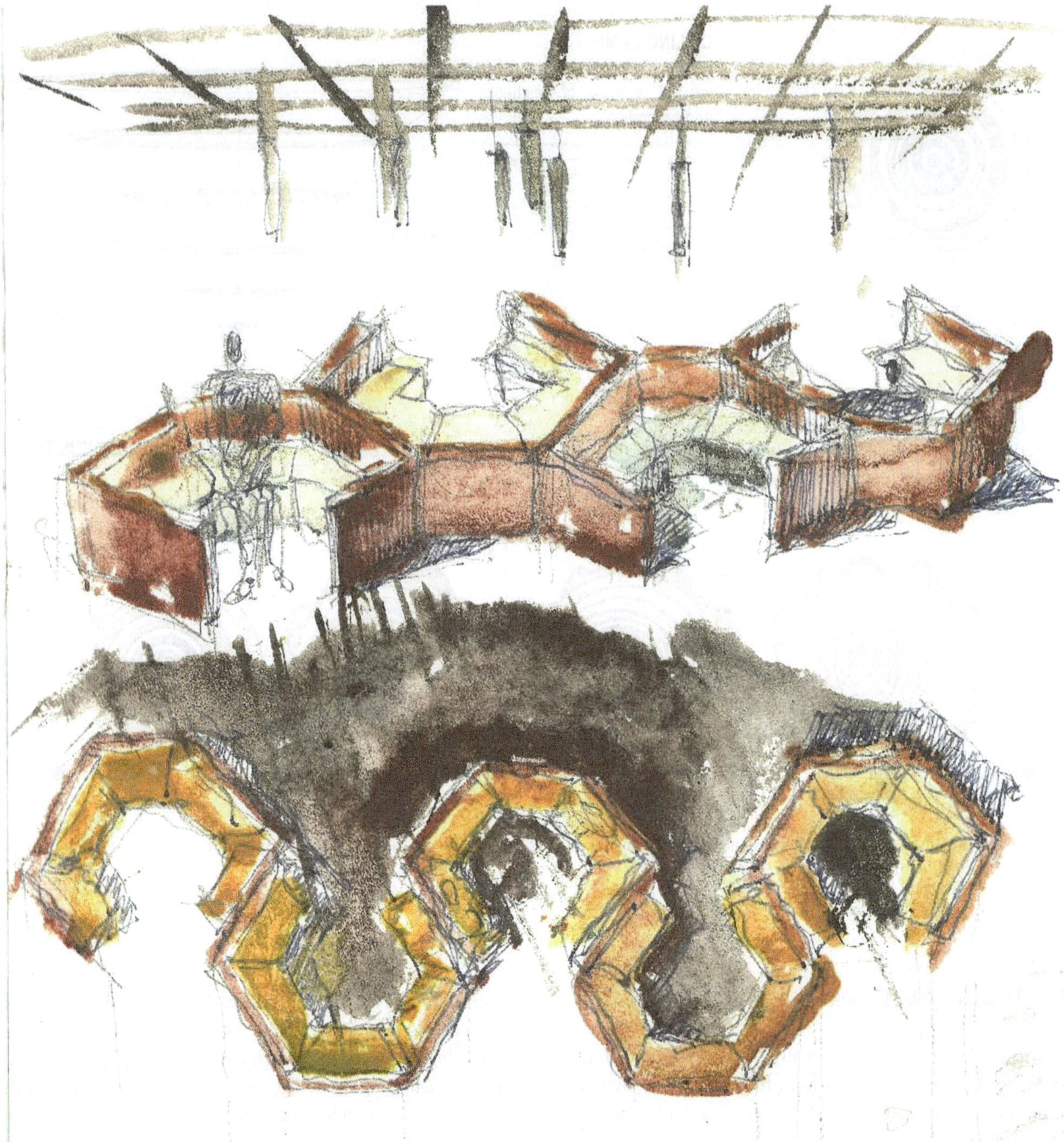

Fig. 7 The project of a flexible multifunctional furniture for a public space or home, inspired by the honeycomb, by the author

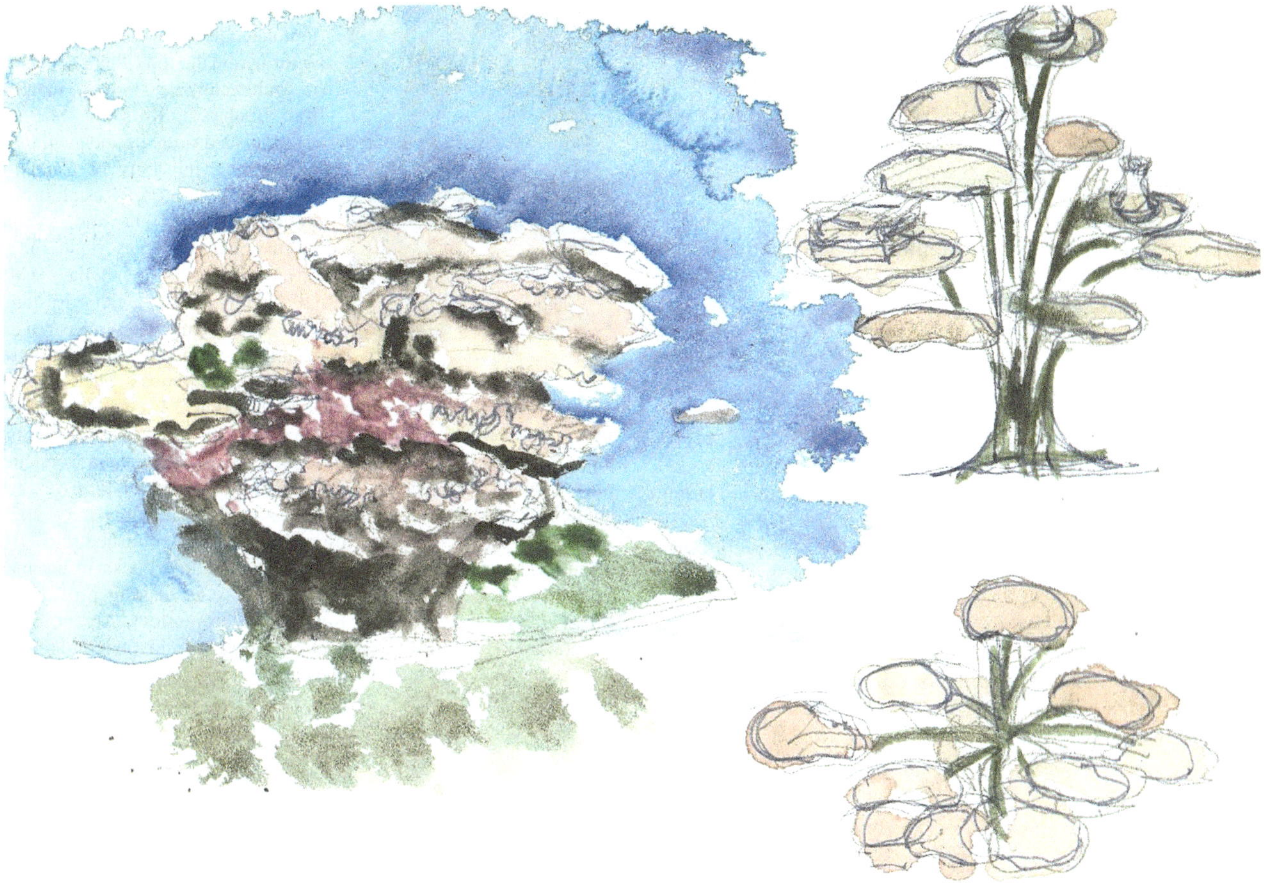

Fig. 8 The project of a flexible multifunctional structure, to be used as a shelf, inspired by the coral, by the author

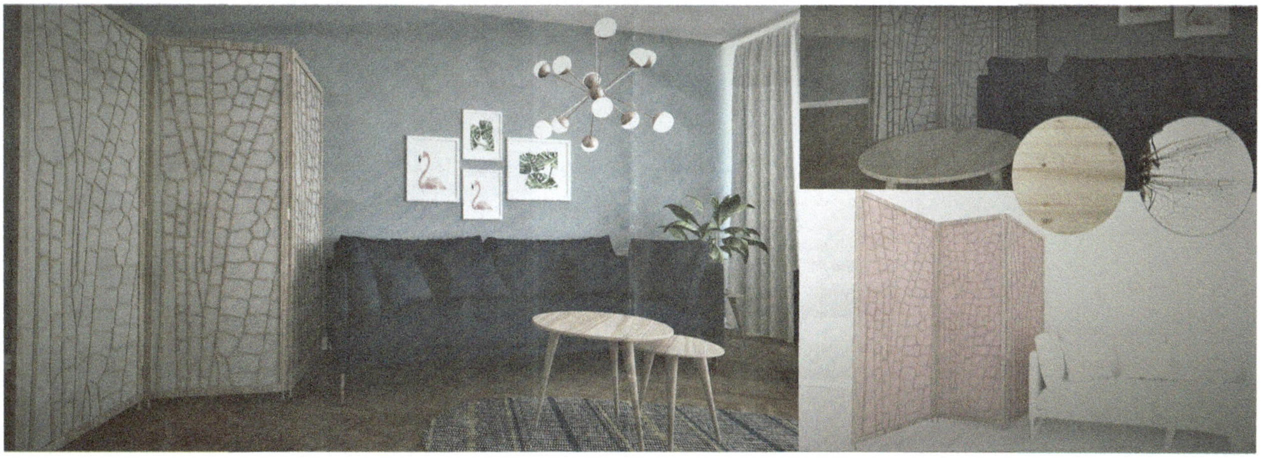

Fig. 9 The project of a wall panel inspired by the texture of a dragonfly wing, by Mădălina Ursu and Eduard Sohan, university project (Faculty of Product Design IMUAU, illustration of the category Interrelation/Network)

dragonfly wing, which generates a resistant irregular network with outstanding stability and special aesthetics. As a general conclusion of the study, the assumed option for one of the four mentioned categories, or the gesture to consciously combine several of them, may lead to a valuable method to be applied in design thinking. Though illustrated with furniture design examples, this method can be translated to several scales, from urbanism to architecture and interior design, being open for further exploration.

References

Benyus, J. (2002). *Biomimicry: Innovation inspired by nature*. William Morrow & Company.

Harding, R. (2006). Ecologically sustainable development: Origins, implementation and challenges. *Desalination, 187*(1–3), 229–239.

Keller, R. S. (2018). *Nature by design. The practice of biophilic design*. Yale University Press.

Laurenza, D., & Taddei, M. (2006). *Leonardo's machines: Da Vinci's inventions revealed*. David & Charles.

Tanov, E. (2018). *Design by nature: Creating layered, lived-in spaces inspired by the natural world*. Ten Speed Press.

Yeang, K. (2006). *Ecodesign: A manual for ecological design*. Wiley-Academy.

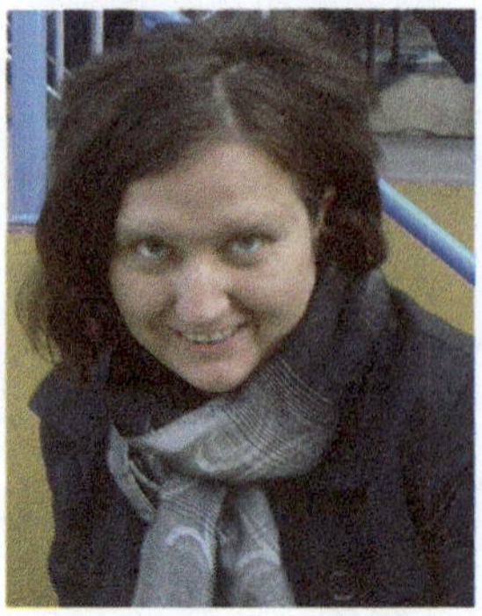

Elena-Codina Duşoiu is an architect (graduated in 2000, IMUAU) and professor within the Ion Mincu University of Architecture and Urban Planning from Bucharest, Romania. She also accomplished study periods within the Polytechnic University of Catalonia, Barcelona (Master in Restoration of Monuments, 2001) and Venice International University (Dottorato di Eccelenza. Storia della Città, dell'Architettura e del Restauro, 2004). Her main fields of research are: rehabilitation and conversion of buildings (published books: The Dynamics of the Sacred Space. The Influence of Function, 3 Breweries and Their Destiny, Ion Mincu Publishing House, 2009), vernacular architecture and ecology (organizer of 12 editions of Spanish-Romanian workshops on vernacular architecture 2006–2018), design and bionics (research studies realized within the IMUAU and the Alcalá University, 2021–2022). She received various research grants and published about 100 scientific articles and 8 authored and co-authored books in the mentioned research fields. Outstanding activity as visiting professor in Spain, Italy, Greece, Czech Republic, Liechtenstein, Argentina etc. Several national and international prizes and nominations in architectural research and architectural design, owner of a personal architecture studio since 2003.

The Bionic Paradigm of Light in Architecture and Design: From Animal Vision to Architectural Conception

Elena-Codina Duşoiu

Judgment is ... the mediate cognition of an object, hence the representation of a representation of it.
Plato

Abstract

The capacity for a multi-sensorial perception is one of the most significant tools in defining the relation of living organisms with the surrounding environment. Studying the world of senses gives constant sources of inspiration in architecture and design, defining space creation and its perception. Going deeper into the study of the mechanisms of senses, we may discover unexpected instruments for creating aesthetic, efficient and bioclimatic architecture.

The study of animal vision, the relationship of living beings, from the simplest species to the most developed ones, with light and color during their whole life, provides solid knowledge, applicable to architecture, design and construction. Taking into account that many species of animals have much more accurate vision than human beings (night vision, underwater perception, 180° or 360° vision, perception of light polarization, etc.), we can investigate how such a relation with light could be applied in the conception of architectural space. A classification of vision systems from the animal world is leading to conclusions that may transform in departure points for future architectural visions, considering light to be compulsory for a correct understanding of space. Original results appear as well while realizing the connection between vision accuracy and life duration of living beings, which seem to be proportional. In a future phase, a study on improved human perception of surrounding space may bring direct applicability into architectural thinking.

This chapter presents as well three study cases of well-known buildings inspired by the anatomy and physiology of the eye, leading to a comparative multi-criterial analysis. The last section of this chapter presents some experimental results of architectural projects, searching an improved and more complex perception of light. The presented projects are selected from the author's academic activity developed with the students of the "Ion Mincu" University of Architecture and Urban Planning from Bucharest and the University from Alcalá de Henares, Spain, in the last years (2021, 2022).

Keywords

Vision system · Light perception · Design with light · Architectural space · Architectural conception · Environment perception

Introduction. Perceiving Space – The World of Senses

This chapter was born out of some simple questions: Which is the relationship between animals and light? Does this relation influence their quality and duration of life? Can the various possibilities of light perception inspire new technological details for our buildings or influence their architectural volume? A parallel can be drawn between the famous definition given to architecture by Le Corbusier: "Architecture is the learned game, correct and magnificent, of forms assembled in the light" (Le Corbusier, 1925, p. 16) and the sense of vision of the living creatures, which for the humans is the

E.-C. Duşoiu (✉)
"Ion Mincu" University of Architecture and Urban Planning , Bucharest, RO, Romania

M. R. Cervera Sarda et al. (eds.), *Architecture Inspired by Nature*, https://doi.org/10.1007/978-3-031-33144-2_5

sense that brings the majority of information and knowledge. However, perception of architecture is not only about seeing. As Peter Zumthor observed: "Architecture needs to be executed. Then the body can come into being. Moreover, the body is always sensuous. To experience architecture in a concrete way means to touch, see, hear and smell it" (Zumthor, 1999, p. 58). It is the same concrete experience that animals have in their complex relationship with the surrounding world. Some have outstanding capacities in dealing with light perception and not only through their eyes. Can these abilities inspire us in the conception on our artificial environment? Because, unfortunately, us, the humans can only imagine "the body as a refuge in a world which would appear to be flooded by artificial signs of life and in which philosophers ponder on virtual reality" (Zumthor, 1999, p. 52).

The present study is bringing together concepts from biology, physics, and architecture, with the purpose of helping a deeper understanding of the surrounding world and translating the knowledge to the construction of our future surrounding environment. The basic bibliography of the study is composed of recent studies referring to the sight mechanisms of various species of animals and their classification. However, architectural interpretations are present as well. A comparative analysis of the species with most longevity from the terrestrial, aquatic, and aerial environment is included as a valuable tool in discovering the most relevant principles of light detection, possible to be translated to the constructed environment. In the end, we present study cases referring to an application of animal vision (horse) to architectural space and to an experiment of "building with light," providing sources of inspiration for future design studies.

Perceiving Through and Without the Eye. Particularities of the Animal Eye

The sight in the animal world covers an enormous diversity of patterns. The most common one is the classical camera-type eye common to humans and other vertebrates. We can still discover the composed eyes of insects, made of a great quantity of tiny lenses, projecting small parts of images and recreating some blurry effects. That is nothing more amazing than the thousands of ocelli spread over the shell of the molluscs (e.g., the clam or the jellyfish), whose way of functioning is rather difficult for our understanding. Some insects have as well their eyes split in various ocelli that help responding to light changes (the tree weta, a sort of spider belonging to the family of *Anostostomatidae*). Some reptiles are provided with a pineal gland with thermoregulation function called "the third eye" (as is the case of the tuatara, a reptile from New Zealand). Other are famous for their outstanding vision at night (salamander). There are even animals that are able to perceive the presence of light with no eye structure at all, due to the visual cells spread all over their body. A unique example is the lobster, provided with so-called "X-ray vision." This type of vision is due to square tubes – natural mirrors that reflect the incoming light, composed of entirely straight walls and right angles, as opposed to the human eye's curved rods and cones (each eye, set on a movable stalk, has up to 10,000 facets that operate like many tiny eyes). This provides the lobster with a 180° field of view. All these examples prove how complex this sense that we generally call "sight" is in fact, even if many times there is a slight difference between it and other senses, as for example the sense of "touch" (Higgins, 2021).

This morphological diversity produces a huge variety in perceiving the world. It is amazing to think how reality can change according to the specificities of visual perception of the various species. The study of the functioning of these various types of systems may lead to the understanding of complex technical principles, possible to translate to architecture and design thinking.

Perception of Color

The most evident presence of light in the animal world is color, an element that is crucial for life: in recognizing their own species, in communicating and being in relation with the environment, in reproducing, in fight for domination between males, or in thermal interaction with the environment. The sight and perception of a variety of colors define, actually, the diversity of the world and life, consisting in a matter of survival. "Groups of animals, for example, can communicate with one another using colours or patterns of light that are invisible to rivals or dangerous predators" (Stevens, 2021a, b, p. 12). Therefore, the perception of color varies a lot, from a very sensitive and deep capacity of defining it (parrots, chameleons, certain types of fishes, etc.) to an infrared vision (snakes), ultraviolet perception (bees), extraordinary sharp night vision (night gecko, cat) and to a rather uncolored (horse, butterfly) or even black-and-white or blue-green vision (sharks, mice). It is also interesting to compare the skin patterns of the animals with their life environment and the accuracy of their sight. Actually, the property of color in the animal body is given by several categories of natural pigments (melanin for dark colors; carotenoids for vivid yellow, orange, and red; psittacofulvin for vivid colors in parrots) and by specific types of structures that allow only short-length light to be seen (green, blue, etc.). Some animals even "borrow" colors from the surrounding environment (the most spectacular case is the one of chameleons, who can change within seconds from a bright green or blue to dull brown, according to their habitat and their dominance). Other species vary their color more slowly from one season to other (arctic lemming, fox, squirrel, etc.)

It may not be hard to understand that the most colored animals have the most developed sense of perceiving color. This fact generates a strong relationship between their physiology, communication with the environment, and their ability to perceive the surrounding reality. For example, the mandrill (an old-world primate), the most colored mammal in the world, has four types of light-sensitive cells in the eye, being able to detect the whole color spectrum, from ultraviolet to red. It seems that the oldest (prehistoric) species developed the most accurate vision of color, much better than the one of humans. However, an outstanding vision of color, due to a great quantity of cone cells, reduces the number of rod cells present in the eye and therefore provokes less night vision, underwater vision, etc. (Stevens, 2021a, b). That explains the fact that contemporary mammals, whose family developed and diversified after the evolution of dinosaurs, usually have less color vision (just two types of cone cells) but better night or gloom vision (this is the case of the horse, which will be further studied in detail) (Kelber et al., 2003).

The quality of eyesight is in a constant relationship with the appearance of the animal. Parrots are enabled with an outstanding capacity of perceiving colors and are extremely colored themselves (e.g., the cockatoo, who is as well provided with the capacity of seeing much more images per second than the humans). The eye of the peacock contains four types of cone cells, allowing it to perceive the entire spectrum from ultraviolet to red. The aesthetic dimension of color (such in the case of the peacock) seems to be useless. However, in the end, it proves to play an important role in reproduction and sexual selection, in the continuity of the species, even if sometimes it seems to prove uncomfortable, disturbing camouflage and providing extra-extensions of the body, as is the case of the peacock tail (Figs. 1 and 2). "The true splendour of his plumage is only revealed in sunlight, and the luckiest (or perhaps more vigorous) male will hold court in the brightest morning sunshine, leaving many of the other males, quite literally, in the shade" (Stevens, 2021a, b, p. 65).

The role of color in recognizing members of the same species and their reproduction is widely spread in the world of: insects (e.g., peacock spider), birds (flamingo, whose color is influenced by their diet), fish (clown fish), reptiles, etc. Many males, especially birds, attract their mates by conscientiously using the effects of light irisation on their plumage, even getting to hypnotic effect (the case of the bird-of-paradise). Some butterfly females are able to detect males belonging to the same species due to ultraviolet perception of the colors and patterns of their wings.

Underwater Vision. Perception of Polarized Light

Water organisms have a different relationship with light and color. The cornea of a whale is less curved than in land mammals, which has the effect of allowing in more light without causing as much refraction. There is a reflective layer within the eye, which redirects incoming light back through the retina, creating double reflection. The obtained result is a clearer image. While coming to surface, above the water, the whale's pupil naturally shrinks, allowing just a small proportion of the light inside. Whales have far more rod cells than cone cells, helping them just to see a bigger quantity of light.

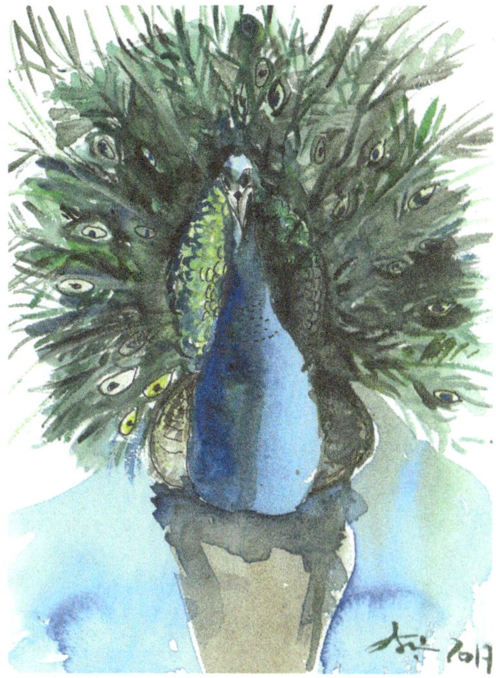

Fig. 1 The opened peacock tail. (Watercolor by the author)

A very special example of animal living in water is the mantis shrimps. Species living in deeper water tend to be bluer and have a vision that developed better to perceive the blue color; the ones living closer to surface see a wider range of colors. An interesting property of their vision is to perceive circular polarized light. The oscillation of the wave of light creates a certain angle of polarization, possible to be distinguished by certain animals (from dragonfly to fish) (Stevens, 2021a, b). Mantis shrimps can detect even more complex direction of polarization (circular polarized light patterns), due to the extreme flexibility in moving their eyes. This fact increases their capacity to perceive color underwater.

Other species have a specific property in relation to the light – their bodies' fluorescence (many of them species of fish). This property functions in darker environments, rich in ultraviolet and blue light, in which fluorescence provides extra signals of the living world.

Some animals benefit of both underwater and sharp above-water vision, due to their vertical pupils, as well as night vision (alligators), and others are able to adapt their sharp vision from water to land as well, being rather terrestrial animals (the turtle) (Fig. 3).

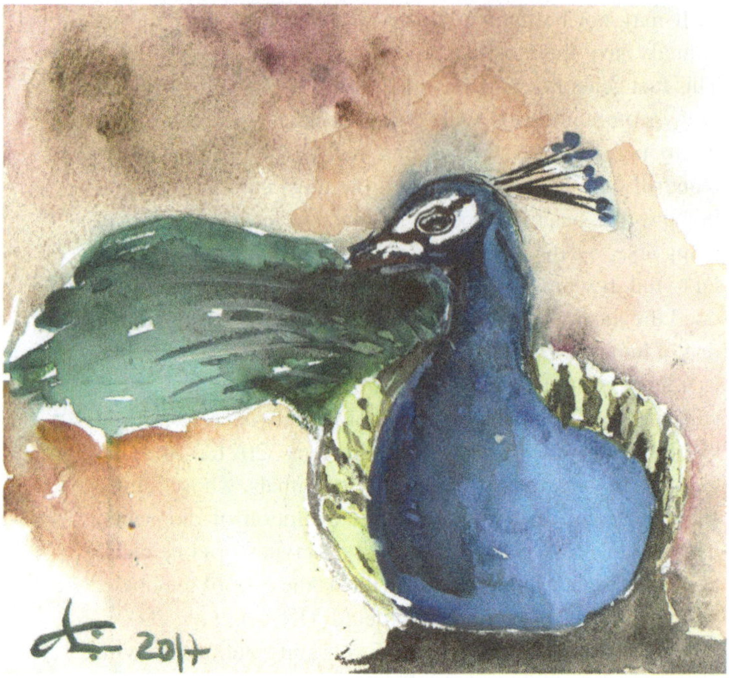

Fig. 2 The amassed peacock tail. (Watercolor by the author)

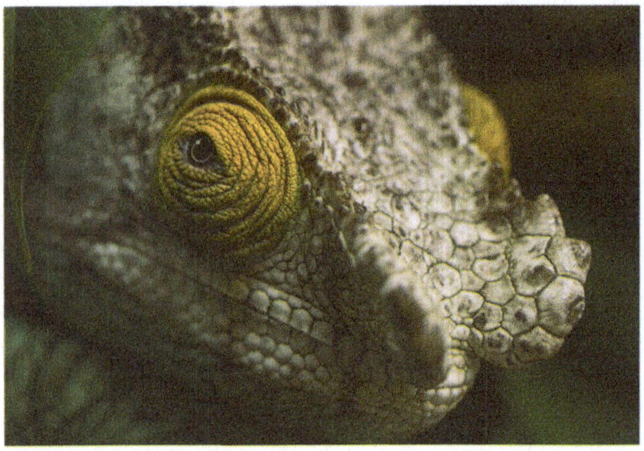

Fig. 3 Chameleon: eye with 360° view. (Source: https://unsplash.com (Photo by S.N. Pattenden on Unsplash))

Angle of Perception. Optics and Geometry

Another aspect of animal vision is the visibility angle and the sharpness of the vision. There are species who see up to 360 degrees (the macaw, who has as well the capability to move the eye bulb inside its socket (Fig. 4), and there are species with an outstanding 3D vision and depth perception (the chameleon, who uses this type of view for recognizing and catching its pride (Fig. 3) (Stevens, 2021a, b). Besides, the majority of species keep their eyes in constant movement for keeping the accuracy of their sight (Land, 1999). One of the most interesting cases is the lobster, provided with so-called "X-ray vision." The eye of the lobster is a multitude of square tubes, natural mirrors that reflect the incoming light, composed of entirely straight walls and right angles. Up to 10,000 facets (having the role of many tiny eyes) are found on movable stalks sustaing each eye. This could inspire some amazing architecture installation able to increase the quantity of natural light penetrating interior space.

Human Sight Versus Animal Sight. Does a Better View Provide a Longer Life?

The basis of human sight is the detection of different wavelengths of light using three receptors from our eyes (cone cells), shortwave "blue" light, mediumwave "green" light, and longwave "red" light. The specific stimulation of these types of receptors provokes the colored ambience we perceive. This has been the basis of the digital, computer-controlled image (RGB spectrum). That makes a diverse perception of the majority of colors, but human sight is not able to perceive very short wavelengths provoked by ultraviolet light, that many species of animals see.

Another ability of seeing that humans lack completely (which is rather common to animals) is the perception of polarized light. This means the way a wave of light travels and oscillates in a certain direction, with its specific angle. Ants, dragonflies, cuttlefish, or mantis shrimp are able to perceive polarization angles (Stevens, 2021a, b). Some animals create patterns of polarization with their own body surfaces (e.g., butterflies living in deep forest areas). The properties of light polarization, referring to angle, intensity, or proportion of light, offer an extended spectrum of vision that is at least as rich as the color spectrum.

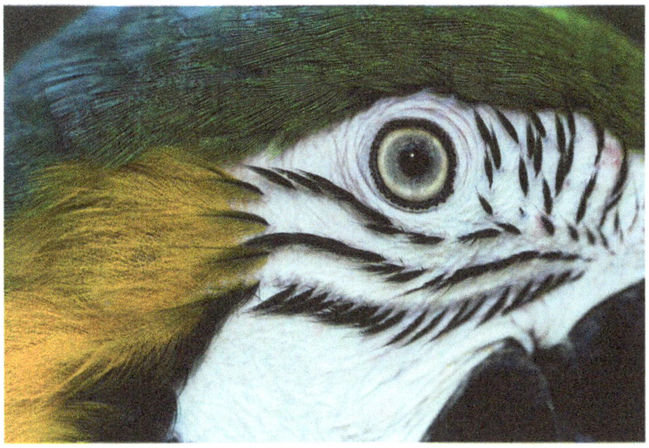

Fig. 4 Macaw: eye with 360° view. (Source: https://unsplash.com (Photo by David Clode on Unsplash))

However, human sight is consistently clearer than the one of the majority of mammals. The human eye can focus on a detail due to the muscles that hold the flexible lens of the eye. This allows focusing on objects of near-so-called visual accommodation, a sense that many mammals lack.

One may study the average period of life of species according to their sight performances, considering view and relation with light in general as the main sense of the living environment. It is amazing to discover a proportional relationship between their life span and the accurate capacities of their eyesight. We can conclude without being wrong that the species endowed with the most efficient view in relation with their living environment are the ones with greatest longevity. The majority live much longer than humankind. Table 1 is a synthetic analysis leading to this conclusion.

Architecture Inspired by the Eye – Some Existing Examples

The parallel between the mechanism of vision and architectural space is an inborn approach, light being a *sine qua non* condition for the existence of architecture. The most common approach observed was the inspiration in the form and mechanism of the human eye. Three examples of famous buildings generated by this concept are presented below.

Kiasma Museum of Contemporary Art, Helsinki

In 1993, the architect Steven Holl was winning the international competition for the construction of the new museum of contemporary art in Helsinki. The design of the building was strongly related with biology from the beginning (Maro Kiris, 2018). Even the name of the proposal comes from the Greek letter "chi" (X) meaning "cross," a concept used as well in anatomy for denominating the crossing of the optic nerves at the basis of the brain (Fig. 5). The form of the building reinterpreted this concept in a large scale, proposing an intersection of two linear volumes. Adaptation to the context and volumetric expression, in addition with light effects, are common features for the creation of Steven Holl, who had designed several museums before and considers abstract interpretation to be "a critical architectural idea" (Carso et al., 2018).

This time, the architectural vision came out through deep scientific research, transferring to the building the property of the anatomic process of sight: the realization of the connection between the stimulation and the knowledge produced in the brain. The input transmitted through the optic nerves is a direct parallel with the dynamic flow generated through the building (Maro Kiris, 2018). The main access is on the southern side, just between the two crossed volumes, beginning an endless virtual axis that connects the building with other cultural poles of the city (Finlandia Hall, by Alvaar Aalto and Töölö Bay). The building is a sensitive insertion in the urban tissue of Helsinki, based on the idea of a spatial and cultural connection, at the same time.

Eye Film Museum in Amsterdam

The new building of the Eye Museum in Amsterdam has been opened about 10 years ago, in 2012, and consists of an extension of the film museum of the city, inaugurated in 1946 (Maro Kiris, 2018). Even if the Viennese architects (Delugan Meissl Associated Architects – DMAA) had the same idea to inspire the building in an anatomic concept related to human vision, the approach is different this time. The architects realized a volumetric essentialization of the form of an eye surrounded by its upper and lower lids (Fig. 6). The resulted space is strongly redirected through the river, building a strong relationship with the urban and natural context. The originality of the approach consists of stylizing in strong snapped lines the form of the eye, giving a geometric, abstract dimension to the volume.

L'Hemisferic from Valencia

Whoever visits today the Spanish city of Valencia is guided to the huge City of Arts and Sciences, a recovered area in the peripheric southeastern part of the city, resulted after the deviation of the river Turia, as an effect of the catastrophe generated by its flood in 1957. The scale of the intervention is impressive, the ensemble consisting in several buildings with a cultural function: Palace of Arts *Reina Sofia*, *L'umbracle* (open-air structure covering a multifunctional space, *L'Hemisferic* (sheltering an IMAX, a planetarium,

Table 1 Synthetic comparison between the vision systems of the most longevity species

No	Animal	Image	Habitat	Expected life duration	Specific property of view
1.	Mantis shrimps	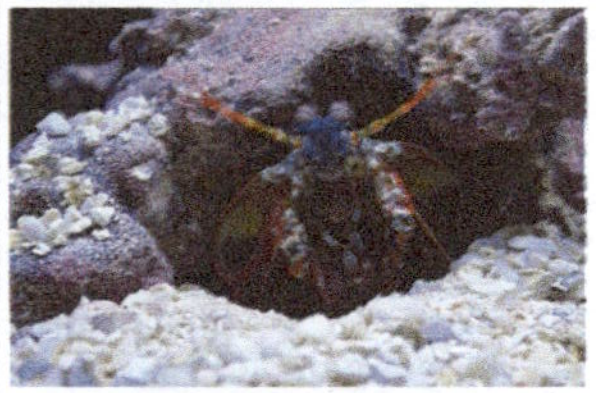	Aquatic environment	20 years (long for a crustacean)	Perception of circular polarized light, increased capacity to perceive color under the water Image source: https://unsplash.com (photo by Amber Wolfe on Unsplash)
2.	Alligator	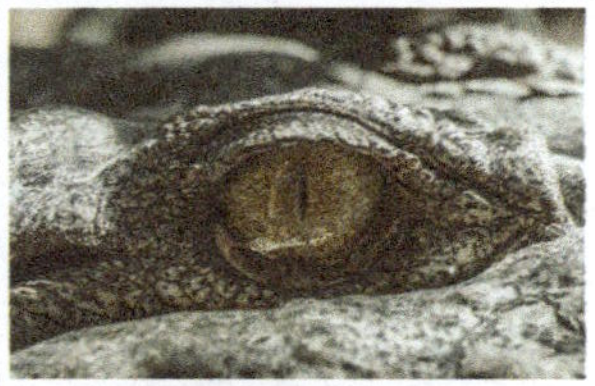	Aquatic/terrestrial environment	>80 years	Vertical pupils, sharp vision above water Image source: https://unsplash.com (photo by Samuel scrimshaw on Unsplash)
3.	Elephant	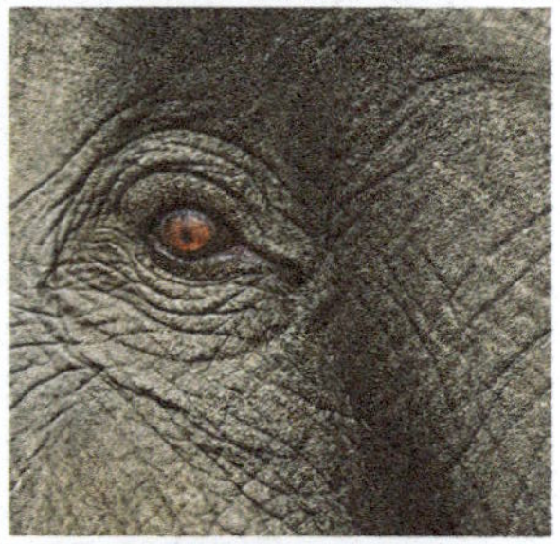	Terrestrial environment	>80 years	Third eyelid, peripheral vision, vision similar to humans but with less color perception Image source: https://unsplash.com (photo by *Inbetween Architects Jerome Charignon* on Unsplash)
4.	Condor	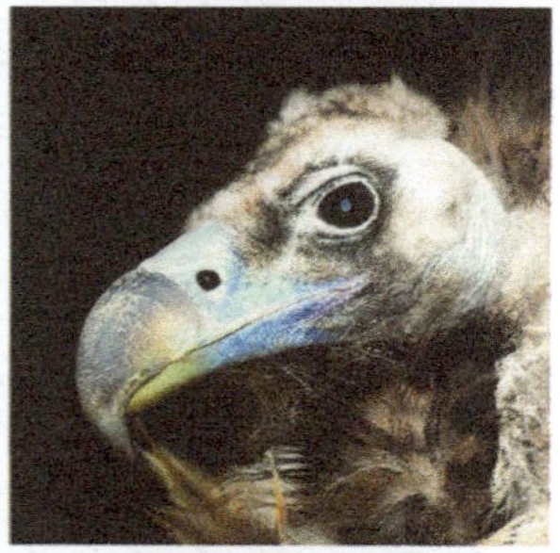	Aerial environment	>80 years	Precision of sight from distance Image source: https://unsplash.com (photo by Marc Grove on Unsplash)
5.	Flamingo	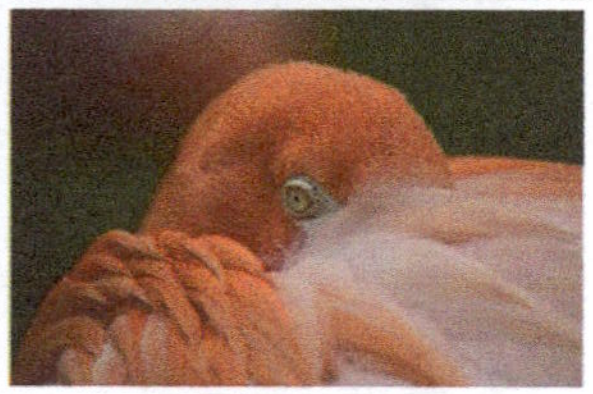	Aerial/aquatic environment	>80 years	Well-developed color perception, in relation with their plumage Image source: https://unsplash.com (photo by Bruno Miranda on Unsplash)

6.	Cockatoo	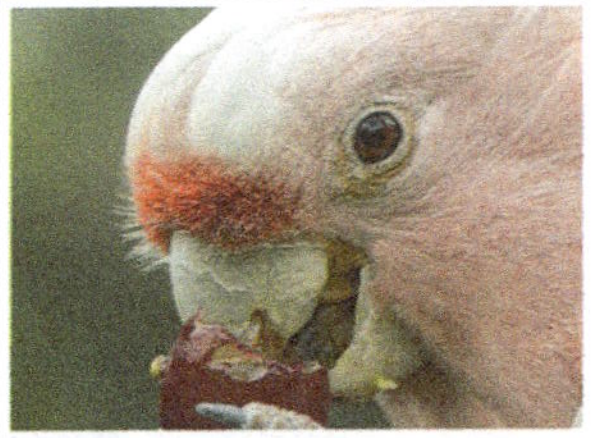	Aerial environment	>80 years	Four light-sensitive pigments (types of cone cells) Image source: https://unsplash.com (photo by David Clode on Unsplash)
7.	Salamander	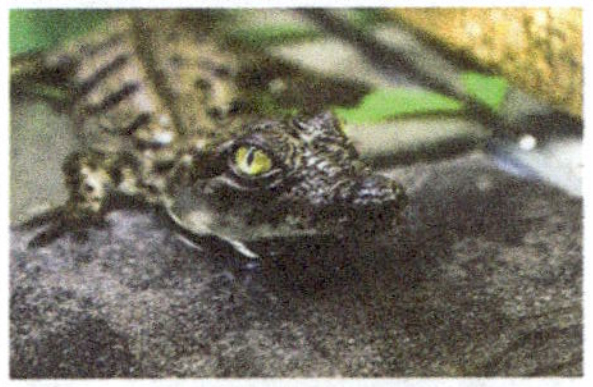	Aquatic/terrestrial environment	>100 years	Adaptation to water view (farsighted) and aerial view (nearsighted), sharp vision at night Image source: https://unsplash.com (photo by Joshua J. Cotten on Unsplash)
8.	Macaw	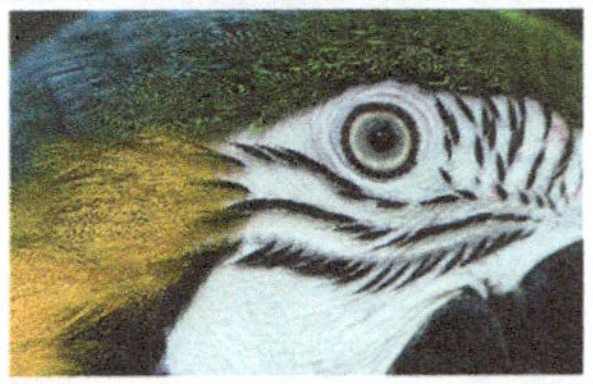	Aerial environment	>115 years	360° view, great movement of the eye bulb Image source: https://unsplash.com (photo by David Clode on Unsplash)
9.	Tuatara	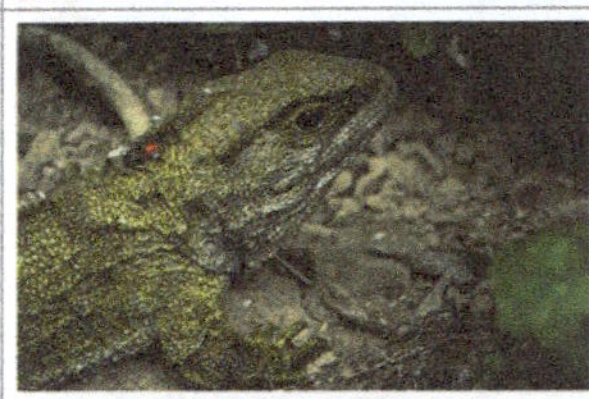	Terrestrial environment	>120 years	Sensitivity for infrared radiation, pineal gland with role of thermoregulation (third eye) Image source: https://unsplash.com (photo by Luca Calderone on Unsplash)
10.	Lobster	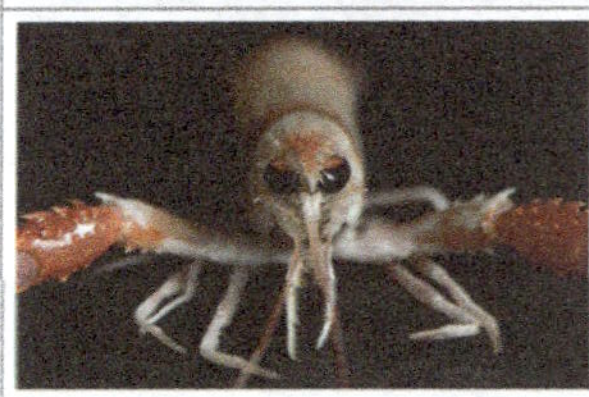	Aquatic/terrestrial environment	>140 years	Straight walls and right angles structure of the eye (reflection), “X-ray” vision Image source: https://unsplash.com (photo by blackieshoot on Unsplash)

(continued)

Table 1 (continued)

No	Animal	Image	Habitat	Expected life duration	Specific property of view
11.	Whale	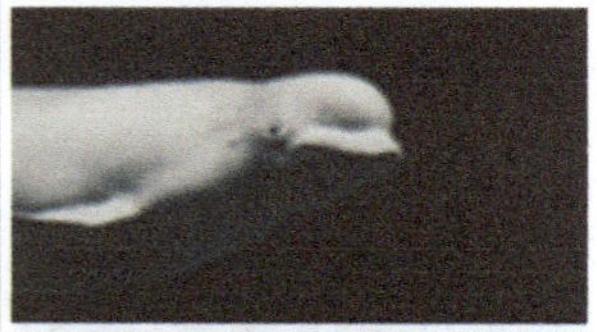	Aquatic environment	>200 years	Rod cells more developed than cone cells, enlarged pupil and reflecting layer allowing light to be reflected twice, reduction of light quantity in above-water view Image source: https://unsplash.com (photo by Mendar Bouchali on Unsplash)
12.	Turtle	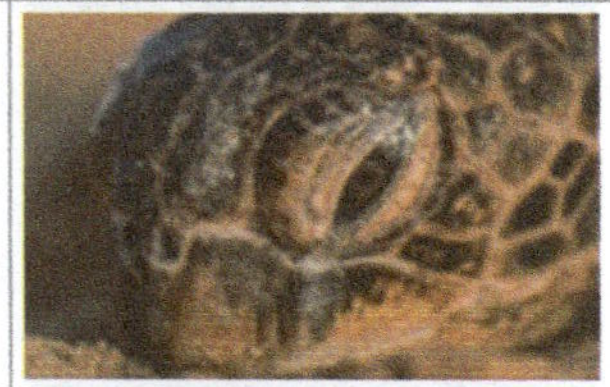	Aquatic/terrestrial environment	>255 years	Adaptation of view from water to land, ability to distinguish patterns, shapes, and colors Image source: https://unsplash.com (photo by Jakob Owens on Unsplash)
13.	Clam (bivalve molluscs)	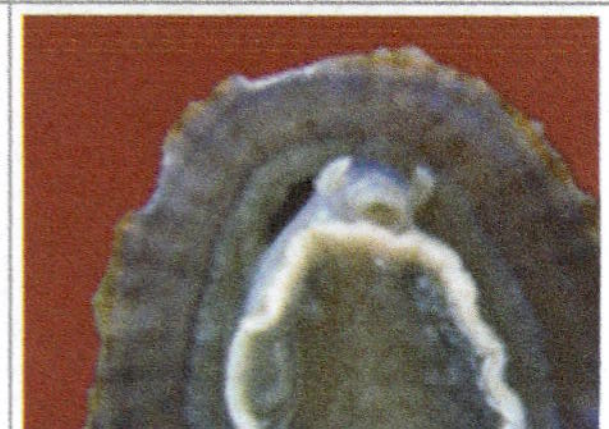	Aquatic environment	>500 years	Hundred small pinhole eyes, detecting changes in light levels Image source: https://unsplash.com/photos/
14.	The tree weta	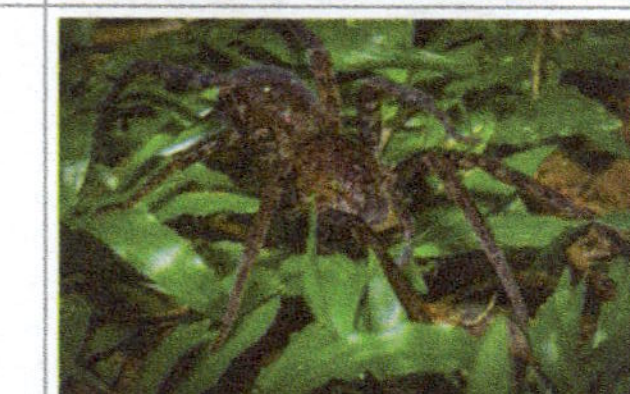	Aerial/terrestrial Environment	Freezes several winters	The two eyes located in front are completed with several *ocelli* Image source: https://depositphotos.com/352834558/stock-photo-tree-weta-endemic-cricket-new.html
15.	Tardigrades	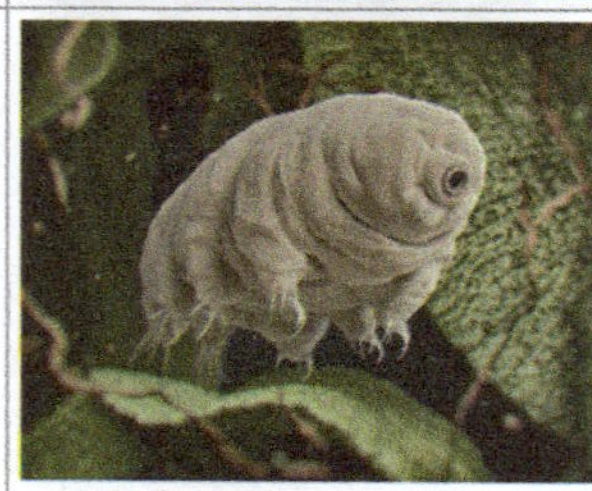	Aquatic environment	Practically immortal	Inverse pigment-cup ocelli, which are located in the outer lobe of the brain, a single pigment cup cell, one or two microvillous (rhabdomeric) sensory cells and ciliary sensory cell(s) Image source: https://depositphotos.com/132045218/stock-photo-tardigrade-water-bear.html

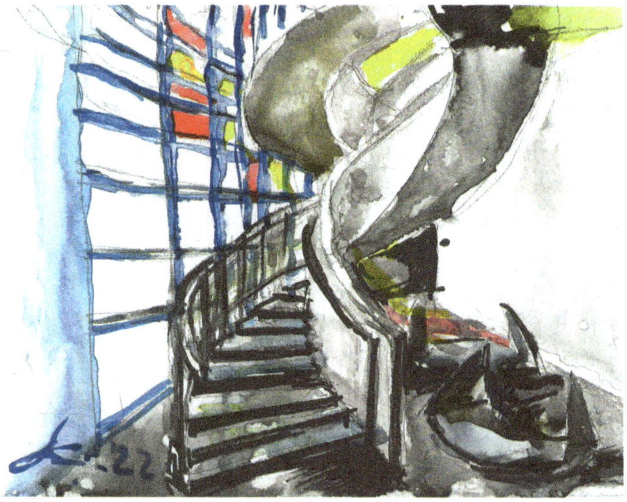

Fig. 5 Kiasma Museum, Helsinki: interior view. (Watercolor by the author)

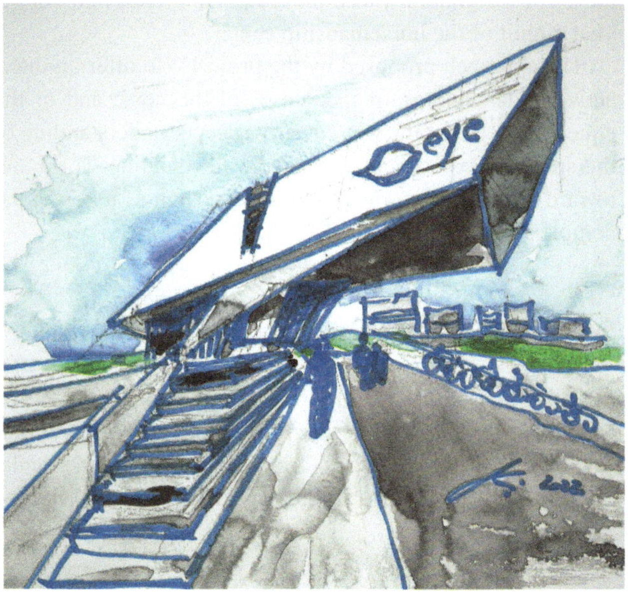

Fig. 6 Eye Film Museum, Amsterdam. (Watercolor by the author)

and several cultural and educational spaces), Museum of Sciences *Principe Felipe*, *L'Oceanográfic* (the greatest museum in Europe dedicated to species living in the oceans) (Luther, 2011). This architectural ensemble is conceived by two architects, well known for their work based on the study of natural structures: Santiago Calatrava and Felix Candela. *L'Hemisferic* (Fig. 7), probably the most iconic building within the City of Arts and Sciences, clearly imitates the form of a human eye and even the process of opening and closing of the lids (the envelope is conceived in a kinetic facade with moving elements that open and close following the necessities). In the center of the building, the volume of the IMAX forms a huge eyeball connected as well with the

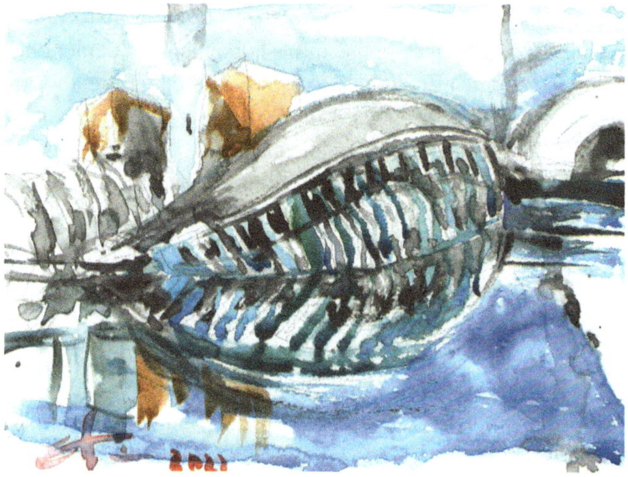

Fig. 7 *L'Hemisferic*, Valencia. (Watercolor by the author)

inside (the media performance) and the outside (an artificial lake consisting in a vast surface of water reflects the whole building).

The comparison between the architectural concepts of the three analyzed buildings brings a survey of multiple possibilities of investigation of the sense of vision, of the morphology of the eye and the visual system in their relation with architecture. It is not casual that the three buildings have a cultural function, being conceived to educate and to inspire. It is hard to define the buildings as "bionic," from a scientific perspective, but all are definitely original and iconic. The first example (Kiasma Museum) is a subtle interpretation of an anatomic mechanism (the transition of the sense of view from the optic nerves to the brain) translated to a building, conceived as a dynamic system orientated through its interior and toward the city. The second one (Eye Film Museum) is a courageous abstractized interpretation of the form of the human eye, with a clear orientation toward the natural landscape. The third building (*L'Hemisferic*) is the result of the adaptation of the morphology of the eye to an architectural volume, from a kinetic perspective. These three examples may constitute the basis for further architectural approach, the challenge being the transition from the formal study to the understanding of view as a complex mechanism, directly applicable to the art of conceiving containers for life.

Characteristics of the Animal Vision – Translations to Architectural Thinking. Some Architectural Experiments

First Study Case. The Horse's Vision

Guiding a complex interior architecture diploma project (created by student Georgiana Voinescu) at the beginning of 2022 lead us toward a detailed study of the horse view. This was because the project was referring to the transformation of a historical manor house into a horsemanship center and a club for people affectionate to horses. Little by little, the concept of the project evolved into exploring the senses of the horse and of the couple horse-equestrian.

Humans and other similar primates have forward-facing eyes and thus have a relatively narrow field of view (about 120° binocular horizontal view and 200° in total). This is completely different for a horse, who has the eyes located on both sides of its forehead. The area seen by both eyes simultaneously, named the binocular field of vision, is quite narrow in comparison with the one seen by each eye separately, the monocular field, and covering in total almost 360°. Accommodation is a specific property, similar to visual acuity, meaning the eye's ability to change its focal length so that objects at varying distances may come into focus. This happens by changing the shape of the lens. Thus, as a rule, larger eyes have a greater range of accommodation than smaller ones, which is the case of the horse. "Large eyes provide long focal lengths and space for a larger lens which in combination gives an ability for higher resolution over a greater range of distances" (Smith, 2020, p. 70). However, this fact is not clearly present in the horse vision. Horses have trouble focusing on objects, especially those that are near to them. In order to improve this blurred vision, the horse completes its information from its other very well developed senses.

As the majority of the mammals, the horse had to adapt to the living conditions in the open, exposing too numerous risks. For these reasons, the horse is in a constant state of wakefulness, carefully monitoring the surroundings and relying on the panoramic view for its safety. The field of view of the horses is almost complete, with a 360° panoramic view in the horizontal plane. The position of the eyes on both sides of the head determines the visual field and the depth perception and gives the capacity to identify even the smallest movements from a very long distance, a property that human eye misses.

The horse's sight is dichromatic (based on blue and green) compared to trichromatic vision (the primary colors of the human color spectrum) (Osorio & Vorobyev, 2008). This fact is due to having only two types of cone cells, in comparison with the human eye that has three.

The interior architecture diploma project I tutored proposed experimental spaces, offering the experience of a horse's vision and of the relationship between a horse and its rider. The project proposed an installation consisting of several cylinders, conceived to reproduce the vision of the horse and the sensations of the rider. The cylinder is a type of surface that allows to experiment the 360° vision of the horse (Fig. 8). Actually, a circular movement brings serenity to the horse and helps him take control of the whole surrounding area. The installation was included as an interest point in the visit circuit of the horsemanship center.

The approach proposed by the project – to offer an interactive understanding of the senses of the horse and of the equestrian – opens the way toward a deeper understanding of the space as experimentation of the surrounding world and toward rediscovering the potential of design as a key in the complex process of human knowledge.

Second Study Case. Building with Light

In November 2021, I proposed to the students of the MUPAAC (*Master Universitario en Arquitectura Avanzada del Proyecto y de la Ciudad*) from the University of Alcalá, Spain, a study of interior space, considering a more accurate

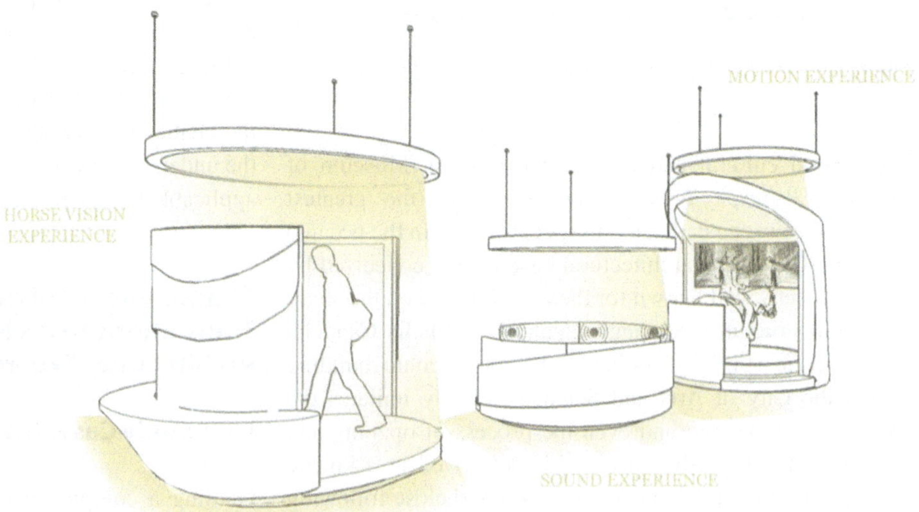

Fig. 8 Experimental installation for experiencing horse vision, by Georgiana Voinescu (diploma project IMUAU, 2022)

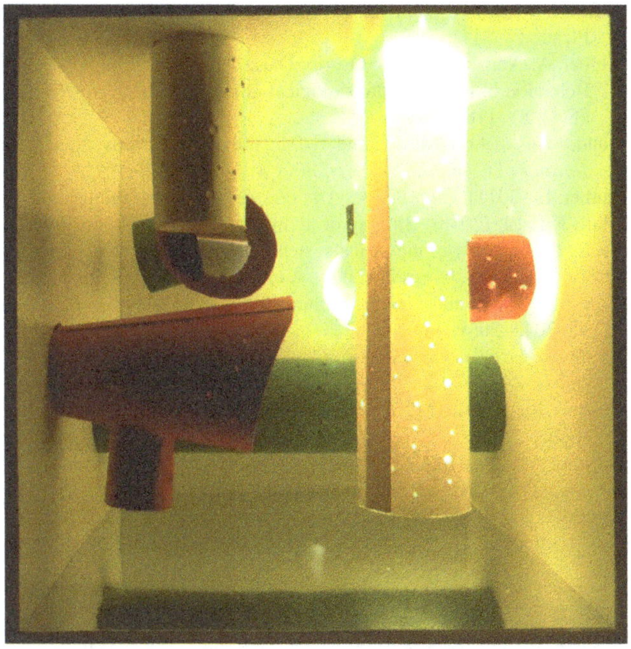

Fig. 9 *Eyes of light*, a study by Maria de Fátima Ugarte Joya and Bernardo Cuellar Medina, MUPAAC, University of Alcalá, 2021

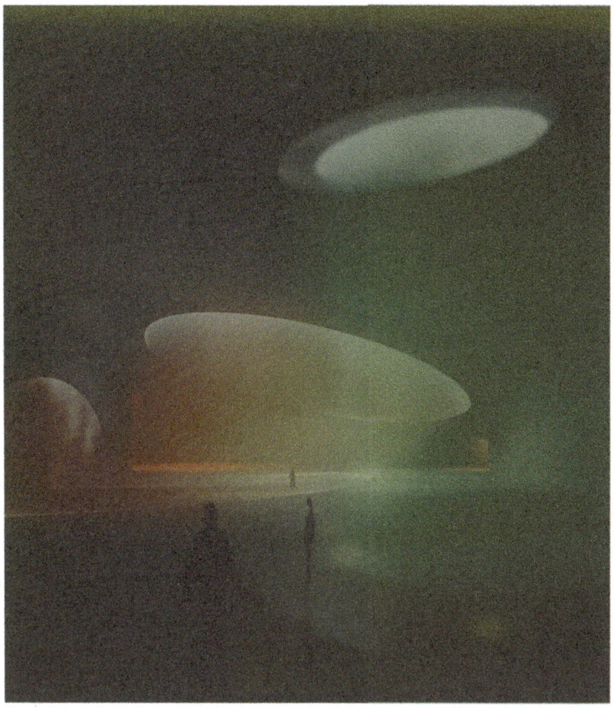

Fig. 10 *The Cave*, a study by Gabriel Palamariu and Lorena Mergea, MUPAAC, University of Alcalá, 2021

perception of light than the normal human one. The results of the brief workshop proposed were amazing. Some proposals departed from the study of the inner consistence of light and from the study of the change of its properties according to the environments and surfaces with whom it interacts (Fig. 9). Others interpreted light as a dynamic membrane or skin, while others attempted to catch and include the ineffable effects of natural light in consistent architectural elements. Another idea was to work with special filters that transform light in completely dark spaces, in a photographic manner. Materials used in interior space can change the wavelengths and can absorb or reflect light (Fig. 10). Shadows can generate space by their own, realizing complementary, abstract volumes. The relation of the skin of living organisms with light and their adaptation to environment can inspire kinetic envelopes (facades) for new buildings. The list is endless, practically.

Conclusions

Let us consider again the questions that generated the study included in this chapter: Which is the relationship between animals and light? Does this relation influence their quality and duration of life? Can the various possibilities of light perception inspire new technological details for our buildings or influence their architectural volume? We have now come across several answers:

1. Generally, the relationship with light is fundamental for living species. However, there is a huge diversity in the possibilities of visual interaction with the surrounding world. Vision of species is perfectly adapted to their life environment and their own physical characteristics. For instance, colored birds and insects have an extremely accurate vision of color, ensured by their four types of cone cells, while water organisms have specific skills in focusing on a sharp, clear image. Many animals are able to perceive polarized light, the infrared and the ultraviolet spectrum, inaccessible to the humans, which gives them a completely different perception of the world etc.
2. The relation with light is undoubtedly fundamental for all living organisms. The table presented in this chapter brings together the species with most longevity from the aquatic, terrestrial, and aerial environment (whale, bivalve molluscs, lobster, species of reptiles, elephant, turtle, species of parrots, flamingo, condor, the tree weta spider, etc.). The majority of them own a prolific vision, developed in diverse senses, much better than the one of the humans. However, there are species whose sense of perceiving light has not been studied enough (such as the tardigrades, which are known for their extraordinary capacity of adaptation and resilience). It seems that some of the tardigrades may even be indifferent or bothered by light. They compensate this fact with other special capac-

ities of adapting to environmental conditions (in hot conditions, they produce heat-shock proteins, which prevent other proteins from warping). Some tardigrades can form bubbly cysts around their bodies that allow them to survive in harsh climates without having to revert into full tun mode. A deep understanding of the sight mechanisms of the most long-living species can undoubtedly help humans in creating instruments and mechanisms for their better adaptation at the surrounding world.

3. As for the relationship between the biological study presented below and the design conception applied to architecture, the analysis of three well-known buildings whose concept bases on the study of the eye and its way of functioning is presented. It seems that by now architects and designers are only considering the study of the human eye, which is not much in comparison with the enormous variety offered by the animal world. This chapter presents as well the study case of a project experimenting the visual sensation of the horse and an experiment of designing interior space with light, which demonstrates a direct application of biological and optical principles into architectural space.

The present work is just a small starting point that could move the avalanche of a new interpretation of space by using new tools inspired from the special skills of animals. This may not lead us to a longer life, as the one of the species presented above, but it will surely help us to obtain a better adaptation to our living environment and a better and deeper understanding of our planet.

References

Carso, K. D., Holl, S., Strizler-Levine, N., Olhavsky, P., Pasti, S., & Swan, K. (2018). *Steven Holl: Making architecture*. Samuel Dorsky Museum of Art.

Corbusier, L. (1925). *Vers une Architecture*. Les Editions G.Crès et Companie.

Higgins, J. (2021). *Sentient. What animals reveal about our senses*. Picador.

Kelber, A., Vorobyev, M., & Osorio, D. (2003). Animal colour vision – Behavioural tests and physiological concepts. *Biological Reviews, 78*(1), 81–118.

Land, M. F. (1999). Motion and vision: Why animals move their eyes. *Journal of Comparative Physiology A*, 341–352.

Luther, E. (2011). *Ciutat de les Arts i les Ciències*. Acu Publishing.

Maro Kiris, I. (2018). Biology of architecture: Two buildings inspired by the anatomy of the visual system. *Turkish Neurosurgery*, 1–5.

Osorio, D., & Vorobyev, M. (2008). A review of the evolution of animal colour vision and visual communication signals. *Vision Research, 48*, 2042–2051.

Smith, A. (2020). Animal vision. In V. Melfi, N. Dorey, & S. Ward (Eds.), *Zoo animal learning and training*. Wiley.

Stevens, M. (2021a). *Life in colour. How animals see the world*. Witness Books.

Stevens, M. (2021b). *Secret worlds. The extraordinary senses of animals*. Oxford University Press.

Zumthor, P. (1999). *Thinking architecture*. Birkhäuser-Publishers for Architecture.

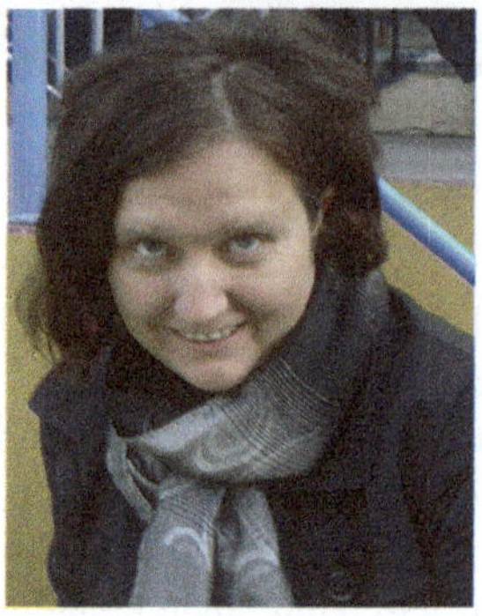

Elena-Codina Duşoiu is An architect (graduated in 2000, IMUAU) and Professor within the "Ion Mincu" University of Architecture and Urban Planning from Bucharest, Romania. She also accomplished study periods within the Polytechnic University of Catalonia, Barcelona (Master in Restoration of Monuments, 2001) and Venice International University (*Dottorato di Eccelenza. Storia della Città, dell'Architettura e del Restauro*, 2004). Her main fields of research are: rehabilitation and conversion of buildings (published books: *The Dynamics of the Sacred Space. The Influence of Function*, *3 Breweries and Their Destiny*, Ion Mincu Publishing House, 2009), vernacular architecture and ecology (organizer of 12 editions of Spanish-Romanian workshops on vernacular architecture 2006–2018), design and bionics (research studies realized within the IMUAU and the Alcalá University, 2021–2022). She received various research grants and published about 100 scientific articles and 8 authored and co-authored books in the mentioned research fields. Outstanding activity as visiting professor in Spain, Italy, Greece, Czech Republic, Liechtenstein, Argentina etc. Several national and international prizes and nominations in architectural research and architectural design, owner of a personal architecture studio since 2003.

Part II

Establishing the Tools: Parametric and Biodigital Architectural Design

Generative Biodigital Architecture and Design: From the DNA to the Planet

Alberto T. Estévez and Yomna K. Abdallah

Abstract

In current times, digital design tools and methods are rapidly evolving continuously introducing deeper integration of Artificial Intelligence and machine learning dominance over the architectural design and construction realm, stressing practitioners and theorists to master these emerging new tools whenever they exist. This acute time-tool equation leads the focus of the architectural design community to the know-how of these technologies whatever their application would be, indication, and justification in an open-end experimental methodology. Despite the significant role of experimental open-end methodology in general and particularly in the design process being defined mainly as a creative process, a rational regulator is needed to guide these modern technological tools to a real sustainable application in the built environment. In the present work, the authors categorize the generative design methods and tools through different biodigital design phases, in a progressive order, from the simplest level of application as only form generation to the most complex interdisciplinary as bioengineered biodigital tissues. These phases include the generative forms phase, the generative behavioral design phase, the generative biomaterials phase, the generative tissues phase, and the generative biocybernetics phase. These various phases mainly focus on the tertiary relation between materials, as the main element in the realization of architectural design, the form that is compatible with the material capacities as well as its functional application and the technology that enables the physical transition of form to reality through materials. The analytical descriptive methodology employed in the current study adopts a case-study analysis approach to exhibit the full experience and the included know-how of the form, material, and technology applied in each case (project). These case studies are manifesting each of these generative biodigital phases. The Biodigital Barcelona Pavilion and the Hyperboloid and Radiolaria Barcelona Furniture Series correspond to the generative forms phase. The Genetic Barcelona Project corresponds to generative behavioral design. The 3D-Printed Biodigital Clay Bricks and the Remote 3D-Printed Clay-Based Façade Unit correspond to the generative materials phase. And the Genetic Barcelona Pavilion and the Biowalls correspond to the generative tissues phase. The generative biocybernetics phase includes a current developing project that will be revealed and published in following studies.

Keywords
Generative design · Biodigital architecture · Biodigital design · Digital architecture · Digital design · Digital organicism · Genetic architectures

Introduction

The evolution in technology has affected architectural design and construction as did in all other aspects of life. And as it is said in the modern architecture movement paradigm of the twentieth century, about the need to design "from the spoon to the city," this has now expanded in our twenty-first century with the possibility to design from the DNA and the BIT to the entire planet, either biological or digital (Estévez, 2021). Within this, questions emerge, almost related to what and how are these new technological advances reflected in the

A. T. Estévez (✉)
ESARQ-UIC Barcelona School of Architecture / iBAG-UIC Barcelona (Institute for Biodigital Architecture & Genetics), Universitat Internacional de Catalunya, Barcelona, Spain
e-mail: estevez@uic.es

Y. K. Abdallah
ESARQ-UIC Barcelona School of Architecture / iBAG-UIC Barcelona (Institute for Biodigital Architecture & Genetics), Universitat Internacional de Catalunya, Barcelona, Spain

Faculty of Applied Arts/Interior Design and Architecture Department, Helwan University, Cairo, Egypt
e-mail: yomnaabdallah@uic.es

M. R. Cervera Sarda et al. (eds.), *Architecture Inspired by Nature*, https://doi.org/10.1007/978-3-031-33144-2_6

design process as methods and tools and how to master their application. And whenever a new method or tool is introduced to design practice, it imposes stress over practitioners and theorists equally, putting all design practitioners and theorists in a competition of time-tool equation, yielding their mental flexibility to the maximum as well as stressing their capacity of learning while partially neglecting the rational implementation of the results of these new tools in the built environment. "How" must not be the first concern in any design case. The right question is "why," to identify objectives before experimentation. This is not to deny the role of an open-end experimental methodology in informing, optimizing, and leading the design process but to organize this experimental practice in the right course. It is not deniable that without the experimental physical models of Antoni Gaudí, we would not have structurally optimized minimal surfaces and minimal materials of architectural marvels as the Sagrada Familia, though accurately tracking this experimental methodology of Gaudí, and others, it is clear that these methods emerged as an answer for "why," not for "how." However, "why" and "how" are closely related in a consecutive relation in the design process. In the current work, "why" is to organize, to categorize, to put the "aim" before eyes and minds, to wisely balance what is for human and what is for a machine, and to reidentify the design process in terms of "what is a tool" and "what is the methodology." Thus, this work aims to identify through each design epoch why use special technical tools and methods and "what is next."

In this work, we focus on materials in the design process as being the only channel of realizing it into tangible reality. And through the "method" of the generative design, the authors opt to detect the reason behind the involvement of most technical advancement in design to construction process in every period. Therefore, the authors propose two twinned branches of advancement, the biological Natural Intelligence + the digital Artificial Intelligence. Hence, the following sections will include these biodigital design phases: *generative forms, generative behavioral design/biomanufacturing (materials and systems), generative biobased materials, generative tissues, and generative biocybernetics.* Figure 1 exhibits a diagram of these five phases and their subcategories exhibiting their interconnected relations.

Generative Forms

Digital design tools emerged as tools of form-making, leading the transition of the design process to digitalization. This involved the integration of CAD technologies as digital drafting tool as well as the BIM (Building Information Modeling) as a design to production workflow methodology and tool (British Standards Institution, 2019). In order to facilitate the translation of design to construction with high accuracy and flexibility in the formal aspects. These form-making tools were informed by the data of the design case or the digital DNA that is comprehended by computers as permutations and combinations of binary code "01." Presenting the concept of "what can be drawn can be built" (Estévez, 2015) accompanied in the same moment with digital fabrication advancement in CAD/CAM technologies and digital strategies of subtractive and additive fabrication methods. Consequently, this fact has opened immense possibilities in the formal complexity and continuity in the architectural design.

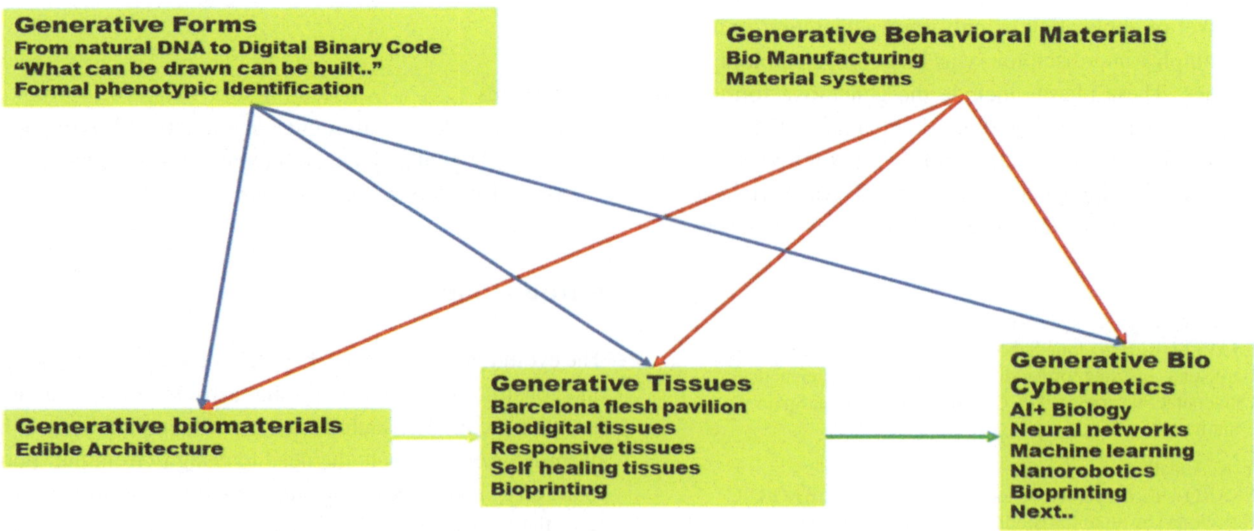

Fig. 1 The five phases of the generative biodigital design. (By authors)

Upon the collaboration between computational programmers and designers, the "parametric design" concept, methods, and tools were presented to the design field to attain maximum control and infinite possibility of modification during all the design to production process, thus saving time and money and optimizing the design process and product. This optimization is based on the algorithmic interpretation of the analog physical form-finding rules as, for example, the hanged chains models of Gaudí or soap membranes of Frei Otto, but rather in the cyberspace in a digital simulation process to optimize the diverse aspects of design including structural, mechanical, material, and behavioral efficiency in terms of interactive and responsive simulation. Each simulation process of the aforementioned optimizes its corresponding aspect in the design case, thus optimizing the design all together in the "form-finding" process based on the genetic optimization process employing genetic algorithms (GA) that are metaheuristics inspired by the process of natural selection and is a group of the larger class of evolutionary algorithms (EA) (Mitchell, 1996). Expanding the design possibilities to the "generative design" is an iterative design process that involves a program that generates a certain number of outputs that meet certain constraints and a designer to fine-tune the feasible region by selecting specific output or changing input values. In this case, the generative designer does not need to be a human; it can be an Artificial Intelligence operator, for example, a generative adversarial network (GAN) (Meintjes, 2018).

Coupling the generative design process of form-finding with the biolearning process, through applying analytic study of biological systems and bioactive behaviors, to use the resulting data into feeding the form-finding process marked the birth of the "biodigital" design era where a biodigital transition occurs from natural DNA to DBC (Digital Binary Code), translated into formal phenotypic identification of biodigitally generated architectural designs in identical formal and functional similarity with their biological peers or DNA twins.

Figure 2 exhibits the manifestation project of the Biodigital Barcelona Pavilion (2008–2009). The project was informed by the data of the imaging study of a pollen (and radiolarian) under a scanning electron microscope (SEM) that enabled the identification of its functional/formal phenotypic traits and their corresponding DNA, being one of the most mechanically resistant, flexible lightweight structures on earth (Seddon et al., 2019). The sporopollenin, the polymer comprising the exine (outer solid shell) of pollen, is recognized as one of the most chemically and mechanically stable naturally occurring organic substances (Qu & Carson Meredith, 2018). Similarly, these phenotypic functional traits were achieved in the Biodigital Barcelona Pavilion.

This biodigital generative forms phase was boosted in terms of formal complexity and fractality, provided by the advancement in additive digital fabrication technologies. The advancement in scaling the 3D printing machines and the development of environmentally friendly filament such as PLA enabled 3D printing of full-scale architectural elements. Figure 3 exhibits the 3D-printed full-scale Hyperboloid and Radiolaria Barcelona Furniture Series, chair and stools, developed from generative form complexity of hyperboloids.

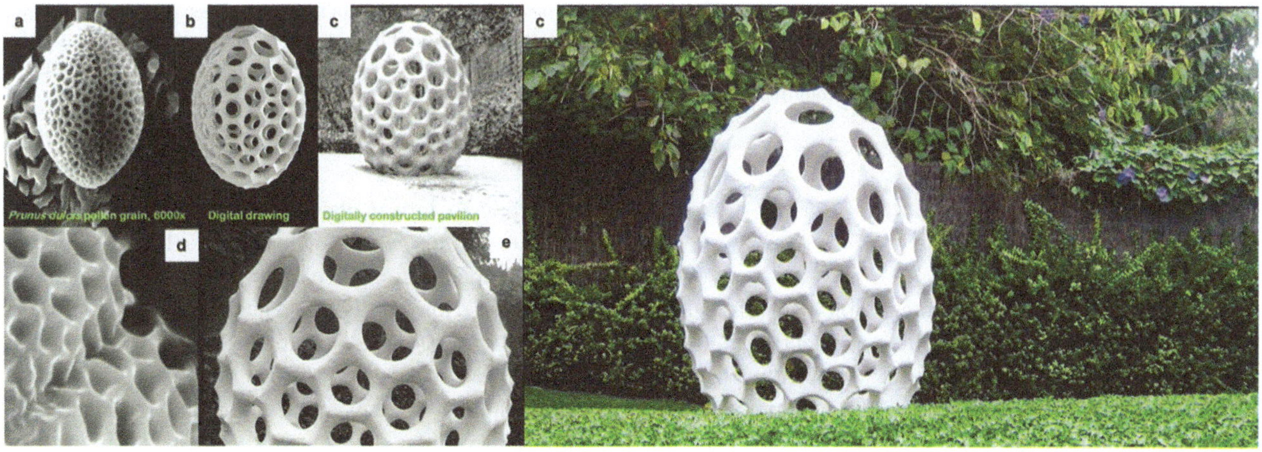

Fig. 2 (**a, d**) Pollen structure, 6.000× and 20.000×, taken with electron microscope, exhibiting pollen structures for extracting (also from radiolarian structures) the genetic rules and the structural parameters for application with digital tools. (**b**) The digital model generated by the algorithmic translation of the natural DNA responsible for the structural efficiency of a pollen grain (elasticity, lightweight, resistance, and material efficiency): this was generated by using algorithmic scripting and CAD/CAM technologies. (**c**) Alberto T. Estévez, Biodigital Barcelona Pavilion, Barcelona, 2008–2009 (col. Daniel Wunsch, computational designer), manifesting the integration of biolearning with generative design tools to achieve the phonotypical physiology of architectural design informed by natural DNA. (**e**) Detail of the digitally manufactured pavilion. (Photos: A. Estévez)

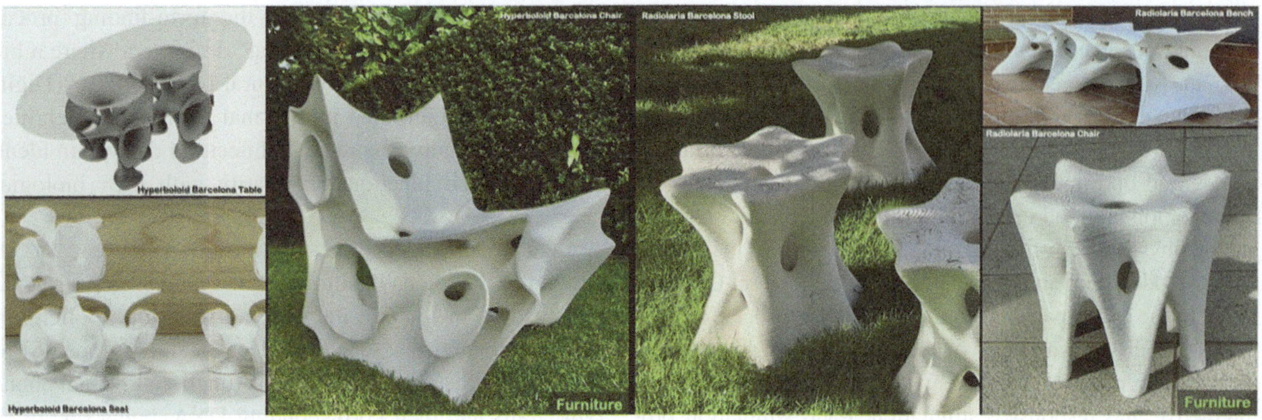

Fig. 3 Hyperboloid and Radiolaria Barcelona Furniture Series, 2019–2020 (col. Pablo Baquero, computational designer), 3D printed at 1:1 scale by Noumena, by Alberto T. Estévez. From left to right, Hyperboloid Barcelona Table, Seat, and Chair, and Radiolaria Barcelona Stools, Benches, and Chair. (Photos: A. Estévez)

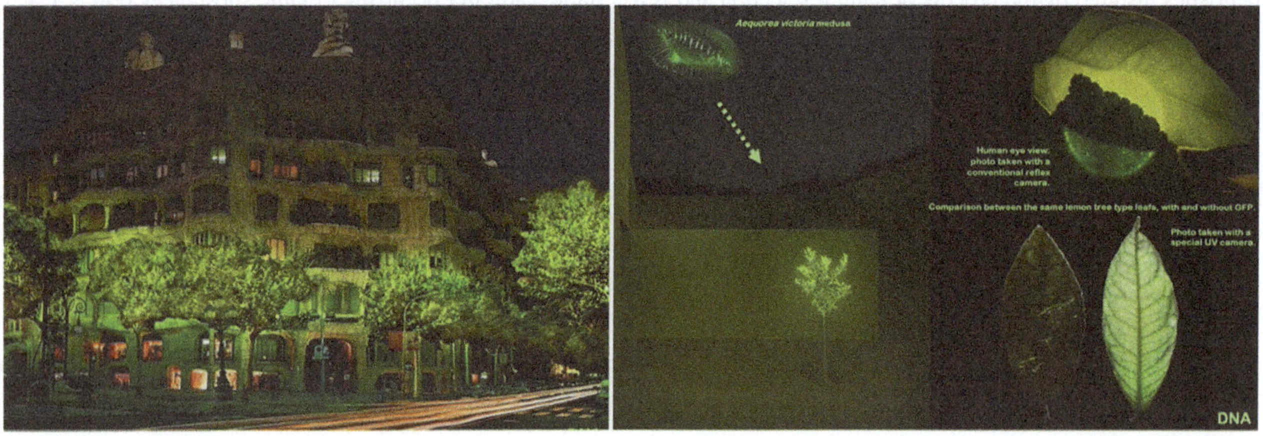

Fig. 4 Genetic Barcelona Project, first phase, 2003–2006 (cols. Agustí Fontarnau and Leandro Peña, geneticists), by Alberto T. Estévez. The project employs genetic modification of natural lemon trees by inserting the GFP gene into their DNA. From left to right, urban landscape image with a possible application of bioluminescent trees, the magical light of bioluminescent trees, and comparison between a real lemon tree leaf genetically modified with GFP and another without GFP from the same lemon tree type. Photos taken with conventional reflex camera (right, above) and with special UV camera (right, below). (Images and photos: A. Estévez)

Generative Behavioral Design/ Biomanufacturing

The generative design tools integration with biolearning methods have triggered the ambition to integrate biological behavioral performance in real time. The objective here was to extend the limits of biolearning to physiological behaviors of biodigital architecture, aiming to achieve maximum autonomous sustainability in the built environment, thus developing systems and materials through exploiting natural biological processes to achieve sustainability by bringing these bioactive agents into action by themselves. The design practice here is coupled with biotechnological experimentation, especially genetic modification. Thus, the natural DNA here was processed two times by two different methods: first, in a generative formal method, translated into digital code to inform the design forms, and the second method, to use this genetic material as the bioactive operator for achieving different ecologic functions in sustainable architecture. This biodigital generative behavioral design was based on two levels of implementation in the architectural built environment: by materials in natural replica method and by systems in a controlled replica method. Figure 4 exhibits the generative behavioral materials achieved by a real genetic modification of lemon trees, by inserting the GFP gene (responsible for the green fluorescence protein production), ensuing their natural bioluminescence by natural replica on the natural fractal scale for a bioengineered urban design, as these genetically modified lemon trees will function as urban lighting units for the streets of Barcelona at night reducing the power demand from conventional nonrenewable sources.

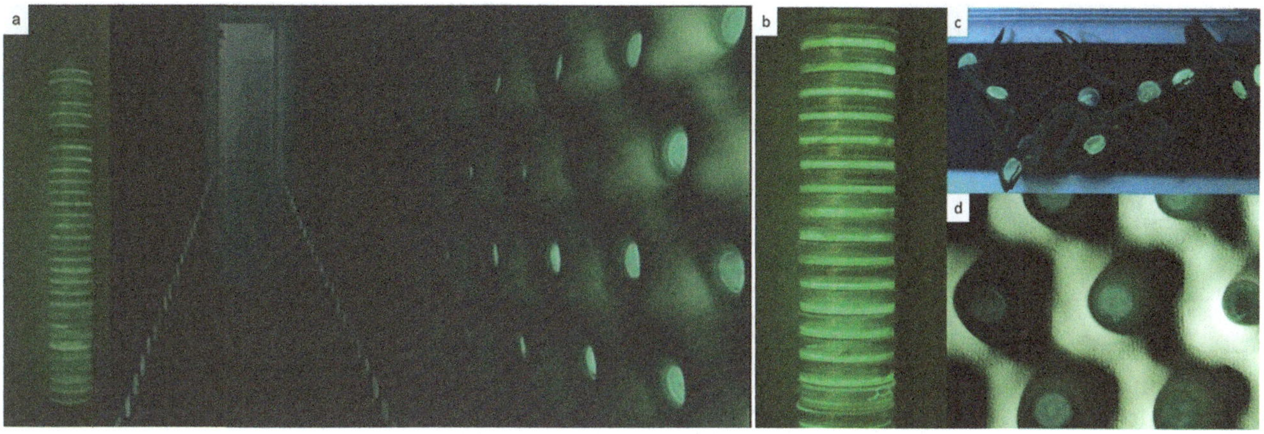

Fig. 5 Genetic Barcelona Project, second phase, 2007–2010 (col. Agustí Fontarnau and Aranzazu Balfagón, geneticists), by Alberto T. Estévez. The project employs *Aliivibrio fischeri* bioluminescent strain to produce natural light in architectural spaces, inside bioreactors that are used as biolamps. (**a**) Biolamps illuminating a full apartment without the need of electricity or installations. (**b**) Biolamp bioreactor of *A. fischeri* bioluminescent strain. (**c**) Biolamp design. (**d**) Biolamps integration in walls and roofs. (Photos taken with a conventional reflex camera as eye would normally perceive: A. Estévez)

The generative behavioral systems in their controlled replica are exhibited in the second phase of the Genetic Barcelona Project (Fig. 5), where bioluminescent bacterial strain was used inside bioreactors as biolamps that successfully were able to light a full apartment without any conventional nonrenewable electrical supply, only requiring the recharging of these systems with fresh media on a weekly basis to continue to grow and to increase the intensity and duration of their bioluminescence. However, this method is less autonomous, as the population of the bioactive agents (the bioluminescent bacteria) is physically constrained by their container or bioreactor that encompass them and their media, as well as the chemical constraints of their growth media constituents and conditions (temperature, pH, etc.), indicating that if the population has stretched over the maximum physical and chemical capacity of this bioreactor, any further replication will not be possible resulting in the decline of the growth rates of the bioactive agents and thus decline in the resulting light intensity.

Generative Biomaterials/3D-Printed Architecture/Fractality/Clay Architecture

Achieving sustainability through material physiology shaped the new biodigital design phase. This phase focused on developing biobased materials that open both possibilities of integrating biodigital generative forms of digital DNA and hosting bioactive agents in an autonomous and sustainable way to attain generative behavioral systems and materials as well. These biobased materials also facilitate the shape-shifting of the architectural element, enabling the design process to include real-time morphogenesis, either by integrating a bioactive agent that grows and interacts with the consortium of materials composing the biobased material + the bioactive agent extracellular matrix. This aspect opens infinite applications of self-healing materials and morphogenetic architecture in real performance.

Composing a biobased material requires concrete knowledge of what should be the base and what should be the filament and in between them what can enhance their coherency and total physical, chemical, and mechanical properties. One of the most prominent candidates for serving as a base material is clay. Clay represents the most available mineral, and the small size of its particles and their unique crystal structures give clay materials special properties, including cation exchange capabilities, plastic behavior when wet, catalytic abilities, swelling behavior, and low permeability, enabling higher application in many industries and processes of clay and clay-based materials (Christine et al., 2017). Building with clay was always a sustainable cost-effective method, due to its very low environmental impact, its availability in nature, easy excavation, and simple processing and production methods (La Noce et al., 2021). However, building with clay according to vernacular methods is not easy or standardized for application on large scale. For example, despite that, the Mousgoum building technique in Africa is a sustainable example of clay buildings having very solid construction. However, these Mousgoum structures require frequent maintenance of the coating, and it is incompatible for application in rainy climates (https://architectureindevelopment.org).

Given this clay architecture in biodigital age and by biodigital means aims to achieve two points: geometrical physiological sufficiency of biobased materials following a biolearning method and the ease of implementation, mass production, and standardization.

This was achieved in the design of the Biodigital Barcelona Bricks Series (2021) as a fractal scale application

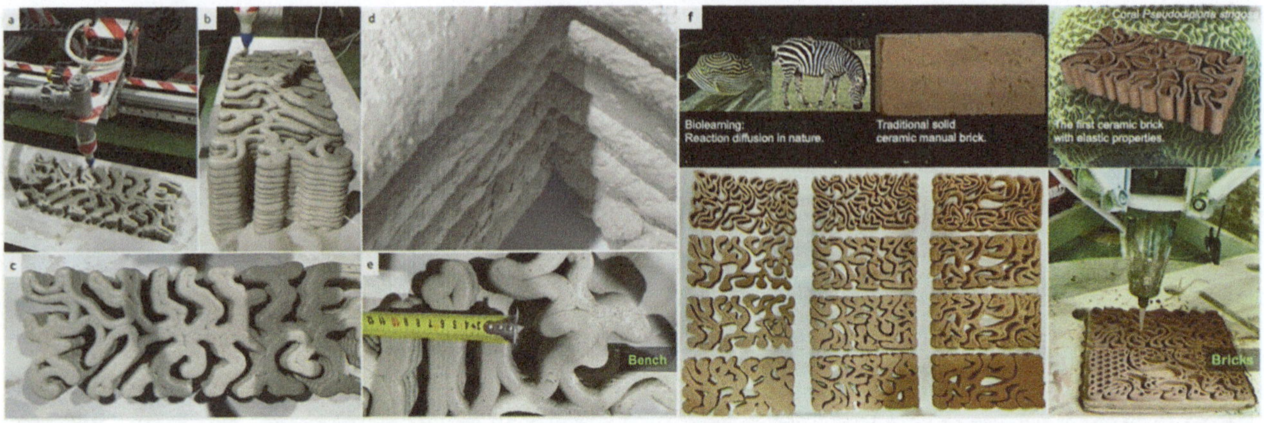

Fig. 6 (**a**) Architectural scale clay printer. (**b**) Remote 3D-printing process of the clay façade unit. (**c**) Fully printed façade unit extracted from the growth logic of corals. (**d, e**) Details of the 3D clay unit showing the coherence and shape fidelity of the print. (Photos by the authors). (**f**) Alberto T. Estévez, Biodigital Barcelona Bricks Series, 2021 (col. Yomna K. Abdallah, computational designer), achieving the creation of the first ceramic brick with elastic properties. (Yomna K. Abdallah, Secil Afsar, Alberto T. Estévez, and Oleg Popov, Remote 3D-Printed Clay-based Façade Unit, 2021, Photos: A. Estévez)

and Biodigital Clay Bench as a larger-scale application. This Biodigital Clay Bench is generatively designed following a form-finding process utilizing branching and reaction diffusion algorithms, which in the same time describes a growth behavioral pattern found in nature. The Biodigital Clay Bench (Fig. 6) was designed using rhinoceros 3D + grasshopper + anemone + kangaroo and fabricated by 3D printing by 3DPlodder.

The project explores two different aspects: one in architectural design and the other in fabrication and material. On the architectural scale, this implementation of biobased materials in 3D-printed architecture was through a remote 3D-printing process in collaboration with industrial entities. This remote 3D-printing process is one category of cyber manufacturing or 4.0 industry. Cyber manufacturing proposes an emergent solution by monitoring and managing the manufacturing process in real time and advancing the feedback/decision-making using digital tools to optimize the sensitive calibrations. Remote 3D printing includes three parameters: the printing material, the printer, and the print design, which are all interconnected. Remote 3D printing gets sophisticated when applied on the architectural scale, experimenting with advanced geometries and new material compositions that are clay/adobe based (Abdallah et al., 2022).

Generative Tissues

The integration of bioactive agents (cells) into compatible biobased materials, in one bioink through 3D bioprinting, aimed to develop autonomous bioactive tissues that can be applied on architectural and urban scale. This process is called biomanufacturing. The biomanufacturing phase of biodigital generative tissues has been developing over years from a mere concept into a tangible reality. When developing living tissues for bioengineering and regenerative medical applications was still in its infancy, the "Biodigital Bioactive Tissues" emerged as a futuristic interdisciplinary research field that promises to fulfill a real living architecture and built environment ambition. In 2007, the project of the Genetic Barcelona Pavilion was proposed as a manifestation of a real soft and eatable architecture, as a genetic reform of the Mies Barcelona Pavilion (Fig. 7), proposing an autosomes bioactive tissue in the built environment that is capable of growth and morphogenesis independently and in real time to achieve various ecological benefits. (The Mies Barcelona Pavilion is a paradigmatic building in the history of modern architecture, located precisely in the same Barcelona where we are. Therefore, in order to propose a "building-manifesto" of genetic architecture, which grows on its own, configured with real living tissues, it was a good opportunity to propose it as a genetic reform project of the same Mies Barcelona Pavilion.)

Enduring on this path, the "biodigital tissues" that were first introduced as a concept in 2000, through the Genetic Architectures Manifesto (Estévez, 2015) (Fig. 8), in 2021, they turned into real experimentation and action, forced forward by the advancement of 3D-printing technologies, giving the ability to print with living cells from various types to attain various materials to be used in the built environment for achieving sustainability through its not contaminating construction process, self-healing properties, and functional and structural efficiency. A current advancement in this research is being conducted in a multidisciplinary collabora-

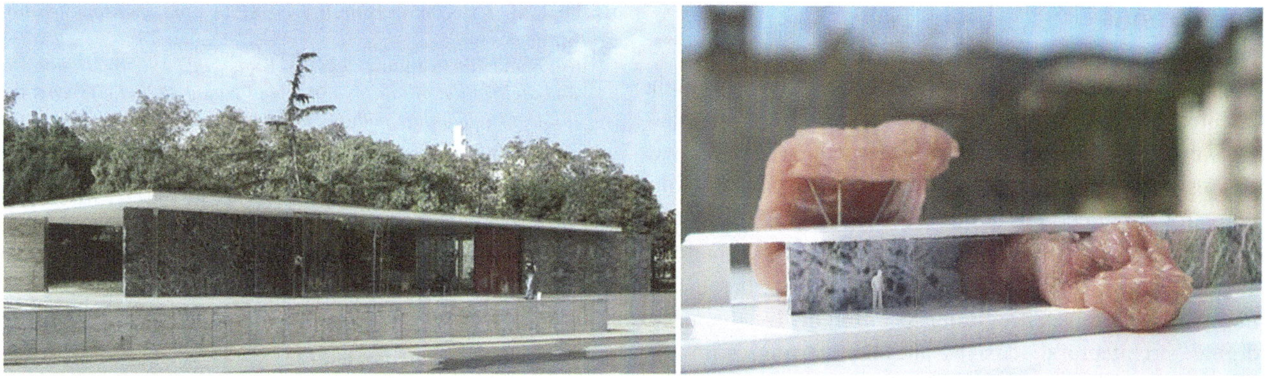

Fig. 7 Left: Mies van der Rohe, Barcelona Pavilion, Barcelona, 1929 (reconstructed 1986). Right: Alberto T. Estévez, Genetic Barcelona Pavilion, Barcelona, 2007 (model, as "manifesto-image," col. Marina Serer): research into genetic control of cell growth, making living tissue grow as a building material, cellular masses that become alive walls that emerge alone, soft and furry architecture that grows. (Photos: A. Estévez)

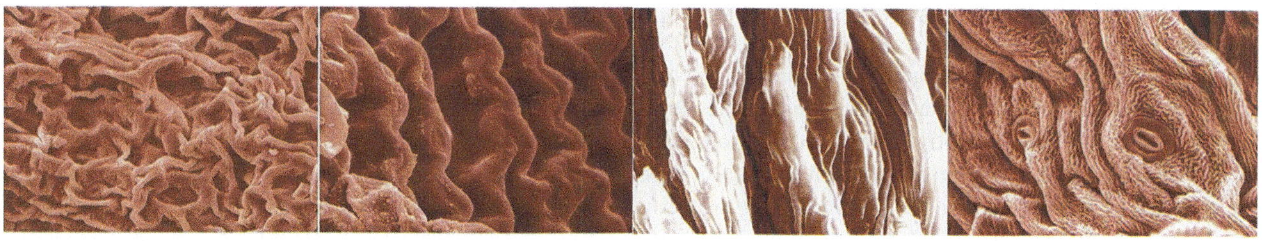

Fig. 8 Biowall systems, 2007 (by Alberto T. Estévez): research into genetic control of cell growth, making living tissue grow as a building material, cellular masses that become alive walls that emerge alone, soft and furry architecture that grows. (Photos: A. Estévez)

tion between academic and industrial entities (Estevez & Abdallah, 2022), combining algorithmic form-finding, biolearning, architectural design, bioengineering, and bioprinting to develop these tissues and study their future application in full-scale architectural built environment. For example, developing biotissues from osteoblast cells, these are the cells that are responsible for the bone tissue formation in human and other mammals.

Generative Biocybernetics

AI integration in the design process have gained much attention lately, due to the possibility of atomization of the design generative process that sufficiently matches the form with the function while evaluating a network of multi-objective/multi-interdependencies in order to reach to the optimized design solution. Pattern recognition is one branch of Artificial Intelligence, which is the first frontier in integrating AI machine learning in the design process. Pattern recognition is the automated recognition of patterns and regularities in data (Bishop, 2006). Consequently, it is employed as a generative design tool that is trained to select the optimum solution based on the feeding data set. Generative adversarial network (GAN) is a class of machine learning frameworks that is employed in further processing of pattern recognition and shape-shifting (Goodfellow et al., 2014). It is composed of two neural networks that contest with each other in a game, where one agent's gain is another agent's loss. Given a training set, this technique learns to generate new data with the same statistics as the training set. The core idea of a GAN is based on the "indirect" training through the discriminator, which is updated dynamically. Thus, in this case, the GAN does not only recognize different patterns from two different input sources, but it is able to combine them in different variations to produce new results. This is one aspect that is involved in the future research of the biodigital generative biocybernetics. The other aspect is identified by the source and type of these image data sets that are used as an input for the GAN. In this case, the images are from a biological source as scanning electron microscope images or fluorescence microscope images. This is the static stage of this generative process. The dynamic stage is still under investigation examining the ability of using dynamic biological microscopy in real time as data sets to feed the GAN for the form-finding generative process, with high expectations of the amount of resulting design iterations and variation that depicts in real time the combinations of biological behaviors. These results are designated to application in hybrid material formation through differential 3D printing (printing with multiple materials, textures, and topologies) that holds infinite functional possibilities.

Conclusions

The integration of digital design and production tools in the architectural design process is intended for different objectives emerging according to each epoch. These objectives are resulting from the need for integral sustainable solutions for architectural design at each epoch while also depending on the available technologies and material advancement to be realized. The present work has identified these objectives according to different five main biodigital architectural design phases: starting with the generative forms phase, mainly employing form-making mechanisms through generative design digital tools such as scripting and CAD, and rhino/CAM technologies for digital fabrication, while depending on biolearning from the SEM microscopy images as the source of the biological DNA to inspire the digital DNA as the form code or script. The generative behavioral design/biomanufacturing phase went further by integrating physically the biological DNA into the built environment through urban biolamps and biobatteries biolamps by genetic engineering and biotechnology as tools in the generative design and production process. Similarly, the generative biobased materials phase went to search for the combination and coupling of form, material, and function by employing form-finding tools informed by the biobehavioral biomathematical modeling, as well as the structural and physical simulation based on the material properties, as well as using 3D printing in a cyber remote manufacturing mode to produce the generated designs. The generative tissues phase has further developed the aspects and limits of bioengineered materials to develop full tissues that will contribute as autonomously living, morphogenetic, and adaptive bioactive materials that will compose the built environment. Finally, the generative biocybernetics phase combines the maximum capacity of generative design tools by employing Artificial Intelligence in form generation through its accurate machine learning generative models as generative adversarial networks (GAN) that can learn the logical patterns of bioagents and predict and generate new creative results while facilitating the futural integration between robotics and bioengineering developing the ultimate combination of biological intelligence and Artificial Intelligence. These pages have also presented the significant impact of the biolearning process as the main methodology for biodigital design through its various phases.

References

Abdallah, Y. K., Afsar, S., Estévez, A. T., & Popov, O. (2022). Remote 3D printing efficiency in clay-based materials integration in sustainable architectural design to production. In J. A. Stagner & D. S.-K. Ting (Eds.), *Renewable energy for mitigating climate change* (pp. 133–152). CRC Press/Taylor & Francis Group.

Bishop, M. C. (2006). *Pattern recognition and machine learning*. Springer.

British Standards Institution. (2019). *BS EN ISO 19650: Organization and digitization of information about buildings and civil engineering works, including building information modelling – Information management using building information modelling*. BSI.

Christine, S., Mana, A., Hanafiah, M. M., & Chowdhury, A. J. K. (2017). Environmental characteristics of clay and clay-based minerals. *Geology, Ecology and Landscapes, 1*(3).

Estévez, A. T. (2015). *Biodigital architecture & genetics: Escritos/writings*. ESARQ-UIC.

Estévez, A. T. (2021). Interscalar fluidity fascinations. In VV.AA., *Informed interscalar fluidity*. European Cultural Centre – Venice Biennale / New York Institute of Technology, School of Architecture & Design.

Estevez, A. T., & Abdallah, Y. K. (2022). Biomaterials & architecture, a possible future: Bio printing architecture. *Journal of Regenerative Medicine, 11*(2), 1000210, 1–16.

Goodfellow, I., Pouget-Abadie, J., Mirza, M., Xu, B., Warde-Farley, D., Ozair, S., Courville, A., & Bengio, Y. (2014). Generative adversarial networks (PDF). *Proceedings of the International Conference on Neural Information Processing Systems (NIPS 2014)*, (pp. 2672–2680).

https://architectureindevelopment.org/project.php?id=68.

La Noce, M., Faro, A. L., & Sciuto, G. (2021). Clay-based products sustainable development: Some applications. *Sustainability, 13*(3), 1364.

Meintjes, K. (2018). "Generative design" – What's that? – CIMdata. Retrieved 2018-06-15.

Mitchell, M. (1996). *An introduction to genetic algorithms*. MIT Press.

Qu, Z., & Carson Meredith, J. (2018). The atypically high modulus of pollen exine. *Journal of the Royal Society Interface, 15*(146), 20180533.

Seddon, A. W. R., Festi, D., Matthew Robson, T., & Zimmermann, B. (2019). Fossil pollen and spores as a tool for reconstructing ancient solar-ultraviolet irradiance received by plants: An assessment of prospects and challenges using proxy-system modelling. *Photochemical and Photobiological Science, 18*, 275–294.

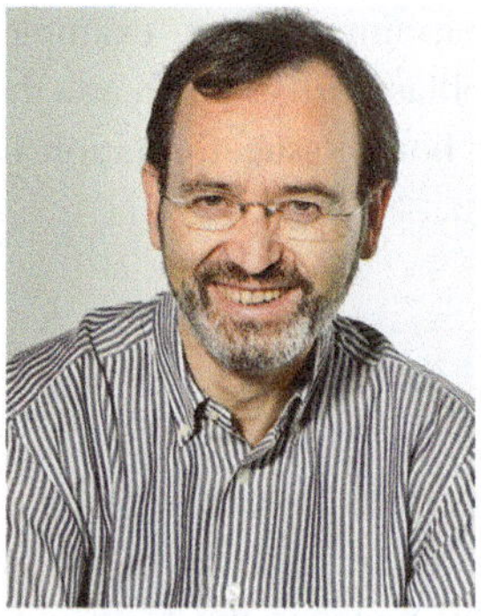

Alberto T. Estévez Architect (UPC, 1983), Doctor of Science (Architecture, UPC, 1990), Art Historian (UB, 1994), Doctor of Arts (History of Art, UB, 2008), with architecture and design office in Barcelona (1983-present). Forty years teaching and researching at various universities. Founder and first director of ESARQ (School of Architecture-UIC Barcelona, 1996), where he works as chairman-professor of Architecture. Creator of the research group, Master's degree and doctorate "History, Architecture and Design" (UIC, 1998-present), and later of the research group, Master's degree and doctorate "Genetic Architectures" (UIC, 2000-present), currently Master of Architecture Biodigital. As well as creator of the Master of International Cooperation with Alex Levi and Amanda Schachter (UIC, 2004-present). With more than 200 publications, dozens of exhibitions, conferences and committees, and invited to give more than 100 international conferences on his ideas and work. He has been director of 25 doctoral theses, and more than 100 master's and degree theses. With six officially recognized six-year research periods. Founder-director of iBAG-UIC Barcelona (Institute for Biodigital Architecture & Genetics) and founder of the Doctorate in Architecture at UIC Barcelona, of which he was its first director. Lately he was also vice-rector-manager of the UIC Barcelona (Universitat Internacional de Catalunya).

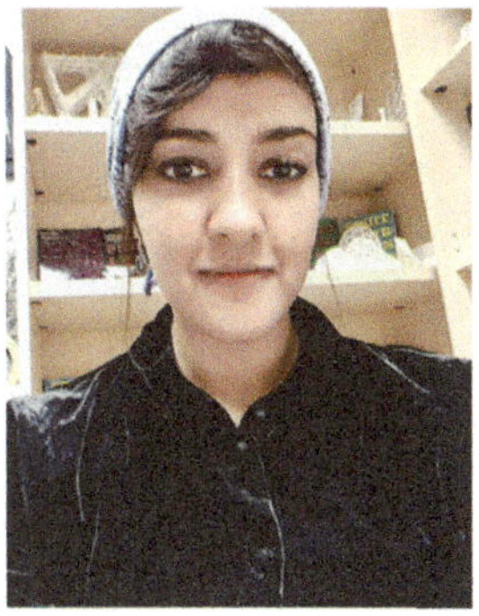

Yomna K. Abdallah is the Deputy Director of the iBAG-UIC Barcelona (Institute for Biodigital Architecture & Genetics), of the GENARQ (Consolidated and Competitive Genetic Architectures Research Group) and of the Master of Biodigital Architecture. She has a doctorate in Architecture (2020), Master in Engineering Science and Technology (2016), Master in Applied Arts-Parametric Design (2016), graduated in Applied Arts (2012). She is also a professor at the Faculty of Applied Arts, Helwan University, Egypt. She has directed multidisciplinary research projects at numerous universities. She has received multiple grants, funds and fellowships, and she has numerous articles published in prestigious scientific journals and publishers.

[illegible]

Bionic Approaches and the Cyborg Culture: Human's Phenotypic and Cognitive Extensions

Mauro Costa Couceiro

Abstract

Architecture, like all human activities, is related to biology. However, architects approach biological analogies in many different ways, often focusing on morphological imitations without bearing in mind the origin of biological processes, although it is common that architects also fail to recognize their "bio-logic," that is, their idiosyncrasies as individual living beings, their role in social and cultural structures, or even the impact of their work on ecosystems.

Under the pretext of the latest research, where digital and analog realities are integrated, this chapter presents some less common aspects of bionic research, the connection of our senses with extended realities (XR), a disruptive approach to perceptions of reality, that has its origins in the Cyborg Culture. This research is only a modest attempt to test technological developments and make them compatible with the slower pace of human's biological adaptation. These are the first steps on a long road we have to travel to hybridize specific characteristics of silicon units from the digital world with the carbon-based human beings.

We are currently developing method, content, and even technology for the dissemination of UNESCO-classified architectural and artistic heritage, exploring the intersections of natural and artificial intelligence. This research aims to create and evaluate digital contents for cultural dissemination, within the scope of Architectural and Urban History, History of Science, and other, supported by significant studies already developed in these fields. To do it, we use cutting edge extended reality technologies to promote universal and democratic access to the architectural history – the history of the most extended and perpetuated phenotype of human civilization. We are also developing foresight exercises on how cultural dissemination will be done and how bionic XR will be used in the perpetuation of memory as well as in progress of human knowledge.

Keywords

Generative Algorithms (GA) · Bionic technology · Lost heritage · Extended reality (XR) · Cognitive revolution · Cyborg culture

Introduction to the Cyborg Shock

At least since the 1970s, intellectuals and philosophers have been warning us that human beings are becoming victims of their own evolutionary success. For example, in the book *Future Shock*, the American futurist Alvin Toffler and Adelaide Farrell (1970) define the term "future shock" as a certain psychological state, both for individuals and entire societies. The shortest definition of the term might be the perception of too much change in a too short period of time. Exponential technological development makes the human being feeling increasingly out of place and obsolete in the artificial environment he creates for himself. Architects and urbanists – massively responsible for the artificial environment that surrounds us – use to approach biological analogies without understanding the origin of biological processes. On the other hand, it is common to witness that architects and urbanists also fail to recognize their own biologic idiosyncrasies as living beings, their impact on ecosystems, or even their role in the cyber-accelerated social and cultural evolution (Fig. 1).

The rule known as Moore's law (Moore, 1965) that states the exponential development of the processing capacity of computers can also be verified in other fields of the technological development of Homo sapiens. We observe that the

M. C. Couceiro (✉)
Centre for Social Studies (CES) of the University of Coimbra (UC), Coimbra, Portugal
e-mail: maurocouceiro@ces.uc.pt

M. R. Cervera Sarda et al. (eds.), *Architecture Inspired by Nature*, https://doi.org/10.1007/978-3-031-33144-2_7

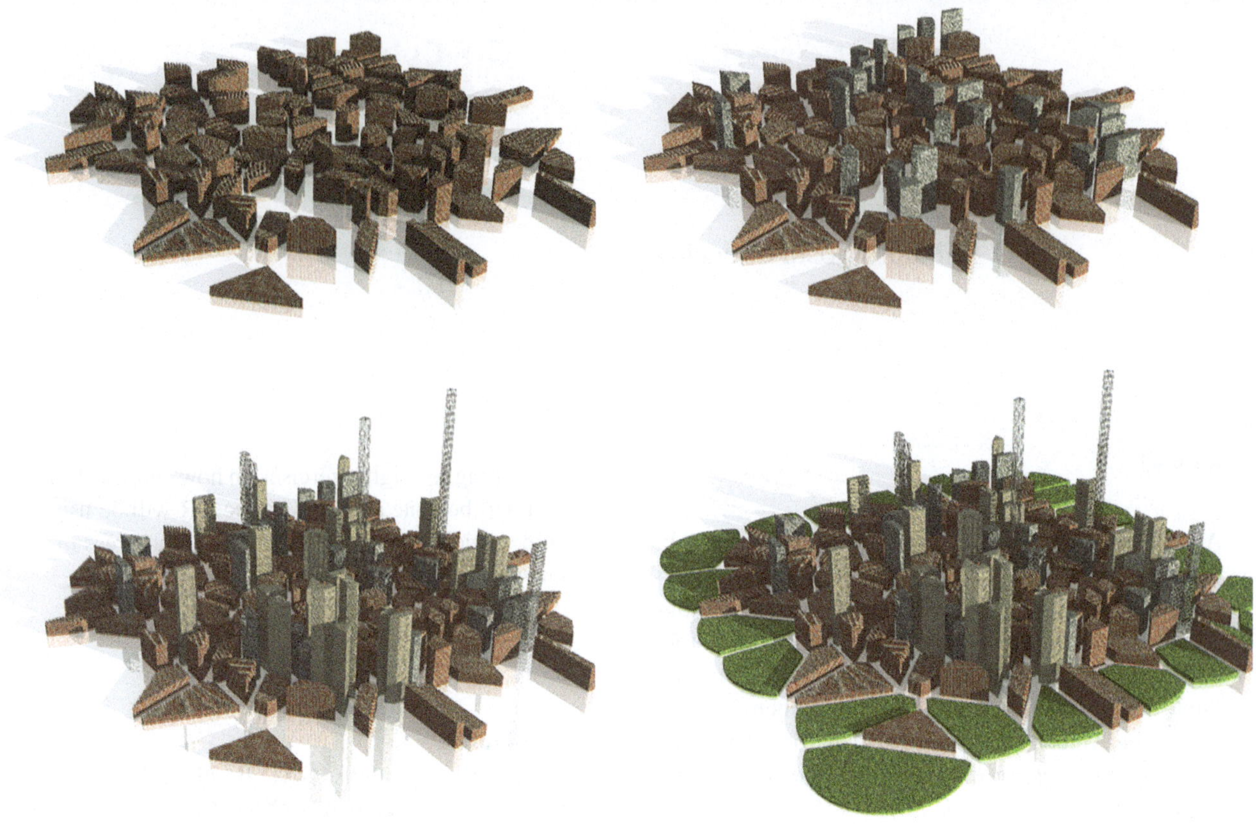

Fig. 1 Exponential growth observed in a Generative Algorithmic City. This software was developed for the first time by Professor Mauro Costa Couceiro in 2012, for a Workshop at the Universitad Complutense de Madrid. This particular simulation is a courtesy of Jessica Martins, former student of architecture at the UC 2014

human being achieves exponential results in any of the directions that engages in scientific exploration (Figs. 2 and 3).

The graphical representation of these exponential developments are asymptotes, and these make the human being more and more skillful and simultaneously more manipulative and destructive. It is because of this destructive character, typical of an asymptotic exploitation of the resources available to human beings, that contemporary philosophers and scientists warn that our civilization may collapse (Hotz, 2020). Thus, we can even hypothesize that these cyclical processes of rise and decline of civilizations may have an increasing amplitude and that our biological elasticity may not be able to withstand the increasing variations. Consequently, we run the risk of civilizational breakdown or even extinction.

We may not even have been the first beings in this galaxy to encounter these complex survival problems (Fig. 4). According to the Astrobiology Copernican Principle, there should be at least 36 advanced civilizations in our galaxy that should therefore communicate with us. However, no electromagnetic signal from our galaxy reveals any kind of intelligence – no intelligent signals are currently reaching our planet. Those signals would only take 100,000 years to reach from one end to the other of our galaxy. Despite the fact that we have the means to listen them for more than a century, nothing has been heard so far. Are aliens silent for fear of being preyed? Are we the incautious ones in a dreadful universe? Or do advanced civilizations simply not last long enough to communicate? Do advanced civilizations have a "natural" tendency for self-destruction? It is not too difficult to imagine a scenario in which, long time ago, in other corners of the universe, advanced civilizations have had their moment of glory followed by their extinction.

Where Do We Come From and Where Are We?

Less than 30 years ago, we were approaching the year 2000. Today's giants like Amazon and Google were taking their first steps as booksellers or developing search engines on a neural-inspired network – the Internet. They were tiny and they were struggling to prove their usefulness. A decade later, social networks were also underestimated in their power. Suddenly, it arises the Arab Spring and the surgical

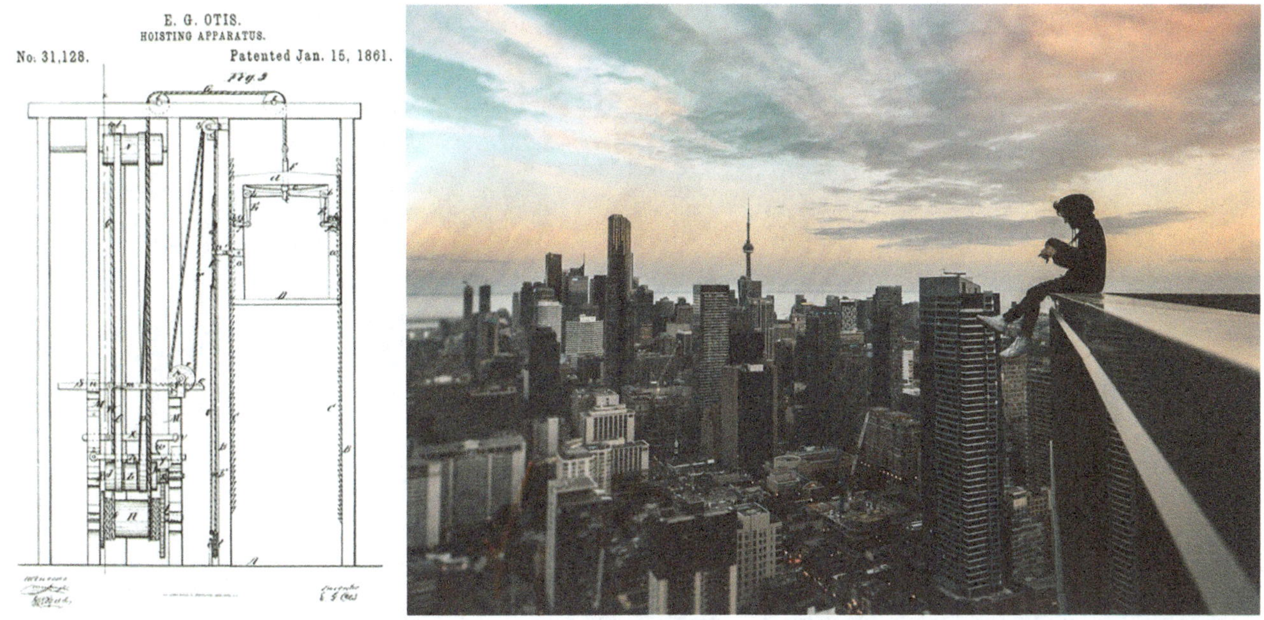

Fig. 2 The exponential and disruptive impact of elevator design on urban ecosystems – a small but revolutionary change in the internal structure of elevators that enabled the high-rise growth of urbanism. On the left: patent US31128A on Google Patents, "Elevator" by Elisha Graves Otis on the safety stop, 1861. At right: photo of Toronto by Yeshi Kangrang on Unsplash, 2017. Otis invented a safety system for the top of the elevator shaft. This ingeniously simple system works as long as there is tension on the cable holding the elevator, keeping the brakes away from the elevator shaft. Otis founded the "E. G. Otis Elevator Company" in Yonkers, New York; nevertheless, he had received only one order in his first 7 months of business. Orders began to come in only after his very showy demonstrations of the brake at the 1854 World's Fair in New York

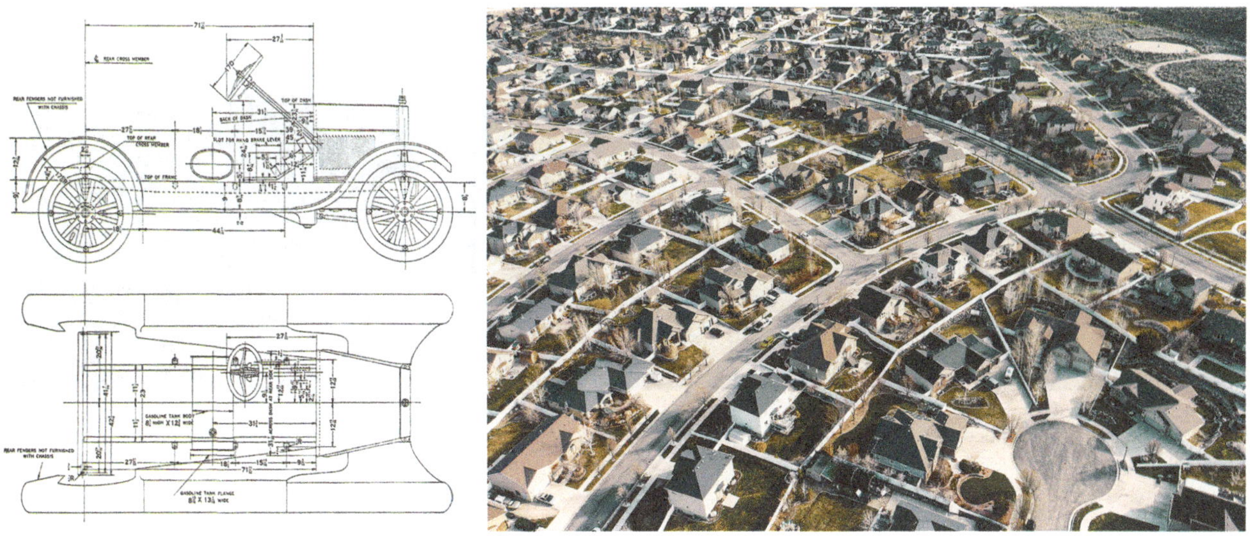

Fig. 3 The impact of car design on the horizontal exponential growth of urban suburbia. Left: the Model T, a vehicle produced by Ford Motor Company from October 1, 1908 to May 26, 1927. It is generally regarded as the first affordable automobile, which brought automobile travel within reach of middle-class Americans. At right: photograph of a suburb of Herriman, a city in southwestern Salt Lake County, Utah, United States, by Michael Tuszynski on Unsplash, 2019

manipulation of Western democracies – metadata knowledge to control the masses, their behaviors, and their ideas. We are moving by leaps and bounds toward the creation of intelligent algorithms that know more about each of us than we do and will be able to manipulate or guide us more effectively than ever before (Fig. 5).

Today, in this third decade of the millennium, we are facing new challenges and creating new socioeconomic struc-

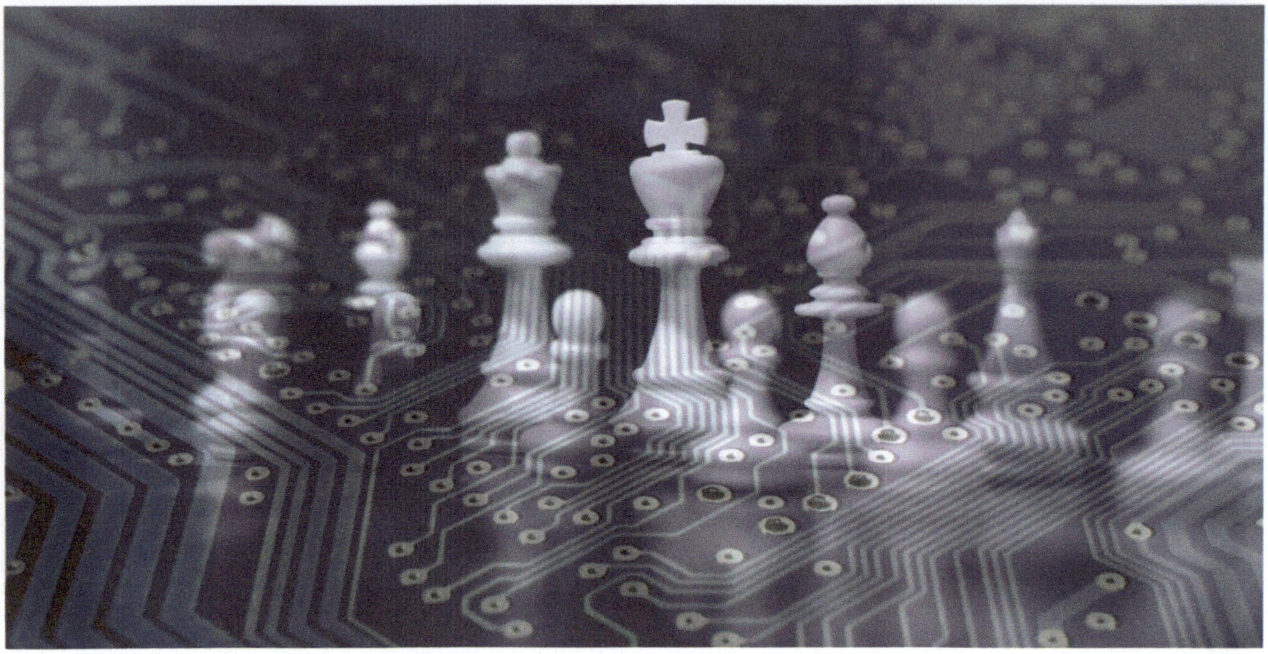

Fig. 4 Photocomposed by the author. We can observe the Homo sapiens' difficulty in the comprehension of the majority of the exponential processes. This has been told ancestrally since the story of the creation of chess: a millinery story that tells that the inventor of chess (Sessa, an ancient Indian minister) asks his king to give him some wheat, agreeing to the principle of squaring the number of grains for each square of the chessboard. The ruler laughs imagining the meager reward for that brilliant invention. Lately, he is informed by the court treasurers that the unexpected amount of wheat grains would exceed the worlds resources. Versions differ whether the inventor becomes a high-ranking advisor or is executed. As with Moore's law, the difficulty we have in understanding exponential growth is exemplified quite well in this story. Moore's law is Moore's observation that the number of transistors on an integrated circuit doubles approximately every 2 years. Moore's law is a projection of a historical trend. Rather than a physical law, it is an empirical relationship linked to the exponential benefits of production experience

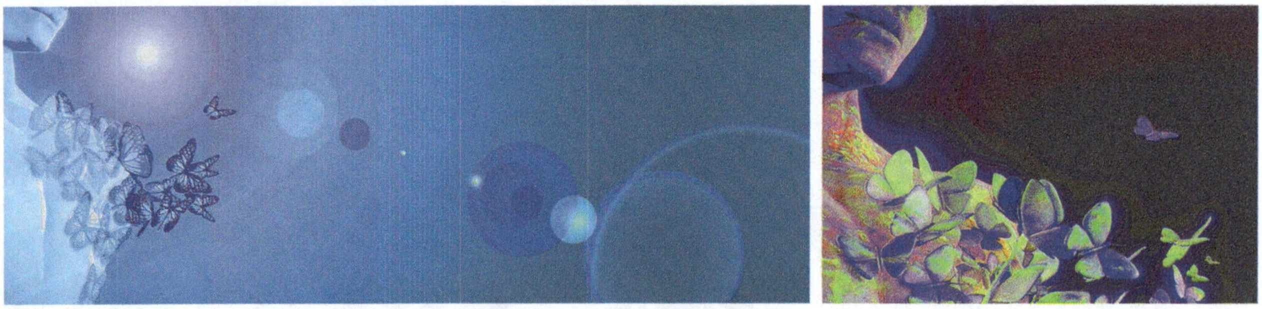

Fig. 5 Illustration title: "Embodying Ideas" – digitally generated and photocomposed by the author for the University of Coimbra (https://www.uc.pt/cultura/ideias)

tures, developed at the pace of artificial intelligence. Global networks are constantly evaluating and testing the potential of prospective futures. It is in these contexts that first attempts to explore intimate connections between diverse intelligences are observed. Human intelligence (HI) and artificial intelligence (AI) are merging into a cognitive bionic whole. These interactions or hybridizations are being carried out by private companies. For instance, microchips are inserted into the human neocortex for subsequent connection with computing devices extrinsic to the user's body (Musk, 2020). As impressive as these intimacies between the silicon and the carbon worlds may seem, this category of research began more than half a century ago, with sturdy advances and also some ethical setbacks. The most emblematic example is already decades old and consists of the bionic application of implants – the direct introduction into the neocortex of filaments to carry incredibly light electrical impulses. These impulses are then absorbed by the neural network to be interpreted as sounds by the human brain. This method, widely explored in audiology surgeries in children up to 4 years of age, is taking its first steps in the visual field. Considering the amount of information processed by the brain in the interpretation of light signals from the world around us, we are still quite far from achieving a system that provides blind people with vision, at least, something similar to the average vision of a human being or a video camera. Scientists in the various traditional fields of knowledge who develop systems of exchange between the digital world and the analog world are confronted with various levels of incompatibility. Therefore, approaches become carefully intrusive, attempting reversibility whenever possible. For the same reason, the content domain also tends to use a balanced mix of externally derived information and amplified interpretations – extended realities.

Technology is becoming more and more intimate with the human body. The cyberpunk culture already glimpsed the cyborg concept with invasive medical procedures, inserting different prostheses, and interacting more and more with our nervous system. There seems to be an unconscious drive behind the technological development. Certainly, there is something because the "restless ape" goes on and on (Bryson, 2004).

Apparently, there is something pushing humans to create and adopt new technologies. We are beginning to mix and/or transfer our living being carbon structural base to a computer-like silicon base as if humans were a kind of caterpillar, building a technological cocoon from which a binary butterfly will emerge, lighter than biological, almost metaphysical, yet more complex and adaptive life form (Rogan & Fridman, 2018).

In the same way that Darwin detected natural selection and Lynn Margulies detected symbiogenesis (Gatti, 2006), we could propose that there are some forces that drive all living things through complexity in an anti-entropic drive, and somehow, digital evolution is connected to that.

"Sentio, ergo sum" (Damasio, 1994)

The latest accelerated transformations based on the intimate connection between digital information and our perceptual capabilities connect us with realities that go far beyond our natural senses. We are giving way to new and prolific fields such as augmented and virtual reality (AR and VR). The integration of these two emerging fields of development gave rise to what has been called mixed reality (MR) or, more broadly, extended reality (XR). Together with accurate 3D surveys, this allows us to capture and simulate architecture as well as other tangible art at unprecedented resolutions. We are reconstructing and delivering experiences that at first glance were impossible to conceive, bringing to life all kinds of architectural experiments in a profoundly unforgettable way, creating new narratives and intimacies (Fig. 6). We allow ourselves to travel virtually in space and time. In this sense, step by step, we digitally reconstruct the vanished architectural heritage, recreating, through multisensory and interactive contents, the way other human beings inhabited this planet and how cultures developed, transformed, and/or perished. Similarly, Noam Chomsky (2019) asserts that there is a human predisposition to symbolic complexities such as language and that there are incredible advantages that only language provides. However, he denotes that this evolution can also be a limitation. Thus, knowledge of built cultural heritage is difficult to convey through oral or literary descriptions. In contrast, the understanding of that same heritage, along with its evolution throughout history, is almost immediate when the heritage is reconstructed and realistically depicted in its historical context.

We focus on recovering examples of architectural design – some of the most tangible and oldest human-extended phenotypes – in which combinations of art and technologies can manifest scientific and aesthetic principles, stimulating the culture of bio-logic. To this end, we test and propose methods to elevate our physical, cognitive, and emotional aptitudes, envisioning (r)evolutions in the learning/teaching fields (Couceiro, 2014). Supported by significant studies already developed to UNESCO, cutting-edge extended reality technologies were used to promote universal and democratic access to culture and world's heritage (Milgram & Kishino, 1994, pp. 1321–1329).

The Tele-Anthropos and the Cognitive Revolution (Conclusions)

As is the case in many other fields, the artificial barrier created by academics between leisure technologies and architectural practices is collapsing. It is almost impossible to be sure that the type of education we promote today will be valid 20 or 30 years from now. We can only assure that the interaction between human beings and digital systems will be increasingly intimate. Biotechnology, associated with digital development, will pose the greatest challenges and opportunities in the coming decades. Consequently, creative technologies will provide interactive experiences in which contemporary events will be nourished by digital information, in order to emphasize certain aspects of present, past, and prospective realities.

> Egyptians worshiped the human eye because they knew the eye was the thing that paid attention.' (...) 'You're gaining access to the real information that's in the world, it's not prepackaged information - that can be false – it's the real information flowing out from the ground of being and if you pay attention to that it will help you move towards the goals that you've already established' (...) 'by absorbing that information – which is learning essentially - you build yourself into a different person' (...) 'The learning that you do along the way transforms you and it transforms the nature of your goals. (Peterson, 2013)

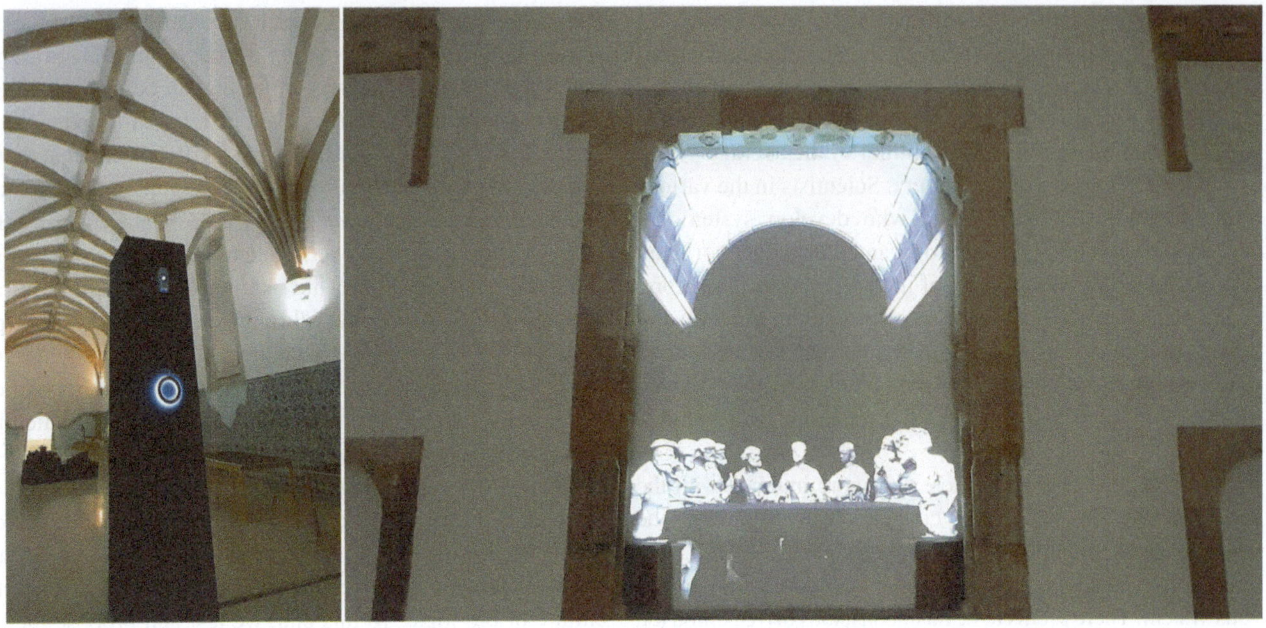

Fig. 6 Author's photographs. Recreation of Hobart's Last Supper Sculptural ensemble through stereoscopic 3D projection and video-mapping (This work was financed by FEDER – Fundo Europeu de Desenvolvimento Regional funds through the COMPETE 2020 – Operacional Programme for Competitiveness and Internationalisation (POCI), and by Portuguese funds through FCT – Fundação para a Ciência e a Tecnologia in the framework of the project SANTACRUZ with reference POCI-01-0145-FEDER-030704 – PTDC/ART-DAQ/30704/2017). A work of digital reconstitution of the vanished sculptural context and the demolished architectonic scenario. All hardware was designed and built for this specific space. The techniques used by the author were developed experimentally in successive stages. The first mixed reality experiments were done with virtual reality glasses in previously surveyed spaces, promoting interactions of the observers with the space and the virtual figures. The attempts to overcome the Turing Test for virtual reality (Renshaw et al., 2016, pp. 2113–2117) were partially or momentarily achieved in several occasions

Digital technologies are therefore primarily responsible for a change in our educational paradigm. We are ending the industrial era of education and making the transition to a different educational system based on tacit (Polanyi, 1966) representations of reality, where virtual and tangible interactions are conceived. Soon, extended reality might be more and more a bionic form of interaction with innumerous contexts, in which communication can only be developed through the fusion of tangible realities – coming from the physical world – and codified elements that expand the sensory experience. It is pretended that the observer perceives the integration of several subjects in a tacitly apprehensible whole. Here, the physical environment is mediated by our brain and gives rise to actions, feelings, and emotions.

Computer extended realities can create environments related to our personality. In that possible near future, artificial neural connections could allow us to play brain games, being tuned to activate our personal fears and/or our psychological reward systems in order to simulate hypothetical psychological realities or stages, as well as giving all the artificial knowledge support to embrace our lives with an ultimately higher level of sensitivity.

The Homo sapiens thus becomes the Tele-Anthropos (Pimenta, 1999, pp. 323–359), capable of experientially knowing things beyond his direct reach, as well as understanding realities through the interpretation provided by his cybernetic cognitive extensions – extensions of his phenotype (Dawkins, 1982). Their existence finds increasing meaning in the scope of accumulated knowledge and experience common to all human beings. In the very near future, architecture, virtual and real, will support our interaction with our environments, ordering the apparent chaos behind our extended phenotypes (Couceiro, 2006).

To conclude, it can be considered that we are still at the beginning of a new era, defined by radical changes in the way we teach and learn. Moreover, it is urgent to find ways to democratize knowledge and involve all citizens of the world in a participatory way through this accelerated paradigm shift.

References

Bryson, B. (2004). *A short history of nearly everything* (pp. 453–458). Anchor Canada.

Chomsky, N. (2019, November 29). Noam Chomsky: Language, cognition, and deep learning | Lex Fridman Podcast #53[Video]. YouTube. https://youtu.be/cMscNuSUy0I

Couceiro, M. C. (2006). Architecture as human phenotypic extension – An approach based on computational explorations. SIGRADI 2006: Post digital: el factor humano. (pp. 56–60). Santiago de Chile, University of Chile: SIGRADI 2006.

Couceiro, M. C. (2014). *Learning (r) evolution – Achieving up consistency between the real and virtual world*. Session 4. ISCTE IUL, Lisbon: Aproged – associação dos professores de desenho e geometria descritiva.

Damasio, A. (1994). *Descartes' error: Emotion, reason and the human brain*. Ed. Crítica.

Dawkins, R. (1982). *The extended phenotype: The gene as the unit of selection*. Freeman.

Gatti, R. C. (2006). A conceptual model of new hypothesis on the evolution of biodiversity. *Biologia, 71*(3), 343–351.

Hotz, G. (2020, October 22). George Hotz: Hacking the simulation & learning to drive with neural nets | Lex Fridman Podcast #132 [Video]. YouTube. https://www.youtube.com/watch?v=_L3gNaAVjQ4

Milgram, P., & Kishino, F. (1994). A taxonomy of mixed reality visual displays. *IEICE Transactions on Information and Systems, E77-D*(12), 1321–1329.

Moore, G. (1965). Cramming more components onto integrated circuits. *Electronics, 38*(8), 114–117.

Musk, E. (2020, August 29). Neuralink: Elon Musk's entire brain chip presentation in 14 minutes. [Video]. YouTube. https://www.youtube.com/watch?v=CLUWDLKAF1M

Peterson, J. (2013, July 18). Potential: Jordan Peterson at TEDxUofT | Public University of Toronto, Canada [Video]. YouTube. https://youtu.be/tLteWutitFM

Pimenta, E. (1999). Teleantropos, a desmaterialização da cultura material, arquitetura enquanto inteligência e metamorphose planetária. (pp. 323–359). Editorial Estampa.

Polanyi, M. (1966). *The tacit dimension* (p. 4). University of Chicago Press.

Renshaw, T., Sonnenfeld, N., & Meyers, M. (2016). *Fundamentals for a turing test of virtual reality*. In Proceedings of the Human Factors and Ergonomics Society 2016 Annual Meeting, SAGE Journals, pp. 2113–2117.

Rogan, J., & Fridman, L. (2018, October 24). Joe Rogan Experience #1188 – Lex Fridman [Video]. YouTube. https://youtu.be/j5FOumrXyww

Toffler, A. (1970). *Future shock*. Bantam Books.

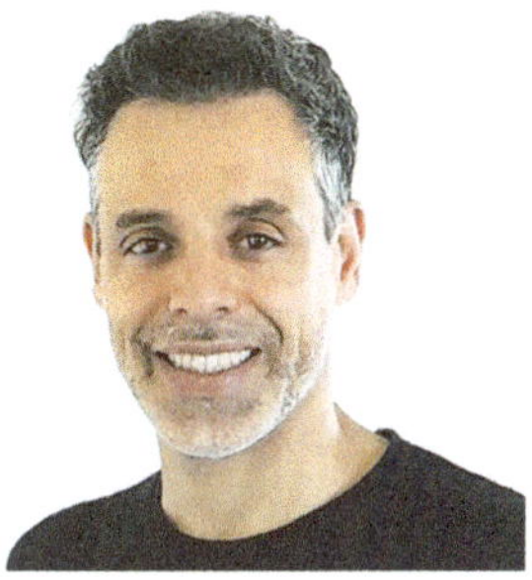

Mauro Costa Couceiro received his Doctor Europaeus degree from UIC's School of Architecture, Barcelona, in 2009. His extensive Ph.D. research was conducted in European and American countries, where he studied biological processes to develop algorithmic analogies for architecture, urbanism and industrial design. Previously, he completed two master's degrees, one in the World History of Architecture, Urbanism and Design and the other in Cybernetics and Sustainability in Architecture. Several institutions in Europe, America and Asia have requested his expertise in using biomimetics to make adaptive and sustainable designs. Research interests include Artificial Intelligence (AI), Generative Design, Building Information Modeling for Heritage (BIM/HBIM), digital fabrication (CAD/CAM) and Extended Reality (XR). Aside from design, architecture and urbanism competitions, he has won awards for doing extended reality (XR/VFX) installations. He has also worked for Sonae, developing and coordinating construction projects for ICT departments and revitalizing industrial heritage.

The Potential of Architectural Forms Derived from the Conoid: Ruled Surfaces in Parametric Design and Construction

Joseph Cabeza-Lainez

Abstract

The conoidal form was grasped in Guarini's treatises around 1660, but due to its elusive nature that is partly based in elliptic curves, adequate knowledge on the surface's properties has not been made available even with the advent of modern mathematics. Fragments of the conoid have occasionally appeared in modern and contemporary architecture, but its use had been discontinued, mainly due to the uncertainties that such construction type posed. The aim of this chapter is to orient the evolution of new architectural forms based on the conoid, offering up-to-date scientific support. The lateral area of such forms has not yet been obtained by differential geometry procedures. As this shape is frequently used in architectural engineering, the inability to determine its surface area represents a serious hindrance to solving several problems that arise in radiative transfer and building construction among others. To address such drawback, we conceived a new approach dividing the surface into infinitesimal elliptic strips of which the area can be easily obtained by means of Ramanujan's second formula. In this complex mathematical process, an evolving set of novel shapes has been derived. The authors demonstrate that the properties of these new surfaces have relevant implications for technology, especially in building science and sustainability, under realms such as structures, radiation, and acoustics. The new knowledge provided by the author, including his own proposals, may help to revitalize and expand such interesting configurations in the search for architectural advances.

Keywords

Conoid · Ellipse · Quadrature · Calculus of surface areas · Number Pi · Parametric design · Structural design

J. Cabeza-Lainez (✉)
University of Sevilla, ES, Sevilla, Spain

Introduction

The first polymath in history to introduce the conoidal figure was Camillo Guarino Guarini (Guarini, 2009) in his famous treatise *Euclides Adauctus et Methodicus* (Fig. 1).

He claimed to be the sole person in discovering the form and also described that it is a sort of cone that ends in a line. (In the ensuing pages of his book, he comes to use the word hyperbolic conoid but for a different body, the paraboloid (Guarini, 2009).)

In another formidable treatise entitled *Architettura Civile*, from 1671, he again refers to this unusual cone, stating that it has limited applicability for the corners of chamber vaults; we surmise that as a kind of squinch, or perhaps he was thinking of some of the imposing vaults in Granada's cathedral that he happened to visit. He goes so far as to calculate the volume of the figure accurately by triangulation, obtaining that it is 1.5 (3/2) times the volume of the equivalent cone (see Fig. 1).

Following these brief appearances, the form remains nearly dormant in architecture until, at the beginning of the twentieth century, it receives a significant impulse by the mastery of art nouveau architects such as Antonio Gaudi, who designed a roof with serially alternated conoids for a school building in front of his celebrated Sagrada Familia Cathedral in Barcelona (Fig. 2).

Around the 1930s, the advent of concrete shell construction favored a revolution of engineering forms and the conoid surface was the recipient of much interest, especially for building hangars, factories, and warehouses (Ramswamy, 2004).

One of the best extant examples is the work of the Bulgarian Engineer Ilja Doganoff who, in 1956–1957, erected a repair workshop for Bulgarian Railways featuring a hundred conoid skylights (Fig. 3).

The shapes were prefabricated in situ and then put on the roof with the help of a crane. They are an example of the extreme feasibility of the surfaces dealt with. Thanks to our recent discoveries, we are able to calculate the daylighting

M. R. Cervera Sarda et al. (eds.), *Architecture Inspired by Nature*, https://doi.org/10.1007/978-3-031-33144-2_8

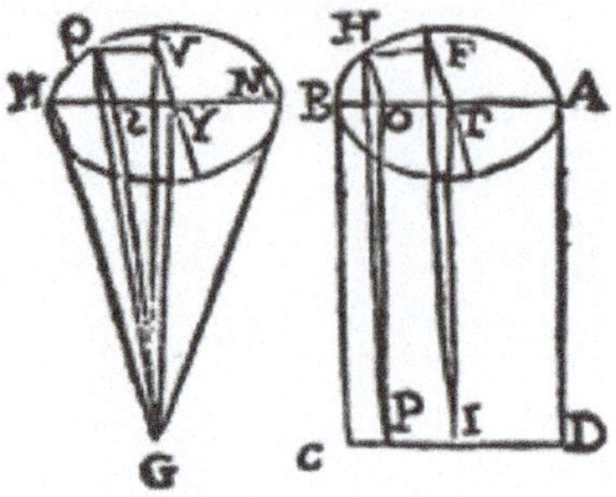

Fig. 1 Comparison between point-cone and the cone that ends in a line according to Guarini

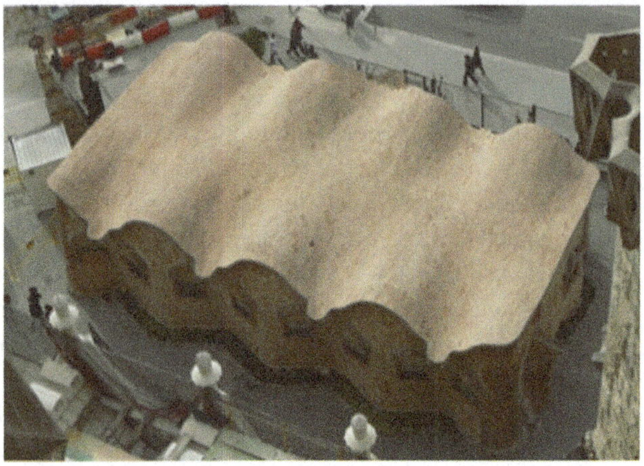

Fig. 2 The Sagrada Familia Schools, rebuilt in 2002. (A. Gaudi)

transmission of similar shapes, an undertaking neither Doganoff nor Ramaswamy (Ramswamy, 2004) were able to perform.

Ramaswamy (2004) and Doganoff (1962) report that owing to the want of knowledge about the surface, structural calculations turn out to be cumbersome, yet engineers still cherish the form because of its many advantageous properties and elegance, citing lighting and economy of construction (Cabeza-Lainez et al., 1997; Cross, 1930) as potential reasons to explain their predilection. Another interesting example of architecture based on this kind of ruled surface is the Atlantida Church (1959) by the Uruguayan Eladio Dieste.

Projects of Conoids Contributed by the Author

The present author has been working in several conoid shapes for more than 25 years, and his experience has ignited in part the present article. Modeling of the characteristic structural, acoustic and lighting properties of conoids has encompassed a significant amount of his career as a researcher. Based on that, he can attest to its sustainability and endurance (Cabeza-Lainez, 2022).

From the structural point of view, as a ruled surface, it can be built directly through straight lines (beams or poles); this fact greatly facilitates the construction and scaffolding, as more natural materials such as bricks or bamboo rods and canes can be used without difficulty, even for reinforcement or repair.

Carbon-fiber coating has become a recent alternative for strengthening even again earthquakes, although moderately expensive for the time being. The arched section of the straight conoid, whether circular or elliptic, presents a vertical tangent. Therefore, if adequately constructed, it is free from horizontal thrusts that might compromise the supporting frame. In other words, it transmits all the loads of the structure vertically and avoids the use of buttresses (Fig. 4).

These consistent and dwindling arches function as girths for most parts of the surface (Dzwierzynska & Prokopska, 2018) and provide increased resistance to a significant degree. It is true that calculation of hyper-static arches is not widely treated in the literature, but we suggest the column analogy method proposed by H. Cross (Cabeza-Lainez, 2009) as a helpful and programming-friendly procedure to solve the problem.

Due to its curvature, the aerodynamics of the roof is excellent for bearing wind loads and other meteorological phenomena such as rain, drizzle, or snow. At the same time, because of the former, it enhances airflow either from the outside or from the internal stack effect with appropriate vents.

Regarding lighting properties, if, as usual, the glazed apertures lie in the curvilinear extremes of the forms, they bring diffused luminance, as we calculated for diverse projects (Fig. 3), and can be easily shaded by eaves protruding from the same brim of the surface.

Acoustic properties stem from the circumstance that the inside surface of the conoid is mostly convex as it can be checked mathematically (Cabeza-Lainez, 2022). Sound waves are diffused in this kind of screen, and consequently, noise and reverberation become dampened. If, through appropriate design, the conoid covers a trapeze or fan-shaped plan (Fig. 4), the effect of an even sound pressure is manifest (Cabeza-Lainez, J, 1997). (In this last case, the surface is not a proper conoid as the forming lines are not parallel to a common plane.)

The aforementioned acoustic benefits are extensive to interior illumination for the same reason of convexity of forms.

The cover in Fig. 11, strictly speaking, is not a conoid because its equation differs from what we have explained in the references. The forming straight lines are not parallel to a plane, but they all coincide along a central vertical axis. However, topping of a fan-shaped plan with this kind of surface offers a very interesting structural property: the larger spans between pillars are covered by arches, while the

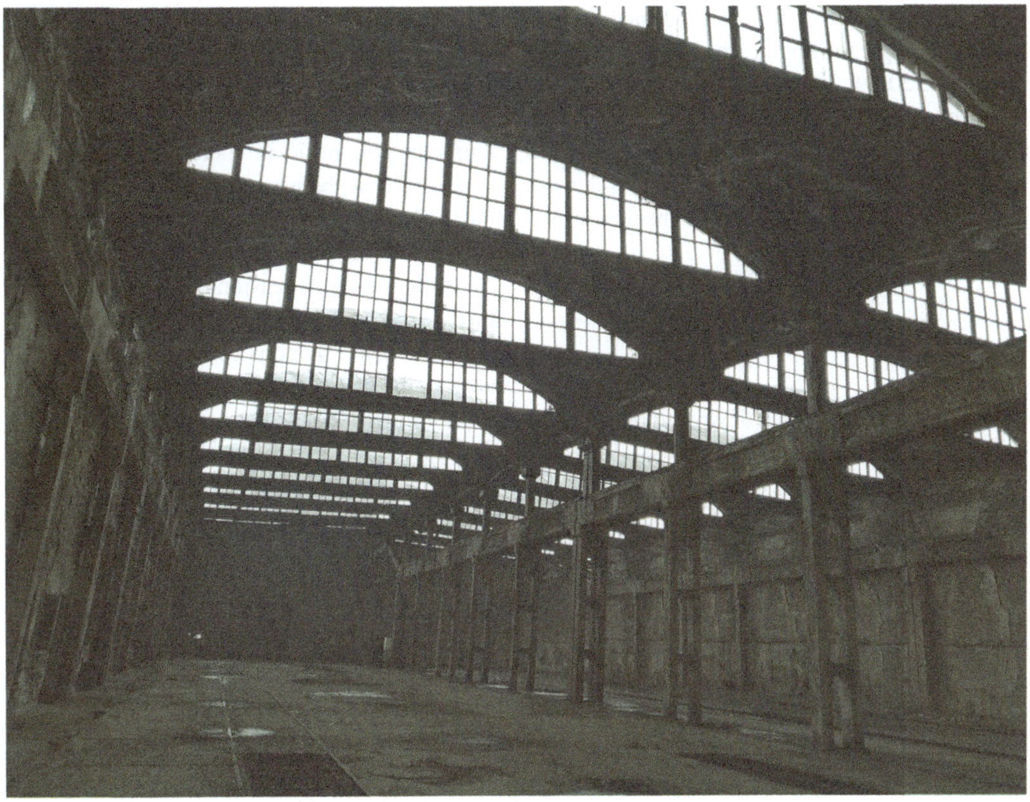

Fig. 3 Ilja Doganoff. current state of the railway depot. (Source: Author)

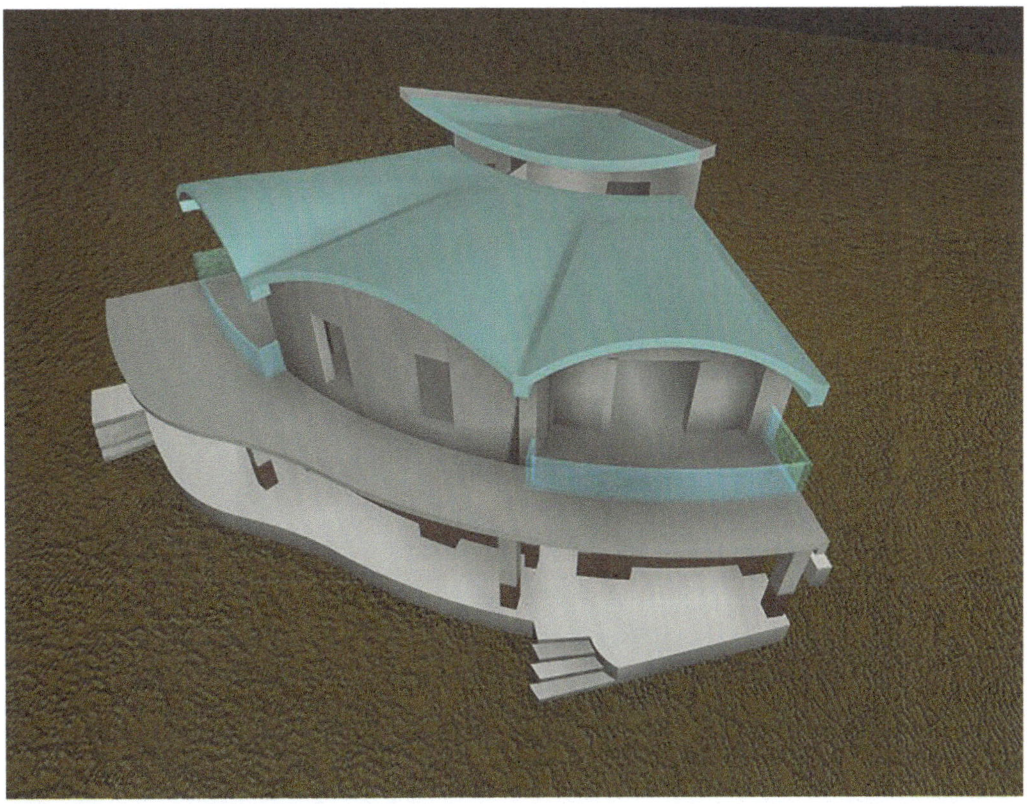

Fig. 4 Three rotated quasi-conoidal roofs designed for a musicians' family. Sanlucar (Seville)

tapered end of the trapeze features a conventional slab or planar beam, which is logical from the constructive point of view (Cabeza-Lainez, 2019).

In this way, the shells' materials can be lighter and smoother. In Fig. 5, we present three vaults consisting of thin layers of hollow brick with steel mesh as a reinforcement. The result has proven to provide increased insulation and adds a variety of light effects.

Future Proposals

As a corollary to the theories elucidated, we will discuss two project forms that we have created with conoids, taking into account evolving technologies. The first one is a system of skylights similar to the one depicted in Fig. 5, but in this case, the glazed parts instead of being planar are also conoids (Fig. 6). Bearing in mind the necessary smoothness on heat and light transfer, this feature presents undoubted advantages (Cabeza-Lainez, 2022). Firstly, the glazing is better shaded and protected by the opaque upper conoid. Secondly, sunlight and heat transmission are modulated by the smooth curves conjoined to the innovative glass properties. In this way, the glass surface becomes load-bearing and collaborates with the general structure. The form can be easily adapted to arrays of skylights as shown in Fig. 6.

The new skylights are more impervious, break-proof, safer, and cleaner in the absence of maintenance, as dust collection is diminished with the curvature.

The second form consists of an innovative proposal for an amphitheater, music, or sports venue (Fig. 7).

In this case, being a double conoid, the advantages previously elucidated are even increased. The sections are closed curves, such as the ellipse and the circumference in the brim. Thus, they work as tension rings or girdles to hold the structure together without severe deformations. The bearing capacity of the shape is extreme. The tiers of the amphitheater are the obverse of the external façade; there is no need to superimpose a conical structure inside a cylinder such as in the colosseum or in Spanish bullrings. Among other problems, the ancient structures were forced to erect giant offload vaults and galleries, which transferred severe thrusts to the outer façade. As a result, we have calculated that savings in building materials of this proposed facility could be massive.

Still, the structure can be easily constructed with straight beam elements and reinforcements. The foundations point to the soil as a kind of arrow, which means that it will be stable, safe, and simple to develop.

The outer surface of the conoid is not vertical but inclined, and so the surroundings of the amphitheater would be self-shaded—an interesting feature in warm or rainy climates.

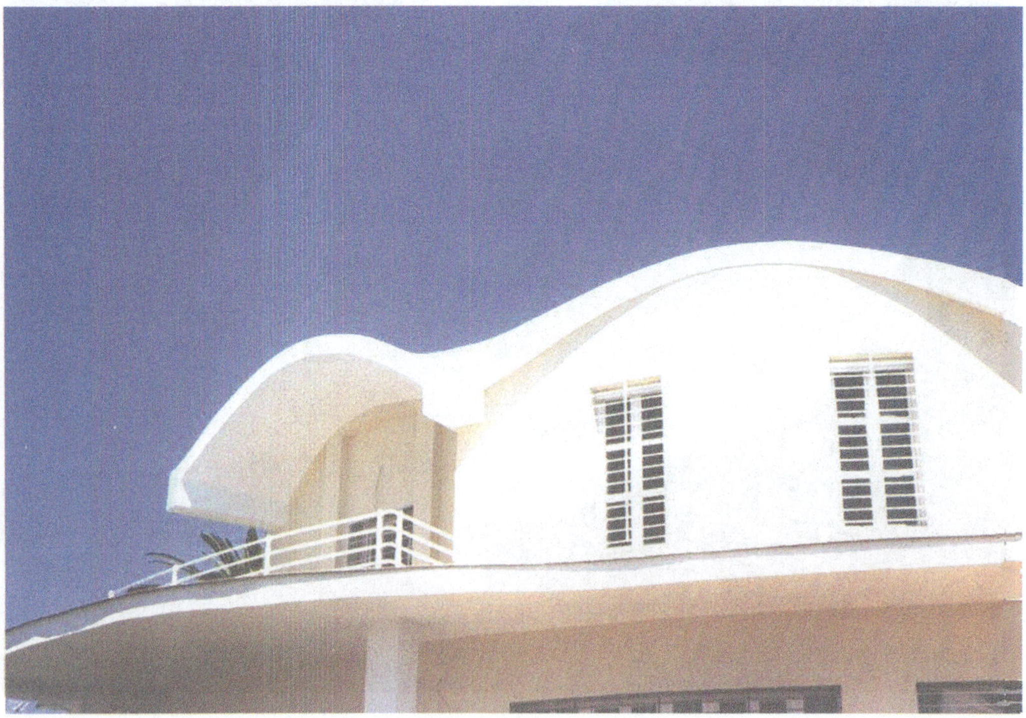

Fig. 5 Detail of pseudo-conoid roofs in which the sun-path produces intriguing variations through the day

Fig. 6 A suggested array of six conoidal skylights

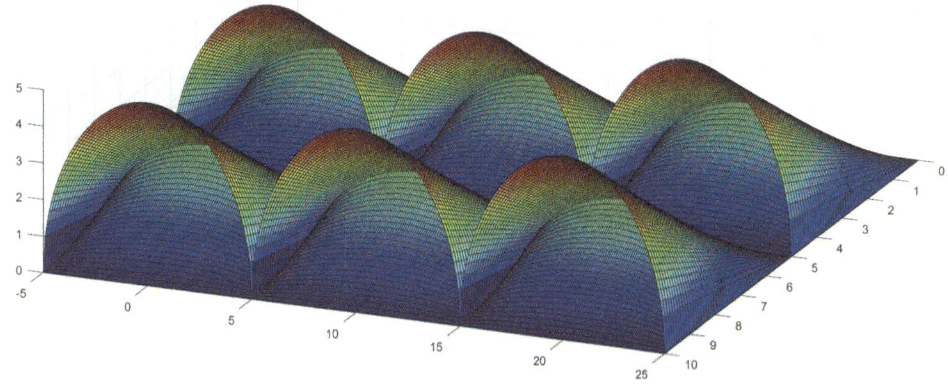

Fig. 7 The amphitheater with proposed retractable shade

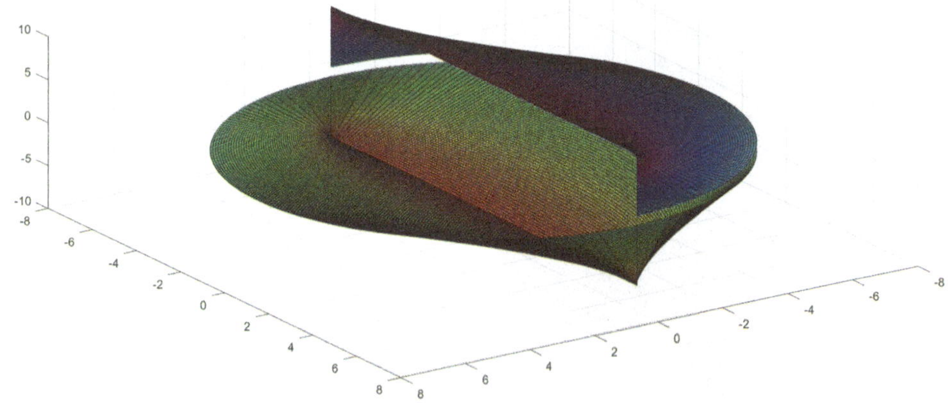

As for the grandstands, it is not difficult to adapt awnings or other shading systems to the inside area in order to protect the tiers from the rain or the sun (Fig. 7).

As mentioned above, the effect of concentration of sound, rather annoying in conventional stadiums, will be almost completely avoided due to the convexity of the surface. In reference [10], we have demonstrated such effect by acoustic ray-tracing procedures. These rely on the finding of the normal to the conoid surface at each point.

The procedure to extract the normal is first-order differentiation of the equation of the surface defined as $F(x, y, z)$. The normal vector is obtained as $N = (F_x, F_y, F_z)$.

We can trace a vector field with the reflected sound rays from an emission point to check that they are effectively dispersed in the air and not concentrated [10].

The only remaining questions would be those of selecting the relative heights of the stages and platforms and other design issues such as ambulation in the venue. Nonetheless, we believe that our proposal could be another good example of how the conoid-based bodies are able to create a significant volumetric space with a comparatively small enveloping surface; the key is that they offer a high spatial compactness, which is usually an added value in terms of heat exchange, costs reduction, and sustainability, in general.

Mathematical Formulation

If we make $L = R$, the regulating equation for Fig. 8 turns out

$$\frac{R^2 z^2}{(R-x)^2} + y^2 = R^2 \tag{1}$$

Generation of New Figures Based on the Previous Findings

We defined the symmetrical figure composed of four conoids, namely, the *Antisphera* (Fig. 9).

Fig. 8 Explanation of the parts of a straight conoid with circular directrix, in this case $R = L = 4$

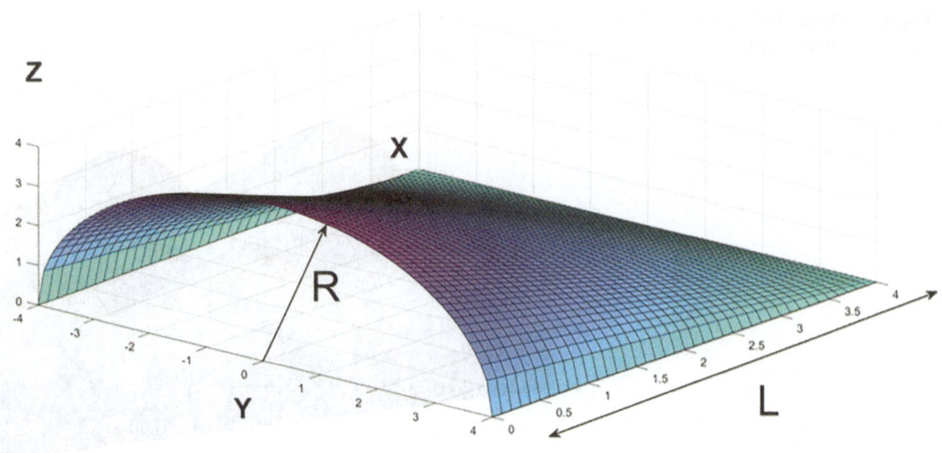

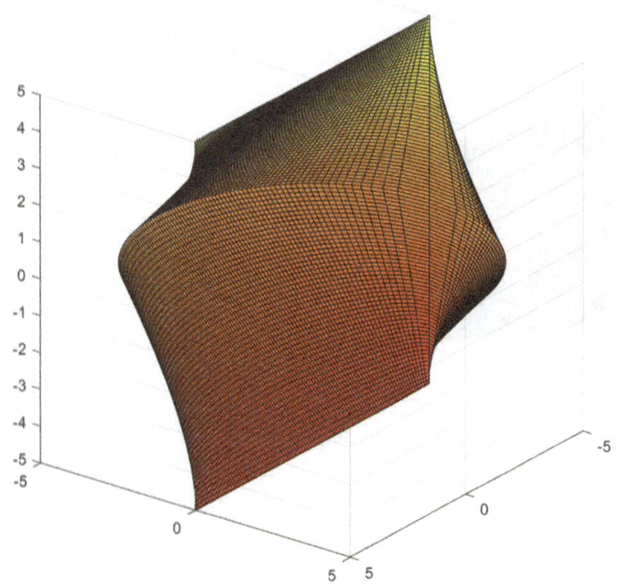

Fig. 9 Depiction of the *Antisphera* for $R = 5$. View from below

Nevertheless, we have conceived that by altering the symmetry and parts of the previous figure, a series of other interesting bodies is derived, maintaining initially the ratio $R = L$.

The first one is opposed geometrically to the *Antisphera* because the edge straight lines intersect at its center plane, which is void. Due to this elusive, and somewhat dual, nature we have coined the name *Dyosphera©* for this shape (Fig. 10).

In total, the *Dyosphera* features eight conoidal sections, organized in groups of four, rotated $\pi/2$ degrees.

The open cavities and sinuous receptacles of this figure make it particularly suitable for aerodynamic architectural creations, monuments, fountains, and so forth. It is also a very stable form because it has four circular bases.

Finally, a very important antisymmetric finding is the *Pterasphera®* (Fig. 11).

This body is tubular in nature. Being internally connected in its entirety, it is apt for conducting all kinds of fluids in an advantageous manner since, for instance, it can reduce the velocity and, at the same time, the noise of transporting the required fluids. Unlike *Antisphera*, it is self-standing and well balanced, which renders it suitable for elongation in the manner of a tower.

Repercussions for Technology

So far, we have presented scientific design developments that we hope will find a myriad of applications in technological areas such as aerospace, building, heritage, and associated industries. Such facts attest to the versatility and feasibility of the solutions presented, which derive from our mathematical investigation. In the last part of the research, due to their complexity, the surfaces have been materialized with the help of 3D fabrication procedures to help decide about some difficult points of the equations or to reflect on future realizations of the proposals. It is undeniable for us that the results attain to the domains of art and design (Fig. 12).

We believe that the implications of this geometrical advance are far-reaching. Due to its internal logic, it would be suitable for biotechnology. Especially the latest development, the *Pterasphera*, being of a tubular nature, would be prone to fluid transportation. As a springlike configuration, it conjoins flexibility and balance. In Figs. 13 and 14, we present examples of possible association and growth in parallel or opposed patterns.

If we analyze the internal section of the tubules (Fig. 14), it is composed of two semi-ellipses of varying sizes, but the span is constant at R; the extreme one is a semicircumference of radius R and the middle horizontal section is a complete ellipse of minor axis R and major axis $2R$.

The dimensions of the section are constant, but its shape is not; thus, the velocity of the fluid inside the tubule can be deftly regulated from the same form. This would offer a clear

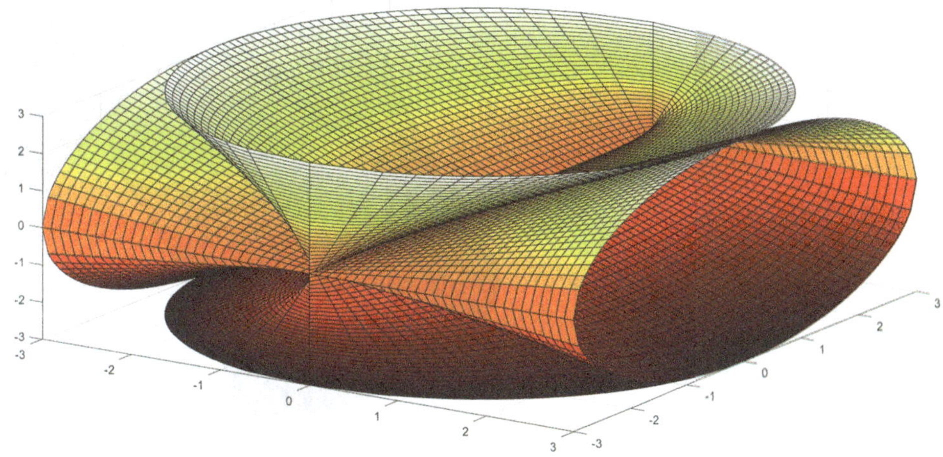

Fig. 10 Distorted depiction of the complete *Dyosphera*

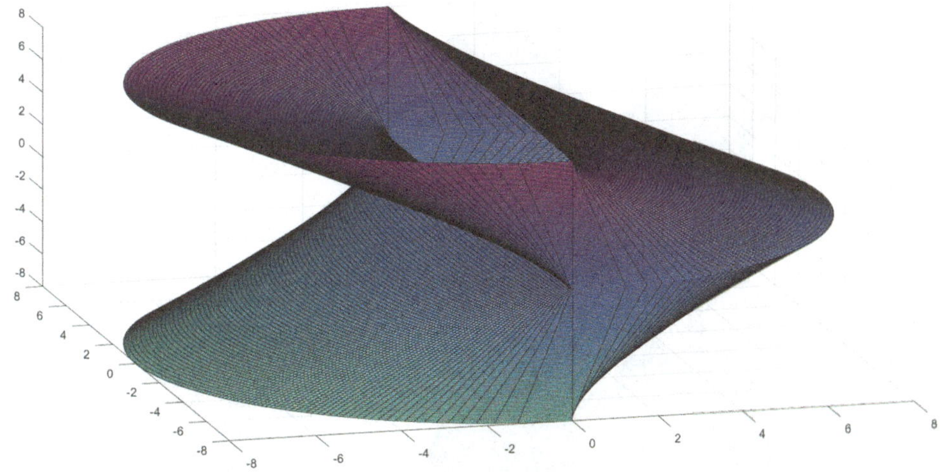

Fig. 11 Initial section of the *Pterasphera*

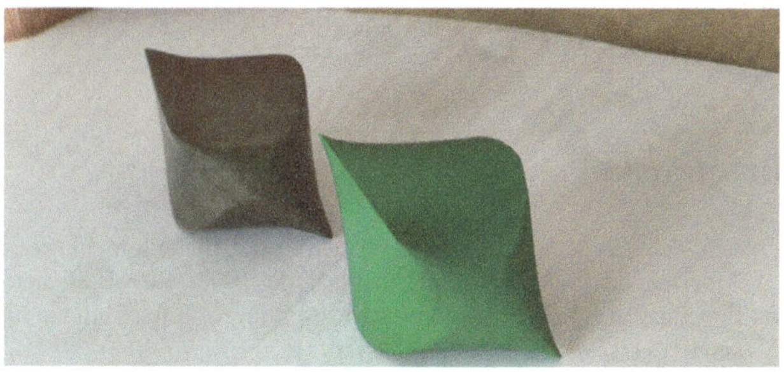

Fig. 12 Antisphera in bronze by the artist Sergio Portela and in plastic 3D print (green)

alternative to reduce the noise level in the ducts or to decant particles in suspension.

In Fig. 14, we present examples of vertical growth of the tubules, resembling vegetal pillars. Such form connects with the art tradition of Coloana Infinitului by Constantin Brâncuşi and with the architectural orders of classical architecture.

We decided to explain this finding because of the possibilities that it showcases, for instance, in tower buildings. Vertical connections are always feasible at the middle plane of the column, but the external envelop will benefit from the sun-tracking or shading properties already elucidated, combined with new photic materials of variable transparency. In consequence, lighting, thermal, and acoustic features will considerably improve the existing conditions.

The structure of the so-conceived tower can be as lightweight as desired due to its inner balance and counterweight. As we have explained previously, alone or better in groups, it should perform adequately under earthquakes, extreme wind conditions, or other unpredictable circumstances (Cabeza-Lainez, 2019).

Conversely, the author, by design operations, has achieved the form of a straight tubule of reversed sections.

Diverse technological fields could have a keen interest in the new forms that we have found, both for the macro- and microscales. It is indeed a leap into a sustainable future, which we hope goes beyond.

Fig. 13 The same tubules opposed

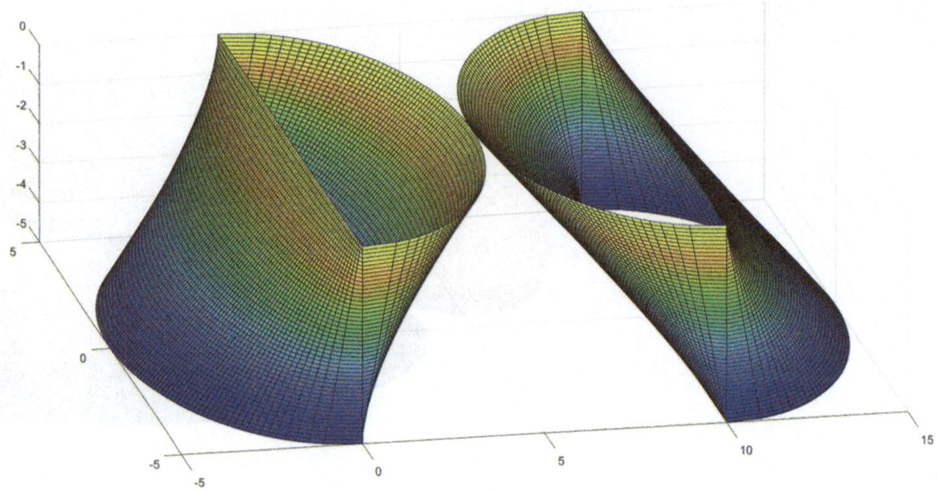

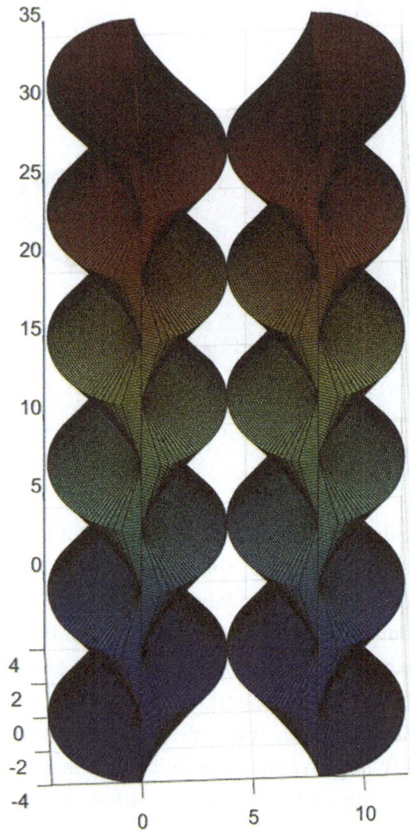

Fig. 14 A symmetric array of twin tubules

Conclusions and Future Aims

We believe to have demonstrated that only science gives form to the formless. That is why some compare it to art or even magic. In the first part of this chapter, we have identified mathematical procedures which greatly enhance our understanding of conoidal shapes. In the process, a new transcendental number has been obtained as a surprisingly accurate approximation of π. Thanks to Ramanujan's conjecture, we have been able to find the lateral area of the conoid, a recurrent form in organic structures whose scientific and technical knowledge was insufficient, although much desired for the benefit of art and architecture.

With such a procedure, we have created no less than four new types of figures that present high potential in many realms, such as aerospace, transport, communication and fluid, light, and sound-conducting devices.

Consequently, we have developed a vast array of revolutionary forms that showcases their utility for the design of sundry elements. Due to their particular geometric properties, they can work, both for heat storage or dissipation, as the case may be. They perform aptly in thermal, luminous, and acoustic radiation domains.

References

Cabeza-Lainez, J. (1997). Rehabilitación del pabellón Plaza de América de la Exposición Universal de 1992. Rev Edificación Ecoconstrucción, 27. Available online: https://hdl.handle.net/10171/16978. Accessed 1 May 2021.

Cabeza-Lainez, J. M. (2009). The key-role of Eladio Dieste, Spain and the Americas in the evolution from brickwork to architectural form. *Journal of Asian Architecture and Building Engineering, 8*, 355–362. https://doi.org/10.3130/jaabe.8.355

Cabeza-Lainez, J. M., Rodriguez-Cunill, I. (2019). The Problem of Lighting in Underground Domes, Vaults, and Tunnel-Like Structures of Antiquity; An Application to the Sustainability of Prominent Asian Heritage (India, Korea, China). *Sustainability, 11*, 5865. https://doi.org/10.3390/su11205865

Cabeza-Lainez, J. (2022). Architectural characteristics of different configurations based on new geometric determinations for the conoid. *Buildings, 12*, 10. https://doi.org/10.3390/buildings12010010

Cabeza-Lainez, J., et al. (1997). Rehabilitación del pabellón Plaza de América de la Exposición Universal de 1992. *Revista de Edificación. Ecoconstrucción, 27*. https://hdl.handle.net/10171/16978

Cross, H. (1930). *The column analogy* (Vol. XXVIII October 14, No.7). University of Illinois Bulletin. Available online: https://www.ideals.illinois.edu/bitstream/handle/2142/4229/engineering-experv00000i00215.pdf?sequence=3. Accessed 20 Sept 2021.

Doganoff, I. (1962). *Berechnung von Kugelschalen über Rechteckigem Grundriß. Mit Berechnungsdiagrammen*. Mitteilungen aus dem Institut für Massivbau Hannover: Düsseldorf.

Dzwierzynska, J., & Prokopska, A. (2018). Pre-rationalized parametric designing of roof shells formed by repetitive modules of catalan surfaces. *Symmetry, 10*, 105. https://doi.org/10.3390/sym10040105

Guarini, C. G. (2009). *Euclides Adauctus et Methodicus Mathematicaque Universalis*. 1671. Torino (Italy).

Ramswamy, S. G. (2004). *Design and construction of concrete shell roofs*. CBS Publishers.

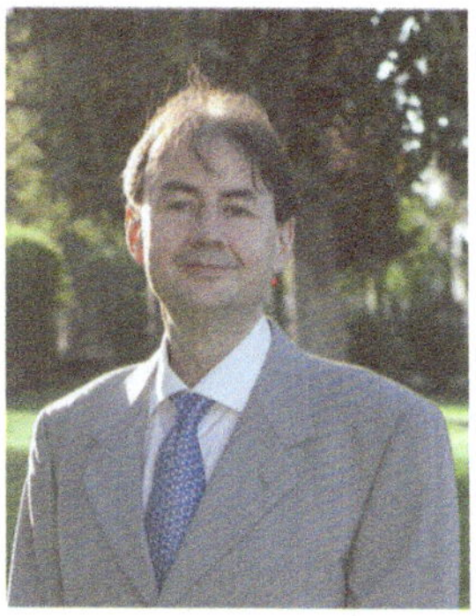

Joseph Cabeza-Lainez (https://orcid.org/0000-0003-1763-9452) holds a Ph.D. in Architecture from the University of Seville (Spain), where he is a tenured Professor of Architectural Composition and Environmentally Oriented Projects. He obtained a Chair-Professorship in 2010. He is a graduate in Oriental Studies and Japanese Language. He speaks more than ten languages, including Chinese and Arabic. He has travelled extensively in Asia, completing a significant research project on bioclimatic architecture. He served as a Research Fellow for the Japan Foundation at Kobe Design University (2004), to whom he is associated, as well as to other Japanese Universities. He has worked in European Research Projects and in cooperation projects with South America. He has collaborated in Architectural Projects with a combined budget of over 90 million Euros, featuring the energy-conscious Retrofit of the Archaeological Museum of Seville, the new Faculty of Engineering at the University of Seville (1998) and the sustainable Railway Hub for El Prat del Llobregat (Barcelona 2009). His activities range from Architecture, Art and Global Culture to Environmental Sciences. He has directed 12 Doctoral Theses. As academic of Fine Arts, he has taught at different universities for over 20 years. He has lectured in the five continents, over 25 countries and published over 100 academic papers and several articles and books including translations from Japanese and Chinese.

The Challenge of Next-Generation Machine Learning Algorithms for Architecture Design and Living Environment

Anca Vitcu

Abstract

In a world powered by an advanced technology relying on stunning machine learning algorithms, with a strong mathematical foundation, integrating its scientific innovations in architecture design process is undoubtedly a requirement for the development of the field, but without obedient surrender, without missing the outstanding historical lessons of built architecture, without losing the essence of architectural thought, and without compromising or neglecting the value of its human content. A blind approach can turn machine learning technology from a valuable tool into a trap by generating solutions to a problem that is essentially ill-defined.

Keywords
Genetic algorithms · Generative adversarial networks · Architecture design

Introduction

While science confronts works, approaches, and results achieved by researchers in different fields, mainly spread over long periods of time, in search of new ways to bridge the gaps, carry forward a particular theory, or formulate a completely new one, art is looking for uniqueness. Both have rhizomatic liaisons with nature and draw their springs in imagination, desire, curiosity, and hard work alike.

Architecture design is an amazing confluence between art and science, a remarkable adventure in real life, a collection of human stories and emotions, a mirror of society changes with solid connections in contextual environments, and a redoubtable mediator among light, ventilation, material performance, and structural behavior. In addition to these, architecture design is algorithmic; it starts with a good idea and gradually evolves from a thorough description of the problem that needs to be solved. If the problem is fuzzily stated, if wrong questions are asked, improper outputs are delivered: "When men misunderstand their own work, they cannot understand the work of others" (Bronowski, 1956).

When Good Science Joins Good Architecture

Algorithms designed to create emerging geometries or generative approaches for topology-driven form-finding processes have in their turn a success history in built architecture when they are founded on arguments, ideas, and logic. Examples of algorithms developed on simple original rules, inspired by Mandelbrot fractal geometry or Robert Ammann tiling discoveries, that generated sound patterns reflected in a light structural complexity can be found in the works of Cecil Balmond: "V & A Spiral Extension" (1996) worked with Daniel Libeskind and "Serpentine Pavilion" (2002) designed in collaboration with Toyo Ito, respectively.

Balmond uses in an inimitable style a broad source of inspiration from mathematics for conceiving novel algorithms to generate forms and search for opportunities of complexity. His advanced thinking system relies on a combination of nonparametric and parametric modeling to translate nature's patterns in a unique way. The emergent forms we discover in Balmond's projects are the outcomes of an enhanced knowledge based on numbers and simple vector rules, on a good understanding of non-orientable surfaces as Klein bottle, on stochastic processes such as random walk, in search to generate a form of order. The accomplishment of his work in collaboration with renowned artists and architects is not due only to the knowledge of complex exploration of mathematics but to the way it was integrated into a multidisciplinary architectural program. In 2000, with a group of architects and engineers, a quantum physicist, and a game theorist, Balmond created at ARUP the Advanced Geometry Unit, laying the foundations of a cross-disciplinary

A. Vitcu (✉)
"Ion Mincu" University of Architecture and Urban Planning, Bucharest, RO, Romania

M. R. Cervera Sarda et al. (eds.), *Architecture Inspired by Nature*, https://doi.org/10.1007/978-3-031-33144-2_9

research group on complex structural geometry, as support for new thinking spaces in architectural vision (Nordenson, 2008).

When Architecture Must Discern Before Engaging a Partnership with Science

The presence of a game theorist in Balmond's team was undoubtedly a wise decision. As well as nowadays, the absence of this knowledge in using generative adversarial networks (GANs) as support for form finding in architectural design is an error, if not an accident.

Scraping the architectural news and going through some academic architectural research articles related to the employment of machine learning algorithms, especially GANs, in architecture design, published since 2014 to now, I noticed a gray assumption, almost a statement, regarding the future of architecture design and architecture as a profession, which is mainly due to a sequence of misunderstandings. In the following lines, I'll briefly remind what generative adversarial networks are and the reason for which they are unable, at least the operational versions available, to create architectural designs but can be reliable partners if are appropriately used.

What Is a GAN?

Created by Ian Goodfellow, a generative adversarial network (GAN) is composed of two neural networks trained in a zero-sum noncooperative game. As the author described, it is a "framework for estimating generative models through adversarial nets" (Goodfellow, 2014). In a general overview, we have the following scenario: a generative model is trained to turn *random noise* into synthetic (false) data; a discriminator model is trained to make the difference between the false data from real examples in a problem known as supervised classification; the generator and the discriminator are parametrized using deep neural networks; and the neural networks learn using the adversarial loss that is a minimax optimization problem. The name of the framework is suggestive regarding the architecture and goals of networks. The three concepts in the title refer to the following (Langr & Bok, 2019):

1. *Generative* addresses the general purpose of the statistical model, which is *creating new data* capturing the training data distribution.
2. *Adversarial* indicates the relationship between the two models in the game, namely, a competitive dynamic between the two models generative and discriminator. In the game, the generator's aim is to create examples that are indistinguishable from the real data in the training set. When the process begins, the generator's outputs are noisy images and the discriminator predictions are random. The central part of GANs lies in the training of the two networks so that the generator becomes more skillful at confusing the discriminator, by maximizing the latter's probability to make a mistake; the discriminator must adapt to get better at classifying a forged distribution from a real one (Foster, 2019). The two networks are in a situation of a zero-sum game. The generator learns through the feedback received from the discriminator's classifications. The training procedure drives the generator to find new ways to outwit the discriminator, and the process continues. In an ideal scenario, the model should reach convergence, an optimal solution or Nash equilibrium.
3. *Networks* refers to the neural networks, a biological-inspired architecture, which can be simple feed-forward neural networks or more complex variants.

The idea underlying of a GAN is sketched in the diagram below (Fig. 1).

An important observation is that the images a GAN will learn to generate depend on the choice of the training set. For instance, if we want a GAN to synthesize images whose content looks like Jean Nouvel's buildings, we have to use a training dataset with images and sketches of Jean Nouvel's works. The two networks are continually trying to cheat each other: the better the generator gets at creating convincing images of Nouvel's complex and complicated buildings, the better the discriminator needs to be at distinguishing real images from false ones.

How We Decide When to Stop?

As we previously mentioned, in a GAN, the two networks, generator and discriminator, have competing objectives acting as *a zero-sum game*, a situation in which one player's gains equal the other player's losses (in other words, when one player improves by a certain quantity, the other player worsens by the same quantity). In generative adversarial networks when one network gets better, the other network gets worse. *Zero-sum games have a Nash equilibrium*, which means a point at which neither player can improve their situation or payoff by changing their actions. A GAN model converges when the discriminator and the generator reach Nash equilibrium, when the following conditions are met:

- The *generator* produces false examples that are indistinguishable from the real data in the training dataset.
- The *discriminator* can at best randomly guess whether a particular example is real or false (have the probability of 0.5 to guess whether an example is real). In our example, if the image belongs to a Nouvel building.

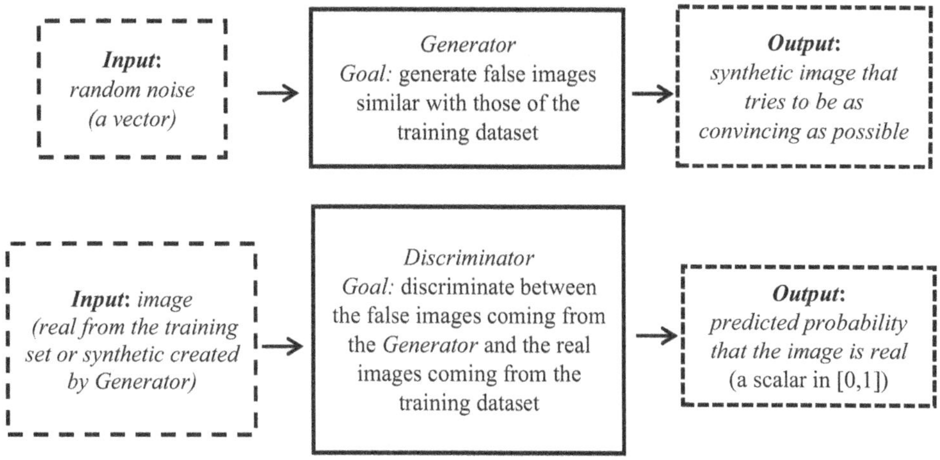

Fig. 1 The core components of a GAN. (Anca Vitcu)

In summary, at the equilibrium point, which is the optimal point in a *zero-sum game*, the generator will model the real data and the discriminator will output a probability of 0.5. Since both networks want to weaken the others, a Nash equilibrium is reached when one network will not change its action regardless of what the other may do.

In practice, there are some technical problems, including the fact that training GAN is not an easy task, that mode partial collapse is one of the hardest problems to solve (the generator repeatedly provides the same images, for which reason networks cannot progress) and the networks sometimes generate behaviors that do not find an explanation in the mathematical foundation. However, when the model architecture and hyperparameters are carefully selected, GAN learning performs well (Goodfellow, 2016).

The standard GAN has been enthusiastically received by the research community, having by now developed several important, competitive variants. GAN is an important innovation based on some simple brilliant ideas, which works very well as a building block for other approaches with sound outcomes. One of them is LAPGAN model that produces samples of images that are more realistic than those originating from a standard GAN. LAPGAN model combines a conditional GAN model with a Laplacian pyramid to generate images containing different levels of detail. A series of conditional GANs is trained to first generate a very low-resolution version of an image and then incrementally add details to the image managing to generate plausible-looking scenes. LAPGAN generators proved able to also deceive human evaluation of samples, not only discriminator networks. The samples from LAPGAN models trained on LSUN categories like "tower" and "church front" are eloquent (Denton, 2015).

Architecture design is about thinking. GANs do not think outside of the training dataset and, when are incorrectly used, manifest a sort of dysfunctional creativity. If we give a GAN several collections of images of Jean Nouvel's amazing projects, well trained maybe, it will be able to imitate the distribution of the training dataset. Conditionally, it will be able to generate examples with similar volumes, shapes, planes, and it will be able to produce new but untruthful content. And all this process has nothing to do with architectural thinking, with "creating a sense of place," about which Frank Gehry said that is one of the most important issues facing an architect (Isenberg, 2009). And worse, if you somehow succeed in solving the ethical and copyright problems, you still find yourself caught in a trap with the false feeling that you are a good architect. Nonetheless, you will always be the second, and as Arthur Rubinstein said in his outstanding interview to Robert MacNeil for WNET/13, on his 90th birthday celebration, in art, "being the second is already wrong."

In a highly technological society, architecture design already proved that has the ability to form novel concepts inspired from art, science, and nature confirming that "originality is the mark of independence of mind" (Bronowski, 1956).

GANs have a promising future in theoretical approaches as well as in practical applications where they have already proved useful in game design, image generation, text-to-image synthesis, face aging, image-to-image translation, video synthesis, high-resolution image generation, and completing missing parts of images.

Exciting topics for computer scientists and mathematicians have been created. A challenging edge for art-related fields, as architecture design, has been envisaged. Even so, these powerful models must first be understood in order to decide how and in what context they can be properly used in art and architecture design. They must not be copied and frenetically applied only because they are fashionable.

Christopher Alexander, one of the noteworthy interdisciplinary architects, who analyzed the structural correspondence between the pattern of a problem and the process of designing a physical form that answers that problem, using set theory to explain methods of algorithmic design

(Christopher, 1964a, b), sensed the danger of unusual magic of computational fundamental algorithms, techniques and tools, for architecture design and architecture as a profession. In his essay to the proceedings for the first conference organized in 1964 by the Boston Architectural Center titled "Architecture and the Computer," he wrote:

> The effort to state a problem in such a way that a computer can be used to solve it will distort your view of the problem. It will allow you to consider only those aspects of the problem which can be encoded and, in many cases, these are the most trivial and the least relevant aspects. (Christopher, 1964a, b, p. 52)

Explorations across disciplinary thinking and practice, as well as inquiries into common boundaries among art, architecture design, nature-inspired mathematical models, computer programming, and artificial intelligence, awakened the interest of architects before John Holland's landmark volume "Adaptation in Natural and Artificial Systems" have been published in 1975 and continues until today. Laying the foundation of a new field of study, based on a novel programming technique, the genetic algorithm, Holland's discourse about the adaptive processes of natural systems, and their use as ground field to design artificial systems, opened a door to several important applications in architectural design, some of them belonging to John Frazer, one of the key founders of generative architecture (Frazer et al., 2002). Frazer's book "An Evolutionary Architecture" stands as an essential methodological document for architects and scientists through the attention paid to the *reasoning* and *care* with which generative models inspired by nature are designed to approach architecture and the living environment as an evolutionary process. *Reasoning* and *care* are also prominent features of surfaces topology research in Frank Gehry (Goldberger, 2011), Zaha Hadid (Judidio, 2016), or Renzo Piano (Goldberger, 2009) works or in machine learning employment to predict how passively actuated façades respond to temperature changes as in Foster + Partners studio technological experiments. These concerns can be read in their own uniquely designed language. Just to remember some of their masterpieces: Walt Disney Concert Hall (Los Angeles, 2003), Stata Center (Cambridge, 2004), Neuer Zollhof (Dusseldorf, 1999) designed by Frank Gehry, Tower C by Zaha Hadid Architects (Shenzhen, in work), The Shard designed by Renzo Piano (London, 2012) or RMK headquarters with a hybrid façade inspired from nature designed by Foster & Partners, completed in 2021. They never repeat themselves.

Alexander knew just like Foster, Gehry, Hadid, Piano, salient leaders in the break from rectilinear geometry, as well as many outstanding generations of talented architects, that in art of design buildings and urban sustainable environments, form search is not a game and that finding the right formulation for a real-life architectural problem is a significant part of solving it and leans on a bold vision and plenty of time dedicated to unconditional investment in learning (Figs. 2, 3, 4, 5). To complete the architectural work, the support from algorithmic thinking and innovative technologies to broad the architectural vocabulary, so that it can then be transposed into a visual and structural optimized or adaptive built form, must be received with discernment, with responsibility.

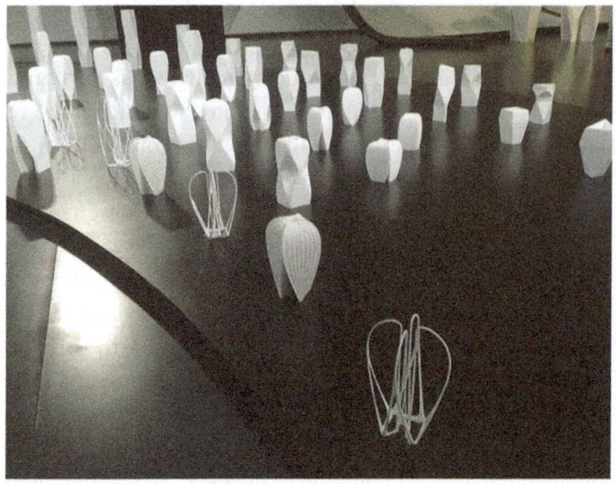

Fig. 2 The Exhibition Zaha Hadid: Form finding study, Institut du Monde Arabe, Paris, 2010. (Photo: Anca Vitcu)

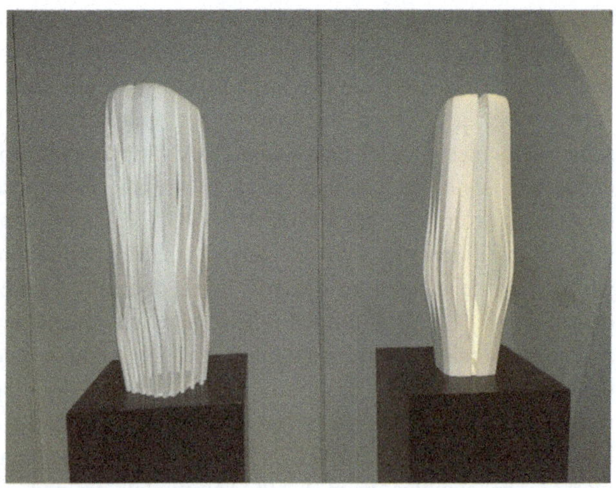

Fig. 3 The Exhibition Zaha Hadid: Form finding study, Institut du Monde Arabe, Paris, 2010. (Photo: Anca Vitcu)

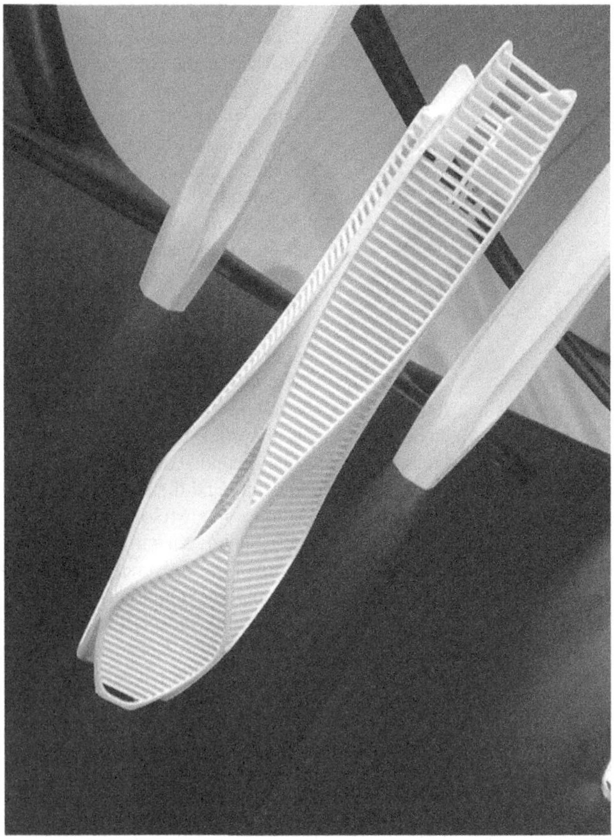

Fig. 4 The Exhibition Zaha Hadid: Form finding study, Institut du Monde Arabe, Paris, 2010. (Photo: Anca Vitcu)

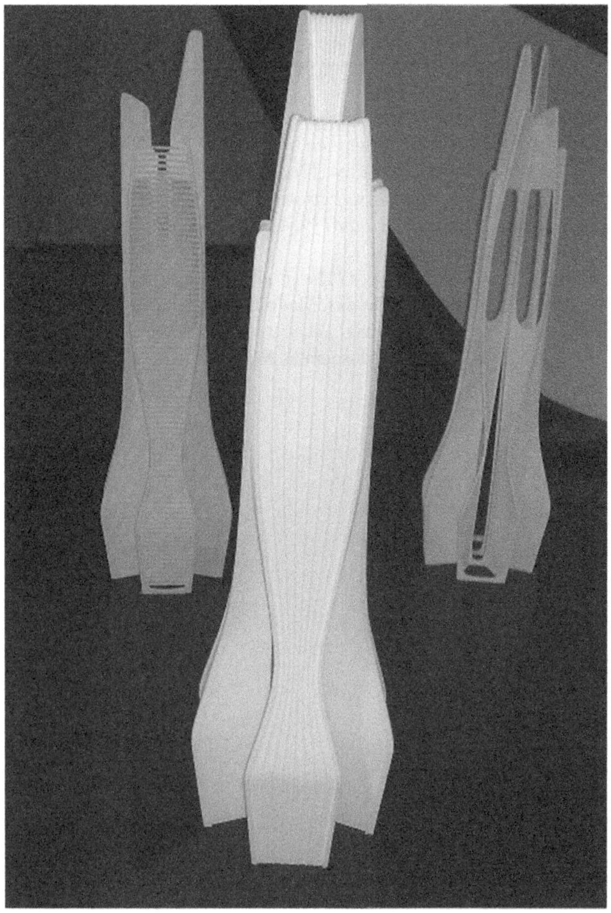

Fig. 5 The Exhibition Zaha Hadid: Form finding study, Institut du Monde Arabe, Paris, 2010. (Photo: Anca Vitcu)

Conclusions

We started this essay with a sample from Cecil Balmond's kneading for discovering new, simple, beautiful generative algorithms tailored for specific architectural programs, a model of how good science serves good architecture when mediated by a brilliant mind and a cohesive professional team. There are so many significant illustrations in which parametric thinking and genetic algorithms support creativity in architectural design or machine learning algorithms the effort of understanding and healing structures, all inspired by nature's divine patterns and processes. All lessons are to be learned. Aicher Otl noticed that "the scope for intelligent solutions is immeasurable, but only if you are in a position to ask questions" (Aicher, 2015). We have to give ourselves time to look at the architectural heritage with modesty, emotion, curiosity, and wisdom, to understand its messages and decently "capitalize" them in our futuristic vision of generating an "architecture without architects."

References

Aicher, O. (2015). *The world as design*. Wilhelm Ernst & Sohn.

Bronowski, J. (1956). *Science and human values*. Harper & Row Publisher.

Christopher, A. (1964a). *Notes on the synthesis of form*. Harvard University Press.

Christopher, A. (1964b). A much asked question about computers and design. In Boston Architectural Center (Ed.), *Architecture and the computer: Proceedings, first Boston architectural center conference* (pp. 52–54).

Denton, E., et al. (2015). *Deep generative image models using a Laplacian pyramid of adversarial networks*. In NIPS'15: Proceedings of the 28th international conference of neural information processing systems (Vol. 1, pp. 1486–1494). MIT Press.

Foster, D. (2019). *Generative deep learning*. O'Reilly Media.

Frazer, John H., Frazer, Julia M., Liu, X., Tang, M., & Janssen, P. (2002). *Generative and evolutionary techniques for building envelope design*. In GA2002: Proceedings of 5th International Generative Art Conference (pp. 3.1-3.16). Milan, Italy: Generative Design Lab.

Goldberger, P. (2011). *Why architecture matters*. Why X Matters Series. New Haven and London: Yale University Press.

Goldberger, P. (2009). *The modern wing: Renzo Piano and The Art Institute of Chicago*. Art Institute of Chicago and Yale University.

Goodfellow, I., et al. (2014). *Generative adversarial networks*. In NIPS'14: Proceedings of the 27th international conference on neural information processing systems (Vol. 2, pp. 2672–2680). MIT Press.

Goodfellow, I., et al. (2016). *Deep learning*. The MIT Press.

Holland, J.H. (1975). *Adaptation in natural and artificial systems*. University of Michigan Press (2nd Ed., MIT Press, 1992).

Isenberg, B. (2009). *Conversations with Frank Gehry*. Alfred A. Knopf.

Judidio, Ph. (2016). *Zaha Hadid (1950-2016): The Explosion Reforming Space*. Taschen.

Langr, J., & Bok, V. (2019). *GANs in action: Deep learning with generative adversarial networks*. Manning Publications.

Nordenson, G. (Ed.). (2008). *Seven structural engineers: The Felix Candela lectures*. The Museum of Modern Art.

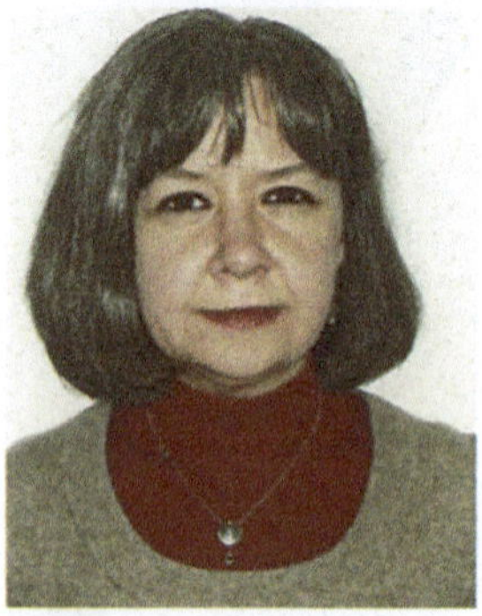

Anca Vitcu Associate Professor, Ph.D., graduated from the Faculty of Mathematics of "Alexandru Ioan Cuza" University (Iasi) in 1993 with a thesis on differential geometry. In 2000 she graduated the doctoral studies in Mathematics at "Gheorghe Mihoc" Center of Mathematical Statistics of the Romanian Academy (Bucharest) with a dissertation on geometrical approach of random algebraic equations. She is part of the academic staff of "Ion Mincu" University of Architecture and Urban Planning (Bucharest) since 2002. In 2013, she founded the courses, seminars and laboratories of Mathematical Models in Architecture and Urban Design, and Fractal Art in Architectural Design, grounded on a transdisciplinary research activity in visual arts, architecture and their connections with branches of science (partnership over time with IMUAU and the Faculty of Computer Science of UAIC Iasi, Romanian Academy, WBI (Belgian Research Fund), Faculty of Architecture of the University of Liège). In parallel, she is teaching topics on 3D computer animation: fundamental algorithms and techniques and Social Media Network Analysis at the Faculty of Computer Science ("Alexandru Ioan Cuza" University of Iasi). For more than twenty years, she has carried out activities in biostatistics research and applied mathematics with leading institutes in Romania and abroad, being since 2010 the coordinator of the Romanian National Group of the International Society for Clinical Biostatistics.

Part III

Studying Scales in Territory: Bionics and Landscape

Landscape as Process: A Transdisciplinary Integrated Approach

Tana Nicoleta Lascu

Abstract

Considering as methodology the transdisciplinary vision, corroborated with the interpretation of a series of concepts rising from a set of interconnected and symbiotic disciplines that transcend traditional academic boundaries such as the holistic systemic approach, the fractal theory and the semiotics, this chapter analyses some aspects within the landscape processes regarding the physical state but also cultural aspects, taking into account the dynamic of the geomorphological state and perception processes.

Given its multiple interconnections with different fields, landscape research offers challenges and opportunities of reflection. Moreover, recent years have brought significant developments in all areas of landscape research, transforming it into an integrated field of study of perceptions and materiality, culture and nature, long-term historical processualities and present states, as a core unity complex for the research area.

Within the framework opened by the European Landscape Convention (Florence, 2000), this chapter aims to take a wide, but not exhaustive, look to the relations and interconnectedness established within the complex natural and human processes that are generating, modelling and influencing the landscape, proposing, from a holistic systemic approach and an eco-semiotic perspective, some paths in order to contribute to the definition of a new strategy, to rebalance a new equilibrium within the landscape system, nowadays under the threat of loss of the usual references regarding the territorial reading, to orient the restarts towards a more sustainable future.

T. N. Lascu (✉)
"Ion Mincu" University of Architecture and Urban Planning, Bucharest, RO, Romania

Keywords

Process · Holistic systemic approach · Landscape perception as sign · Cultural identity

Introduction

More than ever, the crisis we are all experiencing nowadays shows the absolute necessity to understand and to design our landscapes in a more comprehensive perspective. Fredric Jameson considers what we might now call either the disappearance of landscape or, on the contrary, the generalization of landscape: "Postmodern hyperspace has come to exceed the capacities of the individual human body to self-situate, to organize through perception its immediate environment and to cognitively determine its position in a mappable external world" (Jameson, 1991). Nowadays, a total disjunction between the human beings and the built environment has been reached and the continuing sense of defeat is that of the incapacity of our mental, at least for the moment, to map "the immense global, and decentralized network of communication in which we find ourselves as individual and essentially isolated subjects: the map of the immense global network of the Internet".

However, it seems that the loss of the usual references in reading the territory and the daily confrontation with non-places and interstitial or neglected spaces that increasingly characterise our experience, specially escalated by the pandemic context, have heightened the desire to identify, safeguard and celebrate that which escapes or seems to oppose this tendency: the landscape as solution, understood in the sense of a pleasant or wild place, or as a picturesque site.

Revealing the rich character of the landscape, reflecting complex relationships established between the geomorphological and ecological processes, but also as a collective narrative, expression and, at the same time, foundation of communities' identity, this chapter aims to take a wide, but

M. R. Cervera Sarda et al. (eds.), *Architecture Inspired by Nature*, https://doi.org/10.1007/978-3-031-33144-2_10

not exhaustive, look to the processes modelling the landscape, considering the framework opened by the new landscape paradigm, introduced by the European Landscape Convention (Florence, 2000), requiring a perspective beyond the separated objects, to understand their dynamic relationships, considering different types of connection networks that ensure to re-establish the vital connections of the territory in space, between various dislocated events, in time, from the perspective of the historical processuality based on rational analysis of interconnectedness between human and environmental systems.

Methodology of Reference

In his *Transdisciplinary Manifest*, searching out reality as a unity belonging between the disciplines, in a systemic and quantic vision, the physicist Basarab Nicolescu considers transdisciplinary approach corresponding to in vivo knowledge, defined by the correspondence between the external world of the object and the internal world of the subject, as opposed to in vitro knowledge (Nicolescu, 2009).

The opportunity offered by the transdisciplinary approach within understanding the landscape as process is revealed by all characteristics appropriate to this type of knowledge. Rejecting the elitist distinctions, European Landscape Convention focuses on ordinary landscapes, emphasising the values of everyday landscapes (Priore, 2004). Correspondence between the subject and object and including the third logic plays a key role essential when it comes in discussion to issues of values legitimation. There are two main approaches for the landscape interpretation: the first one uses a dichotomy between subjective and objective aspects, excluding human aspects, and is based on the methods of natural sciences, while the second approach is related to social sciences, focusing on values related to liveability, beauty and sensitivity to places. Nowadays, this second approach is being reinforced by the quantum perspective offered by the natural sciences (Lascu, 2013).

The Process of Defining Landscape as an Integrative Concept

In 1993, the anthropologist Tim Ingold defined landscape as "...the totality of human experiences" (Ingold, 2011). However this syntagma seems for many of us quite unfamiliar, it contributed to open new perspectives in the process of defining the landscape.

Once idealised as an Arcadian paradise, the term landscape has been associated, for long time and, even nowadays, for many of us, with an image of nature, reminiscent to a pastoral, idyllic vista or scene – scenery.

Along time, new sciences and theories brought their contribution to enrich the significances of landscape concept: Relativeness Theory, Semiotics, Structuralism, Landscape Ecology, Phenomenology, Chorography etc. The contribution of the new concepts elaborated by different disciplines is relevant within the process of multiple meanings acquired by landscape as an integrative concept (Fig. 1).

Landscape as a System

Although the first attempt to incorporate the whole of human knowledge into one vast scheme belongs to Aristotle (Broadbent, 1987), it was not only until the first half of the twentieth century that the increasing fragmentation and duplication of scientific and technological research and decision-making led the Austrian biologist Karl Ludwig von Bertalanffy to advance his idea of a General Systems Theory (Laszlo & Krippner, 1998). Thus, the systems approach to the phenomena that focuses on the whole instead of the parts, as well as on the complex interrelationships among its essential components, evolved from General Systems Theory, which was developed in 1937 (Laszlo & Krippner, 1998; Pourdehnad et al., 2011).

Process represents the actions carried out at the level of inter-system connections producing that synergistic effect resulting from the functioning of the system, performing its functions and maintaining the living system, by exchanging flows between the social, biologic, ecologic and cultural structures, as matter/energy/information at the level of connections keeping the structure unchanged, in a constant dynamic equilibrium, sustaining the life of the system over a longer or shorter period of time (Fig. 2).

The system comprises the subsystems and the connections between them that work by interacting with each other and with the environment in which they are located, according to algorithms, laws and principles, to achieve a result. The processes give the system its character. The character of a system is different from the character of the component subsystems taken separately.

The structure of the system is given by the way its connections are made. Repetitive processes create repetitive structures. Those repetitive structures that are identical regardless of the scale at which they are studied have been called fractals (Mandelbrot, 1983). A portion of a landscape and all that it contains can be fractal. The study of fractal systems at a holistic interdisciplinary level has proved extremely useful, bringing new sciences such as bionics, biophilia, biomimicry etc., into focus.

The connection between ecology and landscape is also considered pivotal by ecologists, architects and landscape planners (Naveh, 1995; Nassauer, 1995). The concept of landscape is thus intertwined with the different modalities of

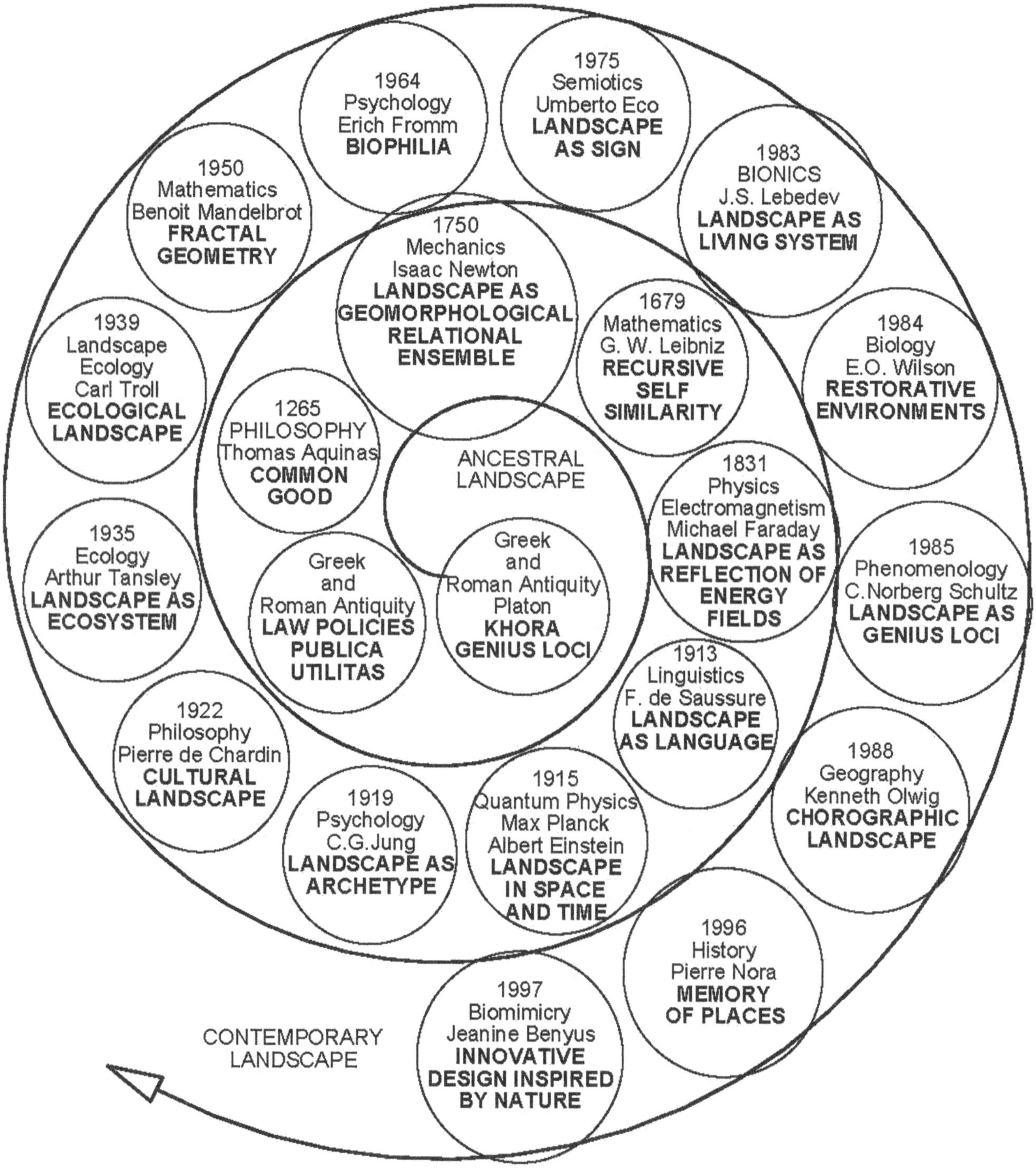

Fig. 1 Natural and social sciences compete enhancing the understanding of landscape meanings, for enriching landscape significance with new concepts opening up new perspectives. (Tana Lascu)

living and perceiving it. Nassauer (1995) proposes a stricter collaboration between aesthetics and ecology, namely, in architecture and design.

The new sciences are responding to the need of human society to restore a balance with its environment by studying and constructing new models of human habitat so that the impact on the landscape/environment/system that keeps us alive does not disrupt the proper functioning of the encompassing system, that unique synergistic effect resulting from multiple and complex processes.

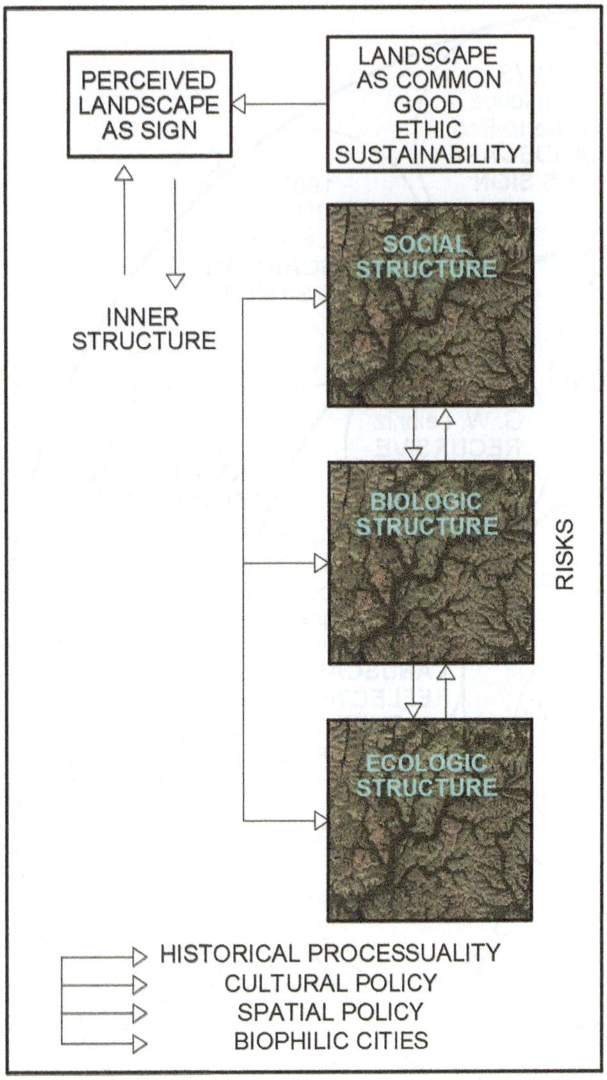

Fig. 2 Layers within landscape: the outer layer as perceived landscape and the inner layers including the social, biologic and ecologic structure. (Tana Lascu)

Cultural Identity and the Historical Signs in Landscape

Culture can be defined as the set of information transmitted along time, from one individual to another, through behavioural means. Besides the system of natural and ecological values, the cultural system of the architectural heritage plays a key role, having different functions, as network system for transport of people and products, but also for transferring ideas, for example, the slow networks, which allow community reflection, contemplation and belonging feeling.

The knowledge of our life environment that we get from our perceptual and cognitive capacities is part of the autopoiesis process not only of the living organisms, but of the whole landscape (Farina, 2006). Human settlements are the places where the information and code system of human culture gathers, adapts and regenerates itself.

As a follower of the New Cultural Geography, which promotes a chorographic approach, Simon Schama analyses from this perspective the sense of character of a place and the appreciation of certain rhythms, the recognition of balance and sensitivity to harmonious relationships, which are inscribed in the genes and adapted to the environmental conditions. Meaningful relationships are established between people and the space in which they live, as the built environment allows the communication of specific cultural values related to thinking, perception, emotional reactions and images. Social and cultural historians continue to explore the interconnections between national identities and human landscapes, focusing, for example, on landscape, myth and memory (Schama, 1995).

Bernard Lassus compares cultural evolution with Darwinian evolution (Lassus, 2010), where both forms of evolution are characterised by a hereditary transmission system. The transmission system (gene versus meme) in biological evolution genes can propagate or disappear (natural selection) paths with multiple and recursive transmission directions as a tree of life, which replaces the dead and broken branches on the surface of the earth, covering the surface with its splendid branches in continual division, based on "affinities between living beings of the same class" (Darwin, 2010).

Within the poetic archaeology defined by Bernard Lassus, in an irreversible process, mementoes follow a new dimension of construction, associating it with the spiritual act of settling, in order to perceive and understand signs and meanings in the landscape, giving them space.

By interacting, landscape and human systems shape each other. Unfortunately, human society has even come to destroy its environment/system by containing faulty subsystems. The system of human society with all its components, including man-made constructions, is still a natural system; it has not been artificially implemented but has evolved through energy exchange processes at matter and information levels.

Landscape Ecology and Landscape Perception as Sign

Nassauer aims to appropriate ecology and aesthetics and the philosophies of perception and aesthetical preference more generally. In her view, one of the key elements needed to analyse and manage landscapes is human perception, between cognition, values and landscape, existing a symbiotic relation (Nassauer, 1995).

Related to landscape, perception process implies associating things through all our senses beyond the according spatial-temporal approach, going far over the boundaries of three-dimensional space to a temporal multidimension, reminding of the intelligence and memory of the things (Farina, 2006). In this way, landscape perception becomes a complex multilayer process connecting us within a frequency continuum. Our interaction with the landscape essence, expressed, beside the common geomorphological formation, by colours, rhythms, materiality, sounds air, memories and imagination, emerges to be the deepest connection with the originating natura naturans. Considering these aspects, landscape valorisation can preserve, enhance and even create new attributes to ensure the balance of the ecosystemic processes providing wellness to human beings.

According to Eugenio Turri, "the landscape becomes such if we manage to give a sense, a cultural meaning to its components, transforming them into signs, through which we communicate with others" (Turri, 1998). These landscapes are simultaneously perceived by our behaviours, closely related with education patterns, own sensitivity, experiences and collective memory (Fig. 3). Any landscape retains the essential function of orientating its users, inhabitants or visitors, not only in spatial perspective but also in terms of meanings and symbols (Lynch, 1971). Education regarding a healthy ecology is in this context essential. People who sense a fundamental sameness between themselves and the natural world feel more empathetic and compassionate with the nature (Dutcher et al., 2007).

Landscape Indicators as Instruments for a Dynamic Landscape Evaluation

The main socio-economic model is based on the increasing use of energy resources, and its environmental consequences have increasingly complex and unpredictable characteristics. As a result, society is in a state of growing uncertainty. New research tools are needed to address this, which can no longer be based on direct measurements, due to the complexity of the phenomena. To meet this challenge, ecological or environmental indicators are used, in response to the high pressure exerted by certain limiting factors on the environment, while also assessing society's reactions, as emergent phenomena having properties of systems that arise from the collective interaction of their components, properties that cannot be predicted by looking only at the structure or behaviour of a small number of constituents subject to fundamental laws.

Despite this need, there is still no European consensus on what these indicators should be. There are still no lists of landscape indicators that are integrated, structured and applied systematically as the landscape quality, for example, is not an inherent datum in the characteristics of the landscape itself, which can be measured scientifically, and others. Instead, it depends on how the landscape is perceived by the population, with appreciation based on a wide variety of physical and material characteristics as well as cultural, emotional and spiritual connections. Indicators cannot therefore be objective; they are rather ambiguous, vague and on the borderline between subjective and objective.

Within the most valuated are the ecological balance of a certain landscape, including hydrogeological stability and security, land consumption and physical environment, human health, quality of life and communities' life and the quality of the landscape: its identity, readability, imageability, perspectives, landmarks, natural and cultural heritage and identity, access to services and sustainable transport, territorial economy efficiency of settlements systems and sustainable touristic enhancement. To identify the ecolands appropriate to an organism, and to plan actions for its conservation and security, Almo Farina presented the eco-field theory as the spatial configuration that, for a given individual, signifies the specific function of accessing a given resource, the set of eco-landscapes that an individual needs in order to secure his basic resources, becoming thus his own "perceived landscape" (Farina, 2006).

Some Aspects Regarding Landscape as Process: Kaštela, Croatia

A landscape of physical superlatives, resulting from karst-limestone rocks of mountain peak of Koziak, part of the Dinaric Alps, overlooking the ancient Roman settlement of Salona, the city of Split and the surrounding bays, offers a synthesis of seven coastal vernacular nuclei, whose history comes back to Illyrian and Roman times, each of them having its own stone castles built in a defensive role during the Ottoman invasion, including summer residences of the nobles of Trogir and Split and the small traditional stone dwellings of local free farmers, owning their land up in the mountains, later complemented by the architecture of the Modern Style and the architecture of the communist and post-communist periods, a scenography of surprising views to the sea, narrow streets between façades and limestone details, pottery roofs, native vegetation and an exceptional intangible heritage, Kaštela coastal landscape (Fig. 4a–k).

The promenade along the seashore is significantly marked by surprising signs such as sculptures revealing legendary characters, ancient occupations or means of transport, stone slabs from the Renaissance period written in Latin, but also from the communist period, while a route up the mountain reveals memorial sites such as ancient churches isolated on the ruins of Illyrian settlements, cemeteries and war memorials linked to recent history, as well as stone quarries and stunning panoramic views towards the coast and islands, rhythmed by the silhouettes of church campanile along the shore. The detailed restoration and renovation of small-scale

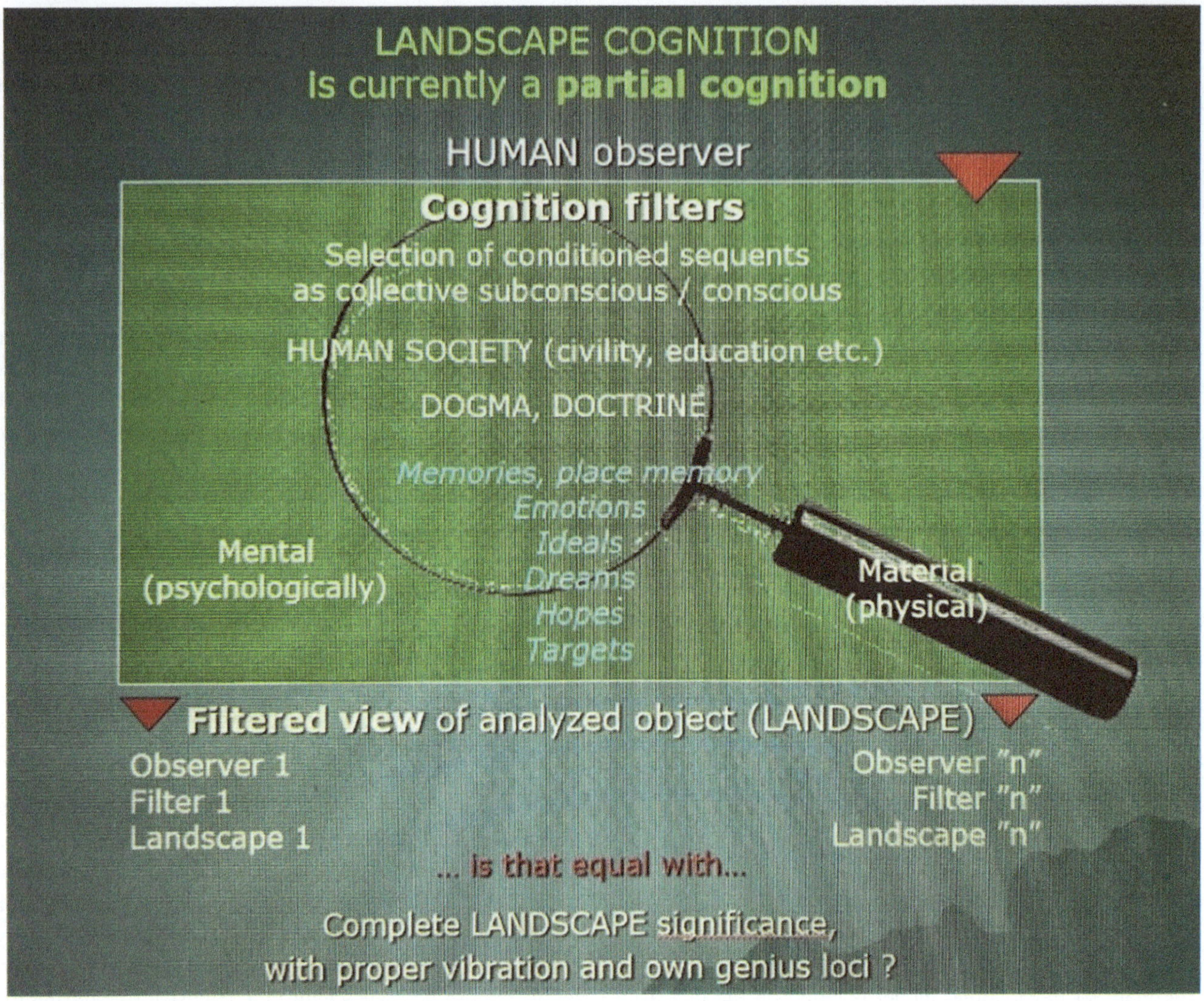

Fig. 3 Filters play a key role in landscape perception as sign. (Tana Lascu)

residential heritage goes in tandem with larger-scale landscaping development projects for the coastal belt, providing new amenities for tourists and locals, extending the existing waterfront. Nowadays, neglected areas of abandoned communist industrial heritage and larger scale architectural heritage, such as hotels and, especially, the many new serial residential interventions that deny and alter the vernacular and Mediterranean styles, represent a discontinuity and a serious threat to the well-established landscape character acquired over many centuries within a complex process of cultural syncretism.

Conclusions

Conclusions can be summarised as follows:

- Integration of perception and aesthetics within the framework of environmental and sociocultural sustainability, interpretative reading of territorial systems, recovery of their historical role and reconsideration of the aesthetic dimension of the landscape are key issues in the project strategy.
- In a transdisciplinary perspective, the focus switches from the individual objects to the history of the processes that generated those objects, paying particular attention to systems networks and relationships that define ordinary or residual landscapes, in which the character shifts, from a formal or historical inventory to a dynamic evaluation generating transformations.
- Critical attention to territorial plots characterised by the multiplicity and stratification of signs rather than by presumed aesthetic or formal homogeneity.
- Centrality of the category of context, understood not as physical contiguity but as the web of significant relationships between assets, even if isolated or fragmented, built both on material connections and, mostly, on immaterial relations.

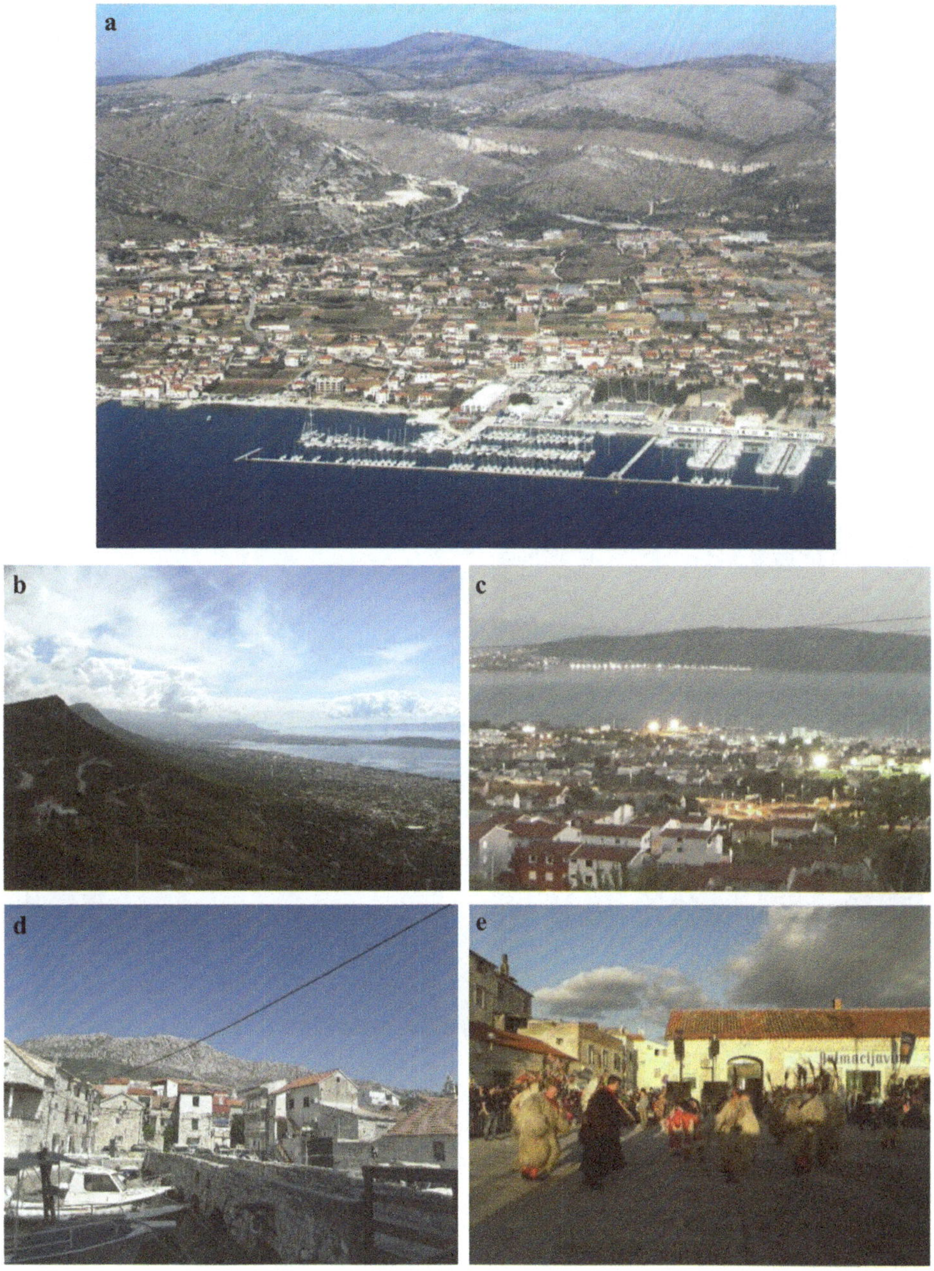

Fig. 4 **Signs and landmarks within landscape processuality.** (**a**) When the combination of natural, irregular fractal geometry of the coastline meets the linear artificial configuration, the landscape acquires complexity and attractiveness: Kaštela, Croatia. (**b**) Perception of the overlapped visual layers of mountain slopes, shoreline, sea and islands, within a consolidated ecosystem. (**c**) Fractal geometry of the rooftops as repetitive structure imposing rhythms and organic coherence within the perception process. (**d**) Vernacular stone architecture, following fractal geometry of volumes, organically integrated within the natural configuration. (**e**) Ereditary transmission systems: bear costumes recalling Vlachs continuity, animating a ritual inserted through Italian tradition: the Carnival. (**f, g**) Landmark as symbol, remembering old transport means in the past, nowadays enhancing sense of identity of place and polarizing community daily life. (**h**) A scenography bathed by filtred sunlight, where transparency and reflections play a key role within perception process. (**i**) Reflections and sequential plans: Kaštilac and Marina in Kaštel Gomilica, in the background industrial architecture in Split and the slopes of Mossor Mount. (**j**) Faulty subsystems inducing a neglected landscape: Palace Hotel, among the oldest hotels in Central Dalmatia, an Art Deco heritage, and its huge park, a natural monument. (**k**) Processes affecting the landscape character: massive alteration and denial of the Mediterranean character of buildings through serial interventions along the coastal area. (Photos by Tana Lascu)

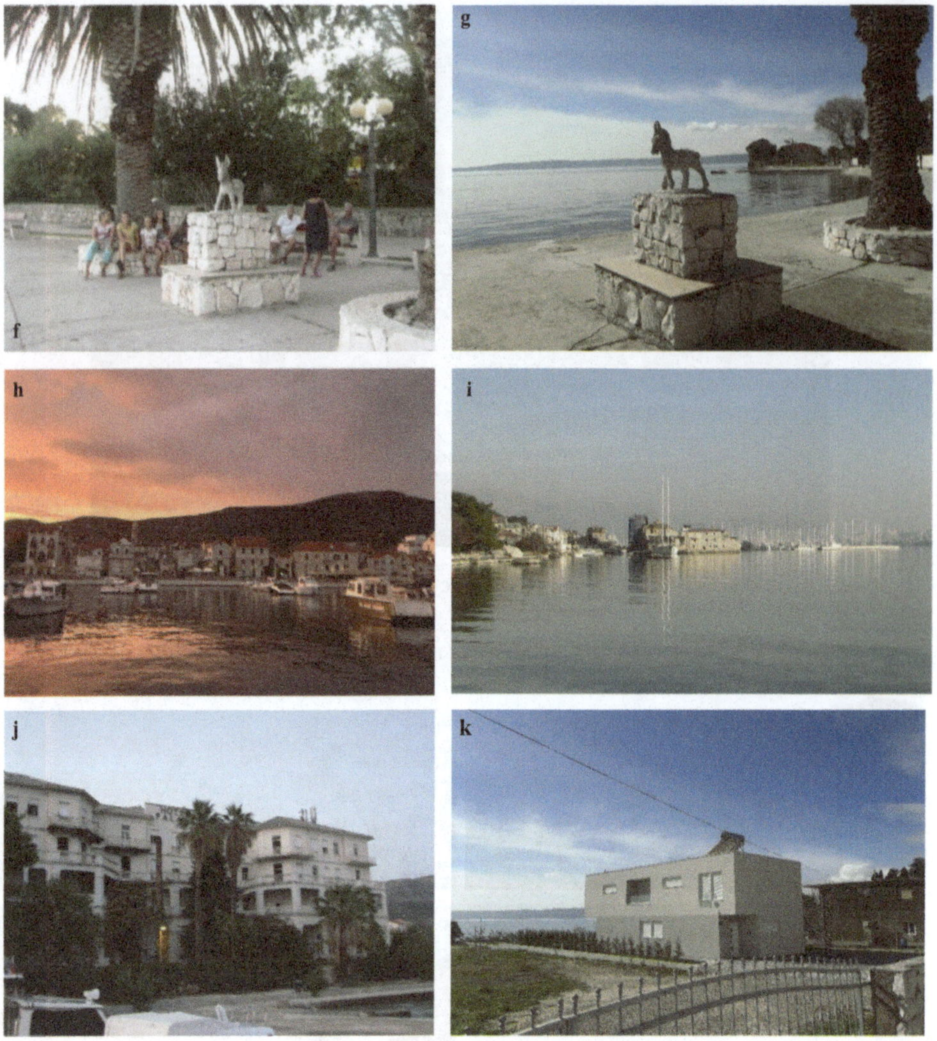

Fig. 10.4 (continued)

From this perspective of historical processuality and intrinsic connection with the place, the relationship between recent architecture, settlement form, and landscape is becoming increasingly relevant in terms of enhancing the attractiveness of community life, the creation of vital spaces and community-friendly spaces, a visual continuity, in line with the (re)-discovery and understanding of the layers of the settlement's fabric read as a palimpsest. From this perspective, the landscape becomes not just an image or a form but a real repository of human experiences, offering different levels of understanding and deciphering it, finding reconnecting and revitalising networks to structure new synergies.

References

Broadbent, D. E. (1987). *Perception and communication*. Oxford University Press.

Darwin, C. (2010). *On the movements and habits of climbing plants*. Cambridge University Press.

Dutcher, D., Finley, J., Luloff, A. E., & Johnson, J. (2007). Connectivity with nature as a measure of environmental values. *Environment and Behavior, 39*, 474–493.

Farina, A. (2006). *Il paesaggio cognitivo* [The cognitive landscape]. Franco Angeli.

Ingold, T. (2011). *Redrawing anthropology: Materials, movements, lines*. Ashgate.

Jameson, F. (1991). *The cultural logic of late capitalism*. Duke University Press.

Lascu, T. N. (2013). Transdisciplinarity in the relationship between the spatial policy and the historical/cultural policy. In *Landscape and imagination - Towards a new baseline for education in a changeing world*. Bandecchi & Vivaldi.
Lassus, B. (2010). *Paesaggio, un'esperienza multiculturale* [Landscape, a multicultural experience]. Editions Kappa.
Laszlo, A., & Krippner, S. (1998). Systems theories: Their origins, foundations, and development. In J. S. Jordan (Ed.), *Systems theories and a priori aspects of perception* (pp. 47–74). Elsevier Science.
Lynch, K. (1971). *Site planning* (2nd ed.). MIT Press.
Mandelbrot, B. (1983). *The fractal geometry of nature*. W.H. Freeman.
Nassauer, J. I. (1995). Culture and changing landscape structure. *Landscape Ecology, 10*(4).
Naveh, Z. (1995). Interactions of landscapes and cultures. *Landscape and Urban Planning, 33*.
Nicolescu B. (2009). *Ce este Realitatea?* [What does reality mean?]. Junimea Publishing House.
Pourdehnad, J., Wexler, E. R., & Wilson, D. V. (2011). *Systems and design thinking: A conceptual framework for their integration*.
Priore, R. (2004). *Verso l'applicazione della Convenzione Europea del Paesaggio in Italia* [Towards the application of the European Landscape Convention in Italy]. Fondazione Benetton Studi Ricerche.
Schama, S. (1995). *Landscape and memory*. Blackwell Publishing.
Turri, E. (1998). *Il paesaggio come teatro, dal territorio vissuto al territorio rappresentato* [The landscape as theatre, from lived territory to represented territory]. Marsilio.

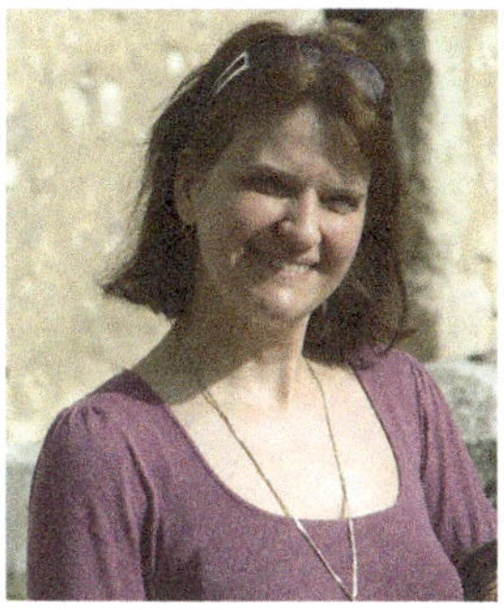

Tana Nicoleta Lascu Ph.D. Architect and Urban Planner, Lecturer at the Basics of Architectural Design Department, Faculty of Architecture of the "Ion Mincu" University of Architecture and Urban Planning from Bucharest. Graduated at IMUAU in 1992 and having a postgraduate specialization in restoration and conservation of monuments and historic sites,, she has been involved in several restoration projects in Romania, France, and Italy, thereafter developing more than 100 projects during the 7-year activity at Cornelis de Jong Architektenburo bna, Middenbeemster, Noord-Holland, the most significant in Urk, De Rijp, Marken and Beemster, elaborating the project for the listing of Beemster polder on the UNESCO World Heritage List. She was a Visiting Professor at the University of Liège (2014), the University of Architecture in Venice – IUAV, and the University of Alcalá (2021). Since 2012, she has represented IMUAU within The Network of Universities for Studies and Education according to the European Landscape Convention – UNISCAPE. Her doctoral thesis "Landscape as an Integrated Concept in the Sustainable Development", finalized in 2011, and her studies published in four books and over 20 international conference Proceedings, as well as the three research grants and more than 50 international curated exhibitions, seminars and workshops, emerged as a result of her constant interest in connecting architecture with landscape, within a transdisciplinary integrated approach.

Disruptive Biotope: The Interstitial Space between Nature and Urbanism

Henry Daniel Lazarte Reátegui

Abstract

Before the arrival of COVID-19 and the mandatory social isolation, the citizens of Lima remembered the existence of that place of collective enjoyment; however, others are witnesses of a positive change in water resources that allows the development of mutualistic interactions of flora and fauna present in the city of Lima. We are referring to the Rímac River along its banks, one of the three most important rivers that the city has, with a length of 140 km from the heights of Ticlio (4818 m above sea level) to the sea outlet, considered in this study as that interstice that generated recreational spaces on its banks and was a source of inspiration for locals and foreigners, capturing those scenarios in works of art, poetry, and lyrics. It continues to be the main source of surface water and the most important aquifer of the Peruvian capital. The Rímac River was considered as an object of analysis in the city of Lima and the principle of explanatory research, based on a case study, being a useful method for the understanding of what is explained. The result of the study was that the reduction of human activity in the pandemic has had several impacts on nature showing us a different ecosystem (biotope + biocenosis) in the Rímac River and a health education along with a civic conscience, through balance and interaction: individual-nature-society; a continuous and erroneous administration of the Rímac River that government political scenarios failed to consolidate through both environmental and urban-architectural recovery projects. Environmental liabilities will continue to increase if we continue to make this natural interstice a place for solid waste, industrial and domestic wastewater drainage, tailings derived from nearby mines and factories, leading to increased pollution and making the interstitial space, for many years, the most polluted place in Peru.

Keywords

Biotope · Interstice · Environmental liabilities · Rímac River

Introduction

For a better understanding of the case study, we consider it is necessary to highlight the meaning of the specified keywords.

The *Disruptive Biotope* is a complex term that has been little addressed in the urban literature. However, we will use it as a framework for this research; *biotope* is defined as that vital space or territory that possesses environmental qualities relevant for a certain community of living beings to develop in it (*Panhispanico*, 2022) and *disruptive* from English *disruption*, which is *breakage or sudden interruption*. Both terms together we define them as that space that was conducive to the interaction, harmony, and coexistence between species through uniform environmental conditions and how it is violently affected and left to oblivion by the accelerated and irregular growth of the city, environmental pollution, and a constant increase in environmental liabilities and passives.

The term *environmental liability* can be defined as that environmental situation that, generated by man in the past and with progressive deterioration over time, currently represents a risk to the environment and the quality of life of people (Aro & González, 2020). In other words, it is a land or place environmentally affected by a historical activity over which there is no control, due to negligence, accidents, or ignorance, being, in addition, potentially places with the presence of contamination.

Regarding the approaches to the definition of *interstice*, we will rely on definitions made by the *Real Academia Española (RAE)* and in some theories or researches that address this definition. In that sense, the term interstitial

H. D. L. Reátegui (✉)
César Vallejo University, Lima, PE, Peru
e-mail: hlazarte@ucv.edu.pe

M. R. Cervera Sarda et al. (eds.), *Architecture Inspired by Nature*, https://doi.org/10.1007/978-3-031-33144-2_11

according to the RAE (from Latin *interstitĭum*) is defined as the space or cleft, which mediates between two parts of the same body, as well as the synonymy related to terms such as free, empty, hollow, gap, interval, opening, hollow, slit, crack, hole, fault, crack, gap, cleft, fissure, slit, and others. The term is applied in various branches of the natural sciences as physics, biology, and geology, among others, and in urban contexts with the Jean Remy's and Ernest Watson Burgess's meanings, with the postulates of the Chicago School where they call it a *transition zo*ne and that proposed by the Belgian School of Urban Sociology that called it an *interstitial zone* (Forero Rodriguez, 2015). This concept arises from Jean Remy's theory called *concentric zones*, which refers to neutral times and spaces, located in the urban center, and dissociated from specific activities, which are little or not at all defined but are arranged for the development of a social and urban life (Delgado Ruiz, 1999).

In 2008, architect Peimbert dedicated his master's thesis to this phenomenon, using the term of *interstices*, leaving aside the idea of *voids*, since emptiness is related to nothingness, supporting the idea of studying them together with their counterpart: full spaces, which he calls interstitial landscapes. He shows his preference for the relationship of interstices to space for art, to which he superimposes his emphasis on spaces: ruins of abandoned buildings, infrastructural and industrial restrictions, etc., which are the preferred spaces for urban art.

> Peimbert acknowledges the existence of what he calls interstices, both in the periphery and in urban areas; he also admits the lack of theoretical or disciplinary lines on the subject of interstices, which makes their treatment difficult (Aguila Flores, 2014, p. 29).

Likewise, the interstice for our case study is the Rímac River, a river that has always been in the territory and remained within the urban structure of the city of Lima due to the accelerated and informal growth of the city, turning its banks year after year into urban remnants, due to the slopes of construction deforestation, solid waste, and land traffic on both sides of its banks.

The Rímac River – The Talking River

The noise produced by the great amount of stones that the waters drag made the natives name it *Rímac* or *Rimaq* (eloquent, speaker, talkative), and with the passing of time, its settlers determined it as Rímac.

Likewise, there are many versions in which they associated the name of the Capital Lima with the denomination of the river. According to Garcilaso de la Vega, the term Lima is the alteration of the voice Rímac, which means *the one that speaks* in Spanish, referring to an oracle worshipped by the natives. Due to its territorial length, the river and the whole valley were named this way (Municipality of Lima, 2020).

The Rímac River begins its journey on the western slopes of the Andes at an altitude of approximately 5508 m above sea level in the Nevado Paca, running through the provinces of Lima and Huarochirí, both located in the department of Lima. Among the most important tributaries of the Rímac are the Santa Eulalia River, the San Mateo or Alto Rímac River, the Blanco River, and the Surco River (Alameda, 2013).

The Rímac River is a water resource that supplies around 9 million inhabitants, who represent 30% of the total population (according to INEI 2019, 32,131,400 inhabitants), as well as generating 550 MW of clean energy resources, allowing 3000 ha of agricultural land and 1500 ha of green areas in the territory to be irrigated (Iagua, 2014).

The Rímac River ended up being located outside the city and later divided it, becoming a natural physical territorial border, due to urban growth and planning, losing a heritage legacy of a collective cultural memory, which, until the closing date of the study, the State Institutions have not been able to recover.

The Bank as a Public Space

The history of Lima is always linked to the coexistence of society and nature, since, in the sixteenth century, the people of Lima used this environment and natural resource (the banks of the Rímac River) as a recreational area, spending an afternoon of entertainment among shrimp and turkey buzzards, participating in popular religious ceremonies and thanksgiving to nature (Fig. 1).

The bank of the Rímac River was the place of meeting and social integration between the inhabitants of the side of the city that carried out activities of celebration daily, especially in the afternoons, next to the inhabitants of the other side of the river (neighborhood of Indian shrimps of San Pedro) that carried out domestic activities (for fishing, this was a place with a strong traditional identity because it was home to people who used to take shrimp from the river, and its streets were named after the activity they carried out, such as the shrimpers' street; Fig. 2). The biodiversity of the flora that existed (among them the plant "pájaro bobo" and the hummingbird representing the fauna that has had this flow) generated a great impression to every visitor who came to the city of Lima (Fig. 3), since this city was surrounded by a natural green belt that later would be replaced by the construction of a wall. The construction of the wall of 5 meters high, 34 bastions, and 10 entrance and exit doors began on June 30, 1684, with viceroy Melchor de Navarra y Rocafull, Duke of Palata, and finished in 1687 (Visit Lima, 2021).

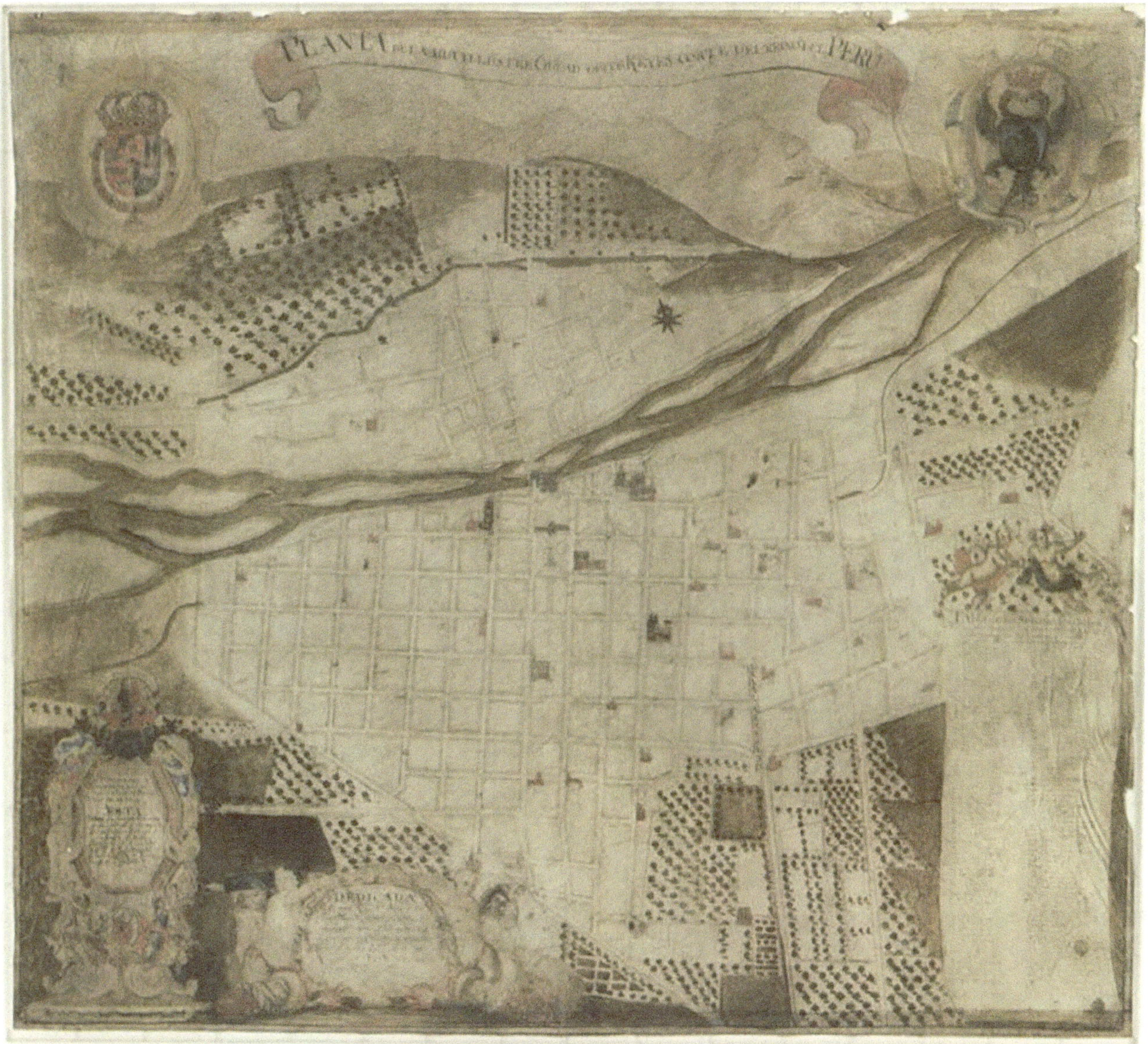

Fig. 1 Plan of the most illustrious city of the king's court of the kingdom of Peru, city of Lima. 1674, Technique Manuscript map, 69 × 80 cm, author Bernardo Clemente Príncipe, Provenance Library of Congress, Washington. ("Rímac, Historia del Río Hablador"/Autoridad Nacional del Agua, Lima: ANA, 2016, Temporary Exhibition (pp. 90–91))

The mere fact of crossing this great river was not an easy task, so a bridge made of sticks was built, located in front of the Church of Nuestra Señora de Montserrat, over which only one person could pass with a respective payment for the right to transit (toll). Later, during the government of the viceroy Marqués de Cañete, it was replaced by a bridge made of stone and brick, solid and robust to resist the force of the water flow due to the heavy rains in the highlands.

Likewise, there are also artistic expressions from the sixteenth and eighteenth centuries showing Lima sharing this natural interstice space.

Study Cases Background

In order to support our study, we will describe some international background information on the importance of the river for the city. This background information has been recovered by way of a synthesis of the master's thesis carried out in 2015 by architect Luis Alejandro Forero Rodríguez (pp. 33–36).

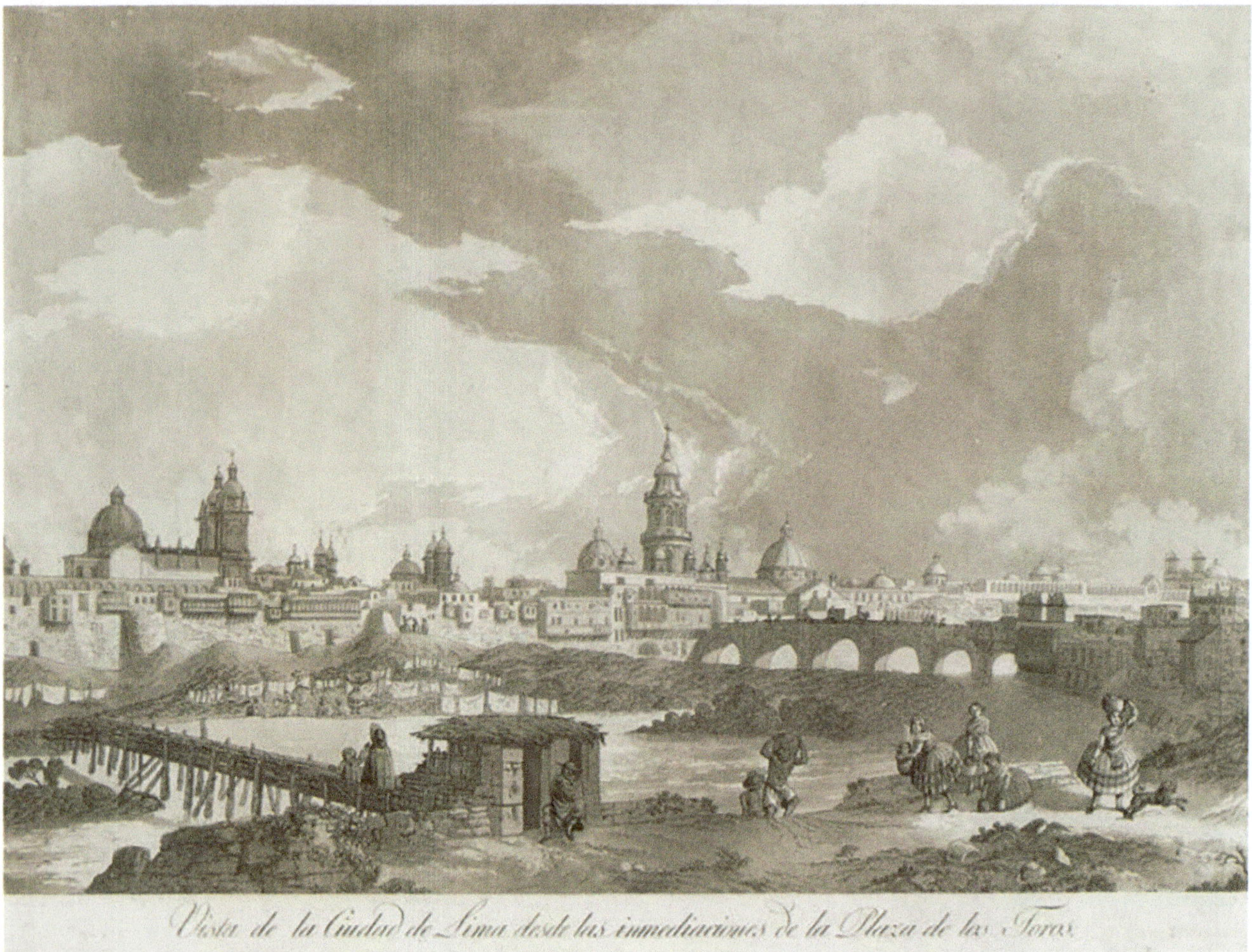

Fig. 2 View of the city of Lima from the vicinity of the bullring. 1791, Technique Pen and ink drawing in sepia wash, 57 × 30 cm. Author Fernando Brambila, Provenance Spanish Naval Museum – Madrid. (Book "Rímac, Historia del Río Hablador"/National Water Authority, Lima: ANA, 2016, Temporary Exhibition (pp. 102 and 103))

Aburrá-Medellín River, Antioquia, Colombia

Dividing the city into western and eastern parts, the Medellín River is 100 km long from its source at the San Miguel peak at an altitude of 2700 m, being fed by the waters of 200 tributaries receiving contributions from 352 other streams.

The contamination of this important body of water for Medellín has been a constant feature of the city's urban development project, as the discharge of sewage into the river has become a mechanism for the expulsion of waste produced by the city.

Aburrá river is a central component in the urban configuration of Medellín, since the development of mobility infrastructure such as motorways and the metro has taken place around the river.

Last March, the administration of the Mayor of Medellín, Aníbal Gaviria, in compliance with the Land Use Plan, presented "Medellín River Park: Equity, Inclusion and Integration" Project, aiming to recover 23 km of the river and including it in the urban public space.

> The park project is oriented towards the incorporation of the river in the solution of mobility problems through the construction of cycle paths and pedestrian corridors. In this way, the case of the Medellín river provides some guidelines for understanding the inclusion of urban rivers in city planning. Some guidelines from Medellín can be taken for use in the case of Bogotá and the Fucha River (Forero Rodriguez, 2015).

The Bodies of Water, Valley of Mexico, Mexico

The insertion of natural elements within cities has, in each particular case, a historical and social background that determines the formation of specific morphological elements. The case of Mexico City is presented as a specific study that

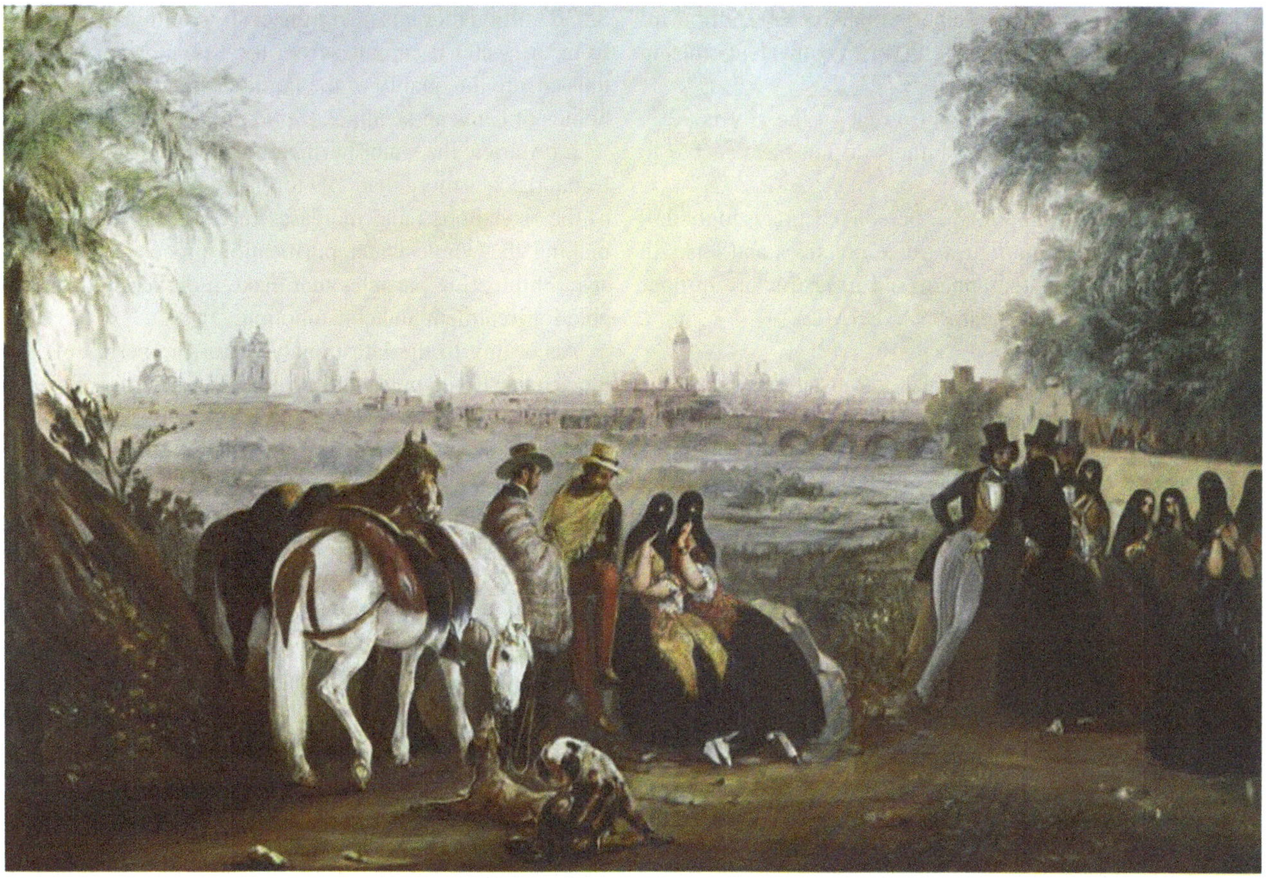

Fig. 3 New Paseo de la Alameda, city of Lima, 1843, Technique Oil on canvas, 56.0 × 89.5 cm, author Juan Mauricio Rugendas, Provenance In: nineteenth century Romantic Peru. Baring Collection, Brother, Lodres (1975). (Book "Rímac, Historia del Río Hablador"/National Water Authority, Lima: ANA, 2016, Temporary Exhibition (pp. 118 and 119))

shows the reasons why this city has also presented a special context that has led to a proposal for the La Piedad River.

In this regard, Fermín Valenzuela (2012) wrote about the transformation of spaces from the interventions to the bodies of water in what is currently known as the Metropolitan Zone of the Valley of Mexico (ZMVM) in which approximately 20 million inhabitants live.

In Mexico City, water bodies have also been inserted through a process of growth, while in the decade of the thirties in the twentieth century, after the Mexican Revolution, the beginning of a series of public works was marked by the piping of the different rivers, since the growing urban expansion toward the south of the city and the establishment of industries to the north began to completely transform the rivers into vertiginous centers of garbage.

In the 1940s, the problem of flooding reappeared and a second phase of river channelization took place. Thus, the La Piedad River was channeled between 1945 and 1960. Once these infrastructure works were completed, the first major avenues and highways were built to connect the city's neighborhoods.

Currently, according to a Conagua Report on the Hydrological System of the Valley of Mexico, "the formerly La Piedad River contributes in the Deep Drainage System that excludes the water captured from the Valley of Mexico and dumps it in the El Salto River" (Forero Rodriguez, 2015).

The Interstitial Space Before and During COVID-19

Having in view the conditions before 2020, we can highlight the following aspects:

- Eighty percent of the drinking water in Metropolitan Lima and Callao comes from the Rímac River, while Lima is the second largest city in the world located in a desert, after Cairo. However, Cairo has the mighty Nile River, which has an average flow of 2830 m^3/s; in contrast, the Rímac has an average flow of only 26.6 m^3/s.

- Lima's water reserves of approximately 330 million m^3 constitute only 30% of the reserves of equivalent cities in population, such as Bogotá.
- The current drinking water coverage in the city is 92%, that is, about 800,000 inhabitants do not have access to drinking water.
- The average consumption per person in Lima is high: 250 liters/day, due to inefficient usage practices and leaks in the distribution system, due to old infrastructure in poor condition ((INEI, Sedapal, ANA, USAID).

The Rímac river and its banks, as a natural interstitial space, are the most deteriorated basin in environmental terms, having as impact scenarios the accumulation of domestic waste, sewage, urban waste, and environmental liabilities (mining), a total of 27 mining operations being located in the Rímac basin, of which seven are still operating and the other 20 are closed or abandoned, the districts of Chicla, San Mateo, Matucana, Surco, Huanza, and Carampoma in Huarochirí having the highest concentration of operations (Alameda, 2013).

The images show a large amount of waste; however, what can be seen in the first figure is not the most worrying thing. What is really dangerous is the mining activity (illegal and informal) at a distance of 160 km from the Rímac River, which is extremely contaminating with hazardous waste that threatens health (arsenic, cyanide, mercury etc.), as well as other worrying factors such as tailings seepage (Fig. 4).

Attempts by the government and the administration of the Rímac River (almost 50 years ago) have failed to recover the river and its banks in its environmental and urban-architectural conception. In addition, the high cost of purifying the contaminated water has to be pointed out, as in 2010, making potable water drinkable for the inhabitants of the capital demanded an investment of 12.7 million soles from Sedapal (La República, 2019), which later "has increased by S/. 30 million more its operating costs to eliminate pollutants from the water it captures from the Rímac River basin and thus ensure the quality of the liquid element that reaches the homes of Lima and Callao" (*La República*, 2019).

La Atarjea, the water purification container for the city of Lima, reaches more than 129 tons of solid waste per day due to the river floods and increases the turbidity of the water, making the whole water purification process more expensive, eliminating particles that make it darker and less suitable for treatment and consumption.

According to the latest analyses for the period of 2019, the Rímac River presented a large number of deficiencies that did not meet the Environmental Quality Standards, due to various metalloids, metals, and thermotolerant coliforms detected (ANA, 2020). This natural interstice has long been at the mercy of contamination by solid waste, domestic and industrial wastewater, and environmental liabilities.

As shown in Fig. 5, according to Francisco Dumler, chairman of the Board of Sedapal, this phenomenon was due to the reduction of solid waste from neighboring factories and citizens by approximately 90%. In other words, 1 TN of debris and garbage was not reached, due to the inoperability of the industries dedicated to extracting material from the quarries (Gestión Newspaper, 2020).

Conclusions

In view of the above, we believe it is appropriate to conclude by way of an interpretative graph (Fig. 6), where we start from a biocenosis that, as a whole represented by living beings, interacted freely and safely in a natural interstitial space. However, it's presence was reduced, and in many cases, there is a disappearance of some living elements in a biotope with vulnerable environmental conditions due to environmental pollution and the development of uncon-

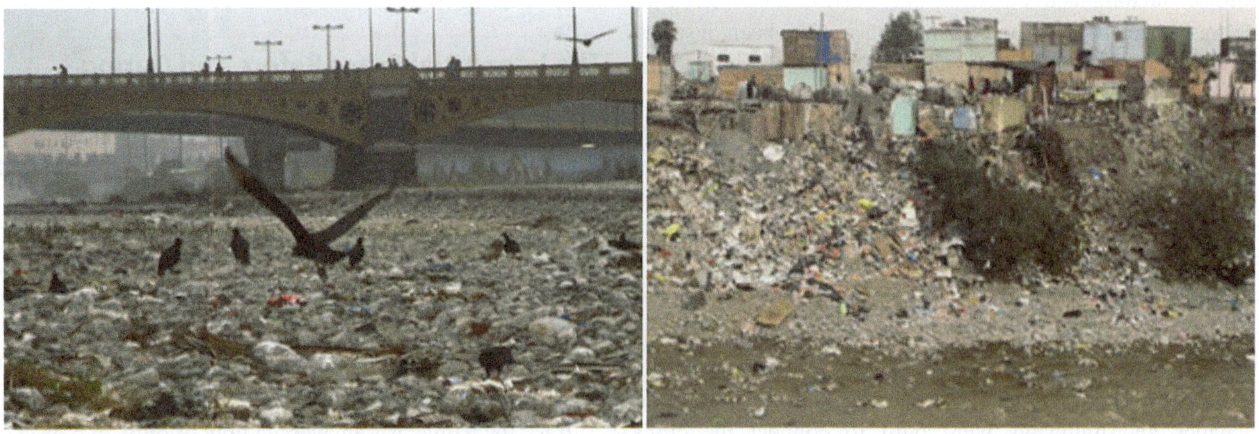

Fig. 4 The Rímac River alarmingly polluted, consequently, the water we drink in Lima. (http://www.aguapureza.pe/rio-rimac-agua-contaminada-lima (Alameda, 2013))

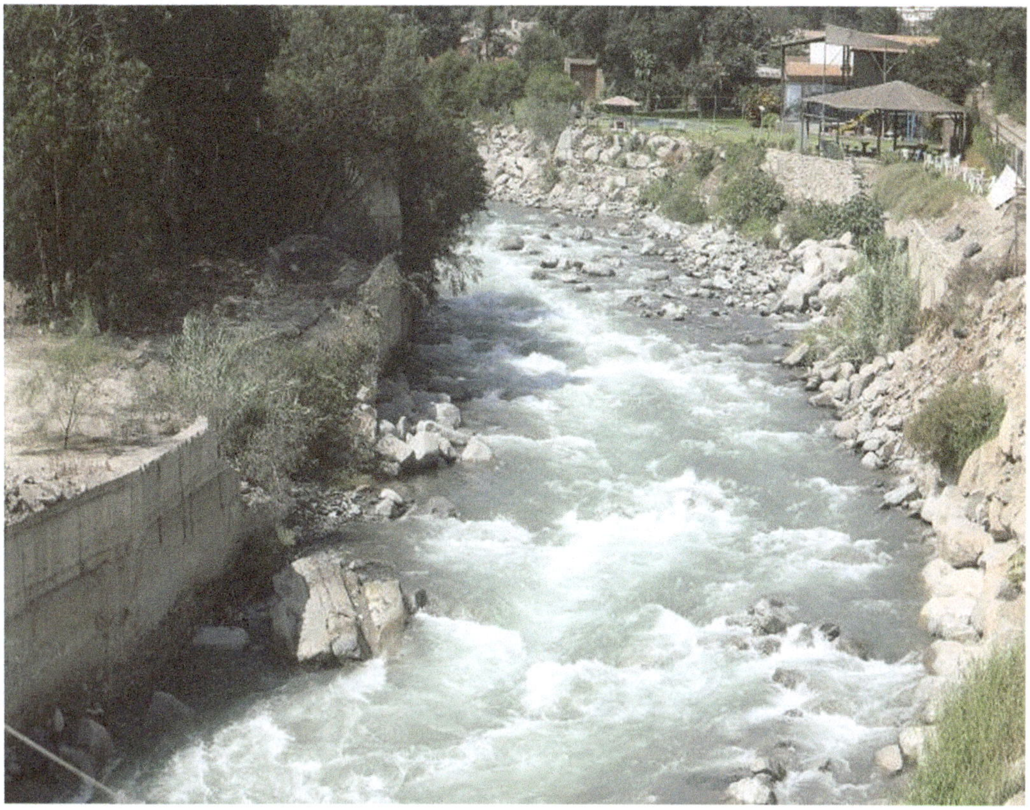

Fig. 5 The Rímac River recovers after mandatory social isolation. Flocks of various species and more vegetation are sighted. (http://www.ana.gob.pe/noticia/minagri-rio-rimac-se-recupera-tras-aislamiento-social-obligatorio. (Tuesday, 21 April 2020))

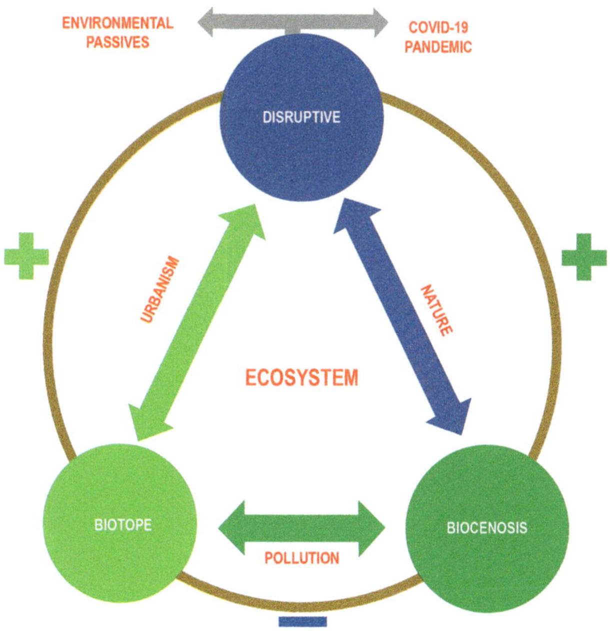

Fig. 6 Diagram of the disruptive biotope – Rímac River. The disruptive biotope will be in the memory of those who have lived and remembered it before an environmental contamination and of those who can now appreciate the potential of an urban image of Lima, making its pedestrianization along its banks safer and turning the Rímac River into a majestic talking river consolidated in a city in constant growth. (Author)

trolled and spontaneous urbanism that ended up relegating to the background, using the natural interstice indiscriminately, progressively turned into a complete garbage dump of the city for more than 50 years, unfortunately leading to the disappearance of a natural ecosystem, where once a symbiosis nature/individual developed and marked a safe place.

References

Aguila Flores, J. L. (2014). *Interstitial space, emergence and transformation*. Universidad Internacional de Andalucia.

Alameda, E. (2013, April 18). *The alarmingly polluted Rimac River – The water we drink in Lima*. Retrieved from http://www.aguapureza.pe/rio-rimac-agua-contaminada-lima/

ANA. (2020, April 21). *Ministry of Agrarian Development and Irrigation*. Retrieved from Minagri: Río Rímac recovers after mandatory social isolation: http://www.ana.gob.pe/noticia/minagri-rio-rimac-se-recupera-tras-aislamiento-social-obligatorio#:~:text=De%20acuerdo%20a%20ultimas%20evaluaciones,y%20ciertos%20metales%20y%20metaloides

Delgado Ruiz, M. (1999). *El animal público: Hacia una Antropologia de los espacios publicos* [*The public animal: Towards an anthropology of public spaces*]. Anagrama.

Gestión Newspaper. (2020, April 21). *Rímac river presents crystal clear waters after 90% reduction of garbage during quarantine*.

Forero Rodriguez, L. A. (2015). *Urban environmental interstices, articulating elements from the public space – Caso Rio Fucha Tramo Carrera 10 – Carrera 30*. Master's thesis on Urban and Regional Planning, Bogotá, Colombia.

Iagua. (2014, April 11). *The Rimac river basin: Looking to the future*. Retrieved from https://www.iagua.es/noticias/peru/14/04/11/la-cuenca-del-rio-rimac-mirando-al-futuro-48235
Municipality of Lima. (2020). *Municipality of Lima*. Retrieved from http://www.transparencia.munlima.gob.pe/component/k2/2-portal/historia
The Republic. (2019, May 25). *Society. Sedapal must spend S/. 30 million to purify contaminated water from the Rimac River*.
Visit Lima. (2021, September 18). *Al Otro Lado Del Rio: Recuperando La Ribera Del Rio Rímac*. Retrieved from https://visitalima.pe/blog/al-otro-lado-del-rio-recuperando-la-ribera-del-rio-rimac/

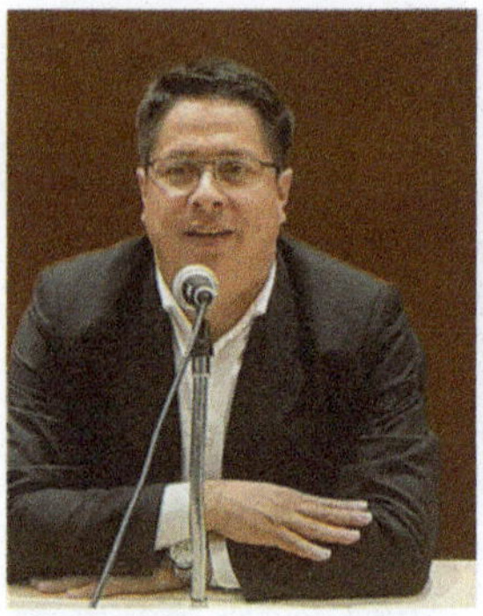

Henry Daniel Lazarte Reátegui Architect by Ricardo Palma University; Master in Digital Home, Infrastructures and Services - Universidad Europea de Madrid. Develops Smart Buildings, Cities and Territories projects coordinating with regional and local governments the management of Public Policies based on Industry 4.0 and Society 5.0 for the benefit of the citizen. Developed the first technology park in Peru (bicentennial park) as a Digital City Node in the district of San Isidro - Lima - 2019. PhD in Public Policy: National Security and Sustainable Development and is a doctoral candidate in Complex Thinking by the Multiversidad Mundo Real Edgar Morin. He collaborates in projects of Sustainable Cities of the Regional Center of Competences in Education for Sustainable Development RCE Lima-Callao, being promoted and certified by IAS/UNU (Institute of Global Reflection and Organization of Graduate Education by the United Nations University).

Life on Earth: Reflections in Times of Crisis

Iván Curiel

Abstract

External agents and changing environmental constants have compromised the formation and development of life on Earth over the centuries. These processes of change have required long periods for the impact to become visible. However, since the Industrial Revolution, these alterations have progressed much faster, threatening our living planet, using natural resources in a predatory way, degrading ecosystems, breaking nature's circular economy and using it as a waste space. In search of a possible solution to all this, new forms in the field of design are studying existing strategies in nature to apply them to the entire development of the human species, always in harmony with the environment so as not to destroy life on Earth.

Keywords

Development · Environment · Circular economy · Linear economy · Climate change · Construction · Design

Formation of Life on Earth

The planet Earth is so far the only living habitat in the known universe. It was formed 4.47 billion years ago – according to measurements by John Rudge, who used various radiometric dating techniques (Kaplan, 2017), with this date being the most accepted by the scientific community. The original planet was nothing like what we know today as Earth. Once the necessary conditions were in place, a small microscopic world began to develop that would be the beginning of the evolutionary history of life and living things on planet Earth.

Life as we know it has its basis in the existence of a molecule that stores information – the information necessary to form a living thing – which we know as DNA (Fig. 1) (Watson & Crick, 1953). It began about 3.8 billion years ago in the deep ocean, saturated with chimneys that expelled sulphur and metals from the Earth's crust. Combined with organic molecules from meteorites, this led to the emergence of the first microorganisms that were able to use sunlight, carbon dioxide and water as a source of energy to produce food through the process of photosynthesis. In return, they emitted oxygen as "waste" and filled the ocean with this element. From this point on, the organisms managed to evolve. These unicellular forms were able to give rise to more complex life forms. At this point, the evolutionary line of living things began, developing into a large family tree and forming an immense variety of living things (Fig. 2).

For millions of years, the development of species has been limited by natural phenomena that, at historic times, almost completely wiped out life, due to the changing conditions of the planet and even to extra planetary objects as evidenced by samples from the Chicxulub meteorite collected by a group of scientists in the Yucatan Peninsula (DePalma et al., 1994). During these periods of mass extinction, some species managed to survive these abrupt variations and are the ones that have evolved to the present day. However, how have species managed to survive on a changing planet for so long? The first answer was elaborated by Jean-Baptiste Pierre Antoine de Monet, knight of Lamarck, biologist and naturalist considered the first evolutionist, and later by Charles Darwin, natural scientist, both reaching, at different historical moments, similar conclusions (De la Peña Seaman, 2018).

Theory of Evolution

Life has adopted multiple shapes and sizes throughout its evolutionary history. The species that live on our planet today have not been the same since their origin, but have developed in the course of biological evolution, depending on whether they were exposed to different climatic conditions, natural phenomena or other circumstances. The evolu-

I. Curiel (✉)
Universidad de Alcalá, Alcalá de Henares, SP, Spain
e-mail: ivan.curiel@edu.uah.es

M. R. Cervera Sarda et al. (eds.), *Architecture Inspired by Nature*, https://doi.org/10.1007/978-3-031-33144-2_12

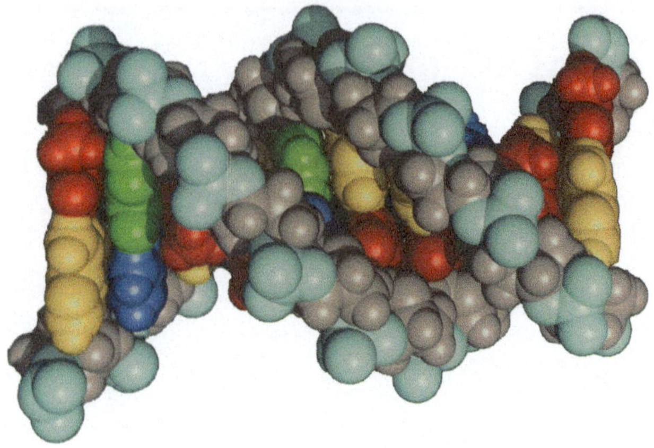

Fig. 1 Representation of a DNA molecule, Vikrazuul. The National Museum of Natural Sciences in Madrid. (Photo I. Curiel)

tionary process is one of slow change over long periods of transformation of organism. Species that fail to adapt because they do not accumulate enough variation are the least likely to survive and reproduce and therefore more likely to go extinct more easily. On the other hand, the species that succeed in adapting through a minimal number of variations are the ones that will have offspring and can survive. These variations can vary greatly within the same species due to the environment. This phenomenon was described in 1859 by Charles Darwin in the theory of the evolution of species, also known as "natural selection" (Darwin & Wallace, 2006).

Natural selection acts against living beings with unfavourable characteristics that cannot adapt to their environment. The main driving force of the evolution of species is genetic mutations in DNA, which Darwin and Lamarck were unaware of and only discovered much later, thanks to the advances of modern science. Mutations are random errors that occur during DNA transcription. The vast majority of these mutations are caused by the characteristics of the environment, which influence the genetic content of every living being. Some other mutations are not due to the environment but are random and may not be unfavourable. These mutations are inherited by the offspring and lead to the genetic differences between organisms (Echeverria, 1994).

In his travels, Darwin observed patterns in the distribution and characteristics of organisms. One of the most important is his study of the "Fringilla coelebs" birds in the Galapagos Islands (Fig. 3). They were not identical on the different islands; they had different morphology, especially the beak, which was due to the different types of food found on the islands. It was concluded that they had a common ancestor but had evolved differently due to environmental conditions.

Lamarck explains the evolutionary process with another well-known example, that of the giraffes (Fig. 4). Those with longer necks were more able to catch food, so they were able to produce offspring (Darwin & Wallace, 2006).

Both based their theory of evolution on the countless evidences they collected. Today, the theory of evolution is still valid, with the additions and clarifications that scientific progress in various fields (genetics, biochemistry, geology, ecology, etc.) has provided us with.

Ecological Closed-Loop System

When we talk about living organisms, we have to ask another fundamental question: How have living things used the planet's resources to survive?

In the natural environment, the residues or wastes of one organism are raw materials for others, thus forming an ecosystem. There are thousands of ecosystems that span all scales, from the ocean or desert to small ecosystems in a decomposing creature. The difference between the two is the number of variables. Large-scale ecosystems such as a marine ecosystem encompass many ecosystems, while a small decomposing insect may only encompass one or two ecosystems.

Seth Galewyrick explains how species interact within an ecosystem to use resources in a closed economic model (Fig. 5) (Galewyrick, 2018).

The tree grows, thanks to solar energy and the photosynthesis of its leaves. Glucose is synthesised, which is used by the aphids to produce a kind of honey that serves as food for the ants. In return, the ants protect themselves from predators.

The ants are the food for the woodpecker, and the woodpecker and the squirrel are the food for the fox. The woodpecker also uses the tree boot to dig its burrow and build a shelter for its young. The excreta of these animals are used by dung beetles, fungi and bacteria, which can obtain nutrients and energy through fermentation reactions. This system is not isolated, but involves many more variables, but in this way demonstrates the circular and efficient use of natural resources (Sommers, A., Learning form nature – circular economies & biomimicry, 2020).

Industrial Development and the Linear Economy

The entire harmony of dynamic systems in vital equilibrium was disturbed with the development of industry around 1760.

The industrial revolution radically changed energy production, forms of work organisation, systems, manufacturing and techniques used. All this led to an enormous growth in the production of goods and thus to a new everyday life that

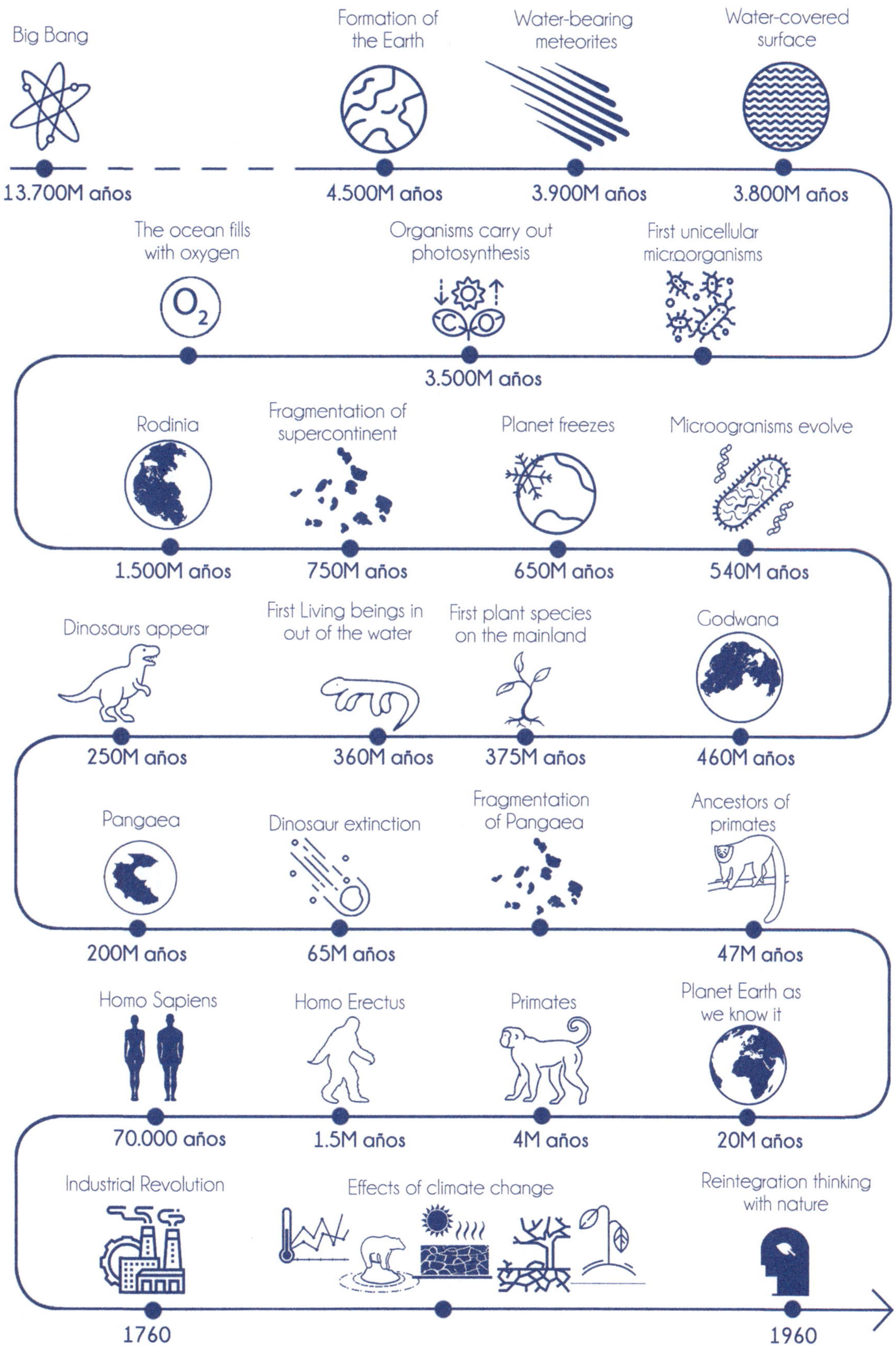

Fig. 2 Timeline of the development of life on planet Earth. (I. Curiel)

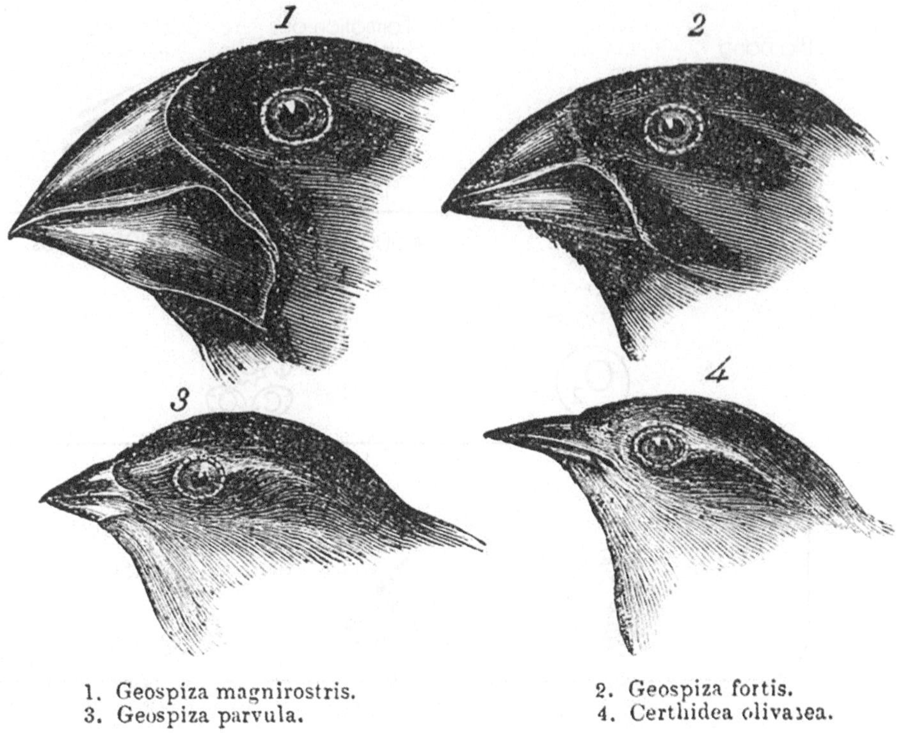

Fig. 3 Differences in morphology of the same *"Fringilla coelebs"* species on different islands. (Darwin & Wallace)

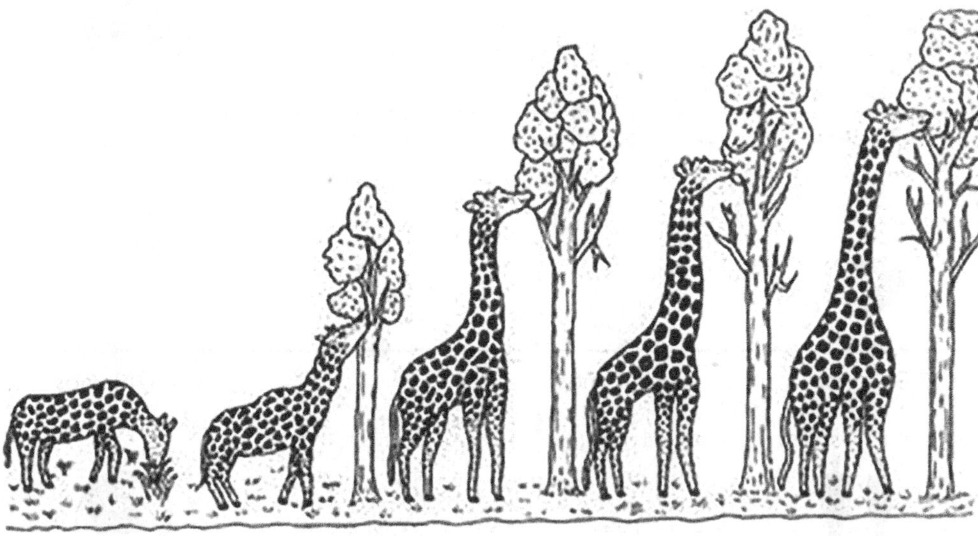

Fig. 4 Differences in giraffe neck lengths. (Darwin & Wallace)

changed from a traditional, agricultural and artisanal society to an industrial and modern society.

Factories meant a radical change in the organisation of work. Large areas allowed the use of machines that displaced manual labour, increased efficiency and shortened manufacturing times (Chaves Palacios, 2004). Key to this development was the invention of the steam engine in 1769 by Watt, which was used in virtually all areas of production; it was the basis for sustaining mining, industrial production and transport. It was powered by the new, non-renewable energy source coal.

The advantages of the invention of the steam engine were obvious: labour savings, product quality, production speed and production capacity. Thanks to large-scale production, prices fell, cities multiplied and the population gained in quality of life and life expectancy. The development of the transport sector, the steamship in 1787 by Fitch and the railway in 1825, used coal as a raw material for their operation.

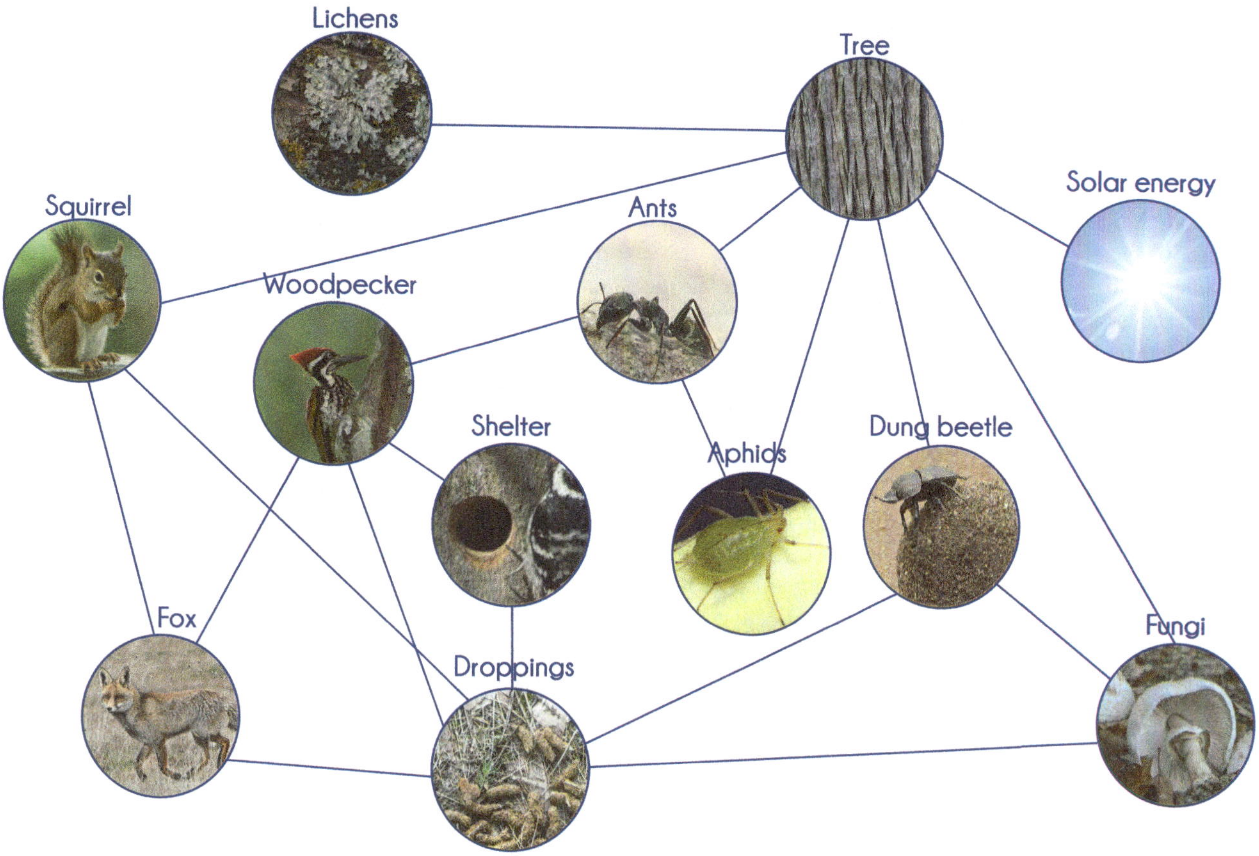

Fig. 5 How the circular economy works. (I. Curiel)

The industrial revolution had a profound impact on the consumption of energy and natural resources. Global CO_2emissions increased exponentially, in dates ranging from the pre-industrial era to 1877 zero to 10 million tonnes of carbon emitted. By 1947, emissions continued to rise, reaching almost 2 billion tonnes of carbon, and, thereafter, there was exorbitant exponential growth, exceeding 7 billion tonnes of carbon in 2003 (Keay, 2007).

Climate Change Due to Human Development

Research carried out by various expert groups shows that the influence of humans on climate change is obvious. The planet is constantly changing, following natural cycles that have changed since the development of industry. But recently, the unstoppable emissions of greenhouse gases that have accumulated in the atmosphere since the industrial revolution are affecting the Earth's natural balance in ways that could become irreversible. The consequences could be devastating if we do not drastically reduce our dependence on fossil fuels and cut emissions (Duarte, 2006).

The exponential consumption of natural resources simply for the greater comfort of the human population since the second half of the twentieth century has meant that the effects of climate change, already seen in the past, are severe and threatening to life on Earth. Climate change in facts:

- Biodiversity loss due to habitat destruction has increased globally to an average of 68% since 1970 (WWF, 2022).
- Rising temperatures are causing the melting of the poles, other ice and snow around the globe and contributing to the disruption of the Gulf Stream. In the last 23 years, global ice loss has increased by 65%. Half of the loss is due to continental ice melt, estimated at 6100 MT, 3800 MT in Greenland and 2500 MT in Antarctica (Rodriguez, 2021).
- The increase in fires is a consequence of the increase in the average temperature on Earth and the concomitant decrease in precipitation. It is estimated that about 55 million hectares burn each year (Castillo et al., 2003).
- Rising sea levels, a consequence of continental melt, pose a serious threat to coastal areas. Sea levels have risen 20 cm since 1880 and are projected to rise another 30 cm to 100 cm by the end of the century (NASA/JPL-Caltech, 2016).
- Lack of rainfall and rising temperatures are leading to land surface desertification, degrading fertile soils and

promoting loss of plant species and soil erosion. By 2025, about 1.8 billion people are expected to face absolute water scarcity. Degradation of arable land is increasing at 30 times the historical rate (Flores, 2022).

- The average temperature increase of the Earth's surface over the last 150 years is 2 °C, according to the WWF report. An accelerated increase has been observed in the last 30 years (Giannakopoulos et al., 2009).
- Material losses due to weather events are increasing. Material losses per year are estimated at €202 million (World Meteorological Organisation, 2021).

The changes in environmental constants do not occur in isolation, but one triggers the other in an accelerated cycle that we are beginning to feel all over the planet. The accelerated degradation of the environment, which is obvious, is triggered by human activities.

Humans extract matter from the environment and convert energy to make products, process raw materials and consume the product. At the end of the product's useful life, it is discarded so that it can no longer be used; it is thrown into the environment in the form of waste and determines a linear economy (Fig. 6) with losses that run counter to natural processes. In clear contradiction, we attribute the word progress to a development process that endangers the life of the planet and the people who live on it, which is paradoxical (Cervera Sardá, 2018).

Impact of Construction on the Environment and Development of New Lines in the Design Field

Buildingis one of the most environmentally damaging human activities. The construction of building is a process that can take years or even decades. In contrast, the lifespan of buildings is not particularly long. Buildings require energy, and much of that energy comes from fossil fuels. The materials used come from all over the world, which also entails energy, mining and production cost. In numbers, construction contributes: One-third of CO_2emissions come from the construction sector, which also consumes 40% of the world's energy, 40% of the world's resources and 25% of the world's drinking water (Fig. 7) (Ritchie et al., 2020).

A rethinking of work, production and human development is necessary; we must turn our gaze to nature, the means by which man can survive. Human needs nature for this survival. Nature has evolved life on earth for 3.8 billion years and has already overcome many of the adversities that we humans still face. Nature has many of the answers we have been looking for. Some schools of thought, design and development, such as bionics or biomimetics, involve a naturalistic conscience to stop endangering life on the planet and try to reserve these harmful effects through energy-saving strategies, optimisation of resources, increased efficiency material conservation and so on.

For far greater optimisation and harmony with the environment, new areas of design, biodesign, have been and are being developed that incorporate living organisms at every scale as a tool and main element of the projects. These new forms of emerging design start from in-depth scientific research applied to design areas, from small objects to cities. The aim is to use ecological design as a tool. These new forms of design therefore reflect a shift in societal priorities towards sustainable approaches to construction, as there is an unprecedented urgency to change traditional methods and rethink goals to counter accelerating environmental degradation (Myers, 2018).

Therefore, there is an urgent need to optimise existing resources and reduce the ecological footprint, starting with the reduction of high CO_2emissions and the reduction of energy consumption.

Since the first greener building approach, a global reduction in the environmental footprint of these early projects has been recorded from 2000 to 2016, saving 55.56 million kWh of electricity and 40.92 million kWh of fossil fuels (Macnaughton et al., 2018). The application of biodesign using digital techniques reveals possibilities for design without constraints in balance with nature, using organisms as energy producers, collectors of harmful emissions, sources of materials, digital storage systems, cleaning agents, etc. The goal is a new multidisciplinary future of construction and design (Kırdök et al., 2019).

These new projects on the table aim to eliminate the use of fossil fuels, such as the genetic project in Barcelona, A. Estevez. By introducing a luminescent protein into the genetic material of trees, it achieves total energy independence, as the leaves of the trees manage to emit light, biolu-

Fig. 6 Linear economy. (I. Curiel)

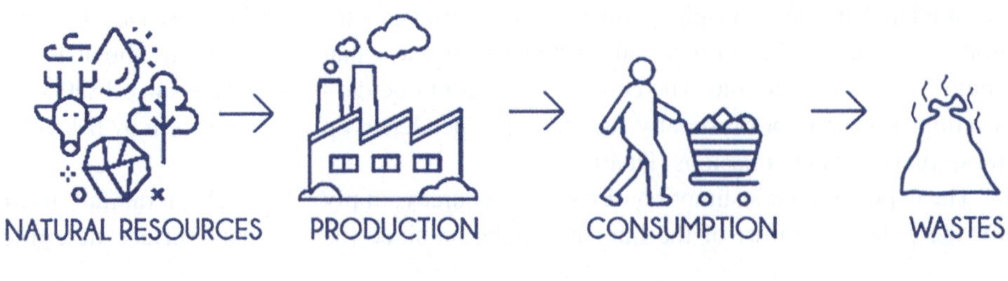

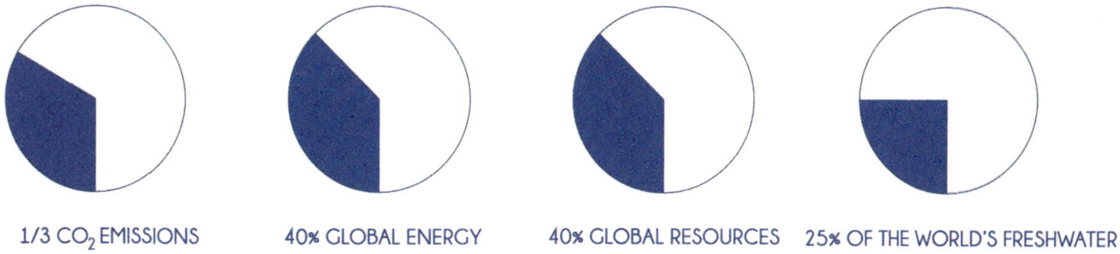

Fig. 7 Global emissions. (I. Curiel)

minescent trees for public and even private lighting. Other future projects, such as the Sahara Forest developed by M. Pawlyn, seek reforestation and repopulation, creating new water, food and vegetation in arid and resource-poor areas and harnessing sun, wind and salt water to develop self-sustaining greenhouses capable of absorbing water from the air, growing vegetation and thereby drawing moisture into the environment and attracting life forms as a result. There are many working groups putting these new projects on the table, from small objects to large infrastructures, to mitigate and reverse all the effects of climate change and reverse the way development takes place in step with nature.

References

Castillo, M., Pedernera, P., & Peña, E. (2003). Incendios forestales y medio ambiente: Una síntesis global. *Ambiente Y Desarrollo De CIPMA, XIX*(3–4), 44–53.

Cervera Sardá, M. R. (2018). *Espacio y tiempo en composición arquitectónica*. Munilla-Lería.

Chaves Palacios, J. (2004). Desarrollo tecnológico en la Primera Revolución Industrial. Norba. *Revista de historia, 17*, 93–109. http://dialnet.unirioja.es/servlet/oaiart?codigo=1158936

Darwin, C., & Russel Wallace, A. (2006). *La teoría de la evolución de las especies*. Crítica.

De la Peña Seaman, C. (2018). *La epigenética y la reivindicación de lamarck*. https://www.cicy.mx/Documentos/CICY/Desde_Herbario/2018/2018-06-21-CdelaPena-la-epigenetica.pdf

DePalma, R. A., Oleinik, A. A., Gurche, L. P., Burnham, D. A., Klingler, J. J., & Echeverria, A. B. (1994). Gene y mutación: una visión historica. LLULL (Madrid), 17(32), 5–24. Periodicals Index Online Segments 1–50 database. https://search.proquest.com/docview/1306824705

Duarte, C. M. (2006). *Cambio global: impacto de la actividad humana sobre el sistema Tierra Consejo Superior de Investigaciones Científicas*.

Echeverria, A. B. (1994). Gene y mutación: una visión historica. *LLULL (Madrid), 17*(32), 5–24. Periodicals Index Online Segments 1–50 database. https://search.proquest.com/docview/1306824705.

Flores, J. (2022). *La desertificación, uno de los grandes problemas ambientales de la actualidad*. https://www.nationalgeographic.com.es/naturaleza/desertificacion-uno-grandes-problemas-ambientales-actualidad_14425

GaleWyrick, S. (2018). Biomimicry for sustainable innovation: The answers are all around us. *Design Management Review, 29*(1), 30–35. https://doi.org/10.1111/drev.12106

Giannakopoulos, C., Le Sager, P., Bindi, M., Moriondo, M., Kostopoulou, E., & Goodess, C. M. (2009). Climatic changes and associated impacts in the mediterranean resulting from a 2 °C global warming. *Global and Planetary Change, 68*(3), 209–224. https://doi.org/10.1016/j.gloplacha.2009.06.001

Kaplan, S. (2017, March 6). Dear science: How do we know how old the earth is? *Washington Post.Com*.https://global.factiva.com/en/du/article.asp?accessionno=WPCOM00020170306ed360040h

Keay, M. (2007). *Energy: The long view (Publisher's version)*. Oxford Institute for Energy Studies.

Kırdök, O., Altun, T. D., Dokgöz, D., & Tokuç, A. (2019). Biodesign as an innovative tool to decrease construction induced carbon emissions in the environment. *International Journal of Global Warming, 19*(1–2), 127–144. https://doi.org/10.1504/IJGW.2019.101776

Macnaughton, P., Cao, X., Buonocore, J., & Laurent, J. G. (2018). Energy savings, emission reductions, and health co-benefits of the green building movement. *Journal of Exposure Science & Environmental Epidemiology, 28*(4), 307–318. https://doi.org/10.1038/s41370-017-0014-9

Myers, W. (2018). *Biodesign nature science creativity*. Thames & Hudson.

NASA/JPL-Caltech. (2016). *Aumento del nivel del mar*. https://climate.nasa.gov/climate_resources/274/aumento-del-nivel-del-mar/#:~:text=El%20cambio%20clim%C3%A1tico%20est%C3%A1%20provocando,ente%201%20y%204%20pies

Ritchie, H., Roser, M., & Rosado, P. (2020). *$CO_2$and greenhouse gas emissions*. Our World in data. https://ourworldindata.org/co2-and-other-greenhouse-gas-emissions.

Rodriguez, H. (2021). *La pérdida de hielo global alcanza un nuevo récord*. https://www.nationalgeographic.com.es/ciencia/perdida-hielo-global-alcanza-nuevo-record_16274

Sommers, A. Learning from nature – circular economies & biomimicry. (2020, April 14). [Video/DVD]. TED Conferences, LLC.

Watson, J., & Crick, F. (1953). Molecular structure of nucleic acids: A structure for deoxyribose nucleic acid. *Nature, 171*, 737–738. https://doi.org/10.1038/171737a0

World Meteorological Organization. (2021, September 2). *Weather-related disasters increase over past 50 years, causing more damage but fewer deaths*. Targeted News Service/HT Media. https://public.wmo.int/en/media/press-release/weather-related-disasters-increase-over-past-50-years-causing-more-damage-fewer

WWF. (2022). *Informe Planeta Vivo 2022. Hacia una sociedad con la naturaleza en positivo*. Ed. R. E. A.. Almond, M. Grooten, D. Juffe Bignoli, T. Petersen. WWF.

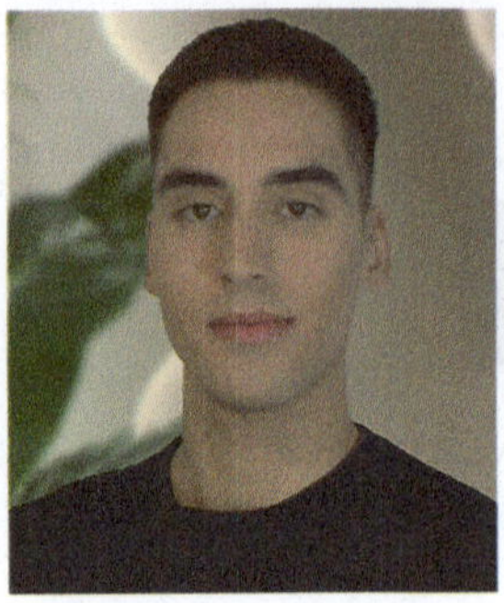

Iván Curiel is an Architect and Urban Planner from the School of Architecture of the University of Alcalá and graduated in the Master's degree in Architecture in 2022. He is currently studying the Master in BioDigital Architecture at the International University of Catalonia, Barcelona. Furthermore, he has participated as support in a couple of projects for the General Foundation of the University of Alcalá de Henares.

Evolutionary Biophilic Concerns in the Landscape and the Relationship with the Future Consciousness Society

Cerasella Crăciun

Abstract

The work aims to make an approach of the history of evolutionary concerns related to the deep connection of human and nature, before the appearance of the term biophilia, with the aim of bringing to present elements/principles/processes/ideas, which can be reinterpreted in the contemporary processes of design, planning, and architectural-landscape and urban-territorial development. From the idea of the tree/garden/landscape as a primordial archetype, passing through the stage of sacralization, to the profane interpretation, from the personalization of nature/life, to the appearance of the first cities, the problem transposed into the human-landscape relationship, the research focusing on a transdisciplinary integrated research method and approach. Sensitive analysis of different forms of biophilic concerns that connect the philosophical and educational ideas, with the development of new urban and architectural morpho-typologies, the evolutionary economic components, overlapped with technological progress, but also subject to some local beliefs and cultures. These generate principles that concern nature/multi-scalar landscape in all its typologies (natural, anthropic/built, urban/architectural, also cultural landscape). The city needs a sensitive analysis of the urban organism and urban planning focused on particular internal urban metabolism, a complex process and a model of conscious design of an integrated biophilic urban organism. Development of urban systems as biophilic organisms involves complex integrated process necessary to introduce a new paradigm practice of consciousness and (meta)ethics, under the conditions of the current transition from the information society, through the knowledge society, to the future society of consciousness. It is necessary to include biophilic aspects as a significant contribution in supporting the life resources of the future, and the nature-human/community relationship, to become a real vector and germ of eco-sustainable development, including contribution of multidimensional education, through knowledge/consciousness applied in urban design.

C. Crăciun (✉)
"Ion Mincu" University of Architecture and Urban Planning, Bucharest, RO, Romania

Keywords

Urban organism · Archetypal garden · Knowledge society · Multidimensional education · Urban metabolism

Introduction

The purpose and objectives of the research are to bring back up-to-date elements, principles, processes, or ideas that can be reinterpreted in the contemporary processes of conscious design on a place/landscape, through a process of design, planning architectural-landscape, and urban-territorial planning, which integrate biophilic attitudes, principles, or multidimensional education elements. These are extracted from the aspects of historical evolution relationship between human and nature along the development and transformation of human society, from the first communities, into the current smart eco-sustainable and resilient urban development.

Different incipient (pre)biophilic ideas appeared in the history of urban planning and the garden. The philosophy of the dominant ideas, as well as the customs of the related principles through (re)interpretation and adaptation to the present, could today (re)generate biophilic approach, thinking, and practice in urban-territorial and architectural-landscape or multidimensional education domains.

The city as an urban organism with a particular metabolism needs to be analyzed and planned in a biophilic perspective, as a conscious design model of a complex process. Without the introduction of related research methods of an integrated transdisciplinary approach on biophilic bases, knowledge still limited at present, new effective solutions for

M. R. Cervera Sarda et al. (eds.), *Architecture Inspired by Nature*, https://doi.org/10.1007/978-3-031-33144-2_13

the future of urban development, but also human, connected with life/nature as vital resources, cannot be found, except through a resilient form and in a truly eco-sustainable connection.

The paradigm of consciousness and (meta)ethics are research methods that must be integrated in educational specialized practice and in planning and designing the conscious design, with preservation of the memory of place/genius loci and local particularity. It's necessary to include biophilic aspects as a significant contribution in supporting and substantiating greatest life resources of the future, and the nature-human/community. Relationship must not be limited by a unidirectional and global knowledge, if necessary to involve ethics and consciousness in planning processes. The act and attitude of planning and design on biophilic bases can and is necessary to become a mediating indicator between development and technology, a true vector and evolutionary germ of future development, including in other related fields, as well as in multidimensional education.

Paradise Garden Typology Design as an Incipient Biophilic Process

Since the earliest times of mankind, long before the appearance of the current terms biophilia or bionics, there is a deep connection between human and nature personified as divinity/big universe, later as Gaia Hypothesis (Lovelock, 2000) and in contemporary understanding as bio-structure and experimental bionics. When people lived in nomadic groups in natural landscapes, there was the idea of the garden as a primordial sacred archetype that represented glorification of the divine nature in its sacred and profane meanings (Eliade, 1987). Initially, sacred education and knowledge began with the development of observations on nature and first biophilic relationship, incipient elements of conscious design, and intervention in nature at the time of the appearance of archaic design processes.

Paradise or Eden Garden (pairi-daeza in old Persian language) can be considered the first historical-archaeological and pre-urban landscape mention of an incipient biophilic process, created by the Divinity, with and through nature. The Babylon Semiramis suspended garden is a biophilic design attitude applied in the first planted garden in history of architectural structure. This denotes an attitude of a green insert, in contrast to the existing desert area in that territory being a lost memory of the divine garden. Ancient Egypt included pleasure gardens, represented in the necropolises, royal funerary, and divine temple gardens, a biophilic sign that symbolically expressed the domineering power and greatness.

The Absolute Landscape and the Beginnings of Biophilia in a Philosophical Approach

Ancient Greece biophilic concerns are related to integration into the macro-landscape and deep respect for the natural landscape. Being an insular people of navigators, the Greeks invent the concept of absolute landscape, harmonious correlation with nature, which brought aspects of ecstasy to the landscape and urban planning of the first democratic cities. The Greeks always created a dialogue between nature and architecture, because nature was considered beautiful and sacred, a stage on which most of the mythical dramas took place, which consisted of a scenography on successive plans from macro- to mezzo-landscape.

The beliefs of the ancient Greeks claimed that only near a tree and a spring human can communicate with the Divinity. They developed the concept of open gardens for all citizens in the new idea of democracy, but also the unique concept of philosophical gardens, dedicated to wisdom, developing new valuable, theories, and aesthetic directions of philosophy in nature (Assunto, 1981).

Aristophanes called the philosophical garden the Empire of the Trees (taken by symbolic interpretation as hypostyle forest of pillars of Greek temples), where the disciples and masters walked, talked, and exchanged ideas. This is a first biophilic multidimensional educational attempt to connect human and nature through thinking and philosophy, considering at that time that only nature can help the thought to be fulfilled and the mind to bear new avant-garde ideas. Epicurus' garden was included in the same educational philosophical typology (Furley, 1996), and the Athena Academy functioned in a public park (Wycherley, 1978) open to all citizens, because only there new philosophical ideas could circulate freely and democratically among people with different status.

Roman Empire come with a new vision, in order to show their dominant and expanding imperial power in nature. Biophilia was still present at the level of the special appreciation of the sites, considered to have a great landscape value. For example, the ancient poet Vergilius supports the abandonment of the Roman City and the return to the domain of the province, considered to be a domain of the joy and virtues of rural life, as well as the beauty of the cultivated plain (Vergilius, 2011). The architectural and urban design was done by integrating into the natural context but transforming the natural landscape. Imperial Park was considered in Ancient Rome a model of disciplining nature as a visualization of power of the empire.

The Scenographic Relationship of Architecture with the Human-Centered Landscape

The Dark Middle Ages brought a marked regression visible including in the urban and architectural development and in the principles approached in relation to the landscape. Later, the Renaissance ideals put human in the center as the measure of all things and rediscovers the ancient idea of the philosopher Protagoras. Compositionally and aesthetically, the principles of an incipient biophilia were used whereby pure but also modified or designed nature contributed to the whole, through symmetry, main and secondary axes focused on the great landscape. It used the compartment into distinct functional areas in new typologies of green spaces with biophilic contribution of closed/secret garden, all of which in fact constitute the history of the architectural-urban evolution of landscape and gardens design (Prévôt, 2006).

In the Baroque era, the interior-exterior spatial relationship, between architectural construction, garden, and great landscape is made through specific composition and the mitigation of the previous schematism and used multiplication principle of visual lines of force and the relationship with the landscape in order to obtain scenographic effects of overall unity and harmony in nature. The green decoration combined with water stage, as well as the processing of a major relief form in a biophilic spirit, is a symbolic role of the cosmic archetypal garden.

Triple revolution context (industrial, urban, democratic) of seventeenth to eighteenth century came with a new biophilic and landscape architectural approach configuration of urban public space, in relation to the public space and a new urban art multidimensional education. Biophilia transformed the idea of archetype garden, which becomes the place where the Roman provinces became the natural landscape, with history, reshape configuration, presence of the motif of the ruins/memories of the past, picturesque, wild but also melancholic. Thus, the newly appeared freestyle promotes nature as a primordial biophilic element in the creation of landscape scenographies, which induce emotions and build biophilic consciousness through design, spaces that contain messages and symbols.

Biophilia in a Complex Integrated and Conscious Urban Organism

Democratic Haussmannian industrial city promotes the idea of unprocessed nature as biophilic idyllic ruralism refuge, correlated with perceptions of the rural and cultural landscape and with new multidimensional education. In the twentieth century, many authors argue that the Earth is a superorganism and proposals to using principles from nature or elements of nature coherently integrated into urban proposals, architectural, and landscape (Lovelock, 2000).

Consciousness society is a desirable future ethical model (Drăgănescu, 2000) determined by a philosophical vision of new paradigms related to education and research with reference to fundamental questions about existence and the relationship between nature/man/technology and education/knowledge through consciousness/ethics.

Consciousness society will emphasize the structural-phenomenological character of information that has become consciousness and will have as functional vectors: grounded and integrated knowledge of existence, management, economy, culture, multidimensional education, and all interacts directly and deeply with the development of future human settlements.

Results – The Planned City as a Complex Process and Conscious Biophilic Design Model

The awareness of the Earth as a living organism has as its starting point the ability of the planet to be self-supporting, similar to the ability of biological organisms but, in contradiction with the city urban system, except through urban planning concerted and approach as a biophilic urban organism (Crăciun, 2008). This must be correlated with a specific and particular image of the city (Lynch, 1998) that must be preserved with a development based on urban models with its related regulations (Choay, 1997), which should also take into account the component of biophilic principles development.

The development of the nature-human relationship through architectural, urban, and landscape conscious design was achieved through evolution along cultural history (Rogers, 2001), in relation to the development of garden design by tracing the development of gardens through history and across social, political, and philosophical boundaries (Turner, 2005).

Urban diagnosis based on a multi-criterial transdisciplinary scientific analysis, including the processes and phenomena at the urban level; components of the place memory; local cultural landscape; economic, social, and political system; management strategies, and urban policies, thus transdisciplinary approach (Nicolescu, 1999), became a tool for integrated research of the natural, anthropic, and cultural landscape of a living multi-scalar and complex organism. Landscape becomes connection link between humans, architectural object, development, and urban planning, including the principles of urban ecology (Forman, 2016; Pickett & Cadenasso, 2017).

Landscape as a fragile resource but also is an asset in development, through the biophilic perspective, is a process that ensures the continuity of the relationship between human and nature, through the design process, as conscious design on a place/settlement who can eliminate vulnerabilities and complex hazards. Likewise, other biophilic components can be seen within landscape strategies, related to the projects at the macro-landscape level on an extended territory of a county or territory, or at the mezzo-landscape level, such as general urban plans of the municipalities, such as the integrated landscape studies.

Conclusions – Biophilia as a Transdisciplinary Approach to Diagnosis, Planning Urban Development, and Multidimensional Education

Biophilic contribution of the landscape planning is important not only for the achievement of a functional mix with eco-sustainable and leisure contribution or for raising the quality of life, but also for the support of human and urban health of cities' complex urban organisms, with minimization of the effects of diseases and urban pathology, with the eco-sustainable biophilic incidence on life in general (Crăciun, 2009). The present/future resilient city needs to be sensitively analyzed and planned as a complex process and conscious design model of an integrated biophilic urban organism in a multiscale-level approach – territorial, administrative, urban, social, community, biological, cultural – while preserving the local morpho-typological specificity and memory of the place/genius loci.

Development of urban systems as biophilic organisms, but also as urban and architecture educational systems, involves a sensitive, complex, and integrated process, a new paradigm of consciousness and (meta)ethics. These need to be adopted including the sense of a multi/trans-dimensional education, absolutely decisive for the current evolution of educational processes in the urban landscape and architectural field (Crăciun & Acasandre, 2015), according to the principles the biophilic basis of life. In the conditions of the current transition from the information society, through the knowledge society, to the consciousness society (Drăgănescu, 2004), only going through a humanological process (Mărincuş, Munteanu, 2009), we can overcome the contemporary discrepancies between development and technological future urban reality.

Biophilic aspects are necessary and have a significant contribution in the transformational education process, in preparation for global transition, only by keeping the particular valuable elements and the local specificity, in connection with applied (meta)ethics, with an essential role in the multidimensional education from society of knowledge (Crăciun, 2020). Only knowledge and consciousness applied in urban design become true vectors and germs of development with significant input in supporting the greatest life resources of the future, a mediating indicator between development and technology, including in other related fields, as well as in multidimensional education.

References

Assunto, R. (1981). *Filosofia del giardino e filosofia nel giardino. Saggi di teoria e storia dell'estetica*. Bulzoni.

Choay, F. (1997). *The rule and the model: On the theory of architecture and urbanism*. MIT Press.

Crăciun, C. (2008). The urban metabolism. Un *unconventional approach to urban organism* (pp. 243, 244). "Ion Mincu" University Publisher.

Crăciun, C. (2009). Urban health and pathology. *Romanian Academic Forum Journal*, 2(FAR XXI), 80.

Crăciun, C. (2020). Manifesto for a paradigm shift in future conscious transformational education. In *Mentoring project – Life quality. technology in retrohumanism. The future with the power of communities* 5. (Vol I, pp. 285–286). Club Romania Publishing House.

Crăciun, C., & Acasandre, A. (2015). "Landscape right" and "right to landscape". Landscape as a tool of meta-ethic and multidimensional education in the knowledge society. In *New approaches in social and humanistic sciences – Multidimensional education and professional development. Ethical values*. Medimond Publishing.

Drăgănescu, M. (2000). Consciousness, frontier of science, frontier of humanity. *Philosophy Magazine, XLVII*(1–2), 15–20.

Drăgănescu, M. (2004). *From the information society to the knowledge society*. Technical Publishing House.

Eliade, M. (1987). *The sacred and the profane: The nature of religion* (trans. W. R. Trask). Harper Torchbooks/Harcourt.

Forman, R. T. T. (2016). Urban ecology principles: Are urban ecology and natural area ecology really different? *Landscape Ecology, 31*, 1653–1662.

Furley, D. J. (1996). Epicurus. In *The Oxford classical dictionary* (3rd ed.). Oxford University Press.

Lovelock, J. (2000). *A new look at life on earth* (p. 35). Devanata.

Lynch, K. (1998). *L'image de la cite*. Dunod.

Mărincuş, N., & Munteanu, F. (2009). *Humanology. The organization of the leadership of mankind*. Odeon Publishing House.

Nicolescu, B. (1999). *Transdisciplinaritatea. Manifest* (Transdisciplinarity. Manifest.) (Contemporary ideas collection. "Plural" series). Polirom.

Pickett, S. T. A., & Cadenasso, M. L. (2017). How many principles of urban ecology are there? *Landscape Ecology, 32*, 699–705.

Prévôt, P. (2006). *Histoire des jardins*. Editions Sud Ouest.

Rogers, E. B. (2001). *Landscape design: A cultural and architectural history*. Harry N. Abrams.

Turner, T. (2005). *Garden history, philosophy and design, 2000 BC–2000 AD*. Spon Press.

Vergilius, P. M. (2011), *Bucolica. The Bucolics of Virgil. With Engl. Notes* [by H. Hayman]. Nabu Press.

Wycherley, R. E. (1978). *The stones of Athens*. Princeton University Press.

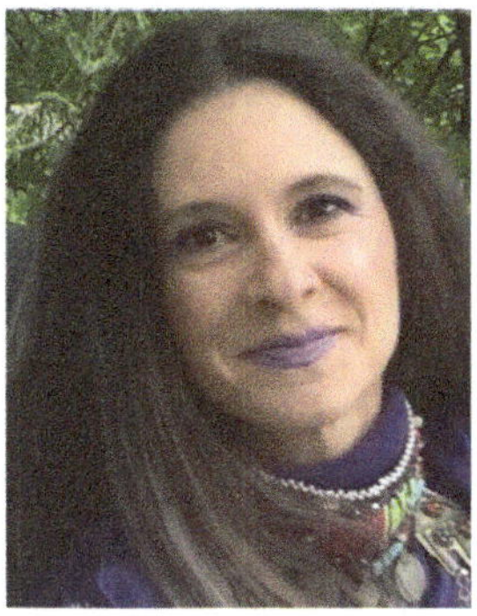

Cerasella Crăciun is an Architect, Urban and Landscape Design Planner – member of the Order of Architects from Romania, Register of Urban Planners, and Union of Romanian Architects. She is Full Professor, Habilitated to Advise Doctoral Research, "Landscape and Territory" Master Director, Faculty of Urban Planning, University of Architecture and Urbanism "Ion Mincu" Bucharest. She is titular of the courses: "Landscape Design," "Protection and Restoration of the Natural Heritage," "Garden History," "Landscape Research, Scenography and Art," "Eco-Development and Landscape," and "Creative and Transdisciplinary Heuristic Research Methods." She is author/co-author/coordinator of over 120 projects, research, studies; published 9 books, 17 book chapters, over 80 articles, 130 conferences, and 80 exhibitions/contests/competitions. She obtained over 25 awards/nominations/selections national/international distinctions as author/co-author – National Architecture Biennale, Urban Planners Romanian Union, European Union Prize for Contemporary Architecture Mies van der Rohe Award, Green Building Council award, etc. She completed research activity in the transdisciplinary fields of Architecture/Sacred – Spirituality/Cultural Landscape/Tradition/Anthropology – Creativity/Transformational Education of the Future, and Border Frontier/Holistic Sciences.

Nature – A Structural Component for Future Human Settlements? Biophilic City Approaches

Codruţ Papina and Cerasella Crăciun

Abstract

Cities have become more like simple inserts in the landscape, not fully conscious of the local context. Essential resources of our daily lives – health, food, energy, protection/shelter (from hazards/unfortunate events) and well-being – are in very high percentages dependent on the external value chains. Only part of our needs is sourced locally, making cities very un-resilient and fragile. How can this be changed? In the current literature, the biophilic city is presented as a new emerging concept that describes an urban structure that includes nature in the spatial configuration, and is oriented towards nature. Biophilic cities can be more than just "nature-inclusive" cities. Authors consider that the potential of the biophilic cities concept can be better exploited through integrated urban and territorial planning. A good understanding of the potential purposes/roles nature can play within urban structures will make cities more resilient, self-sufficient, healthy and sustainable through conscious eco-sustainable design and urban planning. Cities have the possibility of creating a more harmonious relationship with natural elements, through planning, design and adaptation of innovative ecologic solutions. This chapter explores the relationship between people, as a community, and nature, in order to understand what are the important drivers for constructing or regenerating today's cities. From ancient times, where nature was depicted as sacred archetype, to modern times, where the first ideas of purposely planning developments to be in harmony with nature emerged, and to current times of technological and design innovation, this chapter tries to extract and conclude the core values of biophilic nature versus dynamic cities. Nature-based solutions, green infrastructure management, ecological energy sourcing or food production solutions – all have to be part of an integrated territorial, urban and landscape planning and policy-making of how nature can help cities to thrive, reducing the impact on the environment, protecting habitats and encouraging healthy social life and healthy and ecologic behaviour by applying biophilic principles and processes.

Keywords

Biophilic cities · Nature and communities · Landscape planning · Green infrastructure management

Introduction

What is a biophilic city? In the current literature, the biophilic city is presented as a new integrated and emerging concept that describes an anthropic architectural structure, built ensemble or an urban structure that includes nature in the spatial configuration, and is oriented towards nature. It protects the natural elements, being aware of the social and health benefits it has, as well as the benefits on the environmental conditions.

> Achieving the requirements for a biophilic city will go a long way in stimulating social resilience and the landscape, in the face of climate change, natural disasters and economic uncertainty, and other shocks that cities will face in the future. (Beatley & Newman, 2013)

The term biophilia derives from biology, popularized by myrmecologist and sociobiologist Harvard E. O. Wilson. Wilson argues that humans have co-evolved with nature and that we carry with us our ancient minds and our need to connect and affiliate with nature to be happy and healthy. In fact, nature represented a primordial archetype to which humanity related as the sacredness and memory of the place, a direct connection with the planet and the universe. Thus, a definition of biophilia is the innately emotional affiliation of human beings to other living organisms. For Wilson, the concept is a set of rules of learnings developed over thousands of years of evolution and human-environment interaction (an aspect that is taken over by the emerging paradigm of biophilic city). The term biophilia was first used by Erich Fromm to describe the psychological orientation of an individual/existence to be attracted to everything that is alive and vital. The concept was taken up by Wilson, using it in a

C. Papina (✉) · C. Crăciun
"Ion Mincu" University of Architecture and Urban Planning, Bucharest, RO, Romania

M. R. Cervera Sarda et al. (eds.), *Architecture Inspired by Nature*, https://doi.org/10.1007/978-3-031-33144-2_14

similar sense, saying that biophilia suggests "the connections that people subconsciously seek with the rest of the living elements" (transposed into urban and spatial discourse, biophilia suggests the connections that cities/urban systems seek them intrinsically with nature). This chapter is part of ongoing research for understanding what makes a biophilic city, what are core values it should embed and how we can achieve this in current times in order to ensure a sustainable environment. The main aim is to elaborate a set of requirements for biophilic cities classification and a flexible and adaptable road map for making the typical (European) city more included biophilic approach towards nature, creating communities that value, protect and sustainably use nature in their daily lives. For the current paper and research, the objective is to delineate the first set of core values that nature can bring to cities. As a result, this chapter explores the relationship between communities throughout recent strategies or projects and nature, on the idea that biophilic cities are not a new phenomenon or process but only a term resulting from the conscious and intended vision of returning back to nature, in a way that does not diminish the convenience and quality of life of contemporary and future cities.

Contemporary Evolutive Strong Models and Approaches of the Interconnections: Between Communities and Nature

Contemporary Evolutive Models

Since the 1970s, the Gaia Theory (Lovelock, 1991) has been outlined whereby the Earth behaves as a superorganism, made up of all living organisms and the surrounding environment (natural or built – human settlements), giving rise even to new pluri-disciplines of scientific research. The theory is currently being explored that the living matter on Earth – the air, the ocean and the soil – forms a complex integrated and conscious system that has the ability to maintain the Earth as a place conducive to life. Gaia is defined as "a complex entity that includes the Earth's biosphere, atmosphere, oceans and soil; the whole constituting a feedback or a cybernetic system that seeks to create an optimal physical and chemical environment for life on this planet. The maintenance of relatively constant conditions through active control can be conveniently described by the term homeostasis" (Lovelock, 2000, p. 35).

Approaches of the Interconnections Between Communities and Nature

For a complete urban diagnostic process to assess biophilic city potential of the urban-territorial development today, all the city's diseases/existing conditions must be identified and investigated, not only of the built frame itself but also of all the systems that coexist in an interdependent relationship within the city. Thus, the city is seen as a living urban organism in permanent metamorphosis and change, as well as of all the living systems involved in the urban metabolic processes (Crăciun, 2008, p. 244), from the "green waves" existing in the city to the biological systems. Microorganisms, birds and animals must go through a planning process to identify and investigate their dysfunctions/diseases/affections at the level of biological pathology of life – human beings/inhabitants, fauna and vegetation, ecosystems, etc. – but also of urban pathology of the urban, rur-urban or rural settlement.

> The urbanists we need must imagine urban forms that can satisfy the sensibility of the urban world in which the sphere and the curve have a place. They must find a geometry more complex and truer than that of the nest and the straight, a geometry of life, of flower opening and plant growth, of elementary vital structures: the egg and the cell. (Ledrut, 1977)

Ledrut, through his comment, refers to the way/methods/form of development of cities, which is in antithesis to nature, the forms and structure of natural elements. He supports a rethinking of urban morphology, the creation of a more organic, more natural environment, in the sense of being familiar with the living world. This principle, of the functional imitation of the living world, in a utopian scenario, can also be translated to the functioning of urban system, creating new types of developments, more biophilic.

Examples of Biophilic City Approaches in Recent History and Today

Portland (United States): Recognized as one of the greenest cities in the world for innovative green initiatives: Portland showcases advanced storm water management practices by implementing the Green Street Project, reducing the percentage of impervious surfaces and increasing the percentage of green spaces, to allow water to infiltrate into the soil. It was also one of the first cities to implement an urban growth boundary, requiring increased density and compact development within the city while protecting farmland and natural areas outside the boundary. Portland has one of the highest percentages of green space per capita in the country, including large natural areas such as Forest Park and Oaks Bottom Wildlife Refuge. Portland's proximity to nature in and out of the city makes it moving towards becoming a biophilic city.

Singapore: Have built on the motto "Singapore – City in the Garden". The city features a developed network of trails and green corridors that connect to parks and natural areas. Thus, the city opens up to nature, and the inhabitants find it easy to spend time in open, natural spaces. The city has a developed green infrastructure, with a strong local policy for the promotion/adoption of green terraces and facades being implemented – something that contributes to the reduction of urban greenhouse effects (Singapore's urban area has even lower temperatures than the metropolitan average).

European "Garden Cities" are unique variations of the Ebenezer Howard model presented in his work "The Garden City", elaborating a new (utopian) model of development, which represented an innovation for certain urban developments in the world. The author proposed the outline of a new ideal society, formulating a possible solution to the "land problem" (more precisely the problem of sustainable development), which at the time consisted of (1) urban congestion and (2) rural exodus (Howard, 1902). In his vision, the central city is surrounded by rural satellites, in a strong collaborative relationship. His vision triggered a specific urbanization movement in Europe (especially England and Poland), being developed communities in nature – cities with a very high degree of planted spaces, even forests (Fig. 1).

A Different Source of Inspiration? Rural Areas as Biophilic Settlements

The Archetypical Rural Settlements – In Harmony with Nature?

Throughout history, one type of development considered by the authors to be one of the few anthropic systems/human settlements that developed a very unique relationship with nature is the rural settlements.

For centuries, the archetypical rural village represented places where people constructed communities that have a very deep and profound connection with the nature and landscape. Due to the way the rural communities (organically) "plan" the way they produce food, and energy, recycle waste, protect the environment and are aware of a wide variety of environmental and ecological knowledge, they can be considered biophilic communities. These settlements generated unique and organic inserts in the landscape in direct coordination with environmental conditions and the location of resources (Fig. 2). In these situations, the "love for nature" is manifested through high synchronisation between people (as a community) and natural cycles, with all the resources that they bring. These communities' existence and well-being were/are heavily dependent on crop production and all the benefits that the specific landscape can produce. Classic rural settlements managed to reach a relatively optimal level of self-sustenance, due to a long process of trial and error over generations. The "common sense" of rural life was not to disturb too much the natural way of things, understand how to ensure optimal production of resources for the next year and so forth. We can say that rural communities organically followed the principles of sustainable development and circular economy, characterized by "closer interaction between actors, by optimising the use of resources, and between production processes and social and cultural matrices" (McDonald & Brown, 1984). This "love for nature" has been passed down from generation to generation, creating an evolving heritage of knowledge, meaning, identity and a sustainable way of thinking.

Rural Communities as Inspiration for Urban Management and Policies

Several characteristics of the archetypal rural settlement have to be taken as important inspirations for the way we manage cities and urban infrastructures, and also for the way we can promote and support social cohesion and the creation of sustainable communities. For example, rural communities hold important knowledge on how to increase crop production sustainability, applying for centuries solutions or strategies that they are now reconsidered or (re)-innovated as nature-based solutions. The way agriculture is performed in these types of landscapes (considered subsistence agriculture, not to be confused with the idea of living on the verge of poverty) is very ecologically conscious. Crops have a high degree of biodiversity, purposely supported by rural communities, through the cultivation and propagation of pollinator plants, attracting insect predators to avoid diseases and pests. The rotation of cultures is carefully chosen in order to ensure the regeneration of land and preserve/maintain optimal conditions.

If we were to characterize rural settlements, using the lens of current policy objectives and agendas for cities and regions at the European level or worldwide, we can extract valuable knowledge: (i) high-level of understanding of resource circularity; (ii) very little waste; (iii) the value and impact of preserving/supporting biodiversity; (iv) use of ecologic materials for buildings; (v) use of locally sourced materials that create a unique and context-driven cultural landscape, in harmony with the natural landscape in which the settlement has evolved in; (vi) strong community spirit; and (vii) day-to-day contact with nature. These values present in the archetypical village are at risk of being lost or forgotten. Living in an age of technology, speed, globalization and variable economy, rural villages have population shrinking, severe population ageing or the present rapid urbanization processes, "transforming landscapes formed by the rural lifestyles into urban-like ones" (Antrop, 2000). Translating this way of life to the scale and complexity of a typical city is not considered

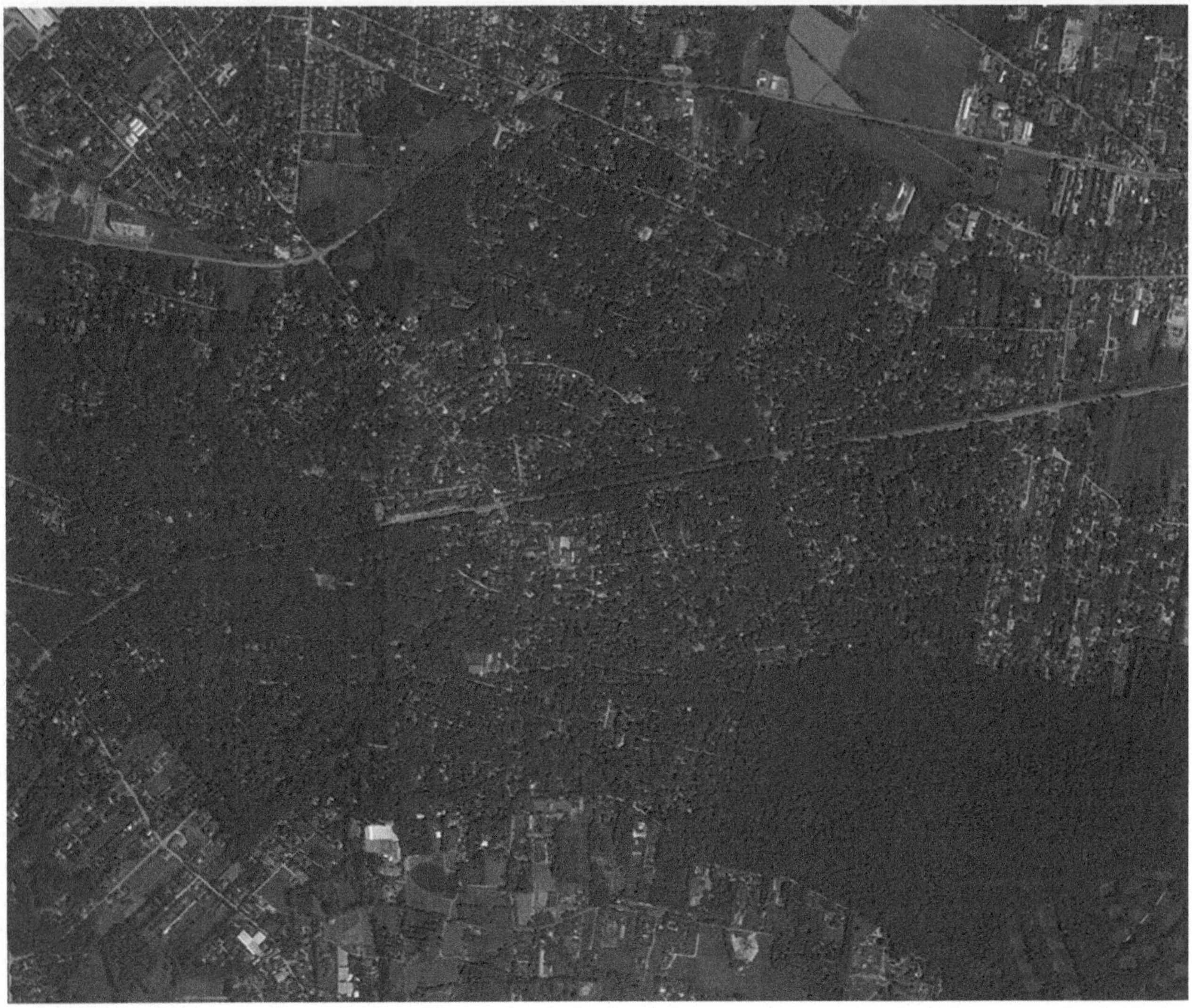

Fig. 1 Podkowa Lesna Garden City, Poland. In the 1920s, a unique housing estate was created here in the natural, forested surroundings, which has attracted many well-known figures from cultural, scientific and political life since the beginning of its existence (https://podkowalesna.pl/). (Photo: https://earth.google.com)

feasible, but nevertheless, important "lessons" can be extracted. For example, practices such as community gardens in cities, under the umbrella of nature-based solutions, have become in the last years very successful interventions that promote production of food locally, social interaction and environmental protection (Fig. 3). In many cases, community gardens were included in strategies/projects of the city, thus triggering a regeneration of neighbourhoods.

Territorial Approach for Biophilic Cities?

By 2050, the total Europe's population in urban regions will increase by 24.1 million, and the rural population to fall by 7.9 million. Currently, rural region represents 28% of Europe's population (ESPON, 2020). This phenomenon of shrinking put to risk relevant assets of the cultural landscape that holds not only a unique and, in many cases, a more sustainable way of life but also a huge potential to create new more biophilic territories, by connecting the rural settlements with the urban centres, ensuring that the specific value of the rural village persists. For this to be possible, it is important to "diversify the local economy and to creatively mobilise endogenous resources, increase resilience and adaptive capacity by selectively downsizing infrastructures and rightsizing the local economy" (ESPON, 2020). Hence, the policies at regional level are important to strengthen the role of rural settlements in an integrated territorial approach. Planning local-level value chains between rural components and urban ones at functional urban area level can reduce the

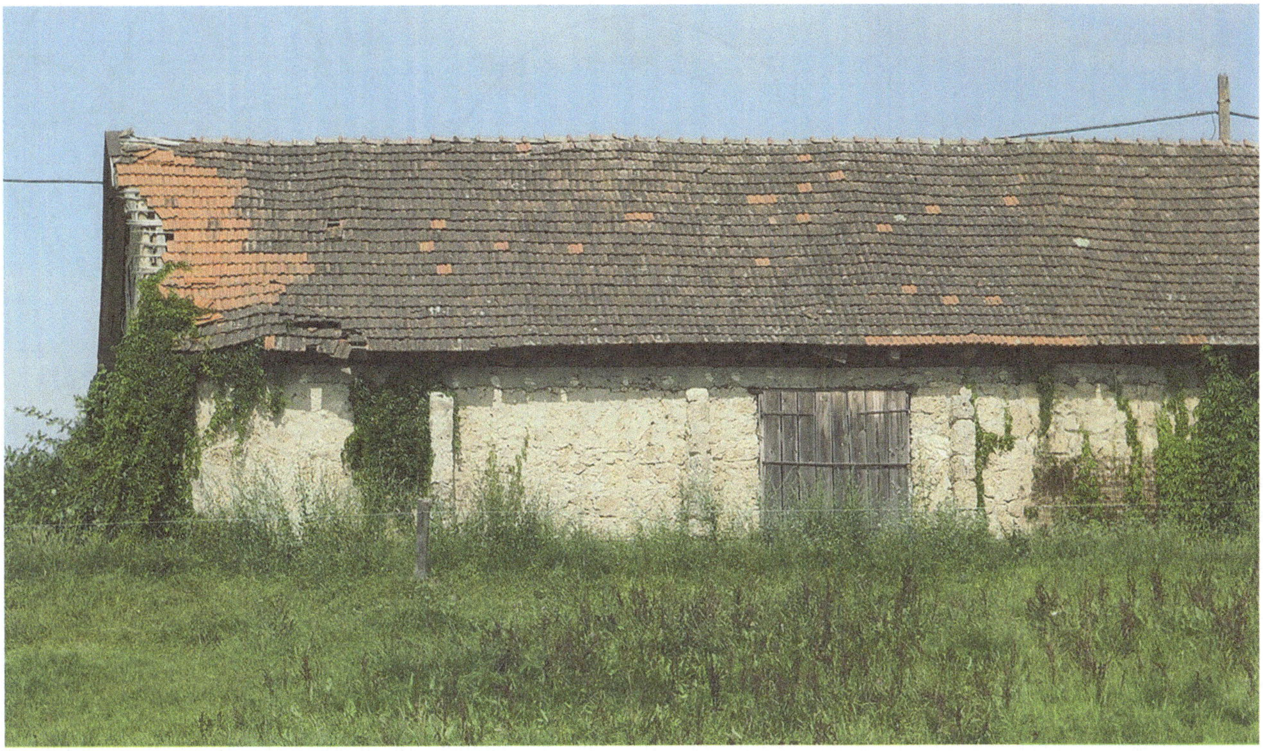

Fig. 2 Rural landscape in Eftimie Murgu village, Banat region, Romania. The village is known for its famous watermills, which are hundreds of years old. Archaeological discoveries attest that this village was inhabited by indigenous communities since 4000 years ago. (Photo: Authors)

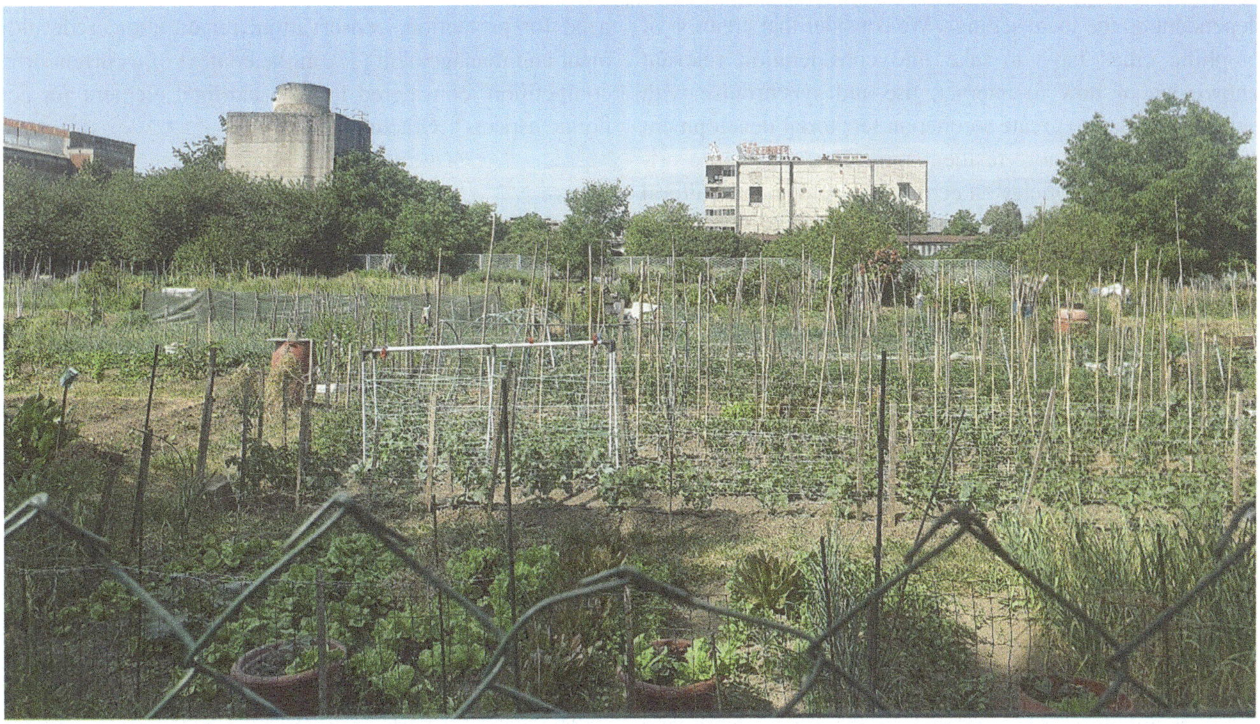

Fig. 3 Community gardens in Sesvete, Zagreb, located in a post-industrial neighbourhood. Community gardens became part of the local identity, as the neighbourhood slowly started to shift from industrial activities. (Photo: Authors)

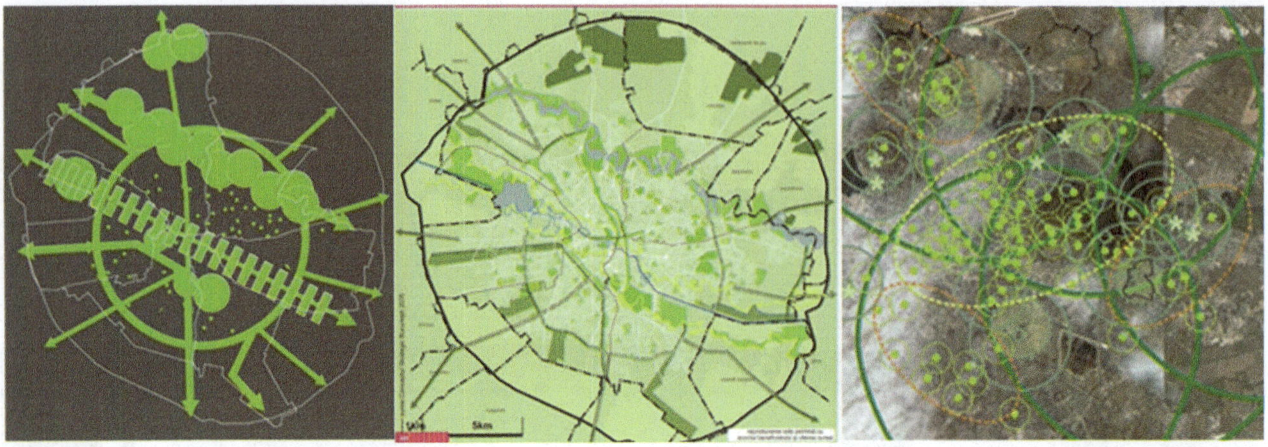

Fig. 4 Approach with biophilic components in studies and real projects. Urban Planning Strategies: "Public Space, Quality of Urban Life and the Landscape as a Fragile Resource. Mezzo-Landscape and the Natural, Anthropogenic and Cultural Landscape in the City of Bucharest", Fundamental Study in the Project: "Integrated Urban Development Strategy of the Bucharest City and Its Support and Influence Territory (BUCHAREST STRATEGIC CONCEPT 2035"/ CSB 2035, IMUAU-CCPEC, 2012)

impact on the environment, improve resilience of the local economy and empower rural communities to continue their way of life, adapted to modern times. Several relevant challenges emerge: (i) how to ensure that the activities performed in the rural areas (either production, agro-tourism, natural resources exploitation) remain eco-friendly/sustainable and follow/respect the co-created knowledge of generations and (ii) how to ensure a territorial coherence and interconnectivity of rural areas with urban centres. The answer is heavily dependent on the local settings. We consider that creation of biophilic cities have to take into consideration relevant knowledge of how to interact, use and synchronise with nature, and also to create a coherent territorial development in order to make more resilient territories and nature oriented. Emerging policies on ecologic development, cultural heritage, territorial mobility and ecosystemic services have to be integrated and carefully monitored. Nature have to be seen as the driving factor and the main guiding system to which other human-made infrastructures and communities have to respond to and align with.

Examples from Current Practices (Authors' Work)

Contemporary Territorial Strategies Approaches

The applied research "*Public Space, Quality of Urban Life and the Landscape as a Fragile Resource*" is a study that develops biophilic ideas related to the mezzo-landscape approach and the Natural, Anthropogenic and Cultural Landscape in the city of Bucharest, within the "*Integrated Urban Development Strategy of the City of Bucharest and Its Territory of Support and Influence – CSB 2035 – Strategic Concept of Bucharest 2035*" (Crăciun, 2012-1). It proposes a biophilic perspective regarding the rehabilitation policy and the establishment of intervention strategies regarding the aesthetic-functional valorisation of an urban public space, as well as the landscape, which becomes an asset, a welcome privilege, to consider as a criterion priority in development and planning, the landscape potential of the territory addressed (Fig. 4). In this approach is important also "the need for protection, conservation, renaturation, redevelopment and management (...), in the context of a current urban competition, considered to be a cardinal element for economic aspects" (Crăciun, 2012, p. 5).

Urban Approaches Project

The integrated landscape study, "*Buzău Green Municipality*", has as its main objective the elaboration of an integrated landscape study based on biophilic principles, which proposes the enlargement of green spaces, for the improvement of the quality of life, the elimination of pollution and the support of an eco-sustainable environment, more clean and resilient to face current climate pressures. Also, the study made proposals related to the development of new functions, activities and leisure/leisure facilities needed by residents, visitors and tourists, by proposing a biophilic green strategy to guide the urban development of the city and the preparation of the process of investments for its future development.

The complex work established the intervention priorities, addressing the following sections/issues: (a) the natural and quasi–/semi-natural framework characteristic of the area; (b) establishing the functions compatible with the status of

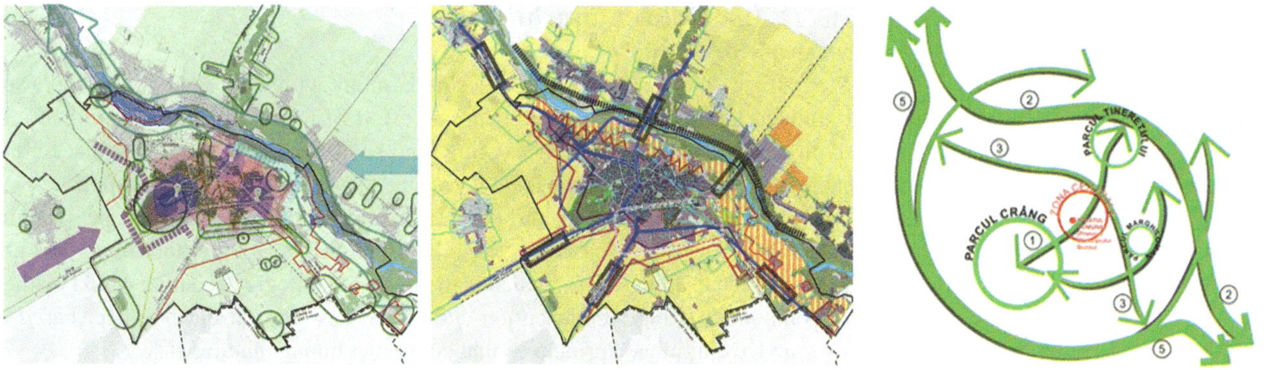

Fig. 5 Approach with biophilic components in studies and real projects. Urban Planning Documentation: Integrated Landscape Fundamental Study: "Buzău Green City", in the project: "Integral General Urban Plan of the Municipality of Buzau", (IMUAU-CCPEC, 2021)

the area and their balanced distribution in relation to the determining/dominant function; (c) environmental problems resulting from the analyses regarding the ratio of natural framework/constructed framework; (d) heritage values that require protection; (e) natural and anthropogenic risk elements, the area's relations with neighbourhood ecosystems; (f) general aspects regarding the implications of neglecting an urban function with an essential effect on the health of the population; (g) the quality of the existing green spaces, including the typologies, as well as their relationship in the territory; (h) the effects induced by the urban planning functions in the vicinity and proposals to improve the observed situation; (i) proposals for solutions that integrate all types of natural, anthropic and cultural landscape in extra-urban and intra-urban areas, necessary for the future City General Urban Plan.

The Strategic Scheme for the Development of the Landscape and the System of Green Spaces proposes the adoption of the landscape and the green system, seen not as a barrier to urban development but as a biophilic promoter and model of an intervention strategy and of a unique institutional form of management.

This approach can contribute not only to increasing the quality of life and the physical, mental, emotional and psychological health of the population and communities in general but also as a "vector" of development in the territory, of the urban economy (tourism, agriculture, forestry, eco-technologies, etc.) and includes several major biophilic strategic elements (see Fig. 5, third scheme): (1) the development of representative structuring green axes; (2) the establishment of major green-blue axes (in areas related to water), as a direction of interest and the permeable limit of relation with the local mezzo-landscape, inside the city; (3) the creation of secondary longitudinal green axes, green pockets and landscaped areas that connect the city's neighbourhood; (4) magnetic biophilic poles/nuclei of focus, attraction and development of different types: productive (urban agriculture, permaculture, seeds and horticultural research), leisure and promenade, sport-leisure and tourism; (5) the major green-yellow-blue belt, the area with the role of protection and permeable relationship with the macro-landscape at the level of the administrative and county territory; (6) the middle inner green belt, by converting some non-functional industrial spaces into areas of landscaped park with integrated activities (sport-leisure, exhibition-cultural, research, technology and IT, etc.) and the redevelopment of existing functional industrial areas, with a percentage of space green of min 30%; and (7) establishing the system of existing and proposed green spaces for development, to support the ecological effect of the green network of biophilic development.

Incremental Steps for Biophilic Cities Innovation (Research Projects and Urban Experiments)

Important steps in the journey of urban management and sustainable development are being made by research and innovation-actions projects. From current practice and projects, we can give two good examples of how nature can help urban regeneration processes: (1) proGIreg project (Productive Green Infrastructure Regeneration, funded under H2020 programme) deployed applied research project with eight nature-based solutions: landfills conversions, soil remediation, community urban farms and gardens, urban aquaponics systems, green walls and roofs, accessible green corridors, polinator biodiversity. In order to prove sustainability and impact, these solutions are experimented with four front runner cities: Dortmund, Zagreb, Turin, Ningbo. Other four cities – follower – build regeneration strategies of (post-industrial) neighbourhood in co-creation approach, aiming to bring positive economic, environmental and social impact on the local communities. Impact of the research project on wider level is that it capacitates and empowers the involved cities to elaborated policies and strategies differently, including nature in different infrastructures and urban systems. (2) UIA SPIRE Baia Mare (Smart Post-industrial Regenerative Ecosystems, funded by Urban Innovative Actions Programme) experiments with remediation of heavy-metal polluted land, with the use of nature, through a process called phytoremediation. The innovative aspect of the project is that phytoremediation plantations represent the trigger to new local productive value chains, biomass resulted from plantations being used for clean energy. At the same the time, within the new local value chains, the community represents the main component, being involved in environment-oriented activities and reclaiming the polluted land. Regeneration is considered on long term, through creating a new identity of the Baia Mare city, with communities oriented towards nature and care for the environment. Instead of mining and metallurgic work, the city remediates the historic pollution and creates affordable and clean energy sources.

Results

Biophilic urbanism is the practice of management, design and planning that makes it possible to create a city that is oriented towards nature. From the perspective of the current chapter, there are two important aspects of biophilic urbanism: (1) To create the context/settings for communities-nature interactions, through urban and architectural design, and also through territorial planning. People being in contact with nature provides wide variety of health and psychologic benefits, together with the ecosystem services that nature brings into urban environments. (2) To promote nature-oriented urban management, planning and integrated policy-making for sustainable development. If present in urban environment, nature does not have to be a "static" component of the local (eco)system. It is important to assess how nature can be taken as a resource than can produce, regulate, catalyse certain processes or mitigate, resolve or ameliorate different types of challenges that cities are facing. An integrated approach is needed, besides technological and engineering innovations. Going back to rural settlements, in order to be efficient and sustainable, technologies can play a part but is more important to have an integrated approach. Ecological technical solutions, biophilic architecture and other such solutions currently have a very advanced technological readiness level. The current paper notes the lack of urban management and planning practices that address these design techniques in an integrated way. Natural systems, or in short nature (including green infrastructure, nature-based solutions, natural areas, natural phenomena, biodiversity, etc.) must be seen both as a valuable resource, able to generate other resources, to catalyse certain processes and address a wide range of urban issues: productivity (energy, food), local environment and microclimate, urban resilience.

Conclusions

Creating cities in harmony with nature, which are more sustainable, is not an effort of innovation but an effort of integrating existing (and past) knowledge and design according to the local landscape conditions and level within it.

Anthropic urban and architectural landscape planning based on biophilic principles is an important integrated process, which is the basis of strategic decisions with a resilient contribution and which can prevent development based on principles that contradict human nature.

Policy instruments are considered optimal approaches for triggering long-term impact, but for that, there has to be concrete proof of the benefits the nature brings.

In this sense, an approach in integrated inter-, multi- and trans-disciplinary teams of research, conservation, design and urban planning specialists is necessary, for an efficient approach to the city and the landscape.

Incremental advancements are realized through applied research, and strategic and planning documents. As much as the political and administrative component represents the turning factor of urban and regional development, also the active and vocal involvement of experts in multi- and inter-disciplinary environment is needed to steer development directions and promote coherent and sustainable ways of development.

The biophilic perspective is a truly eco-sustainable approach and a preoccupation since ancient times of mankind, regarding the relationship between man and nature. This represents, at the same time, a chance for the future to capitalize on principles that are centred on life and living, in cities and territories that can get closer to the inhabitants and the community, at the expense of an anthropic, technological society that brings problems of urban and human pathology, alienation and disease.

The city is a gigantic processor not only of the anthropic, built, technical, mobility and infrastructure components, of capital and power, but also processors of people's lives, binder of communities, ecosystems, ecotone and general life.

References

Antrop, M. (2000). Changing patterns in the urbanized countryside of Western Europe. *Landscape Ecology, 15*, 257–270.

Beatley, T., & Newman, P. (2013). Biophilic cities are sustainable, resilient cities. *Sustainability, 5*(8), 3328–3345.

Crăciun, C. (2008). *The urban metabolism. Un unconventional approach to urban organism*. "Ion Mincu" University Publisher.

Crăciun, C. (2012). Research study: *Public space, quality of urban life and the landscape as a Fragile resource*. Mezzo-landscape and the natural, anthropogenic and cultural landscape in the City of Bucharest in the project: "Integrated Urban Development Strategy of the City of Bucharest and its Support and Influence Territory – CSB 2035 – BUCHAREST STRATEGIC CONCEPT 2035", p. 5.

Howard, E. (1902). Garden cities of to-morrow. London: Swan Sonnenschein & Co.

Ledrut, R. (1977). *L'espace en question ou le nouveau monde urbain*. Editions Anthropos.

Lovelock, J. (1991). *Gaia. The practical science of planetary medicine*. Gaia Books.

Lovelock, J. (2000). *A new look at life on earth*. Devanata.

McDonald, G. T., & Brown, A. L. (1984). The land suitability approach to strategic land-use planning in urban fringe areas. *Landscape Planning, 11*(2), 125–150.

Policy brief: Shrinking rural regions in Europe | espon. Available at: https://www.espon.eu/rural-shrinking (Accessed: 31 May 2023).

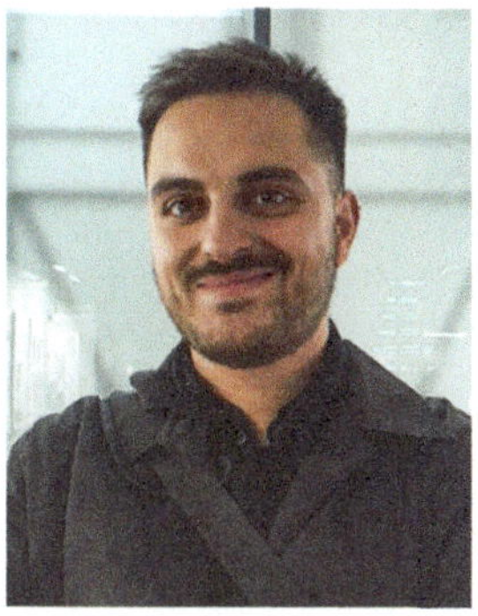

Codruț Papina (Ph.D. candidate) is an Urban and Landscape Planner and Designer, doctoral student, project manager of applied research projects, holding a degree in Landscape Design and Planning with a master's degree in Urban Management for Competitive Cities.

His doctoral studies at IMUAU Bucharest focus on the development of urban peripheries, biophilic cities and sustainable approaches to how cities can grow and transform. He has valuable experience in co-creation processes for strategic planning, urban design and urban regeneration.

In the last years, he was involved in various urban planning projects, sustainable development strategies and various consultancy or strategic documents, along participation and winning in urban and landscape design competitions. Currently he is project manager of H2020 project proGIreg (Productive Green Infrastructure for Post-Industrial Regeneration) and project leader of JPI Positive Energy Districts Call project KINETIC (Knowledge Integration for Neighbourhoods in Energy Transition led by Inclusive Communities).

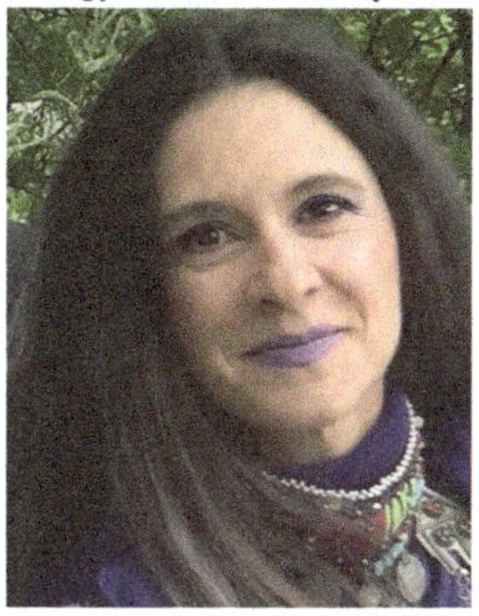

Cerasella Crăciun is an Architect, Urban and Landscape Design Planner – member of the Order of Architects from Romania, Register of Urban Planners, and Union of Romanian Architects. She is Full Professor, Habilitated to Advise Doctoral Research, "Landscape and Territory" Master Director, Faculty of Urban Planning, University of Architecture and Urbanism "Ion Mincu" Bucharest. She is titular of the courses: "Landscape Design," "Protection and Restoration of the Natural Heritage," "Garden History," "Landscape Research, Scenography and Art," "Eco-Development and Landscape," and "Creative and Transdisciplinary Heuristic Research Methods." She is author/co-author/coordinator of over 120 projects, research, studies; published 9 books, 17 book chapters, over 80 articles, 130 conferences, and 80 exhibitions/contests/competitions. She obtained over 25 awards/nominations/selections national/international distinctions as author/co-author – National Architecture Biennale, Urban Planners Romanian Union, European Union Prize for Contemporary Architecture Mies van der Rohe Award, Green Building Council award, etc. She completed research activity in the transdisciplinary fields of Architecture/Sacred – Spirituality/Cultural Landscape/Tradition/Anthropology – Creativity/Transformational Education of the Future, and Border Frontier/Holistic Sciences.

Bionic Criteria for Sustainable Labelling of Buildings

Ana Mohonea

Abstract

This chapter proposes a strategy for sustainable labelling of buildings, on the basis of bionic principles resulting from the interconnection between bionics and the criteria for labelling sustainable buildings. It presents bionic principles that are based on principles of life and labelling criteria through BREEAM and LEED systems. An exploratory analysis is performed and two methods of correlation analysis are proposed: at level of terms and at level of relations between bionic principles and labelling criteria listed.

Bionic architecture represents building smart and sustainable, using resources responsibly. Following extensive research, I concluded that bionics, through its principles, can promote sustainability in construction. Also, sustainable developments have the same purpose and, in most cases, use bionic principles. The challenge of the conducted study was to examine whether bionic principles are found in sustainable development and if, by correlating with sustainable development in architecture, sustainable labelling criteria based on bionics can be achieved.

It is noted that this study is a proposal for completion and improvement of the existing labelling systems.

Keywords

Architecture · Buildings · BREEAM · LEED · Bionic principles · Renewable criteria · Sustainable labelling · Innovation in architectural certification

A. Mohonea (✉)
"Ion Mincu" University of Architecture and Urban Planning, Bucharest, RO, Romania
e-mail: ana.mohonea@uauim.ro

Introduction

Bionic architecture builds smart and sustainable, using resources responsibly. At the same time, durable development has the same purpose and, in most cases, uses bionic principles. The challenge of the study was analyzing if the principles of bionics are also present in durable development and to find out if through correlation with the criteria of durable labelling they can become bionic criteria for architectural sustainability.

Environmental changes nowadays have serious consequences that affect mankind. At architectural level, solutions are sought to promote sustainability. Based on this search, there are two approaches. The first approach is one of durable development, based on nature. It means to analyze the labelling criteria of buildings in order to prove their sustainability and how green they are. The second approach is found in bionics, which uses a technique principle rooted from nature that guides life. Both approaches have the same purpose, that of finding the necessary elements for sustainable architecture to answer actual consequences that happen to the natural environment.

This analysis proposes to show the similarities and differences between bionic principles (which have been taken, analyzed, and synthesized from life principles) and sustainability criteria (the criteria have been taken and analyzed from the labelling criteria of green buildings) and what elements can be found from one's category into the other's, but also what relations can be made between them.

One aspect of novelty of the article refers to finding ways to measure the integration in nature of a building. The architect strives to integrate architecture *in tandem with nature* and approaches this by demonstrating a correlation of bionic principles with criteria for sustainable labelling.

The methodology was focused on defining principles of bionic architecture. This was achieved by using analogy and systematization of the concepts contained in the principles of life and principles of bionics. The correlation method was based on the comparison of bionic principles and criteria for

M. R. Cervera Sarda et al. (eds.), *Architecture Inspired by Nature*, https://doi.org/10.1007/978-3-031-33144-2_15

labelling sustainable systems in order to achieve bionic criteria of sustainability, thus making a correlation analysis at the level of terms and relations.

For this study, the chosen labelling criteria is Method for Assessing the Impact of Buildings on the Environment (BREEAM) in Europe and Leadership in Energy and Environmental Design (LEED) in the USA. The chosen two systems are the most important and utilized systems in labelling sustainability worldwide. The reporting of bionics to each of these was also pursued.

LEED and BREEAM represent the most accepted evaluation systems from the industry of sustainable buildings. They have similar means, approaches, and structures of evaluation, but the methodology differs. These evaluation systems take into consideration a big number of parameters in evaluating the performance of buildings, BREEAM having a bigger number of managed buildings than LEED.

Labelling Criteria of Sustainable Buildings

BREEAM is an evaluation system that was made in 1990 in Great Britain, being the oldest analysis and certification method of sustainable building. The points are given for each criterion and sub-criteria depending on the analysis domain. It is applied in different architecture programs: offices, schools, houses, industrial units, etc. The building can be categorized as Pass, Good, Very Good, Excellent, or Outstanding. The BREEAM program confers a sustainability strategy that takes into consideration all zones from the construction branches (BREEAM, 2019).

Adopting the BREEAM program has many advantages both from an environmental and economic point of view. Some of the advantages are implementation of low energy requirements and operating costs, higher rental value at property level, improvements in the level of productivity due to workers' access to a comfortable working environment, an improved reputation of the construction industry by showing its commitment to protect the environment, short time to sell buildings, as well as other advertising advantages.

The building sustainability assessment categories are Management, Health and Well-Being, Energy, Transport, Water, Materials, Waste, Land Use and Ecological Aspect, Pollution, and Innovation (Fig. 1).

The BREEAM system is used in more than 70 countries. In most countries in Europe, a new system is developed that is tailored to the specifics of each respective country in collaboration with the National Scheme of Operators. The collaboration between them is in advanced stages in countries such as Holland, Spain, Norway, Sweden, and Germany. Achieving BREEAM standards through design helps to create sustainable buildings with an increased quality level.

LEED is the most popular evaluation system at global level for building's sustainability. LEED aims to promote design and construction that increases profit and reduces the negative effects of buildings on the environment, improving the health and comfort of occupants. LEED applies an environmentally friendly method even in terms of managing their communication (which is done mostly online) and minimizing paper waste. To accredit buildings, LEED evaluates engineering, design, and construction (LEED, 2019).

The LEED labelling system approach can give even projects in the design stage a score, but they can only receive certification at the end of construction. LEED grades evaluated buildings in four categories: Certified, 40–49 points; Silver, 50–59 points; Gold, 60–79 points; and Platinum, +80 points.

The LEED assessment categories are as follows: Location and Transportation, Sustainable Sites, Water Efficiency, Energy and Atmosphere, Materials and Resources, Indoor Environmental Quality, Design Process and Innovation, and Regional Priority (Fig. 2).

Bionic Principles and Sub-principles

The principles of bionics can be defined by means of principles of life, specific to all organisms. Life always evolves and adapts, managing to create favorable conditions for living. By analyzing the bionic principles and its sub-principles, bionic design methods can be deduced, leading to improved sustainability of future architecture projects.

Nature never wastes and never uses more than it needs. The principles of nature formulated by Janine Benyus emphasize how nature (Benyus, 2002), as a life-generating element, relates to the components outside it:

- Nature works on the basis of sunlight.
- Nature uses only the amount of energy it needs.
- Nature adapts the form to the required function.
- Nature reuses everything.
- Nature supports cooperation.
- Nature relies on diversity.
- Nature implies the conditions of the place.
- Nature weighs the excesses from within.
- Nature has as a principle the power of limits.

Everything can be related to the environment, an environment in which the Earth represents the catalytic element so that the principles of life can also be defined in relation to it. We can say the Earth is defined by limits, but these do not stop it from relating correctly to the environment, to capitalize on the interdependencies between its elements, thus cultivating relations of cooperation and self-organization.

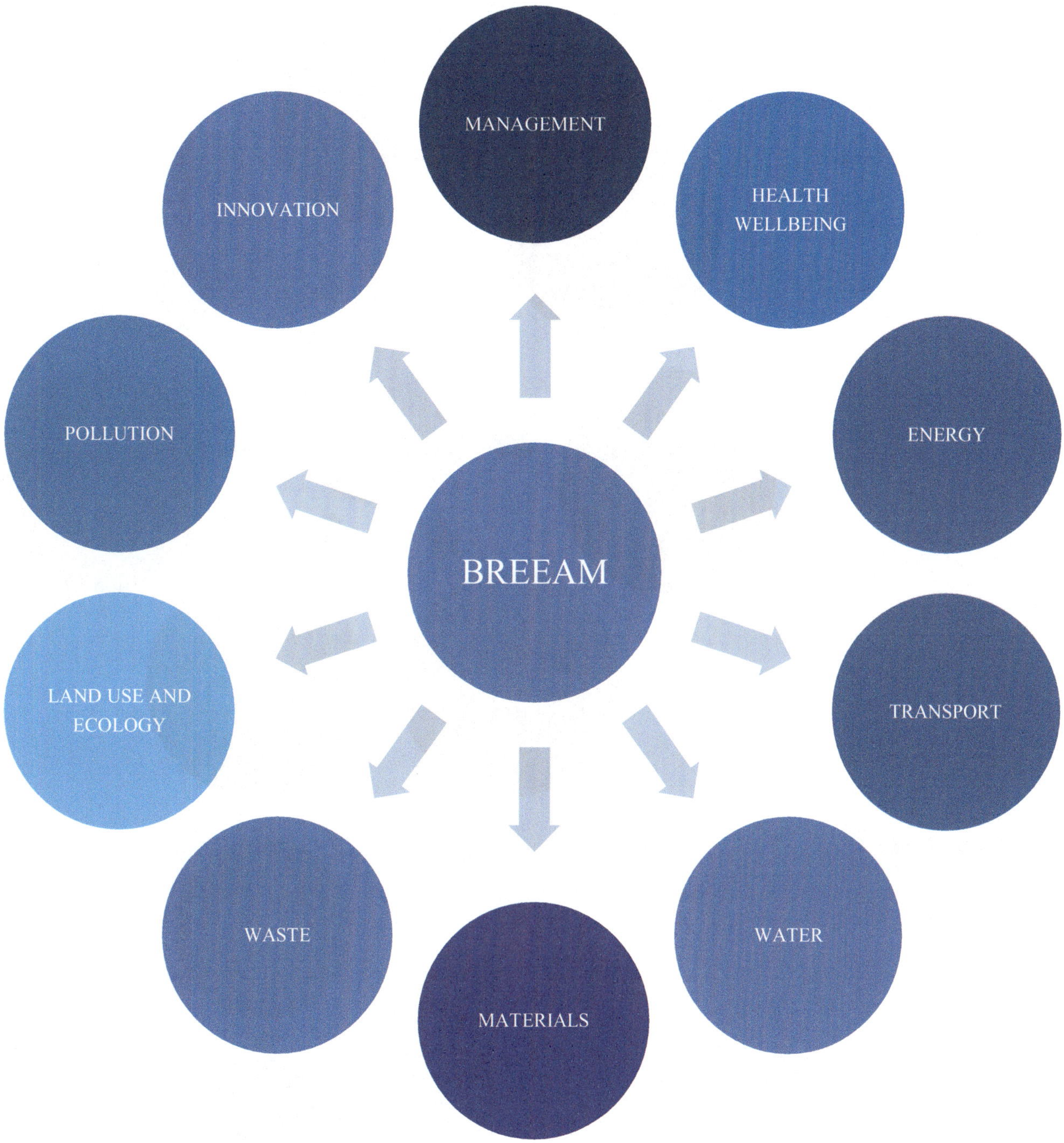

Fig. 1 BREEAM System. Evaluation categories for sustainability. (Ana Mohonea)

Earth optimizes living conditions, rather than maximizing, through the process of recycling all natural materials, adopting a multifunctional design and adapting form to function. Earth's primary resource is water, and it promotes materials that are not harmful to life, relying on self-assembly and chemical processes occurring in water.

Earth is characterized by redundancy, decentralization, distribution, and diversity, which allow the cyclical processes of life to be maintained (Wines, 2000).

The principles of life (Fig. 3) can be defined and understood through guidelines that promote sustainability. Bionics can respond to these needs; therefore, the six principles of life (De Garrido, 2012) defined as bionic principles form the basis of the analysis in the proposed study.

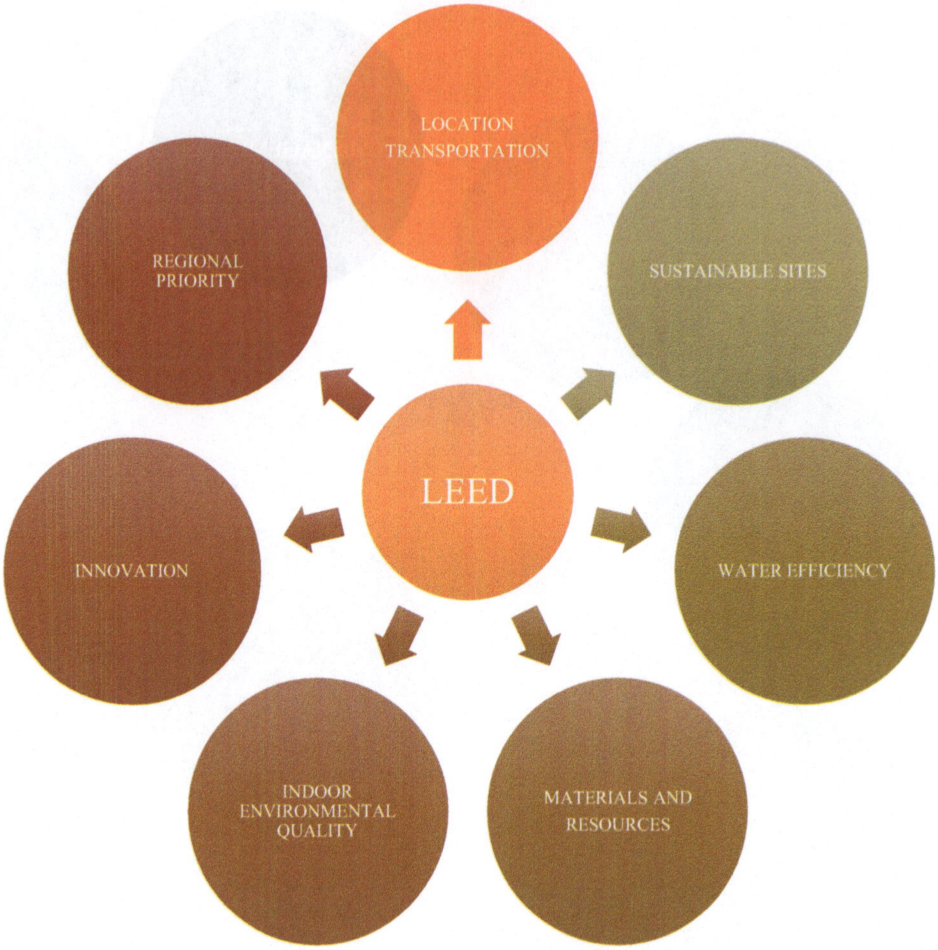

Fig. 2 LEED System. Sustainability evaluation categories. (Ana Mohonea)

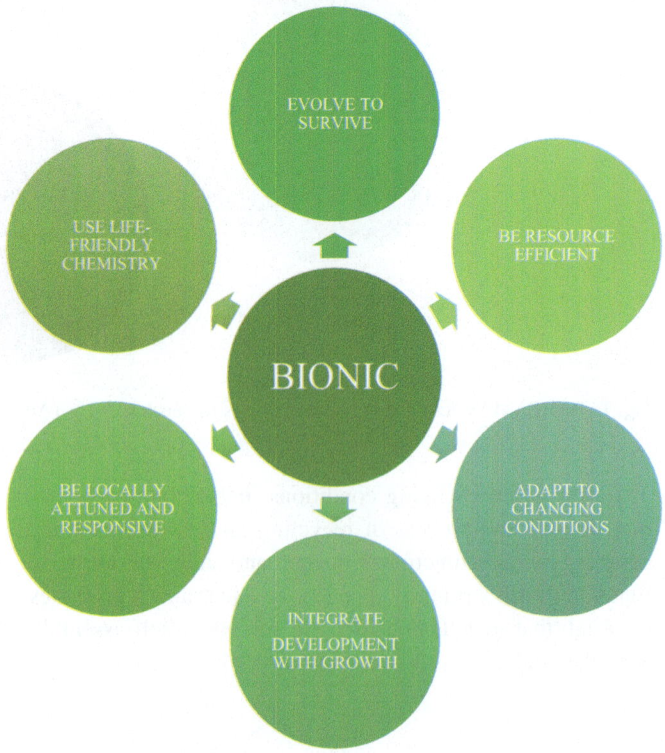

Fig. 3 Scheme of bionic sustainability. (Ana Mohonea)

The Interconnection Between the Labelling Criteria and Bionic Principles

The analysis was carried out basing on bionic principles and sub-principles, derived from the principles of life, as well as basing on BREEAM 2018 and LEED V4.1 (2019) criteria. For the criteria, the last application options were chosen, the field being for *New Constructions and Major Renovations*.

The principles and criteria were structured in order to carry out the analysis. An attempt was made to correlate the six bionic principles with the criteria to which they correspond, but a link was also made between BREEAM and LEED at the criteria level. Following the relationship of bionic principles with the labelling criteria of sustainable development, a series of advantages that favor sustainability can be deduced.

Both the principles and the criteria have as stated objectives: the creation of indoor comfort through indoor air quality, the creation of both renewable and efficient energies, and the creation of methods for capturing water and minimizing its consumption, managing emissions, optimizing and separating waste, and creating a resilient system for the use of sustainable materials.

Nature has the same goals, and through its mechanisms, it has managed to create solutions (Pawlyn, 2011). Building with a low environmental impact and implementing cutting-edge technologies can also succeed in creating sustainable solutions. An analysis was carried out at the level of aspects and indicators present both at the level of BREEAM and LEED, as well as at the level of BIONICS (Fig. 4).

Any new method of designing in close relation with bionics could improve the sustainability performance of a building. It can bring demonstrable benefits against already existing sustainability criteria if they can represent viable solutions for sustainable development in the current context. Through bionics, which currently offers the most innovations, new solutions can be created, but also a bridge between bionic principles and labelling criteria promoted by sustainable development.

Bionic Criteria of Sustainable Labelling

The study that was carried out was based on an exploratory analysis, at the level of keywords and at the level of relationships between the three analyzed systems.

The first method aimed to perform a correspondence analysis at the level of terms, and the second aimed to generate a correspondence analysis at the level of relationships between the bionic principles and the sustainable criteria of BREEAM and LEED. The analysis program Atlas.ti version 7 was used to carry out the study.

Regarding the used language in the study, a few clarifications are imposed:

- The principles of life are defined as bionic principles, and the generated documents are related to the name BIONIC.
- Bionic principles are defined by bionic sub-principles.
- Sustainable criteria are defined for each labelling system separately by sub-criteria.
- The sustainable labelling criteria are simply called criteria and are represented by LEED and BREEAM, these naming the related documents.
- The criteria and principles represent the three categories of analysis.
- Writing lines of code to help generate solutions was based on the analyzed principles and criteria as well as their correlation; codes and subcodes define the keywords in the analysis categories.
- The analysis was carried out in English.
- The analysis methods refer to correspondence analysis at the level of terms and correspondence analysis at the level of relationships.
- In creating the schemes, tables, and graphs, the same colors chosen to represent the criteria and principles at the level of visual representation in this study were kept every time. Thus, blue was chosen for BREEAM, red for LEED, and green for BIONIC.

For the two analysis methods proposed, the creation of codes was sought, which generated a series of links based on the contents of the subcodes. This approach was used because it was desired to obtain the most conclusive results. After the first stage of the study, represented by document analysis, followed the second stage of analyzing the results and generating solutions.

Through the application, the codes and subcodes of the search-analysis variables were generated, obtaining after the analysis the "network" of the study, a map of each code. The competition of the codes and their interconnectivity were realized starting from the concrete results obtained in the case study generated with the help of the documentation in English. Behind all the generated codes are words and paragraphs considered relevant for the analysis, belonging to several documents entered within the project. Subcodes were generated to search for the core elements characteristic of each of the LEED and BREEAM criteria – correlated with BIONIC.

After introducing several types of documents into the program (doc, docx, rtf, html, mht, pdf, jpg, avi), with the three categories of analysis as the subject, words and phrases were selected that defined important concepts in the realization study, and codes and subcodes were generated. The generated codes were named according to the mentioned principles and criteria. To generate them, the documents introduced in the first stage were analyzed (Fig. 5).

ISSUES	INDICATORS	BREEAM	LEED	BIONIC
ECOLOGICAL	Environmental emission	YES	YES	YES
	Resource materials	YES	YES	YES
	Waste	YES	YES	YES
	Water	YES	YES	YES
FUNCTIONAL	Area efficiency	NO	NO	YES
	Conversion feasibility	NO	NO	NO
TECHNICAL	Fire protection	NO	NO	YES
	Durability	YES	NO	YES
	Cleaning and maintenance	NO	NO	YES
	Endurance(hail, storm, flood, earthquakes)	NO	NO	YES
ECONOMIC	Cost of life cycle	YES	NO	YES
	Assessing value	NO	NO	NO
SOCIO-CULTURAL	Security	YES	NO	YES
	Accessibility without boundaries	NO	NO	YES
	Regional and social aspects	NO	YES	YES
PROCESS-MANAGEMENT	Planning process	YES	NO	YES
	Construction site process	YES	YES	YES
	Applying use	YES	YES	YES
	Exploitation	YES	NO	YES
SITE LOCATION	Micro location	YES	YES	YES
	Public transport conditions	YES	YES	NO
	Cycling accessibility	YES	YES	NO
	Neighbourhood	YES	YES	YES
	Ways of work legal	NO	NO	NO
	Possibility of expansion	NO	NO	YES
	Area consumption	YES	YES	YES
	Protection of nature and common areas	YES	YES	YES
	Biodiversity	YES	NO	YES
ENERGY	CO2 emissions	YES	YES	YES
	Energy efficiency	YES	YES	YES
	Renewable energy	YES	YES	YES
	Equipment for technical constructions	NO	YES	NO
	Monitoring	YES	YES	YES
	Under metering	YES	NO	YES
	Energy efficiency of the electric building equipment	YES	NO	YES
COMFORT AND HEALTH	Thermic comfort	YES	YES	YES
	Air quality indoors	YES	YES	YES
	Acoustic comfort	YES	NO	YES
	Visual comfort	YES	YES	YES
	Comfort in operation	YES	YES	YES
DESIGN AND INNOVATION	Architecture	NO	NO	YES
	Art indoors and outdoors of the building	NO	NO	YES
	Innovation	YES	YES	YES

Fig. 4 Issues and indices present in BREEAM, LEED, and bionics. (Author: Ana Mohonea)

First Method – Correspondence Analysis at the Level of Terms

Through the first method, a correspondence analysis of common terms or keywords was proposed within the three categories of analysis: BIONIC, BREEAM, and LEED.

In the first stage, the analysis of terms of the three previously mentioned categories was carried out, at the level of the entire document. The next stage assumed the relationship between the content of the BIONIC principles and the content of BREEAM and LEED sub-criteria. Also, the content of the BREEAM and LEED criteria was taken into account in relation to BIONICS sub-principles, with the aim of establishing to what extent the compared contents refer to the same notions and name the same ideas.

The source of the codes taken in the study is represented by keywords found in English codes and subcodes. They are words that are the basis of the analyzed codes (approximately

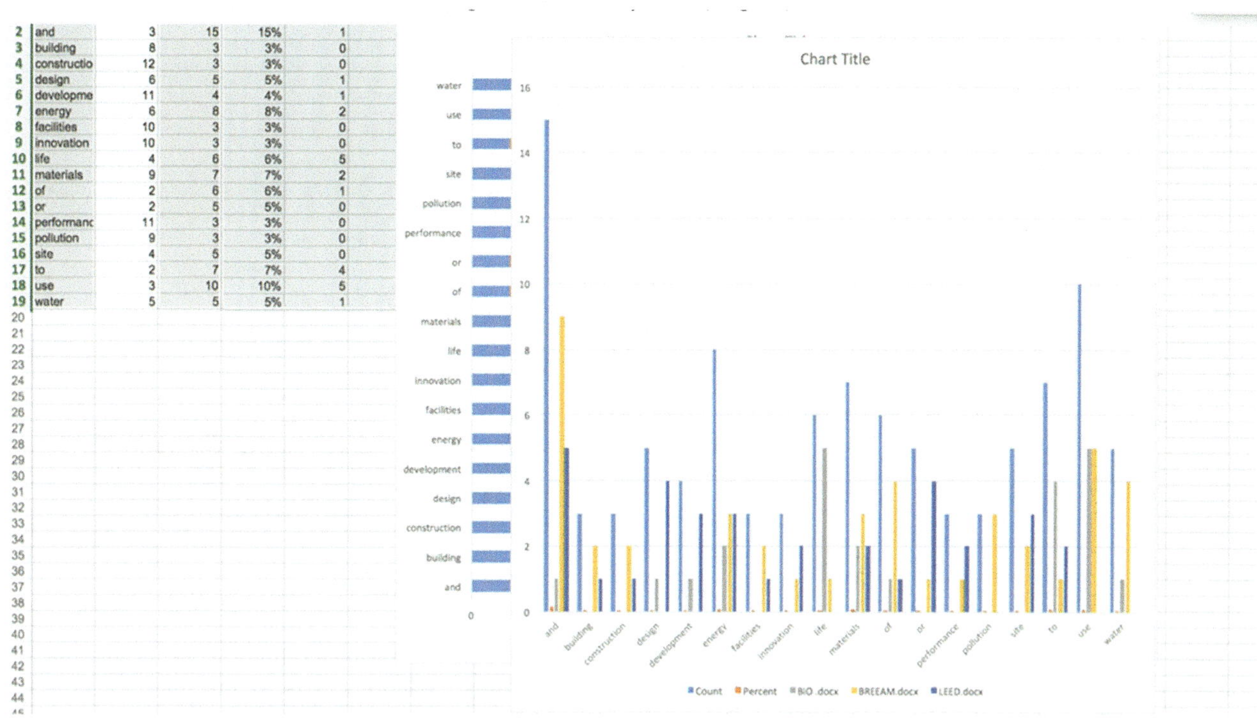

Fig. 5 Analysis study first stage. (Ana Mohonea)

140 words were the basis of the analysis imported within BIONICS). Filtering of relevant words was done through the program, finding that there are some keywords that were the basis of several codes or subcodes within the application (e.g., the word construction is found 10 times in LEED, 4 times in BREEAM, and 2 times in BIONIC).

The analysis highlights the presence of relevant words within the criteria and principles. Statistically, the same relevance per word or relevance per code was observed, within the analyzed criteria, both within the analysis performed. Following the analysis, the presence of bionic principles was observed both in the LEED criteria and in the BREEAM criteria, not necessarily at the level of the same keywords, but the final weight was the same. The results obtained were recorded in the presented tables, being counted by means of databases generated by the applications used in the analyses (Figs. 6 and 7).

Second Method – Correspondence Analysis at the Level of Relationships

Through the second method, a correspondence analysis was carried out between the relationships established between the three categories of analysis (Fig. 8). First, the links between codes and principles, respectively, between codes and criteria, existing within each category were generated. Thus, the relationships between BIONIC and its codes (codes that were named after the bionic principles) first resulted, then between BREEAM and its codes (codes that were named after the BREEAM labelling criteria), then LEED and its codes (codes that were named after the LEED labelling criteria), and then the links between the common codes of the two criteria were created (Fig. 9).

The second part of the analysis consisted of generating relationships between BIONIC and BREEAM, and the third part of the analysis between BIONIC and LEED. The relationships obtained were relationships of association, causality, opposition, and membership. For example, between BREEAM and its codes, as well as between LEED and its codes, the relationship of belonging (is part of) was generated. The association relationship (is associated with) was generated between some of the BREEAM and LEED elements. Two types of relationships were generated between BIONIC and its codes: association and belonging (is associated with and is part of). Three types of relationships were generated between BIONIC, and BREEAM/LEED based on association, membership, and opposition codes (is associated, is part of and contradicts).

Results Interpretation

It is observed, following the analysis, the finding of the same types of relationships between the bionic principles and the labelling criteria, but in different proportions.

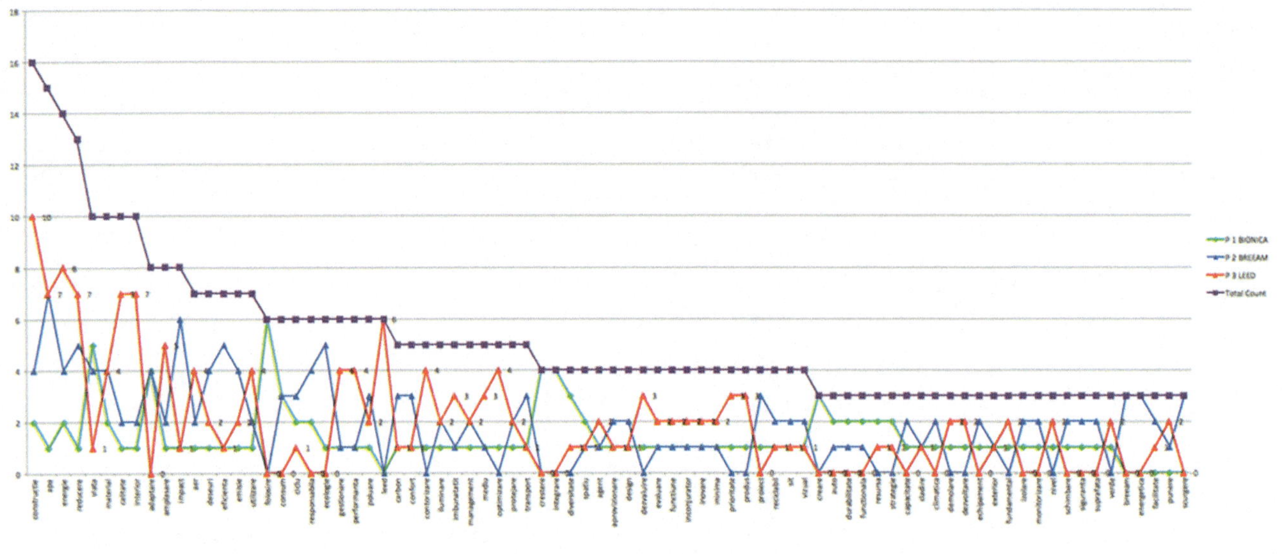

Fig. 6 Obtained results after conducting the analysis with first method. (Ana Mohonea)

The study results help to develop architectural-bionic design methods, adapted to the current requirements of sustainable development, defined as evolutionary method, regenerating and smart.

Initially, it was proposed to measure the integration in nature of a building by comparing bionic principles with the basic requirements for the construction of buildings, but later in the development of the thesis, those were replaced with labelling schemes of sustainable construction.

It was found that such labelling schemes and certifications, known worldwide, are based on a number of criteria that allows their relationship with bionic principles and demonstrates the possibility to measure integration in nature of a building.

The study showed that bionics can be implemented on several levels, and therefore, the section of case studies was larger than was originally thought, being added to the case studies based on the presentation of a new method of analysis to those already proposed.

This study does not propose an entirely bionic approach on built space but highlights the link between bionic and sustainable architecture in order to enhance the comfort of living and to understand the importance of nature in achieving sustainable development. Not everything must be bionic because it would be a world with which the people could not identify themselves. It is important for architects to propose harmonious bionic solutions that do not contradict the cultural-spiritual matrix that defines contemporary society.

The innovative value of this analysis is represented by bionics impact in architecture, achieving a comprehensive content obtained through an interdisciplinary research methodology, by analyzing specific issues and generating multiple connections and interactions between areas. Anticipating the risk factors and implementing environmentally responsible solutions is a conscious approach by the architect who is proposing solutions based on the principles of sustainability.

It can obtain *certificate of integration into nature* through the application of bionic criteria for labelling the sustainability of a building and can develop a method for architectural-bionic design who will allow to measure the percentage of integration of the building in nature. Future generations of architecture and design students can learn to apply the bionic criteria, developing sustainable projects with a high degree of integration in nature.

Conclusions

The study of the relationship between natural habitats and ecosystem can be a starting point for the development of sustainable urban habitats. Analysis, comparison, and replication of principles used in nature, at their level of functioning, represent the chance to design urban habitats to be in harmony with the natural ones, and to join in the effort to protect the planet, the biodiversity, and, not least, the life of future generations.

In this study, bionic principles and BREEAM and LEED labelling criteria were analyzed and compared. We conclude that several parameters from bionics can be implemented in sustainable labelling criteria in order to improve the quality of buildings for a more durable architecture and greener environment.

The article proposed a strategy (correlation analysis at level of terms and at level of relationships) in order to achieve criteria of sustainable labelling based on bionic principles, aimed at reducing the negative impact of building on the natural environment. Filling standards for evaluating sustain-

WORDS	P 1 BIONIC	P 2 BREEAM	P 3 LEED	Total Count		WORDS	P 1 BIONIC	P2 BREEAM	P 3 LEED
construction	2	4	10	16		construction	0,125	0,25	0,625
water	1	7	7	15		water	0,066667	0,466667	0,466667
energy	2	4	8	14		energy	0,142857	0,285714	0,571429
reduction	1	5	7	13		reduction	0,076923	0,384615	0,538462
life	5	4	1	10		life	0,5	0,4	0,1
materials	2	4	4	10		materials	0,2	0,4	0,4
quality	1	2	7	10		quality	0,1	0,2	0,7
interior	1	2	7	10		interior	0,1	0,2	0,7
adaptability	4	4	0	8		adaptability	0,5	0,5	0
location	1	2	5	8		location	0,125	0,25	0,625
impact	1	6	1	8		impact	0,125	0,75	0,125
air	1	2	4	7		air	0,142857	0,285714	0,571429
waste	1	4	2	7		waste	0,142857	0,571429	0,285714
efficiency	1	5	1	7		efficiency	0,142857	0,714286	0,142857
emissions	1	4	2	7		emissions	0,142857	0,571429	0,285714
utilization	1	2	4	7		utilization	0,142857	0,285714	0,571429
use	6	0	0	6		use	1	0	0
consumption	3	3	0	6		consumption	0,5	0,5	0
cycle	2	3	1	6		cycle	0,333333	0,5	0,166667
resposability	2	4	0	6		resposability	0,333333	0,666667	0
ecology	1	5	0	6		ecology	0,166667	0,833333	0

Fig. 7 Analysis – percentage of retrieval of each relevant subcode within each of the analyzed categories (% P1 BIONIC + %P2 BREEAM +% P3 LEED = 100%). (Author: Ana Mohonea)

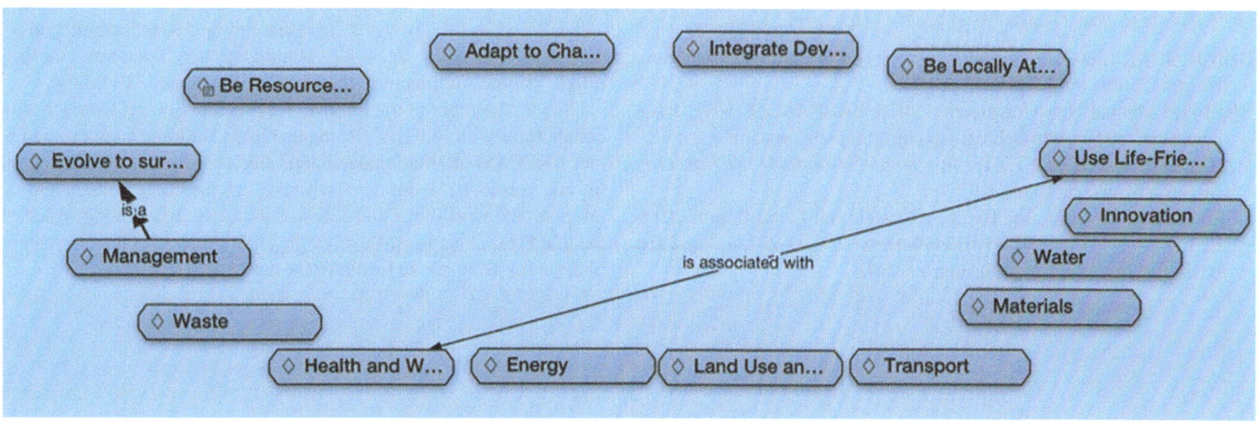

Fig. 8 Correspondence analysis at the level of relationships. (Ana Mohonea)

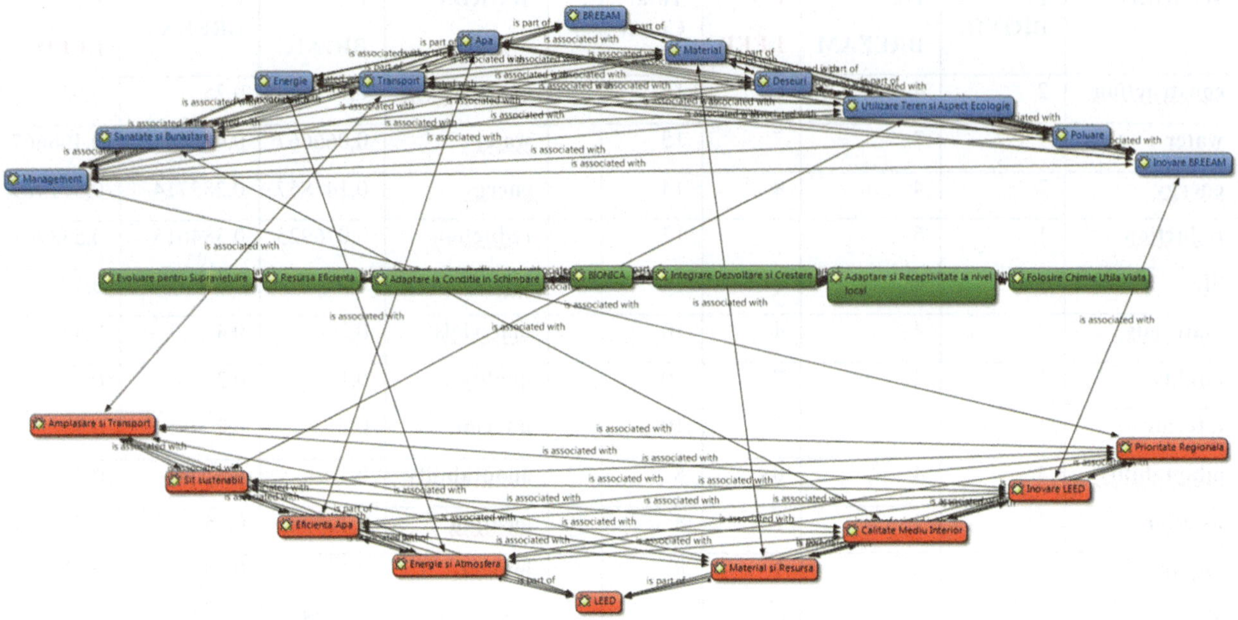

Fig. 9 Analysis of the relationship between BREEAM, BIONIC, and LEED. (Ana Mohonea)

ability in the built environment could lead to the creation of an architectural-bionic design method that will become a useful working tool in the design of sustainable buildings.

One could conclude that the implementation of bionic in architecture can help to reduce negative impact on the naturalenvironment, on one hand, and to redefine standards for assessing sustainability in the built environment on the other. A new way to approach the criteria for labelling buildings through the principles of nature transposed by bionic may be proposed, and a higher level of ambition may be reached in trying to get the balance between the built environment and the natural one.

References

Benyus, J. M. (2002). *Biomimicry – Innovation inspired by nature*. Harper Collins Publishers.

BREEAM Manual New Construction 2018. BREEAM, SD5078, Issue 3.0 (31.07.2019), https://bregroup.com/products/breeam/

De Garrido, L. (2012). *Green in green: Sustainable architecture*. Instituto Monsa de Ediciones.

LEED Reference Guide for Building Design and Construction 2014. LEED v4, ISBN:978-1-932444-18-6 (01.11.2013/ updated 08.2019). https://www.usgbc.org/leed/v41

Pawlyn, M. (2011). *Biomimicry in architecture*. RIBA Enterprises.

Wines, J. (2000). *Green architecture: The art of architecture in the age of ecology*. Taschen GmbH.

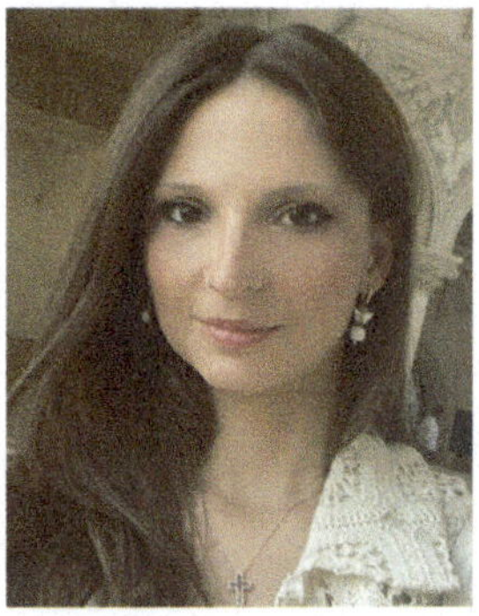

Ana Mohonea is an Interior Architect, graduated the Faculty of Interior Architecture of the "Ion Mincu" University of Architecture and Urban Planning in Bucharest, Romania.

She studied her undergraduate, master's and doctorate at the IMUAU, where she is currently working as an Assistant Prof. Ph.D. Arch. She was awarded the "Bene Merenti" for the outstanding results of her academic studies and the Medal of the IMUAU for her diploma project.

Ana Mohonea was a pioneer of research in bionic architecture in Romania; her doctorate thesis is titled: "Bionic Architecture. Premises for Sustainable Building".

She is a member of the Interior Architecture Society from Romania, collaborator with ROGBC (Romania Green Building Council) and former BREEAM international assessor. She participated in many competitions, projects, summer schools, architecture biennales and international workshops and received many awards, among which was the first Prize at National Level of the eighth edition of the Design competition for avant-garde furniture for her *Guitar Chair*.

Part IV

Researching on Materials: From Macro to Nano

Microalgae Biotechnology: The Bioindustry of the Future

María Segura Fornieles

Abstract

AlgaEnergy is a Spanish biotech-based company exclusively focused on microalgae, photosynthesis microorganisms with high biomass productivity and high capacity for CO_2 biofixation. Microalgae biomass is natural source of proteins, fatty acids, carbohydrates, vitamins, and other compounds such as minerals or carotenoids, having application in many sectors, including agriculture, food and feed, cosmetics, aquaculture, energy, and pharmaceutical, among others. AlgaEnergy develops R&I projects that result in products and solutions for all those sectors that can contribute to improve the quality of life of population through a circular and sustainable business model.

Keywords
Microalgae · Phytoplankton · Biomass · Biostimulants · Food · Feed · Cosmetics · CO_2 biofixation · Biotechnology · Circular economy · AlgaEnergy

Introduction

Prokaryotic microalgae (blue-green algae or cyanobacteria) and eukaryotic microalgae, essential components of phytoplankton, are microorganisms endowed with the ability to develop oxygenic photosynthesis, releasing oxygen (O_2) as a result of photolysis of water (H_2O) and synthesizing organic matter to the oxidized forms of the primordial bioelements (C, N, P, S), such as carbon dioxide (CO_2), at the expense of an inexhaustible source of energy, the sunlight. They are highly efficient organisms in the use of solar energy, with high biomass productivity and high capacity for CO_2 biofixation (Esteban Clares et al., 2014).

Due to their excellent biochemical profile and high growth rates, microalgae and cyanobacteria generate a multitude of products, having applications in many sectors.

The industrial development of microalgae and cyanobacteria, based on intensive cultivation and controlled in photobioreactors, represents a valuable tool to help alleviate the three major crises that humanity faces, namely: food, energy, and the environment.

Global Context

Mankind must overcome the main twenty-first century challenges including food security (the world's population is growing, meaning food security is a priority), environment (mitigating the climate change is one of the modern society's greatest concerns), and energy (our future will depend on obtaining a more efficient and sustainable energy source).

Microalgae are well positioned as a clear opportunity for improving these scenarios. Regarding food security, microalgae increase the crops' productivity and are the raw material with the highest protein content on the planet. In addition, they are essential to increase production in aquaculture. Concerning environment, microalgae are the most efficient CO_2biofixation systems and the greatest producers of oxygen on the Earth. In terms of energy, microalgae are a source of fourth-generation sustainable and inexhaustible biofuels.

The Company: AlgaEnergy

AlgaEnergy is a biotech-based company, established in the year 2007, exclusively focused on microalgae.

The company stands out for its experience and passionate team that consolidates almost 5 decades of research in microalgae, its technologies and culturing facilities developed "in-house" (the most advanced in the world), and its innovative products that are efficient, competitive, and sustainable.

M. S. Fornieles (✉)
AlgaEnergy, Madrid, Spain
e-mail: msf@algaenergy.com

M. R. Cervera Sarda et al. (eds.), *Architecture Inspired by Nature*, https://doi.org/10.1007/978-3-031-33144-2_16

AlgaEnergy is a B Corp company, becoming the first company in its sector to obtain the certification worldwide, after having demonstrated in a rigorous audit that its activity generates a positive impact on society and the planet.

What Are Microalgae?

About 3000 million years ago, the cyanobacteria, and some 1500 million years later, the eukaryotic microalgae populate the terrestrial and aquatic systems of the Earth constituting, together with the higher plants, the support of life and the oxygenation of the atmosphere. Microalgae and cyanobacteria are responsible for half of the planet's primary productivity and the generated new oxygen into the atmosphere (Shestakov & Karbysheva, 2017).

There are tens of thousands of species, which exhibit a multitude of shapes, sizes, and colors.

They are the first step in the food chain in the sea and responsible for more of the 50% of the oxygen on Earth. They do not compete with the production of other raw materials, they grow in all types of water, they are extremely productive, and they are able to biofixate up to 2 kg of CO_2 for every kg of biomass produced.

Microalgae Composition

As a result of their excellent biochemical profile and high growth rates, microalgae and cyanobacteria are microscopic biofactories that generate a multitude of products of interest because of being a natural source of proteins, fatty acids, carbohydrates, vitamins, and other compounds, such as minerals or carotenoids. Due to their rich and complete composition, they have application in many sectors including agriculture, food and feed, cosmetics, aquaculture, energy, and pharmaceutical, among others (Fig. 1).

Microalgae Culture

The intensive and controlled culture of microalgae and cyanobacteria is developed in photobioreactors, operated to obtain biomass, or extracts from it, as final product.

Photobioreactor is understood as those systems in which there is a physical separation between the culture and the medium that surrounds it, normally transparent, and that allows better control of the culture conditions, thus increasing productivity.

Inside them, the division and cell growth of the cultures take place, in an efficient, sustainable, and controlled manner. Therefore, the photobioreactor, as an essential component of the production process, must be designed in compliance with a series of requirements that allow these premises to be fulfilled.

The main parameters to control during the development of microalgae cultures are solar radiation, temperature, pH, culture medium, mode of operation, including dilution rate, and photobioreactor design, including mix-up.

Once the culture reaches the optimum concentration inside the photobioreactor, it must be harvested, which consists of separating a part of the microalgae culture to collect the microalgae biomass contained therein, to stabilize it, and process it based on the business application of interest (Fig. 2).

Results

AlgaEnergy Facilities

Having exclusively dedicated our first 10 years to applied research sourcing from more than 3 decades of basic R & I, today we can confidently say that one of the greatest accomplishments of AlgaEnergy is the successful scale-up of microalgae cultivation processes, which allows us to put innovative, competitive, sustainable, and highly effective products on the market.

AlgaEnergy owns two microalgae cultivation plants. One is our Technological Platform for Experimentation with Microalgae (TPEM), located at the Adolfo Suárez Madrid-Barajas airport (Madrid). The other is the Arcos de la Frontera plant in Cadiz, whose secondary objective is the biofixation of CO_2 through microalgal cultivation, to turn this greenhouse gas into products of high commercial interest.

We also operate the UPT® Production Center, exclusively dedicated to the production of our range of agricultural biostimulants, marketed as DynaMix®.

In operation since 2011, TPEM is probably the most powerful and flexible private R&I tool known in the international field of microalgae biotechnology. It was built in collaboration with AENA and IBERIA. Located at the Adolfo Suárez Madrid-Barajas airport, TPEM combines different types of photobioreactors, which allow the simultaneous cultivation of several different strains of microalgae. TPEM's mission is to generate valuable knowledge about technologies and processes on a scale preceding the industrial level, as well as to serve as a R&I and experimentation tool (Figs. 3 and 4).

ARCOS plant, located in the South of Spain (Arcos de la Frontera, Cádiz), became operational in 2014 and is on the site of the largest combined cycle power plant in Spain, owned by Iberdrola, world leader in renewable energy and technological partner of AlgaEnergy since 2009.

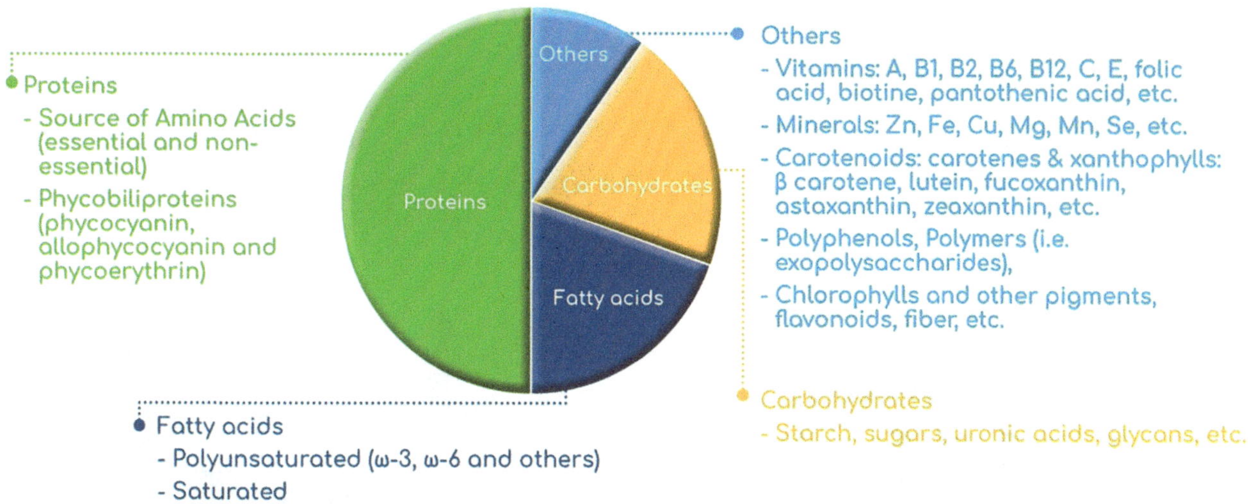

Fig. 1 General microalgae composition

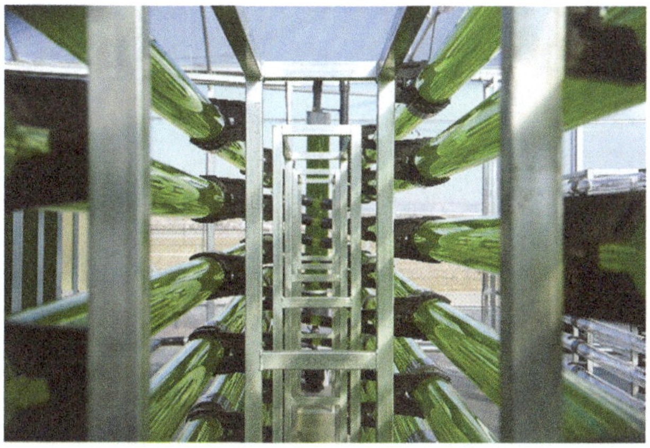

Fig. 2 Microalgae photobioreactor for biomass generation

The main nutrient of the microalgae, CO_2, is taken directly from the combustion gas chimney of the power plant and pumped into microalgae cultures. So ARCOS plant produces microalgae biomass through CO_2 biofixation from flue gases, transforming this harmful gas into products of high commercial interest (Figs. 5 and 6).

R & I Projects

AlgaEnergy, using basic research coupled with its own know-how, has the mission of promoting new lines of basic research and applying and improving performance achieved in the lab at industrial level (applied research).

R&I's project portfolio ranges from genetic engineering applied to the modification of certain microalgae's metabolism to the development and implementation of new and more efficient systems for cultivation and processing of microalgal biomass.

AlgaEnergy leads and participates in international R&I projects with a total investment of more than EUR 80 million. AlgaEnergy leads the way with more than 150 consortiums (universities, research centers, and companies), all firmly committed to progress and sustainable development (Fig. 7).

Applications and Sectors

At AlgaEnergy, R&I is the seed of all the high-value products we produce for a variety of market sectors. Our firm commitment to innovation propels us to develop innovative, sustainable, and, above all, highly effective solutions for industries in which microalgae have an immense application potential.

Currently, AlgaEnergy has products and solutions, based on different microalgae, for agriculture, nutrition (both human and animal), cosmetics, and aquaculture. In addition, we are studying the immense application potential of microalgae for the pharmaceutical, biomaterials, and biofuels sectors (Fig. 8).

Conclusions

Microalgae and cyanobacteria are not only our origin but also our future. AlgaEnergy is the first company in the world taking advantage of the CO_2emitted by other industries to cultivate its microalgae as part of a circular and sustainable business model (Fig. 9).

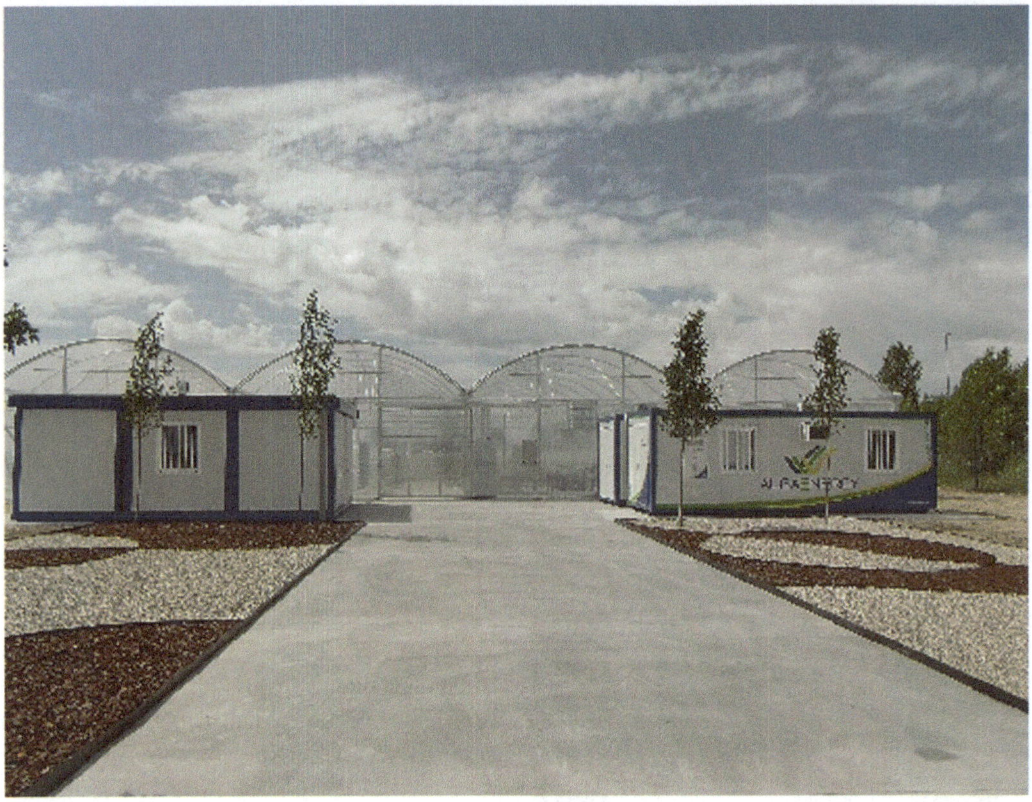

Fig. 3 TPEM facilities, including laboratories, offices, and greenhouse

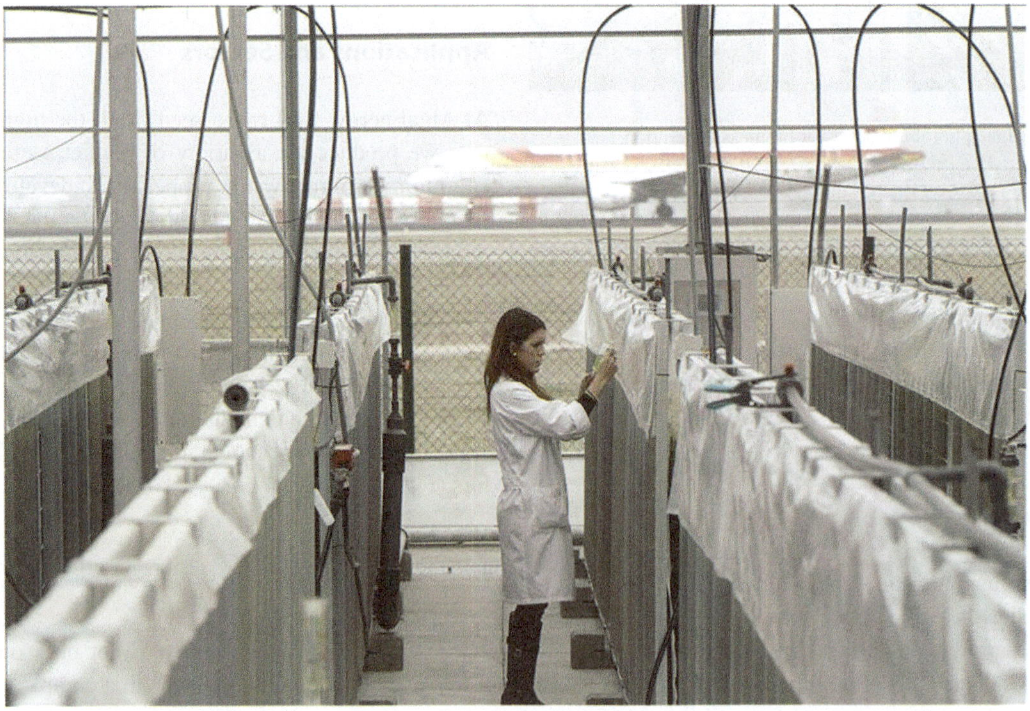

Fig. 4 TPEM flat panel photobioreactors

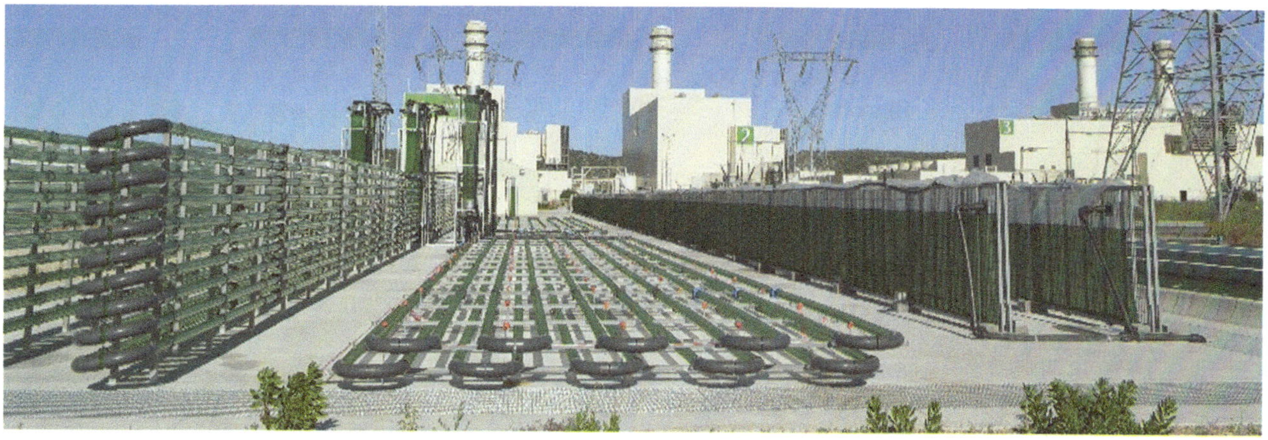

Fig. 5 ARCOS plant, flat panel, and tubular photobioreactors' view

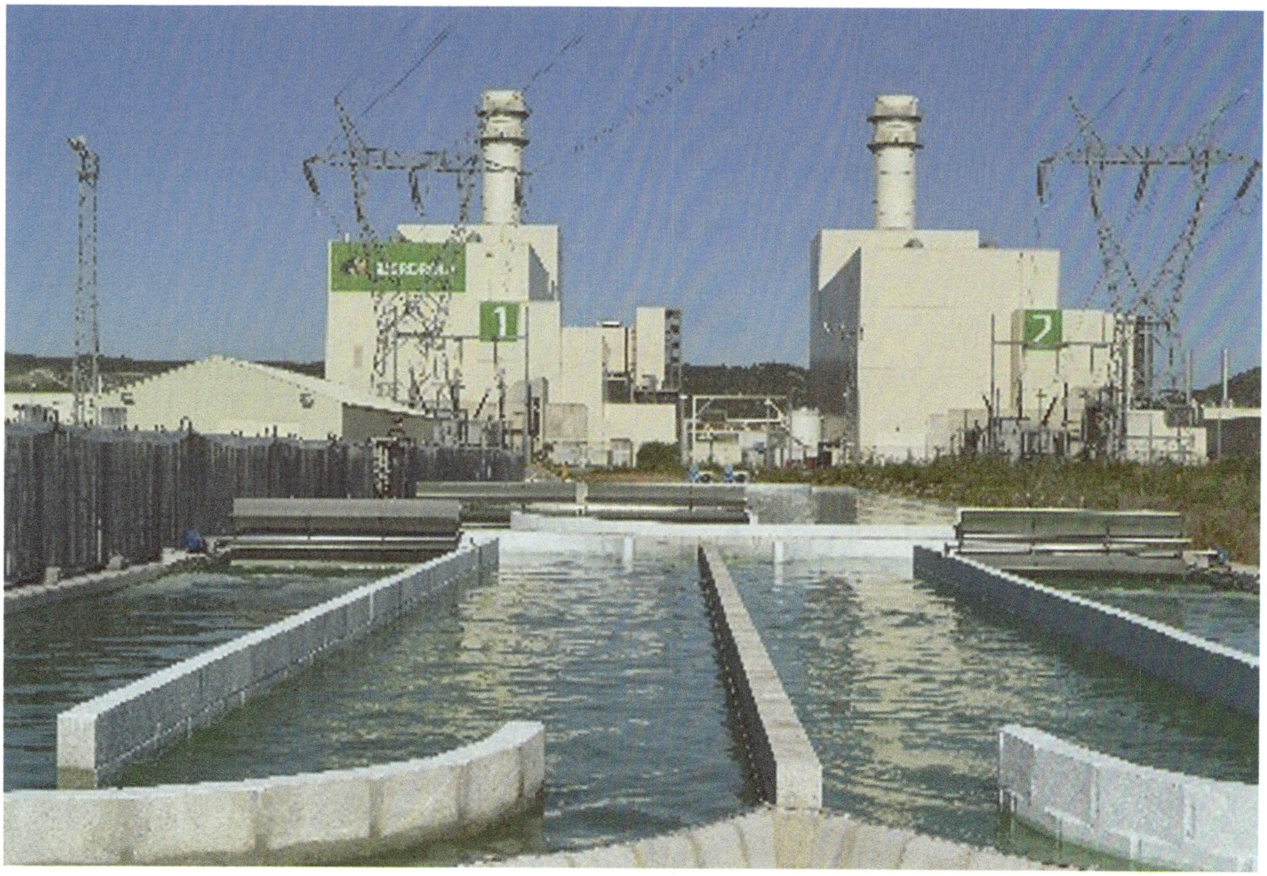

Fig. 6 ARCOS plant, raceway reactors' view

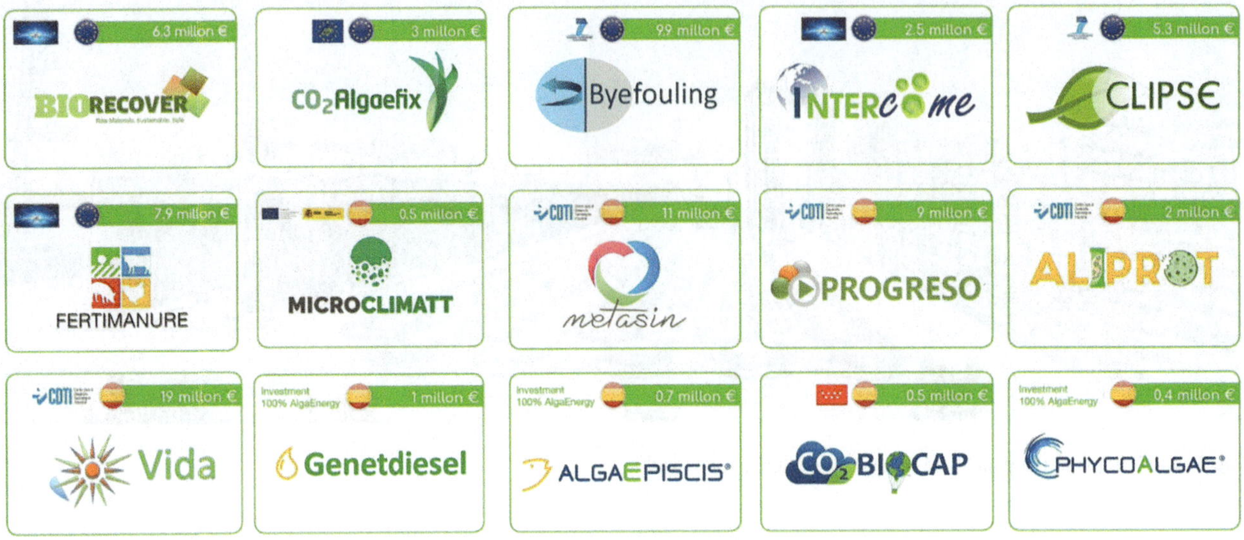

Fig. 7 AlgaEnergy's R&I projects

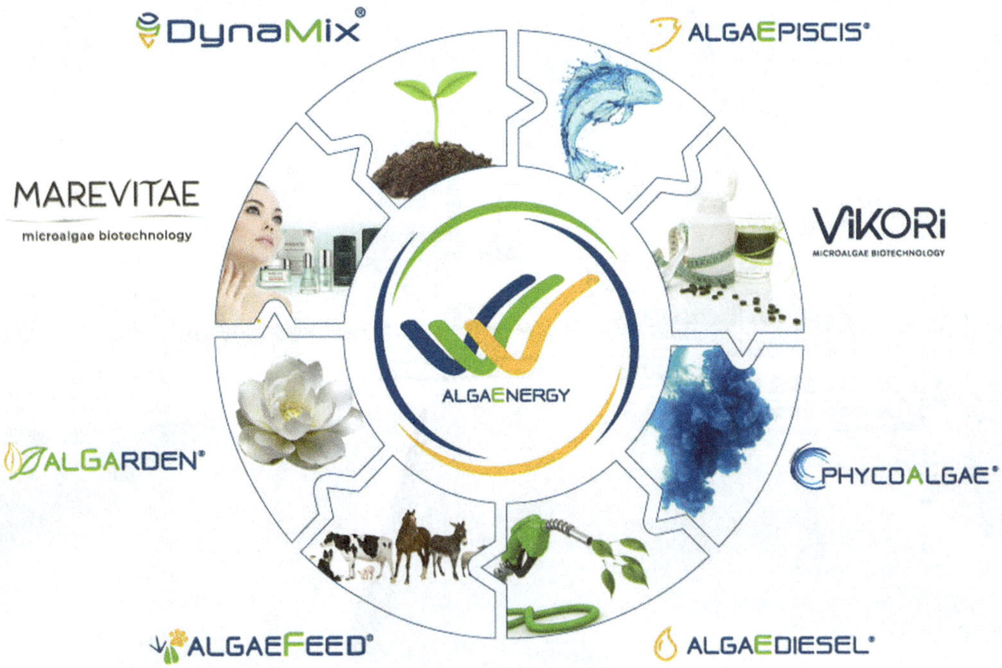

Fig. 8 AlgaEnergy brands for several market applications

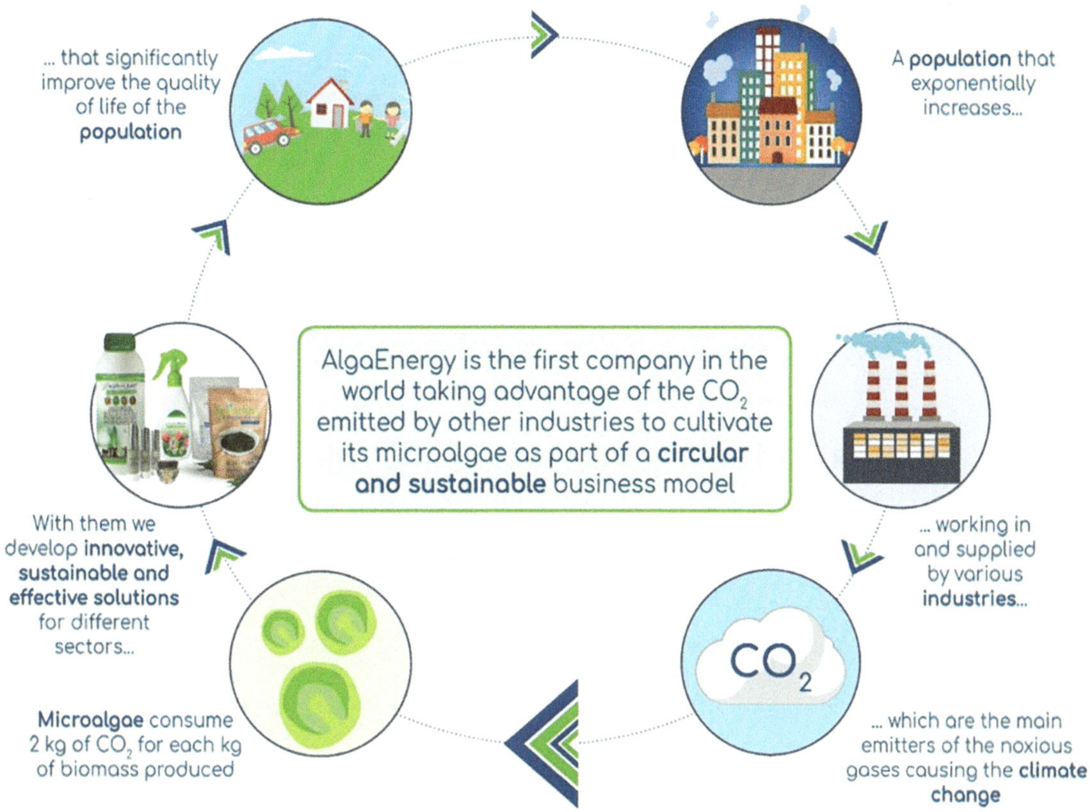

Fig. 9 AlgaEnergy circular economy business model

References

https://www.algaenergy.com/

Esteban Clares, M., Moreno Fernández, J., García Guerrero, M., & García González, M. (2014). Assessment of the CO_2 fixation capacity of Anabaena sp. ATCC 33047 outdoor cultures in vertical flat-panel reactors, *Journal of Biotechnology, 187*, 51–55.

Shestakov, S. V., Karbysheva, E. A. (2017). The origin and evolution of cyanobacteria. *Biology Bulletin Reviews, 7*, 259–272.

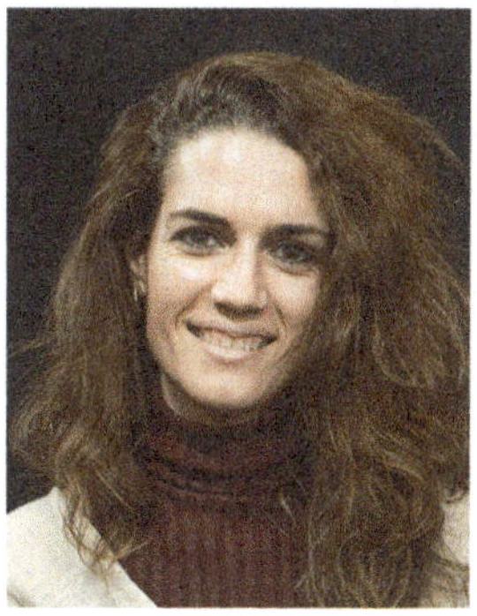

María Segura Fornieles is entitled in Chemical Engineering and Industrial Engineering. She also has a Master in Renewable Energies.

As *AlgaEnergy*'s Deputy General Manager and Technical Director, she has developed, since 2010, her professional activities in the engineering field. Most especially her activity focused on the design of construction and installation of facilities for the culture of microalgae, such as AlgaEnergy pilot and industrial facilities, including the development and improvement of culture processes and photobioreactor designs, development of control and instrumentation systems, as well as in the direction and management of several R & D projects.

Cultivating Microalgae in Architecture: Toward a Zero Carbon Print: A CO_2 Sequester Analysis

Ma. Rosa Villalba and María Rosa Cervera Sardá

Abstract

The need to implement projects that contribute to environmental conservation and prevent global warming by reducing greenhouse gas emissions is increasing. The objective of this work is to transfer the technology of microalgae photobioreactors to the architecture, in such a way that the façades or rooftops of buildings could became biomass production plants and CO_2 collectors through natural resources. This will mean that the architecture would assume a change of approach, both in an aesthetic and functional way. A new active and efficient urban landscape is generated and the bioclimatic and metabolic behavior of the building is improved by introducing in the construction process living organisms in relatively small ground space. Focusing on the CO_2 sink effect of a flat photobioreactor integrated in a building located in the city of Madrid (Spain), we pretend to demonstrate that a façade could act as a CO_2 collector despite of the carbon footprint produced by its elements. Taking into account all the parameters needed for the microalgae to grow, we can estimate the amount of biomass that is produced to get the CO_2 that is fixed during the photosynthesis process and compare it with the carbon footprint of the materials of the façade.

According to the calculations of the photobioreactor proposed, it will take almost 8.5 years to fix the manufacturing CO_2 footprint of the materials used. However, in 50 years of lifespan of the materials, a photobioreactor of 338 m^2 will subtract from the atmosphere more than 166 tons of CO_2, which is equivalent to the drive of almost 768,000 km of a gasoline car. However, it must be taken into account that after studying the carbon footprint of the photobioreactor materials, the final amount of CO_2 effectively subtracted will be more than 166 tons.

M. R. Villalba (✉)
Alcalá University, Madrid, Spain
e-mail: mrosa.villalba@edu.uah.es

M. R. Cervera Sardá
Universidad de Alcalá, Alcalá de Henares, Spain
e-mail: rosa.cervera@uah.es

Keywords
Architectural photobioreactor · Green façades · Sustainable architecture · Bioenergy · Biomass · CO_2 · CO_2 sink, Sustainability

Introduction

Currently, the need to implement projects that contribute to environmental conservation and prevent global warming by reducing greenhouse gas emissions is increasing. According to the United Nations, global carbon dioxide (CO_2) emissions have escalated by nearly 50% since 1990, causing the global average temperature to rise by 0.85° between 1880 and 2012 (United Nations, 2020a). These changes in temperature lead to multiple environmental problems ranging from rising sea levels to increasingly extreme weather events. Among the main factors that generate CO_2 emissions, energy production is one of the most important as it accounts for 60% of global greenhouse gas emissions (United Nations, 2020b).

In response to this urgent need, it has been proposed to build a system that includes the absorption of solar energy with the fixation of CO_2. Photosynthesis is the only natural phenomenon that involves these two processes, so the goal is to benefit from it by capturing sunlight through cells composed of living organisms such as microalgae.

Microalgae are a relatively easy and versatile crop to grow. They have numerous applications not only in the biofuel industry but also as biomass, as fertilizers, for nutritional purposes, and many other uses. The main advantage of their use is based on its capacity as fast-growing photosynthetic microorganisms that can double their volume over a period of 1–10 days if the medium allows it; their high lipid content, which can reach 50% by weight of dry matter in some cases (an oil production of 15–300 times higher than soil plant species per unit area); and the use of less surface area for cultivation as they can be grown in vertical structures. Finally, the most important feature that signifies a usable advantage is its

M. R. Cervera Sarda et al. (eds.), *Architecture Inspired by Nature*, https://doi.org/10.1007/978-3-031-33144-2_17

greater CO_2 fixing capacity compared to other plant species (Santos Montes et al., 2014).

Microalgae have traditionally been grown in closed or open photobioreactors. Photobioreactors (PBRs) are containers that allow these organisms to grow in an aqueous medium that contains the elements necessary for their development: nutrients, a source of CO_2 and the exposure to sunlight. The purpose of this document is to outline design options and strategies for integrating flat photobioreactors into the façade of buildings. Acting as an architectural structure, PBRs combine beauty with use. In cities, flat-plate PBRs will be a design option that can remove excess CO_2 from the atmosphere and produce useful algae products.

Therefore, the purpose of this document is to outline a simple design of a PBR to analyze the amount of biomass and CO_2 that could be sequestered and compare it with the amount of CO_2 needed to produce the materials that build it.

Photobioreactors in Architecture

So far, algae cultivation has been done industrially. So the objective of our work is to transfer it to the architecture, in such a way that the façades or rooftops of buildings could became biomass production plants through natural resources.

The combination of elements as an architectural enclosure with the biomass production gives them a unique aesthetic and iconic value. This will mean that the architecture would assume a change of approach both in an aesthetic and functional way. A new active and efficient urban landscape is generated and the bioclimatic and metabolic behavior of the building is improved. Living nature is incorporated into the construction (Fig. 1).

The façades would become something active and dynamic, through the conduction of fluids with microalgae, that adds functional value to the purpose of protection of the building. The aesthetics of a living element change over time, so this new façade would change accordingly. Therefore, the aesthetic perception is transformed to be the source of other designs that adapt to the new functionality and the temporality that this involves. This approach provides new opportunities to generate a versatile architecture that develops according to the changes in temperature and sun exposure of the year.

Within this research, we have been working on different designs that combine the functionality with different aesthetic approaches that work with the diverse forms of industrial cultivation of microalgae:

Photobioreactor Inspired on the Microstructure of a Butterfly's Wing

This photobioreactor design mimics a thin scale of the microstructure of a butterfly wing placed as tiles, generating a fickle structure that is able to lay over any surface or façade. The photobioreactor modules are held from a primary structure and can be easily hung or taken down as needed. The modules' suspension pieces allow these to be articulated so that they can be positioned at will for maintenance, design, or user convenience. They can be of a tubular shape or located within a "glass-like sandwich" with all microalgae fluids running inside (Fig. 2).

Tubular Photobioreactor on Façades

This type of photobioreactors is composed of acrylic tubes exposed to sunlight. A pipe system is formed to move the culture medium (fluid that concentrates the elements necessary for the growth of microalgae) so the microalgae could grow. Through the rhythm of the tubes, their orientation and color, the system could be managed in such a way that they constitute an aesthetic element for the building (Fig. 3).

This is a new way to incorporate nature into the urban environment of today's cities. Photobioreactor façades offer the possibility of combining different types of vegetation that coexist symbiotically to promote their growth. In this case, the photobioreactor system would produce natural fertilizers that promote the growth of the vertical garden that structures the façade. Consecutively, through this design, CO_2 is captured both by the plants we include and by the microalgae embedded in the piping system. Therefore, it not only becomes a versatile way to cover the building but also provides an attractive aesthetic element.

Apart from all the abovementioned benefits, one of the most important advantages of this kind of façades is that they present an effective thermal insulation within a cost equal to or less than the existing coatings (Cervera Sardá & Vicente, 2016).

Designing a Flat Photobioreactor as a CO_2 Sink

The main objective of this research is to calculate the CO_2 sink effect of a flat photobioreactor. To this end, it has been decided to work on an existing prototype building, located in the city of Madrid (Spain) and considering the climatological characteristics of that city (Fig. 4).

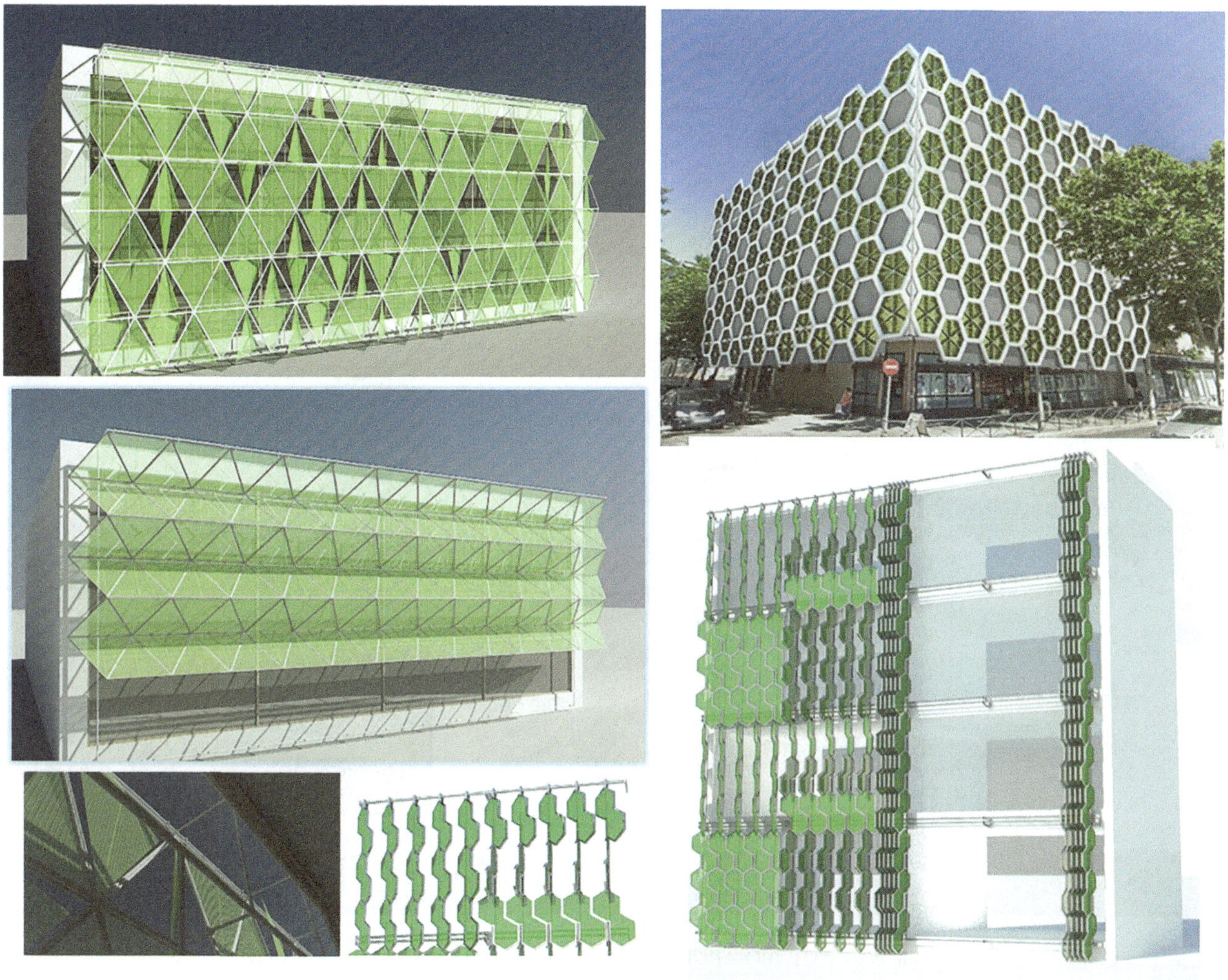

Fig. 1 Vertical bioreactors on building façades. (Cervera & Pioz Architects, 2014)

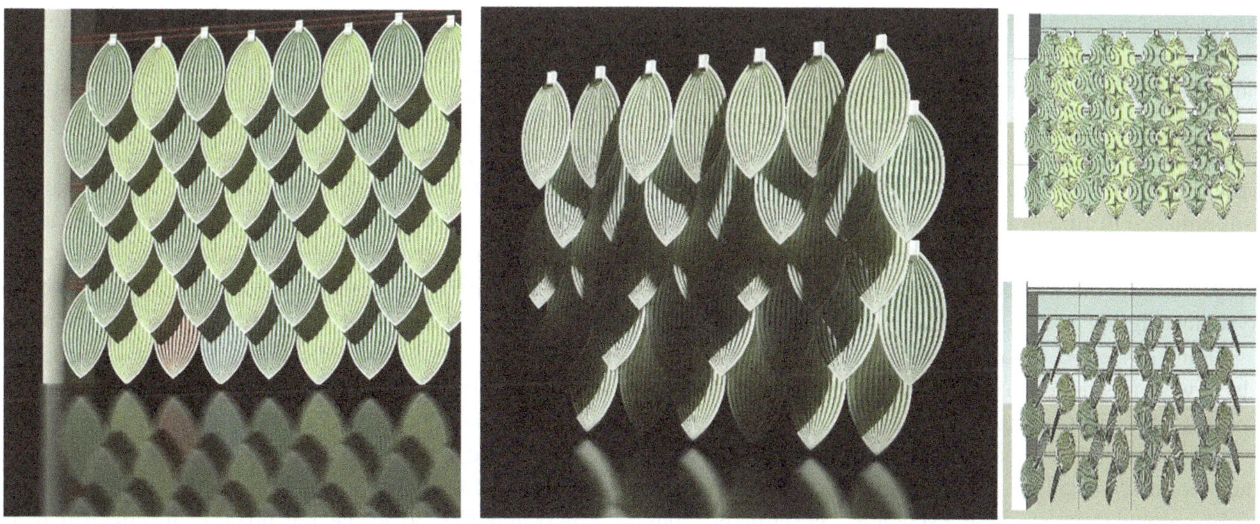

Fig. 2 Butterfly-inspired façade. (Cervera & Pioz Architects, 2014)

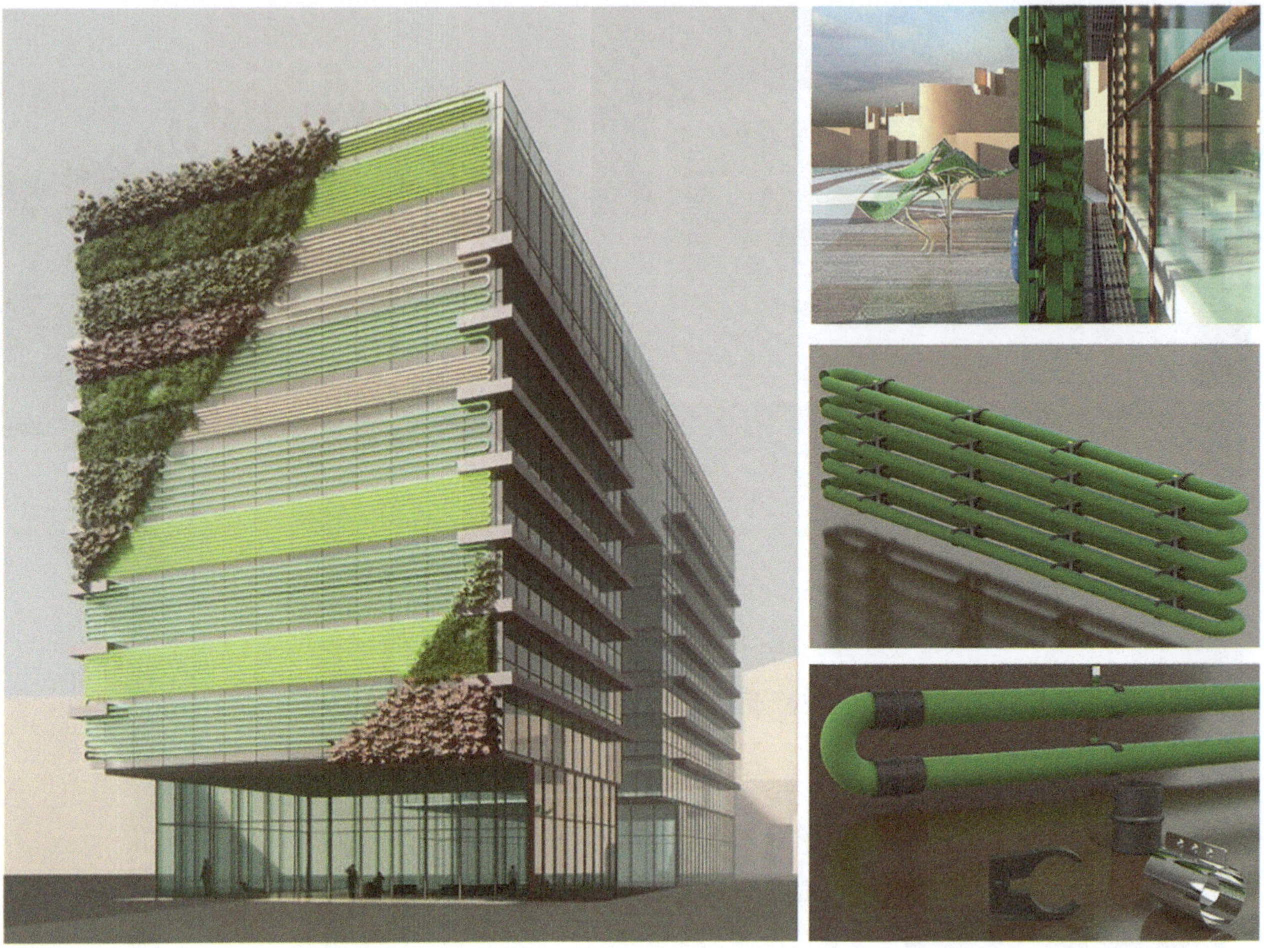

Fig. 3 Design made for Alcalá Mobile Week 2021. BIO-ARCHITECTURE: Towards zero impact with Microalgae. (Cervera Arquitectos)

As a practical exercise, we choose an existing building with a south-facing façade of 13.00 m in height and 26.00 m wide. These measures define the dimensions of the modules that make up the photobioreactors cells and the location of the pumping system necessary for the operation of the façade.

To determine the dimensions of the project module, multiples of three were used to avoid material waste, being the base measurements 1.20 m × 3.00 m. However, the module was adjusted to 1.00 m × 2.60 m in width to cover the entire existing façade and to give space for the attachment of the base structure, the water distribution and biomass collection systems (Fig. 5).

Stainless-steel C metal profiles, anchored to the building slabs, were positioned to support each of the panels and give space to the photobioreactor operating system. Over this structure, a metal plate is attached to support the panels and the pivot mechanism.

The module is made up of a stainless-steel structure composed of C-type profiles that house an acrylic container that will be inserted using the profile as a channel. The container will not have a top cover to allow the natural release of oxygen produced by the photosynthesis of the microalgae. In the same way, the upper closure element will have open spaces for this purpose (Fig. 6).

Inside the container, it is intended to place a plastic element that produces a slope of at least 1% to avoid the accumulation of material at the bottom of the panel. The gas mixture distribution system will be fixed to this element, which must cover the entire width of the panel to guarantee the continuous movement of the culture medium.

To optimize the distribution of the growth medium through the photobioreactor, it was decided to locate the pipes for the supply and collection at the slab's level, so the framework structure provides a channel for the pipes to be laid. Also, the dimension of the modules coincides with the height of the building's mezzanine to allow the pivoting movement of the panels.

It is intended to locate the pivoting mechanism in the transversal axis of the panel to keep the hydromechanical and gas mixing connections fixed while the panel can be

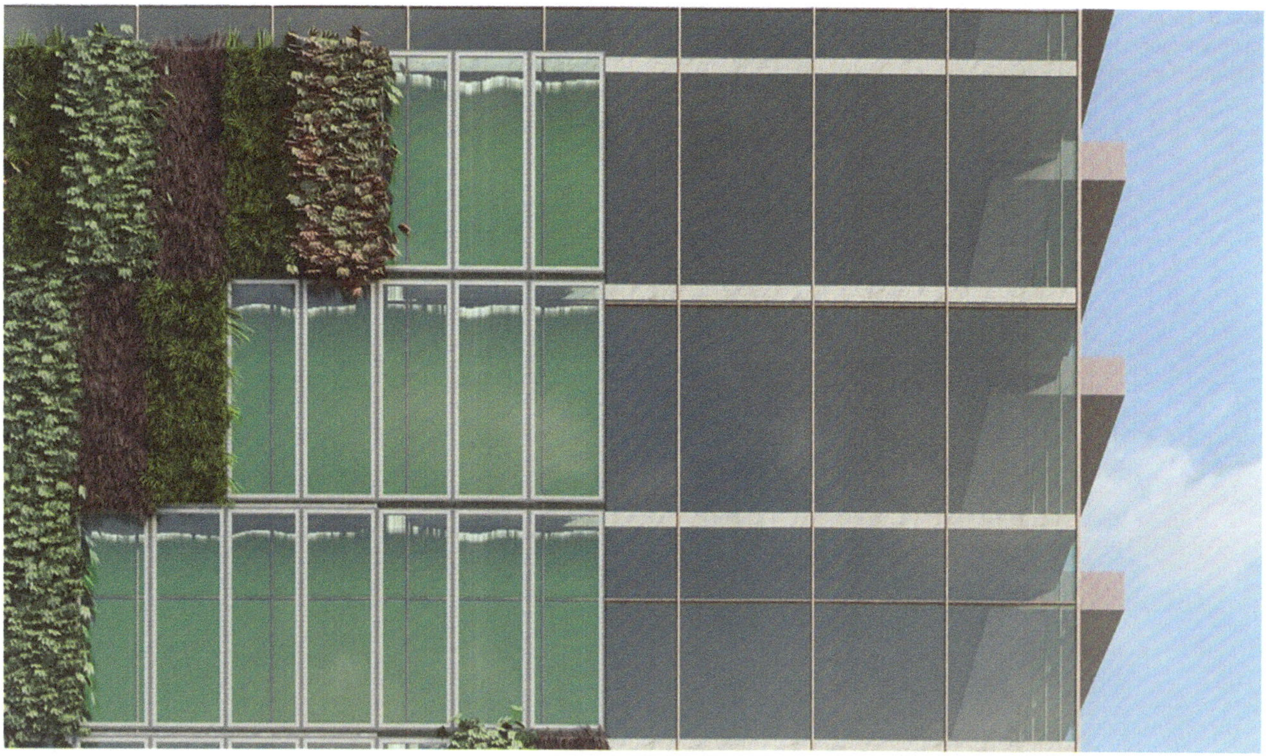

Fig. 4 Flat photobioreactor as an alternative of the previous design in Fig. 2. (Cervera Arquitectos)

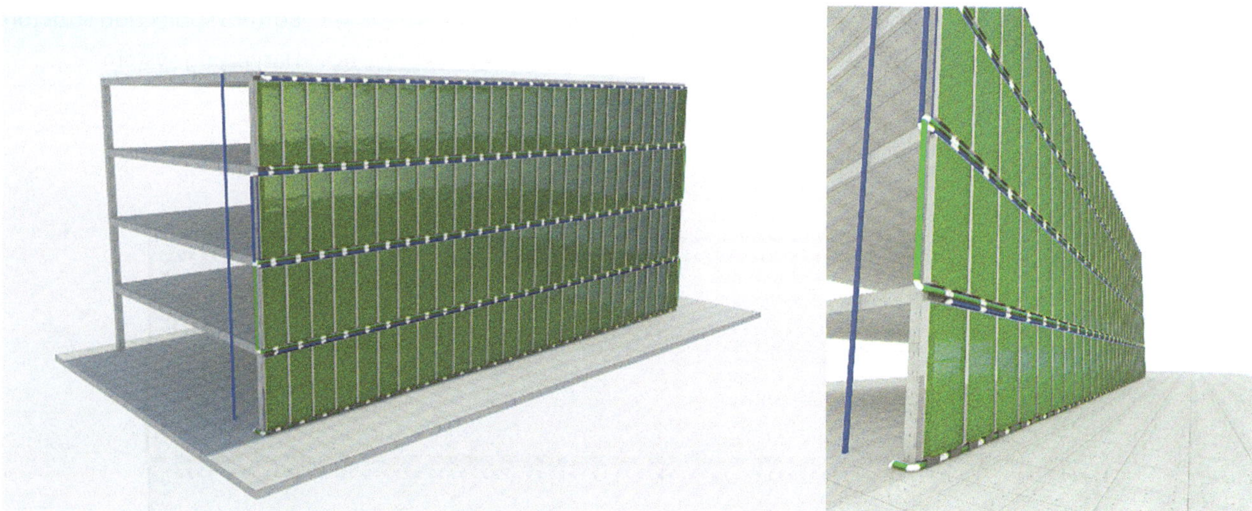

Fig. 5 Schematic design of a flat photobioreactors on the façade of an existing building. (Cervera Arquitectos)

rotated to take advantage of the solar incidence or in turn to cool the crop in the summer months.

The distribution of the fluids is done by gravity to optimize the use of pumps that move the fluid along the façade. It is intended to collect rainwater from the roof of the building and store it in an underground compartment to take advantage of the thermal insulation that the earth provides against extreme cold and heat (Fig. 7).

The PBR will be semi-open to allow the release of O_2 from the top. Microalgae produce oxygen, as a product of the oxygenic photosynthesis, which in excess would decrease optimal growth. Therefore, some O_2 degassing system is required. The simplest of these systems is to leave the top of the PFRs open and allow oxygen to exit through the opening.

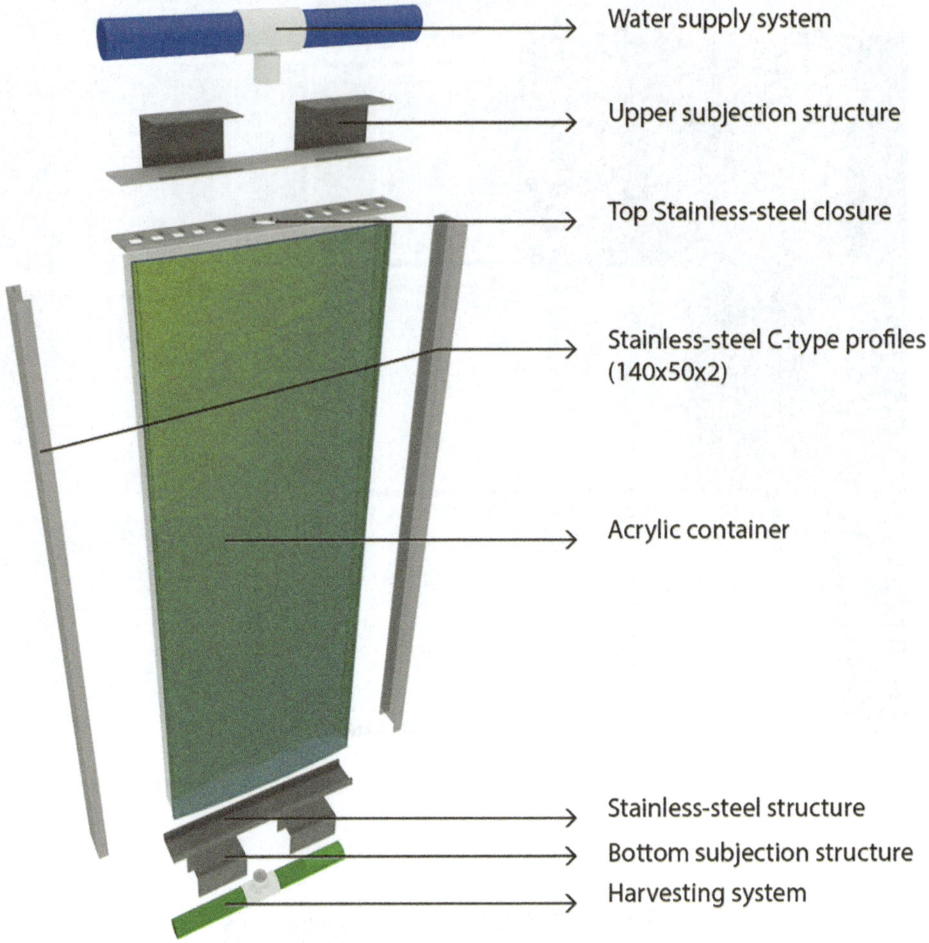

Fig. 6 Elements that form the photobioreactor's module. (Cervera Arquitectos)

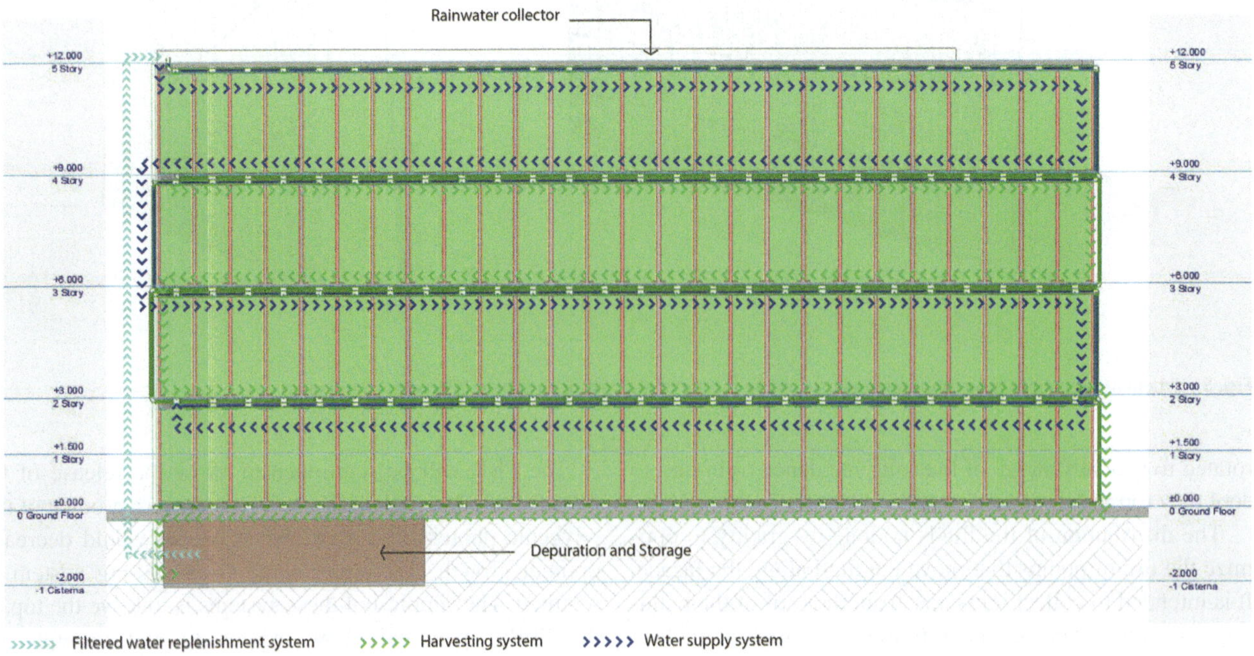

Fig. 7 Harvesting system. (Cervera Arquitectos)

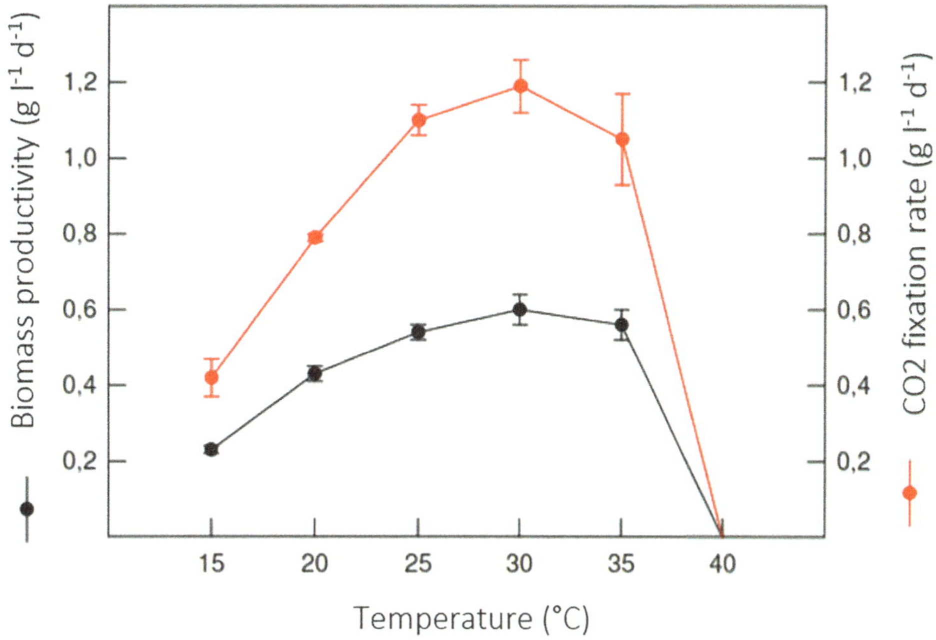

Fig. 8 Effect of temperature on biomass productivity and CO_2 fixation rate of *S. vacuolatus* in continuous cultivation. (García Cubero et al., 2017)

Necessary Parameters for the Operation of Photobioreactors

For the proper operation of the photobioreactor, the following parameters were taken into account in consideration to achieve the highest possible productive activity in biomass production and CO_2 fixation.

Temperature and Light

Both temperature and light are important factors in determining the growth rate of algae. Sun irradiance is one of the few conditions that is largely out of experimental control, and temperature is largely too. Many PBR studies are carried out under ideal conditions, usually around 30 °C; however, ideal conditions are generally not possible in external environments. Heating water before pumping it to PBR is an energy-intensive effort and will require more energy than it will produce. Therefore, the temperature of the water will be determined by the outside temperature, as well as by the sunlight in the photobioreactors.

The average monthly temperature in Madrid ranges between 5 °C and 25 °C (Climate-Data.org, 2020). At first, we consider to use different strains of algae to adapt to the high variation in temperature. However, the Scenedesmus strain has shown a great adaptability to temperature change, as shown in a research done by Xin, Hong-ying, and Yu-ping in 2011. They summit the *Scenedesmus* sp. LX1 strain to controlled temperature of 10, 20, 25 and 30 °C and realized that the growth rate between these temperatures were no significant (Xin et al., 2011, p. 3100).

Even though this type of strain of Scenedesmus have high adaptability to cold environment, there weren't enough data of the growth rate related to the CO_2 fixation rate of the *Scenedesmus* sp. LX1. For this reason, we choose to use the data (Fig. 8) from the DIGITAL.CSIC of García Cubero (2014), which compares de biomass productivity (g L^{-1} d^{-1}) at determinate temperatures with the CO_2 fixation rate (g L^{-1} d^{-1}) of the strain *Scenedesmus vacuolatus*:

There was no data to temperatures below 15° to stablish the biomass productivity, so there is going to be too much uncertainty in the calculations of the growth rates of the months with an average lower than 15°. Although we estimate the growth gate on these months to be minimal, still there are some studies that indicate that a type of Scenedesmus strain can be found in extremely cold regions of the Arctic and Antarctic (Varshney et al., 2015). So we can assume that this strain can survive in extreme weather conditions.

Nevertheless, for the estimations needed to calculate the biomass production of this exercise, a linear regression analysis was made from the given data and a linear extrapolation was formulated from the trend line obtained from Fig. 8 to get the growth rates for the lowest average temperatures (Fig. 9).

This model is used to determine the annual biomass production and the amount of CO_2 sequestered. Temperature was the only variable accounted for in this growth rate model, but the intensity of sunlight is also an important variable. The two, temperature and irradiance, are related since the most direct sunlight raises the temperature (Table 1).

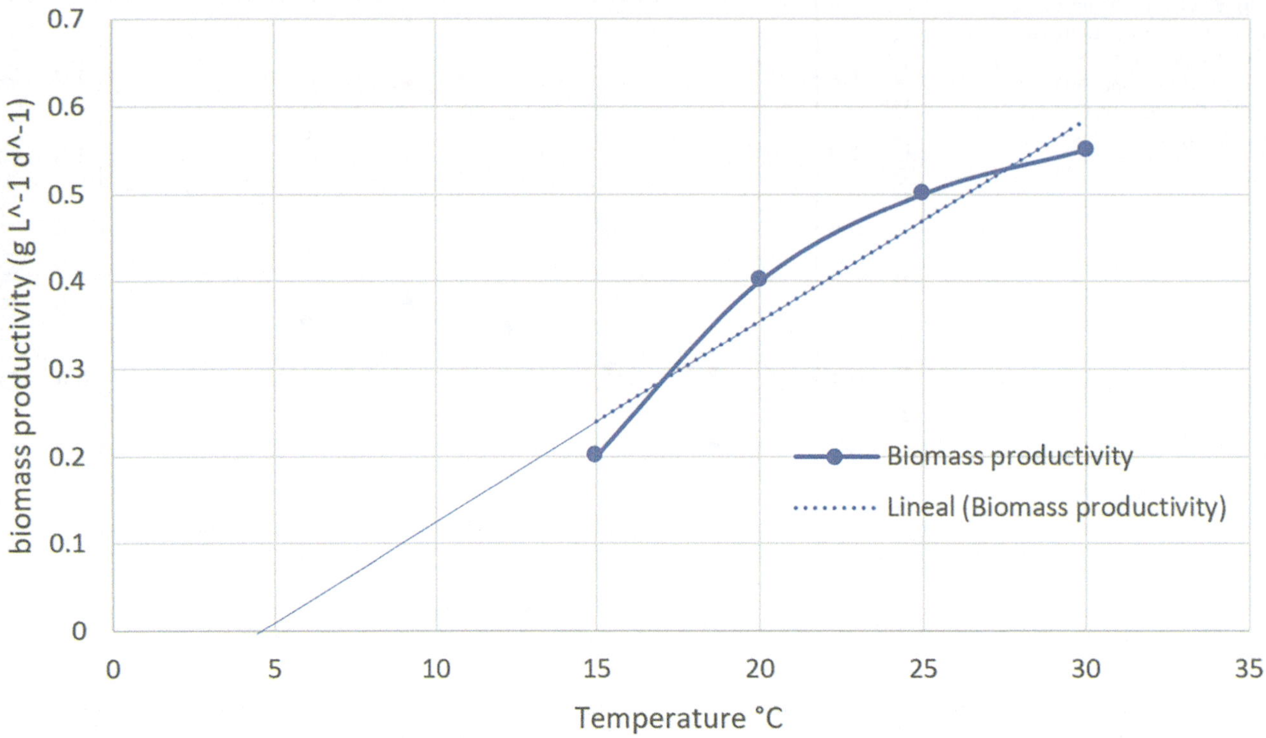

Fig. 9 Linear extrapolation of the effect of temperature on biomass productivity. (Authors)

Table 1 Growth rate versus average temperature (Authors)

	Average temperature (°C)	Growth rate (g L^-1 day^-1)
November	8.9°	0.09
December	5.4°	0.02
January	5°	0.01
February	6.4°	0.042
March	9.6°	0.115
April	12.2°	0.175
May	15.8°	0.33
June	20.4°	0.42
July	24°	0.485
August	23.2°	0.472
September	19.6°	0.388
October	14°	0.217

In order to optimize the light exposure and thermal control that microalgae have available, PBRs are designed on a vertical axis that can rotate. This means that the bioreactor plates can be tilted in the direction needed to optimize exposure based on light need and temperature regulation.

Harvest

It is important to determine what percentage of the mixture should be harvested and how often. In a semi-closed system, as the microalgae grow, the nutrients in the aqueous medium are consumed until reaching its total capacity. This means that after reaching a critical concentration, the algae will begin to die if they are not harvested and the culture medium is renewed. We found this critical concentration (g L^{-1}) for Scenedesmus and determined the harvest periods based on them (Shelef et al., 1984) for Scenedesmus reaches this concentration around 1.2 g L^{-1} (Fig. 8).

Our goal is to harvest around 20–50% of the microalgae mixture each time. This means that the harvest frequency should be every other day. Using the growth rates determined above and the critical concentrations, we found the percentage of the microalgae mixture to harvest during each month (Table 2).

It is needed to clean the entire system, after the coldest period of the year, to make sure that the algae could have its optimal conditions to grow after being exposed to lower temperatures. Therefore, in March, we have a different harvest frequency to let the algae grow for 15 days in order to get its maximal concentration.

Harvesting is set as a certain percentage of the mixture that is extracted by the pipes at the bottom of the photobioreactor into a harvesting mechanism. The actual harvesting

Table 2 Percentage of volume harvested per month. (Authors)

	Max conc. (g L^{-1})	Harvest frequency (times per month)	Days in between harvest	Percentage to be harvested (%)
November	1.2	6	Once a week	30
December	1.2	2	Twice per month	25
January	1.2	1	Once a month	25
February	1.2	3	Every 9 days	32
March[a]	1.2	4	15 days of initial growth + every 3 days	29
April	1.2	14	Every 2 days	29
May	1.2	15	Every other day	28
June	1.2	15	Every other day	35
July	1.2	15	Every other day	40
August	1.2	15	Every other day	39
September	1.2	15	Every other day	32
October	1.2	15	Every other day	18

[a]The month of March has a different harvest frequency due to the initial growth of the microalgae

process is still being determined, but it is likely to combine some techniques such as sedimentation, filtration, and centrifugation. Centrifugation is a great way to separate algae biomass from water, but it's an energy-expensive mechanism (Dassey & Theegala, 2013).

Water

The PBR façade requires no supplemental water in addition to the rainwater collected from the roof of the building. Using climate data from Madrid, we estimated the average rainfall per month, subtracted by the amount of water lost from the centrifugation process. We estimated around 20% of the water from the centrifuge would be lost each harvest. In total, there is a surplus of around 48,000 L of water each year.

The area on the roof for collecting water will be 22.1 m in length by 12.9 m in width. The length was reduced from 20 m to 12 m because the surplus of water was too large to store in a storage tank. Even though we are only saving 22,000 L of extra water, that is exactly the right amount to refill the entire PBR façade once. The 13,347 extra L could be used as fertilizer for surrounding green areas. In March, there is a need of three times of the total volume of the PBR to clean the entire system, due to a restitution of the algae after the coldest period of the year.

CO_2

An essential component to plant growth is CO_2. Through photosynthesis, energy from light is used to convert water and carbon dioxide molecules into glucose. Our source of CO_2 for the PBRs comes from the flue gas of boilers and heating systems of the building. Typical flue gases from natural gas-fired power plants may contain 8–10% CO_2, and typical flue gases from coal-fired boilers may contain 12–14 vol% CO_2 (Song et al., 2004). This CO_2 is generally released into the atmosphere and acts as a greenhouse gas, which can be avoided if it is pumped into the PBRs and turned into glucose.

To determine the amount of CO_2 sequestered by the façade, the daily growth rate of each month is multiplied by the days in the month, multiplied by the total number of L, and converted to kg to find the amount of biomass produced per month. That is summed to find the total biomass per month, which is multiplied by 1.8 kgs CO_2/kg of microalgae (Zhang, 2015) to find the total kilograms of CO_2 sequestered by the PBRs over a year (Table 3).

Flat Photobioreactor Energy Balance

Materials in Flat-Plate Photobioreactors

Algae cultivation in open spaces comes with a few major downsides including low light utilization, loss through evaporation, diffusion of CO_2 to the atmosphere, the requirement of large areas of land, and face contamination issues. Since closed systems don't face these issues, much research has been devoted to developing closed systems, such as flat-plate, tubular, and vertical-column reactors (Ugwu et al., 2008).

Flat-plate reactors make sense to integrate vertically into the façades of buildings. Because of their thin width, they have a high surface area to volume ratio and make good use of sunlight (Tamburic et al., 2011). Placed vertically, there is an issue of mixing in the reactor, which is why CO_2 and air will be bubbled in.

Table 3 Amount of biomass produced and amount of CO_2 fixated per month. (Authors)

	Daily growth rate (g L^{-1} day^{-1})	Monthly biomass produced (kgs $month^{-1}$)	Amount of CO_2 fixated (kgs)
November	0.1	59.14	106.45
December	0.1	13.58	24.44
January	0.1	6.79	12.22
February	0.1	25.76	46.36
March	0.35	78.08	140.55
April	0.35	114.99	206.98
May	0.35	224.06	403.31
June	0.36	275.97	496.75
July	0.36	329.30	592.74
August	0.36	320.48	576.86
September	0.35	254.94	458.90
October	0.35	147.34	265.21
	Total	1850.42	3330.76

Table 4 Life span and CO_2 produced from materials (kgs of CO_2/kgs of material). (Authors)

Type	Materials	Life span	Production of material	Units	References
Support structure					
C profile (160 × 60 ×)	Stainless steel	50	2.91 ton CO_2/ton	kg	ISSF Stainless Steel (2015)
Metal platen (140 × 6)	Stainless steel	50	2.91 ton CO_2/ton	kg	ISSF Stainless Steel (2015)
Collection system					
Biomass collecting tube	PVC pipe	50	18 kg/3 m section of pipe	m	Recio et al. (2005)
		50	18.00	m	Recio et al. (2005)
Water tube	PVC pipe	50	18.00	m	Recio et al. (2005)
		50	18.00	m	Recio et al. (2005)
Water after harvesting	PVC pipe	50	18.00	m	Recio et al. (2005)
T pipe with valve	PVC	50	67.00	Kg	Alsabri & Al-Ghamdi (2020)
90° elbow pipe	PVC	50	67.00	Kg	Alsabri & Al-Ghamdi (2020)
Modules					
C profile 1 (140 × 50 × 2)	Stainless steel	50	2.91 ton CO_2/ton	Kg	ISSF Stainless Steel (2015)
	Stainless steel	50	2.91 ton CO_2/ton	kg	ISSF Stainless Steel (2015)
Type 1 of module: 2.60 × 0.90	Acrylic	10	5.50	kg	EnergyMan (2020)

To select the materials that structure the photobioreactor cells, the CO_2 produced in their manufacture and the life span of each one were studied (Table 4). Once the materials with the lowest environmental impact were determined, the CO_2 generation data were contrasted with the amount of CO_2 sequestered by the façade to establish the period of time that each one needs to balance the CO_2 necessary for its production (Table 5).

One of the most important materials to consider is the microalgae container. The panels will be made of 3 cm thick acrylic sheets for safety reasons. Acrylic was selected over polycarbonate or glass because of its transparency and lesser environmental impact (A & C Plastics, 2020).

Afterward, the use of stainless steel for the structure and frames of the modules was chosen because it has one of the highest recycling rates of any material. On average, about 50% of stainless-steel scrap is used to produce one ton of new stainless steel, according to the 2019 International Stainless-Steel Forum (ISSF Stainless Steel, 2015).

Finally, we decided to use PVC pipes to compose the water distribution and biomass collection systems, since they have the lowest CO_2 emissions between high-density polyethylene (HDPE), polypropylene (PP), concrete, and ductile iron. This data was taken from a report of the Estimation of energy consumption and CO_2 emission associated with the production of these materials carried out by the Environmental Modelling Laboratory of the University of Cataluña (Recio et al., 2005).

Results

To determine the time that would take for this façade to fix the CO_2 produced by the materials necessary for its construction, we calculate the amount of biomass produced to establish the CO_2 that could be sequestered (Table 3) and compare it with the amount of CO_2 needed to produce the materials needed to construct the façade. The results were as follows (Table 6):

Table 5 CO_2 output from containing material. (Authors)

CO_2 outputted vs. sequestered	CO_2 output from materials				
Type	Materials	Material production (kgs of CO_2/kg of material)	Total material used (kgs)	CO_2 produced from materials (kgs)	References
2.60 m × 0.90 m	Laminated glass	3.08	3699.1	11,393.1	Liu et al. (2014)
	Polycarbonate	6	1752.2	10,513.2	Boustead (2005)
	Acrylic	5.5	1650.0	9074.9	EnergyMan (2020)

Table 6 CO_2 output from materials compared to photobioreactor CO_2 sequestered. (Authors)

Type	Life span	Units	Total material used	CO_2 produced from materials (kgs)	CO_2 sequestered per year (kgs)	Years to balance carbon cost of materials:	CO_2 sequestered in 50 years (kgs)
Support structure							166,538.18
C profile (160 × 60 × 2)	50	kg	260.61	758.38	3330.76	0.23	
Metal platen (140 × 6)	50	kg	1379.68	4014.88		1.21	
Collection system							
Biomass collecting tube	50	m	104.70	628.08	3330.76	0.19	
	50	m	8.40	50.40		0.02	
Water tube	50	m	104.70	628.08		0.19	
	50	m	8.40	50.40		0.02	
Water after harvesting	50	m	11.80	70.80		0.02	
T pipe with valve	50	kg	47.38	3174.46		0.95	
90° elbow pipe	50	kg	4.76	318.92		0.10	
Modules							
C profile 1 (140 × 50 × 2)	50	kg	2405.31	6975.40	3330.76	2.09	
	50	kg	859.04	2491.22		0.75	
Type 1 of module: 2.60 × 0.90	10	kg	1677.02	9223.61		2.77	
			Total	28,384.63	3330.76	8.52 years	

It will take 8.5 years to fix the CO_2 produced by the façade. However, the life span of the materials reaches mostly up to 50 years. For this reason in 41.5 years, this façade can remove 138,226.69 kg of CO_2 from the atmosphere before it is necessary to change the materials that compose it. These 138 tons of CO_2 will be equivalent to 767,926.07 km of a running gasoline car.

According to the photobioreactor model proposed in the present study, it will take almost 8.5 years to fix the manufacturing CO_2 footprint of the materials we are using. However, the life span of the materials allows the façade to remove more than 138 tons of CO_2 from the atmosphere before it is necessary to change them, constituting a great contribution to reducing greenhouse gas emissions.

Conclusions

Microalgae façades are an aesthetically pleasing solution to generating energy and sequestering CO_2. Taking into account algal growth conditions such as temperature, pH, light, mixing, and concentration, microalgae are a productive and beautiful crop that can be implemented in relatively small ground space.

Discussion

It is necessary to find and test recycled materials that have a lower carbon footprint to make more viable for us to reach a zero footprint with this project in less time. Furthermore, the main issue discovered with this research is the importance of continuing the work on the calculation of the CO_2 that will be produced while the system is working, since in the present study the use of energy for the operation of the pumps and other collection systems was not taken into account.

Finally, it is important to account for the CO_2 cost of growing, separating, and transporting microalgae along with the energy cost of materials and pumps. Larger areas of photobioreactors can make better use of the energy cost of operating. In the future, scaling up this technology would be an efficient way to capture more CO_2 and increase production-biomass. Overall, PBR façades are a creative way for cities to become greener and will likely be a leading technology in the future.

Acknowledgments This work had the support of the program MIT-Spain Universidad de Alcalá (UAH) Seed Funds 2020.

References

A & C Plastics, I. (2020). *Polycarbonate vs Acrylic: Learn about the difference between acrylic & polycarbonate material.* A & C Plastics, Inc. https://www.acplasticsinc.com/informationcenter/r/acrylic-vs-polycarbonate

Alsabri, A., & Al-Ghamdi, S. G. (2020). Carbon footprint and embodied energy of PVC, PE, and PP piping: Perspective on environmental performance. *Energy Reports, 6*, 364–370.

Boustead, I. (2005). *Eco-profiles of the European plastics industry.* A report prepared for PlasticsEurope, Bruxelle, Belgium.

Cervera Sardá, R., & Gómez Pioz, J. (2014). Architectural bio-photo reactors: Harvesting microalgae on the surface of architecture. In *Biotechnologies and biomimetics for civil engineering* (pp. 163–179). Springer. https://doi.org/10.1007/978-3-319-09287-4_7

Cervera Sardá, R., & Vicente, C. A. (2016). Case studies on the architectural integration of photobioreactors in building facades. In F. Pacheco Torgal, C. Buratti, S. Kalaiselvam, C. Granqvist, & V. Ivanov (Eds.), *Nano and biotech based materials for energy building efficiency* (pp. 457–484). Springer. https://doi.org/10.1007/978-3-319-27505-5_17

Climate-data.org. (2020). *Climate Madrid (Spain).* Climate-data.org. https://en.climate-data.org/europe/spain/community-of-madrid/madrid-92/#climate-table

Dassey, A. J., & Theegala, C. S. (2013). Harvesting economics and strategies using centrifugation for cost effective separation of microalgae cells for biodiesel applications. *Bioresource Technology, 128*, 241–245.

EnergyMan. (2020). *Eco-friendly alternatives to acrylic (plexiglass) – 2020 guide.* Ubuntu Manual 2021. https://ubuntumanual.org/eco-friendly-alternatives-acrylic/

García Cubero, R. (2014). *Producción de biomasa de microalgas rica en carbohidratos acoplada a la eliminación fotosintética de CO2.* (Tesis doctoral). Universidad de Sevilla. Sevilla.https://digital.csic.es/bitstream/10261/101928/1/TESIS%20DOCTORAL%20RGC.pdf

García-Cubero, R., Moreno-Fernández, J., & García-González, M. (2017). Modelling growth and CO2 fixation by Scenedesmus vacuolatus in continuous culture. *Algal Research, 24*, 333–339.

ISSF Stainless Steel. (2015). *Stainless steel and co2: Facts and scientific observations* (Paper presented at the International Stainless Steel Forum).

Liu, M., Zhu, S., & Li, W. (2014). CO2 emission of building glass production. *Energy Education Science and Technology Part A: Energy Science and Research*, 1203–212.

Recio, J. M. B., Guerrero, P. J., Ageitos, M. G., & Narváez, R. P. (2005). *Estimate of energy consumption and CO2 emission associated with the production, use and final disposal of PVC, HDPE, PP, ductile iron and concrete pipes.* Universitat Politécnica De Catalunya.

Santos Montes, A. M., González Arechavala, Y., & Martín Sastre, C. (2014, 01/02/). *Uso y aplicaciones potenciales de las microalgas.* Revista: Anales de Mecánica y Electricidad, Periodo: 1, Volumen: XCI, Número: I, 20–28. Madrid, España.

Shelef, G., Sukenik, A., & Green, M. (1984). Microalgae harvesting and processing: a literature review. United States: 10.2172/6204677. https://www.osti.gov/biblio/6204677

Song, C., Pan, W., Srimat, S. T., Zheng, J., Li, Y., Wang, Y., Xu, B., & Zhu, Q. (2004). Tri-reforming of methane over Ni catalysts for CO2 conversion to syngas with desired H2/CO ratios using flue gas of power plants without CO2 separation. *Studies in Surface Science and Catalysis, 153*, 315–322.

Tamburic, B., Zemichael, F. W., Crudge, P., Maitland, G. C., & Hellgardt, K. (2011). Design of a novel flat-plate photobioreactor system for green algal hydrogen production. *International Journal of Hydrogen Energy, 36*(11), 6578–6591.

Ugwu, C. U., Aoyagi, H., & Uchiyama, H. (2008). Photobioreactors for mass cultivation of algae. *Bioresource Technology, 99*(10), 4021–4028.

United Nations. (2020a). *Goal 13: Take urgent action to combat climate change and its impacts.* https://www.un.org/sustainabledevelopment/climate-change/

United Nations. (2020b). *Ensure access to affordable, reliable, sustainable and modern energy.* https://www.un.org/sustainabledevelopment/climate-change/

Varshney, P., Mikulic, P., Vonshak, A., Beardall, J., & Wangikar, P. P. (2015). Extremophilic micro-algae and their potential contribution in biotechnology. *Bioresource Technology, 184*, 363–372. https://doi.org/10.1016/j.biortech.2014.11.040

Xin, L., Hong-ying, H., & Yu-ping, Z. (2011). Growth and lipid accumulation properties of a freshwater microalga Scenedesmus sp. under different cultivation temperature. *Bioresource Technology, 102*(3), 3098–3102.

Zhang, X. (2015). *Microalgae removal of CO2 from flue gas.* IEA Clean Coal Centre.

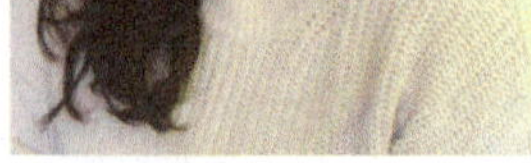

Ma. Rosa Villalba is an Architect, Ph.D. Candidate at the University of Alcalá with the theme: "Implementation of photobioreactors in Architecture" (2020-present). She accomplished the Master in Advanced Architecture and City Projects, with double specialty in Environment and Architecture (2018–2019) and Urban Design (2019–2020). She has been awarded with the mention for the Best Master's Degree record (2019–2020). She had published scientific contributions in IOP Conference Series: Earth and Environmental Science "Will the Covid-19 Pandemic Transform Our Urban Habitat? A Microalgae Photobioreactors Labyrinth-Garden as an answer for The Post-Covid Sustainable City" (2021) and I Congreso Edificios Energía Casi Nula: "Fotobiorreactores urbanos de algas para la ciudad verde: reinterpretando el Laberinto" (2020) – online digital article.

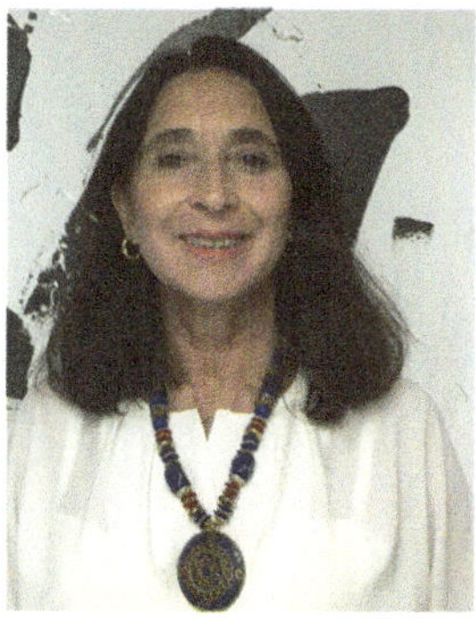

María Rosa Cervera Sardá Ph.D. Architect, Professor and former Dean of the School of Architecture at the University of Alcalá. Current Director of the Master's degree in Advanced Architecture and City Projects. She is the author of writings and books on architecture, among which we highlight: Bionics, Biomimetics and Architecture (2019); Space and Time in Architectural Composition (2018); Recycling Mumbai. Re-imagining the Slum (2012); Madrid, Recycled City (2011); Iron in 19th Century Madrid Architecture (2006). A regular speaker in professional and academic circles, she has given lectures in Spain, China, India, USA, Bolivia, Peru, El Salvador, Venezuela, Romania, Italy, etc.

As an architect, Rosa Cervera has received several awards – the "Antonio Maura" award, the COAM award, the "Transfer of Knowledge" award – and has won several international architecture competitions.

Rosa Cervera is a pioneer in the research of Bionics, Biomimetics and the application of biological structures to innovative and efficient architecture and urban design. A direct result of these studies is the Self-Sustainable Vertical Garden City, Bionic Tower.

Nano-modified Materials for New Construction Technologies: Self-Compacting and 3D Printing

Hugo Varela, Gonzalo Barluenga, and Arnaud Perrot

Abstract

Self-compacting concrete (SCC) and 3D printing (3DP) have become two of the most innovative and promising construction technologies for the XXI Century Architecture with cement-based systems (CBS). Both applications require specific fresh material properties, fluid in the first case and firm in the second one. However, material cast-in-place in both technologies shows similar issues to be addressed regarding material fresh state rheology and its control. The main rheological parameters to consider for fulfilling cast-in-place requirements are shear yield stress, viscosity and structural build-up. Besides, mechanical properties as compressive yield stress, Young modulus and critical strain must also be considered at early age, when setting begins. Nano-modification can be the answer to modify fresh state rheological and early age mechanical parameters of CBS. The aim of this study was the evaluation of CBS rheological behaviour with nanocomponents and VMAs for SCC and 3DP applications. The combined use of nanocomponents (NC), as nanoclays and nanosilica, and viscosity modifying admixtures (VMAs) has shown the ability to tune CBS rheology. Moreover, the addition of a high-range water-reducing admixture (HRWRA) is essential to produce CBS with an appropriate consistency and workability, although some incompatibilities with other components have been described. CBS rheological and mechanical parameters can be studied using specific experimental test methods, depending on the properties and the consistency required by each technology and application. Nano-modified CBS showed an increase of shear yield stress while the addition of VMAs increased viscosity. The combination of both components produced synergetic effects in the material over time, overcoming issues observed for both technologies.

Keywords

Self-compacting concrete · Digital fabrication · 3D printing · Rheology · Thixotropy · Nano-modified materials · Construction technologies

Introduction

Cement-based systems (CBS) are a primary construction material in architecture applications. Nowadays, innovative cement technologies such as self-compacting concrete (SCC) and 3D printing (3DP) with cement-based materials (CBS) are solutions to reduce the amount of material, cost and energy used, to improve structural and design efficiency and, in the case of 3DP, to eliminate the requirement of a formwork (Wangler et al. 2016; Okamura & Ouchi, 2003; Ferron, 2008).

Although CBS fresh properties are different in both technologies, both SCC and 3DP show similar issues regarding the rheological properties required for their cast-in-place process (Wangler et al., 2016; Roussel, 2018; Perrot et al., 2018).

Rheology of CBS is an essential point to control fresh state of this material (Roussel, 2018; Perrot et al., 2018). Parameters as shear yield stress, viscosity and structural build-up are used to evaluate CBS rheology. Besides mechanical properties at early age, as compressive yield stress, Young modulus and critical strain have to be also considered (Perrot et al., 2018). SCC has to be fluid to be pumped and to fill a formwork without requiring compaction. Then its consistency should evolve at rest to avoid excessive formwork lateral pressure (Kim et al., 2010; Barluenga et al., 2017).

H. Varela (✉) · G. Barluenga
Universidad de Alcalá, Departamento de Arquitectura, Alcalá de Henares, Madrid, Spain
e-mail: hugo.varela@uah.es

A. Perrot
Université Bretagne-Sud, IRDL, Lorient, France

M. R. Cervera Sarda et al. (eds.), *Architecture Inspired by Nature*, https://doi.org/10.1007/978-3-031-33144-2_18

On the other hand, 3DP has to be pumped, extruded and placed. Then, layer by layer, it has to be vertically constructed, reaching enough mechanical capacity to keep in place during setting and hardening. During this process, rheological and mechanical properties have to develop and adapt to the required values (Perrot et al., 2018; Roussel, 2018).

Nano-modification has been described in the literature as an excellent opportunity to improve microstructural behaviour and to control CBS rheology. The small particle size of nanocomponents produces a high specific surface area and fill the small gaps within the matrix of cement (Sanchez & Sobolev, 2010). Besides, some nanocomponents (NC) as nanoclays and nanosilica have demonstrated to be useful to modify CBS rheology (Kuder & Shah, 2007; Kim et al., 2010; Hou et al., 2013; Varela et al., 2020). Moreover, the use of VMAs is another way to reduce CBS fresh state issues. VMAs are able to modify viscosity and control water drainage inside fresh CBS (Palacios & Flatt, 2016; Varela et al., 2021). A combination of nanoparticles and VMAs is an effective way to control rheology and to reduce CBS cast-in-place problems (Varela et al., 2021).

The aim of this study was the evaluation of rheological properties of CBS with NC and VMAs for SCC and 3DP applications. Synergies between NC and VMAs to control rheological parameter were assessed.

Self-Compacting Concrete Technology

Context

New innovative and advanced concretes were designed during the 90s, being self-compacting concrete one of the most relevant new materials (Okamura & Ouchi, 2003). SCC is a fluid material that can adapt itself to any imposed shape. SCC increases quality control and reliability regarding conventional concrete and reduces cost, energy, noise and labour risk during its cast-in-place (Ferron, 2008).

SCC cast-in-place can be divided into three main stages: material pumping, pouring into the formwork, and rest and set inside the formwork. Some issues have been described concerning SCC, as flow restrictions during the pumping process, due to its rheopectic behaviour, increasing viscosity under pressure (Barluenga et al., 2017). Besides, the high SCC fluidity generates high lateral formwork pressure that would require stronger and tighter formworks (Ferron, 2008; Kim et al., 2010).

One strategy to avoid these cast-in-place issues is controlling the rheology of the material in fresh state throughout the use of nano-modification (Sanchez & Sobolev, 2010). Rheological parameters such as shear yield stress, viscosity and structural build-up are essential to control the rheology of SCC (Roussel, 2018; Perrot et al., 2018; Perrot & Rangeard, 2019). Several experimental test methods have been used in this study to evaluate these rheological parameters (Varela et al. 2020, 2021).

The addition of VMAs and NC showed the ability to change SCC rheology during the different phases of its cast-in-place process. Nanoclays and nanosilica are able to increase initial yield stress and structural build-up of this material (Kim et al., 2010; Hou et al., 2013; Varela et al., 2020, 2021). VMAs change material's viscosity and also change yield stress (Palacios & Flatt, 2016; Varela et al., 2020). The combination of both NC and VMAs produces synergies that can improve SCC behaviour over time (Varela et al., 2021).

Figure 1 summarizes the stages of the process, the design parameters, requirements, issues and components related to SCC technology. The main aim of this study was to evaluate fluid pastes and SCC samples with NC and VMAs to assess their effects and synergies on the material rheology.

Material

Figure 2 presents CBS compositions used in this study, which were designed at two scales: cement paste and concrete. A cement paste with a cement type CEM I 42.5R blended with limestone filler (2:1) was designed in order to reduce cement content and to produce sustainable pastes. A low water to binder ratio was selected (0.35) and a HRWRA was incorporated to reach a fluid consistency. Then nanocomponents and VMAs were added to the paste to control rheology. The nanocomponents included in this study, 2% by cement weight, were nanosilica and four nanoclays, an attapulgite, a bentonite and two sepiolites (in powder form and dispersed in water). Three types of VMAs were added a polyacrylamide-co-acrylate based (VMA1), a polyether-methylcellulose-based (VMA2) and a synthetic co-polymer (VMA3), separately and combined with nanocomponents to evaluate synergies between them.

SCC compositions used the selected pastes with aggregates, gravel with a maximum particle size of 16 mm and a river sand of 0–4 mm.

Experimental Test Methods and Evaluation

Flowability and rheology of SCC technology were experimentally evaluated. The effects of NC and VMAs on the paste rheology were evaluated on cement pastes and afterwards on SCC. Figure 3 shows the testing methods carried out in this study on cement pastes and on SCC samples.

SCC cement paste was evaluated with a mini-cone slump test (MCS) and a dynamic shear rheometer test (DSR). Mini-

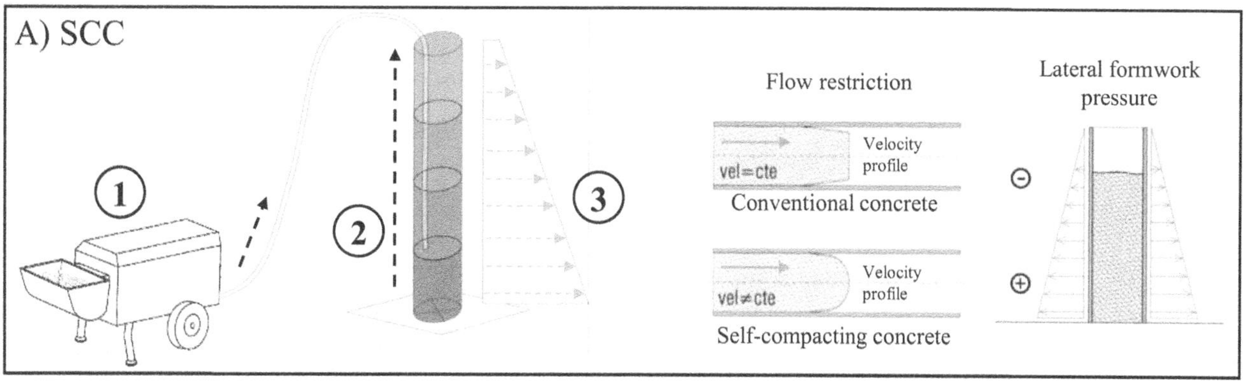

Phases		Rheological	Issues	Requirements	Components
(1)	Pumpability	Shear yield stress; viscosity	flow restrictions; segregation	slip layer; low yield stress; viscosity	HRWRA, w/b
(2)	cast-in-formwork	Shear yield stress; viscosity	Segregation; not self-compacting	low yield stress; viscosity; fluid consistency under pressure	HRWRA, VMAs
(3)	Reduction of pressure	Shear yield stress; viscosity	Formwork lateral pressure; paste leaks	High structural build-up; firm consistency at rest	Nanoclays and nanosilica

Fig. 1 Summary of the main features of SCC technology. (Source: Own illustration)

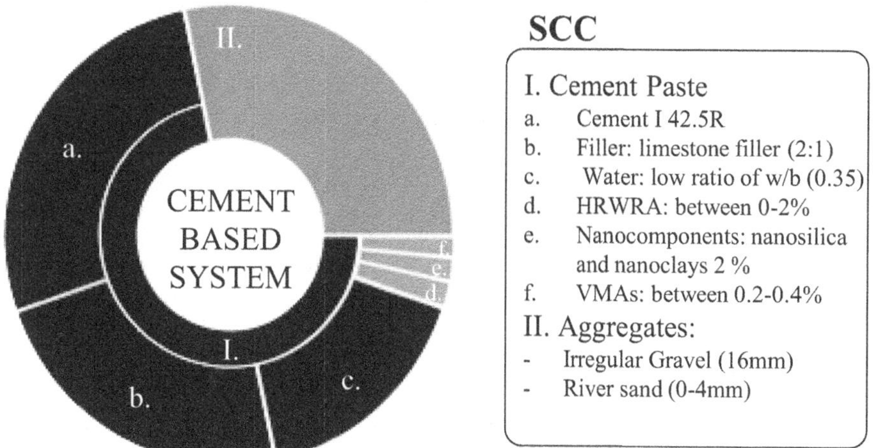

Fig. 2 Material compositions of SCC samples: cement paste and concrete SCC. (Source: Own illustration)

cone test consists of a 3D-printed mini-cone that is filled with cement paste and is lifted allowing free spreading of the paste sample and measuring the final spread diameter and final spread time of cement paste (Fig. 3a). These parameters were used to calculate paste rheological parameters, yield stress (τ_0) and viscosity (μ_0) (Ferron, 2008; Varela et al. 2020). Besides, τ_0 was measured over time, evaluating paste structural build-up (A_{thix}) of mixtures. Reversible and irreversible mechanisms involved in τ_0 values increased were measured in samples left at rest and stirred before performing the test. These parameters were also measured with DSR tests with a coaxial cylinder geometry (Fig. 3b). However, only dynamic (irreversible) τ_0 values were measured with DSR. Two different protocols were used: a time curve (increase of shear) and flow curve (constant rate). In the first one, τ_0 and μ_0 were measured and afterwards structural build-up (irreversible) was calculated (Ferron, 2008; Varela et al., 2021).

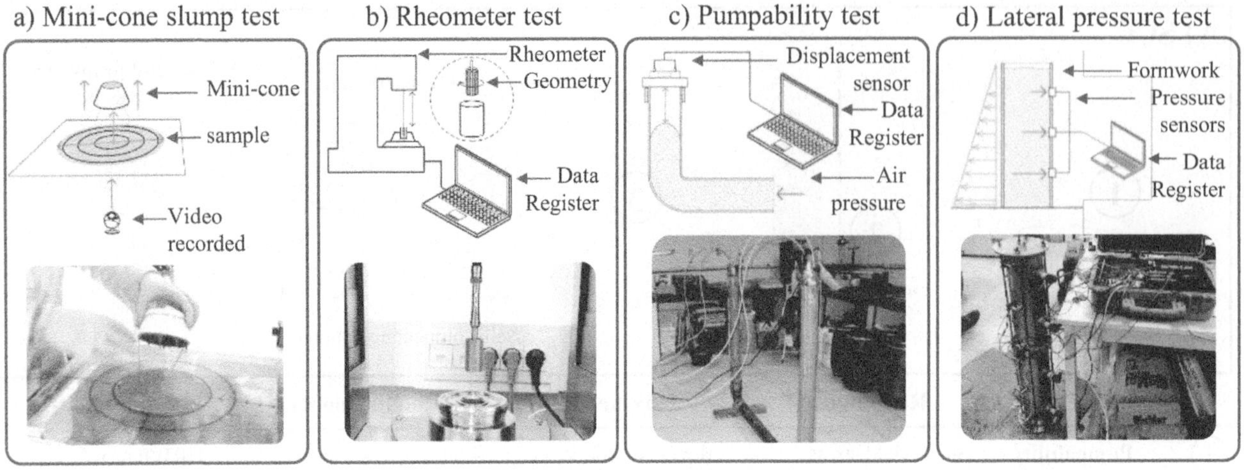

Fig. 3 Experimental test methods used to evaluate SCC: cement paste and concrete tests. (Source: Own illustration)

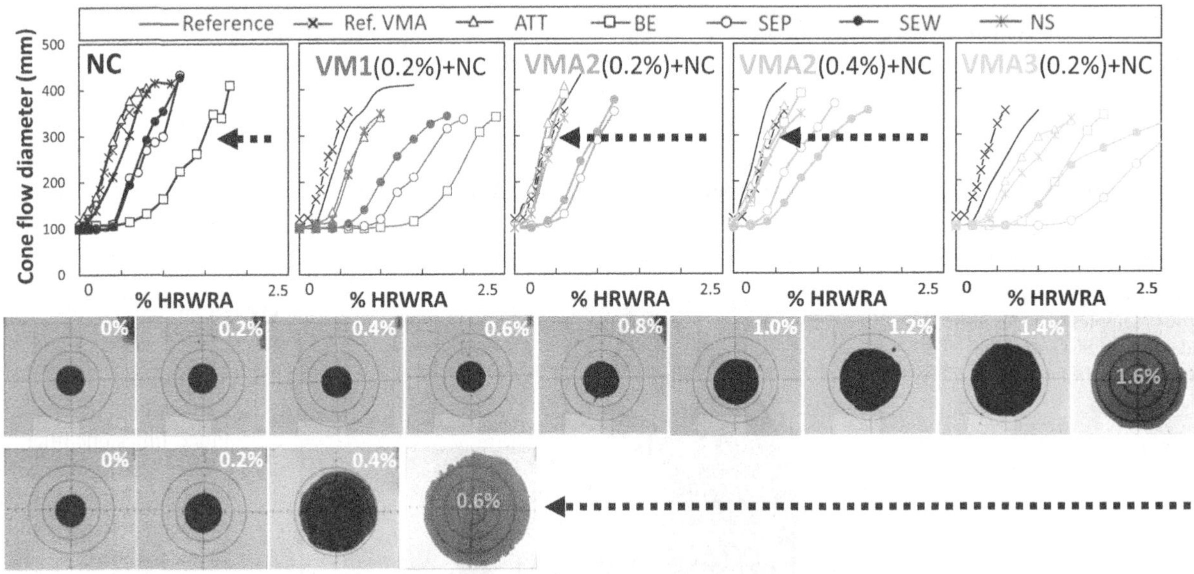

Fig. 4 Increase of HRWRA on NC and VMAs combination cement paste samples: mini-cone slump test. Comparison of BE and VMA2+BE sample evolution increasing HRWRA percentage. (Source: Own illustration adapted from Varela et al., 2021)

SCC was evaluated using a pumpability test apparatus (PU) and a lateral formwork pressure test (LFP). In the PU test, a SCC sample was tested in a U-shaped pipe connected on both ends to an air pressure circuit (Fig. 3c). That air pressure moves the SCC in both directions alternatively. Then, displacement and air pressure needed to move SCC sample were measured. Those parameters can be related to τ_0 values of the sample. Besides, the front profile of the moving sample can be observed, showing the characteristic rounded velocity profile SCC rheopectic mixtures (Barluenga et al., 2017). On the other hand, LFP was performed in a vertical formwork instrumented with wall and capillary pressure sensors (Fig. 3d), designed to measure the lateral pressure exerted by SCC samples on the formwork (Kim et al., 2010). The difference between hydrostatic and lateral pressure can be related to τ_0 values.

Main Results

Figures 4 and 5 show the main results obtained on cement paste for SCC technology. A high demand of HRWRA was required for NC to achieve the target flowability, especially for pastes with Bentonite (BE) (Varela et al., 2020). However,

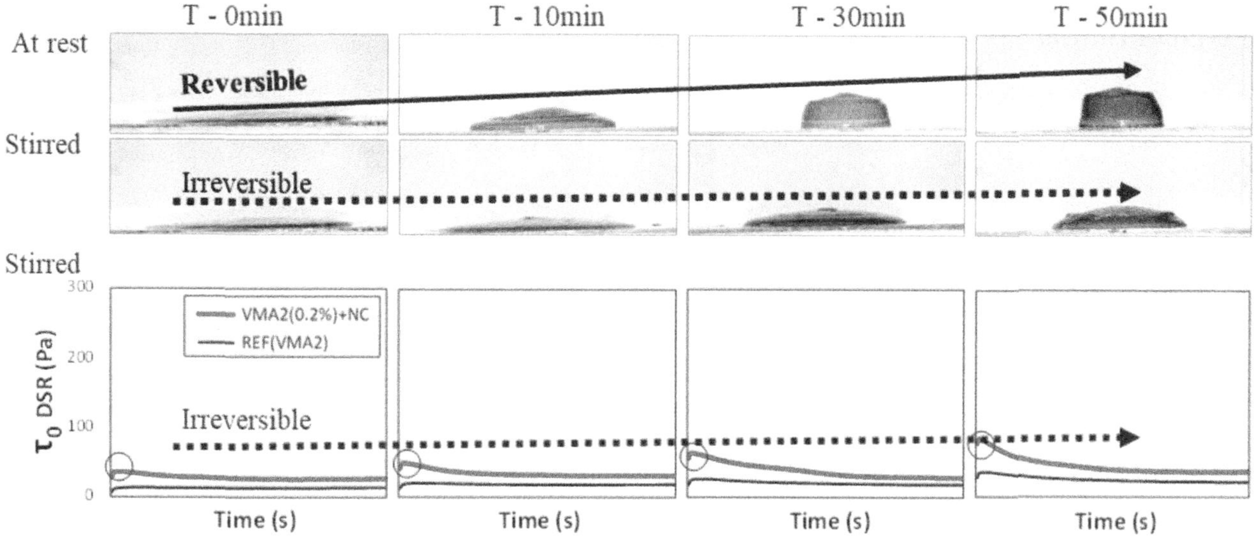

Fig. 5 Reversible and irreversible structural build-up effect (thixotropy effect) on VMA2+BE cement paste sample: mini-cone slump test and DSR (CCR) test over time. (Source: Own illustration adapted from Varela et al., 2021)

the combination of VMAs and NC exhibited positive synergetic effects on paste rheology. VMA2 with BE cement paste sample (Fig. 4) showed a synergetic effect, reducing the amount of HRWRA required to reach the high flowability required by SCC samples regarding to the paste sample only with BE (Varela et al., 2021).

The structural build-up (effective thixotropy) of the SCC pastes was assessed over time (Fig. 5). Reversible and irreversible structural build-up was evaluated on samples left at rest and stirred, respectively. Paste with VMA2 and BE showed a remarkable increase of reversible structural build-up on samples left at rest, whereas stirred samples showed a low increase of the irreversible part of the structural build-up. Hence, when sample was left at rest, a quick increase of yield stress over time was produced (Varela et al., 2021). This type of synergetic effect could be effective to reduce lateral pressure on SCC cast-in-place.

3D-Printing Technology

Context

Digital fabrication consists of the generation of a physic object from a 3D digital model and corresponds to a set of the construction technologies that are receiving more and more interest in the last decades, due to its free shaping possibilities and highly sustainable performance. Among digital fabrication techniques, 3D printing (3DP) is characterized by creating objects superposing layers of extruded material. This new cast-in-place technology allows generating architectural elements with freedom design and structural efficiency. Moreover, 3DP increases construction speed and reduce labour, energy and formwork costs (Wangler et al., 2016).

However, CBS 3D printing presents some issues regarding material's technological requirements for pumpability, extrudability, layer deposition and buildability. To fulfil these requirements, printable CBS should show first low yield stress and semi-liquid state to be pumped and extruded. But then, it should increase yield stress quickly and change to a semi-solid material to maintain its shape after deposition and to be able to bear the weight of the upper layers (Perrot & Rangeard, 2019; Perrot et al., 2018; Roussel, 2018).

To obtain the material appropriate consistency in each phase of the cast-in-place process, it is essential to control rheological and mechanical parameters of fresh CBS (Perrot et al., 2018; Roussel, 2018). Shear yield stress, viscosity and structural build-up are the rheological parameters, and compressive yield stress, Young modulus and critical strain are the mechanical material that have to be experimentally measured (Perrot & Rangeard, 2019; Roussel, 2018).

Nano-modification was again used as a design method to control CBS rheology. In this study, nanoclays (NC) and VMAs were used to control initial drainage issues at extrusion (increase of force) (Kuder & Shah, 2007) and then, nanoclays could produce an increase of shear and compressive yield stress (Varela et al., 2020, 2021). Actually, some NC could be able to replace cellulose ether-based VMAs to control 3D-printing parameters (Kuder & Shah, 2007).

Figure 6 plots the process, design parameters, requirements, issues and components related to 3D-printing technology. The aim of this study was to evaluate the effects and synergies of NC and VMAs on cement paste rheology for 3D-printing technology.

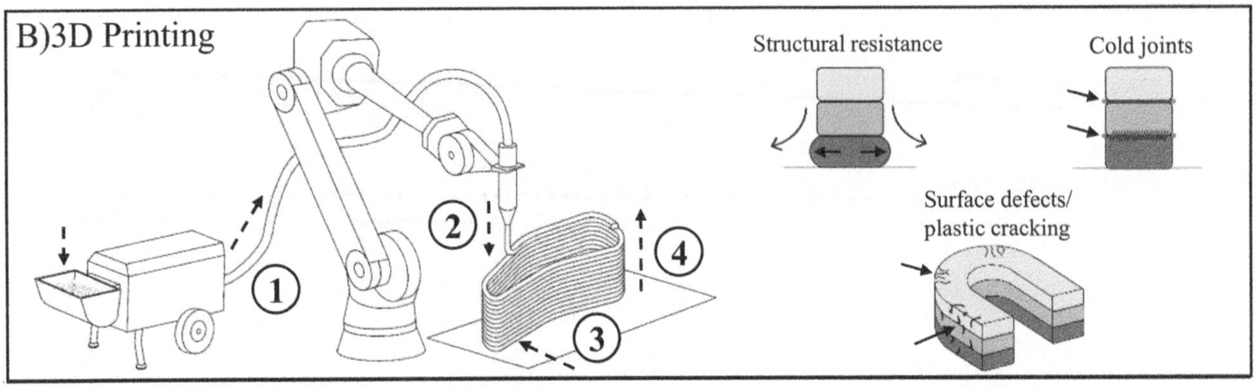

Phase process	Rheological	Mechanical	Issues	Requirements	Components
(1) **Pumpability**	Shear yield stress; viscosity	-	Plug; segregation	slip layer; low yield stress	HRWRA, w/b VMAs
(2) **Extrudability**	Shear yield stress; viscosity	-	Plug; drainage; time windows	Avoid drainage; low yield stress	HRWRA, VMAs
(3) **Deposition first layer**	Shear yield stress; viscosity	Compressive yield stress; Young modulus	Discontinuity; surface defects; Geometry control	low yield stress; high viscosity; high young modulus	Nanoclays; VMAs
(4) **Buildability**	Thixotropy index	Compressive yield stress; Young modulus	Cold joints; Buckling	High structural build-up; high compressive yield stress; low young modulus	Nanoclays

Fig. 6 Summary of the main features of 3D printing technology. (Source: Own illustration partially adapted from Perrot & Rangeard, 2019)

Material

Figure 7 describes the CBS compositions proposed for the 3D-printing study. The 3D-printing material was evaluated at two scales: a cement paste phase and a mortar phase.

A reference cement paste (cement type CEM I 52.5R) and another blended (3:1) with fly ash (FA) were designed. A low water to binder ratio was selected (0.27) and HRWRA was used to reach the target consistency for 3D printing, ranging between 0 and 5% by binder weight. The target yield stress measured using the cone-penetrometer test was fixed at 1 kPa. Then three NC, an attapulgite, a sepiolite and a bentonite, were added in a 2% by binder weight and two VMAs, 0.1% of a polyacrylamide-co-acrylate-based (VMA1) and 0.5% of a cellulose ether-based (VMA4), were included by binder weight to the reference paste with FA.

On the other hand, mortar was designed incorporating river sand with a maximum size particle of 1mm. Besides, natural fibres could be added to improve material buildability.

Experimental Test Methods and Evaluation

Figure 8 plots the four experimental tests that were carried out on cement paste in this 3D-printing evaluation.

A cone penetration test (CPT) was used to evaluate τ_0 of CBS. The tip cone fall into the sample and depth penetration was measured to calculate τ_0 value (Mazhoud et al., 2019). The test was repeated three times for each measurement. This method was carried out over time, from 10 to 90 minutes, to evaluate the τ_0 evolution and the structural build-up of cement paste. A DSR controlled constant rate test (CCR) was also carried out to assess τ_0 of cement paste. A rate of 0.5 s^{-1} and a four-vane geometry was used. The torque value was used to calculate τ_0 of sample through a formulation described in the literature (Perrot et al., 2018). This test was also performed over time (10–90 min) to obtain structural build-up of mixtures. On the other hand, squeeze test (SQT) carried out in a DSR was used to evaluate compressive strenght (σ_0) at fresh state. A plate geometry was used to squeeze a cylinder cement sample with a constant velocity of

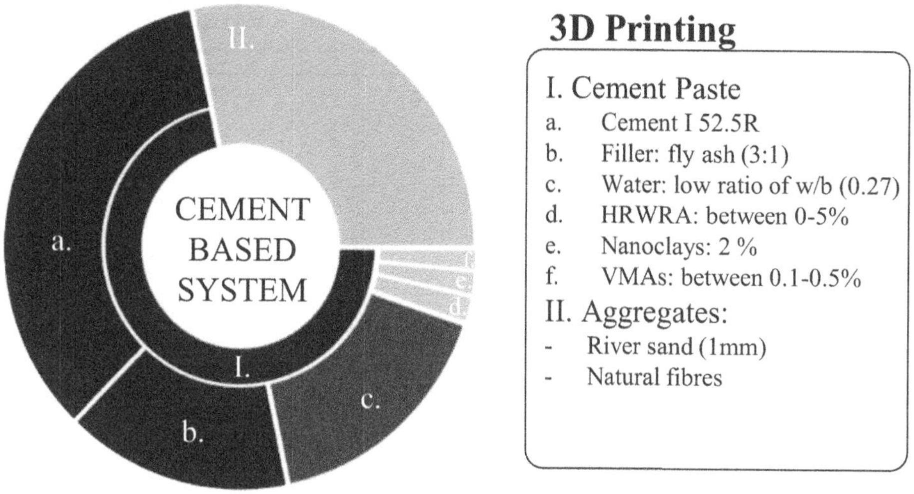

Fig. 7 Material compositions of 3D-printing samples: cement paste and mortar SCC. (Source: Own illustration)

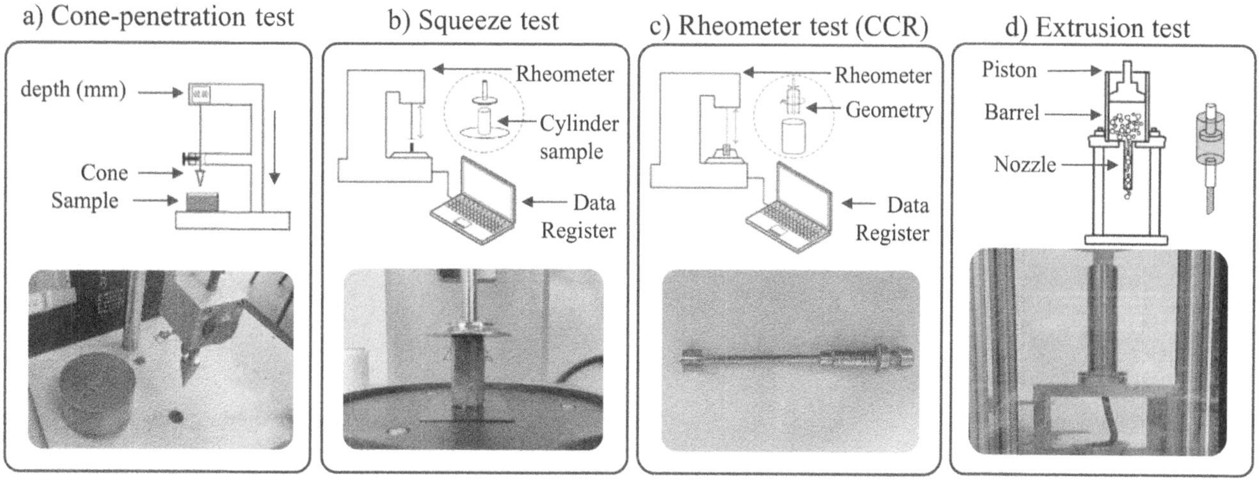

Fig. 8 Experimental test methods used to evaluate 3DP: cement paste and mortar tests. (Source: Own illustration)

0.1 mm/s. The normal force obtained in the compression was used to compute σ_0 value. SQT was performed over time (10–90 min) to measure σ_0 evolution over time. In addition, a paste extrusion test (EXT) was carried out at 10 min. The extrusion force and the displacement were measured to calculate both τ_0 and σ_0 (Perrot et al., 2018).

Main Results

Figure 9 summarizes the main results of cement paste with NC and VMAs for 3D printing. All samples in this study showed an initial τ_0 stress of 1 kPa. The use of FA reduced τ_0 over time. However, FA did not reduce much the force extrusion. That reduction was reached in all samples with cellulose ether-based (VMA4) being the samples with lower shear yield stress. Nevertheless, sepiolite (SEP) also produced a force reduction effect avoiding drainage in the piston of extrusion test, although it showed a higher increase of τ_0 over time. Other NC were not able to produce the same effect than SEP. The combination of VMA4 and SEP showed a synergetic effect, being able to be extruded appropriately but produced higher τ_0 values over time at EXT and CCR than pastes without SEP.

On the other hand, VMAs reduced σ_0 evolution over time. VMA4 showed a low increase of σ_0 over time but sample with SEP and VMA4 increased σ_0 values. In general, SEP increased initial σ_0, but over time (from 70 min), samples with lower amount of HRWRA showed a higher σ_0 evolution. That is the case of FA reference, which setting started probably earlier than SEP samples setting due to they have high amount of HRWRA.

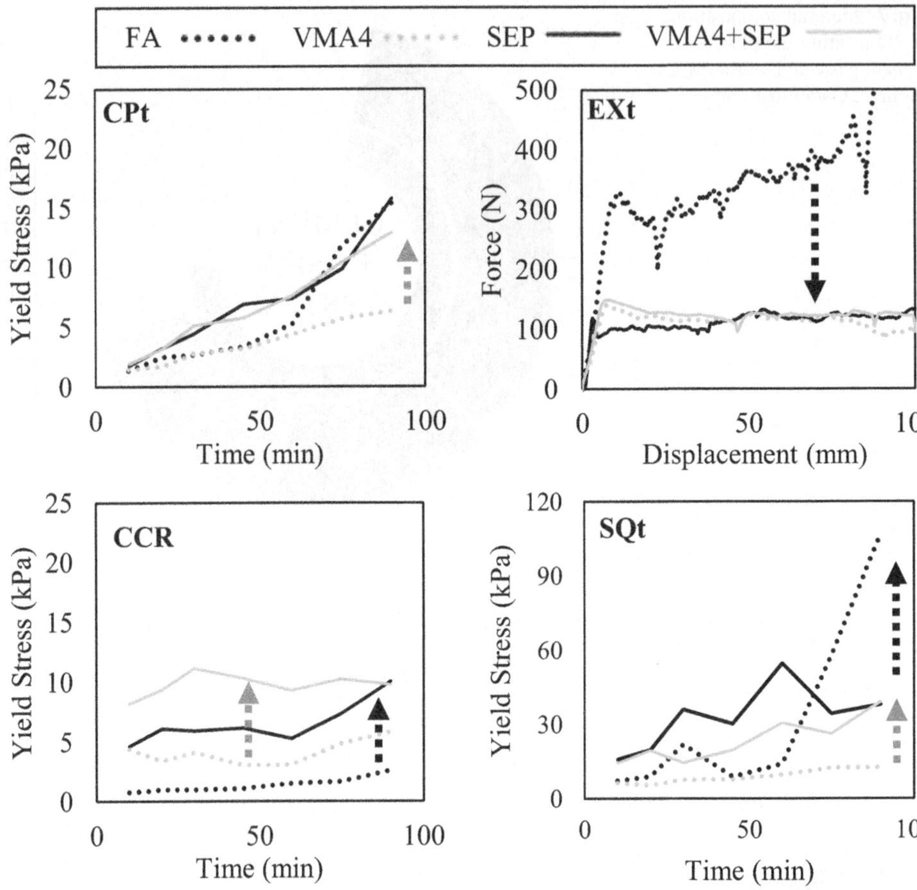

Fig. 9 Main results of cement paste samples of 3DP study. FA reference, SEP, VMA4 and VMA4+SEP samples as a resume of 3DP samples evaluation. (Source: Own illustration)

Conclusions

In this study, the rheological and mechanical properties of cement-based system (CBS) with nanocomponents (NC) and viscosity modifying admixtures (VMAs) for self-compacting (SCC) and 3D-printing (3DP) cast-in-place technologies were evaluated. The experimental test methods used to measure rheological and mechanical parameters were described. From these results, some conclusions can be drawn:

The study of rheological and mechanical parameters is necessary to understand the change of material consistency and solve the cast-in-place issues of SCC and 3DP.

The use of experimental test methods in laboratory is essential to carry out an appropriate evaluation of rheology of CBS. Each technology needs its own specific evaluating test methods.

Nano-modification is able to adjust the rheological and mechanical parameters of CBS in the fresh state. Especially, NC (as nanoclays and nanosilica) and VMAs are able to modify these parameters on CBS.

The combination of NC and VMAs can produce synergies to control the initial yield stress and the structural build-up of CBS.

BE combined with VMA2 presented a synergetic effect on SCC cement pastes that produce a reduction of HRWRA percentage regarding samples only with BE. Besides, the BE and VMA2 sample showed a huge increase of reversible build-up over time (effective thixotropy).

Opposite to FA reference, SEP and VMA4 produced a good extrusion in 3D-printing cement samples. Also, SEP samples showed an earlier increase of τ_0 over time but did not show the same effect with σ_0 due to probably their high amount of HRWRA required.

References

Barluenga, G., Giménez, M., Rodríguez, A., & Rio, O. (2017). Quality control parameters for on-site evaluation of pumped self-compacting concrete. *Construction and Building Materials, 154*, 1112–1120. https://doi.org/10.1016/j.conbuildmat.2017.07.223

Ferron, R. (2008). *Formwork pressure of self-consolidating concrete: Influence of flocculation mechanisms, structural rebuilding, thixotropy and rheology* [Doctoral Dissertation, Northwestern University]. ProQuest Dissertations & Theses Global.

Hou, P. K., Kawashima, S., Wang, K. J., Corr, D. J., Qian, J. S., & Shah, S. P. (2013). Effects of colloidal nanosilica on rheological and mechanical properties of fly ash–cement mortar. *Cement and Concrete Composites, 35*, 12–22. https://doi.org/10.1016/j.cemconcomp.2012.08.027

Kim, J. H., Beacraft, M., & Shah, S. P. (2010). Effect of mineral admixtures on formwork pressure of self-consolidating concrete. *Cement and Concrete Composites, 32*(9), 665–671. https://doi.org/10.1016/j.cemconcomp.2010.07.018

Kuder, K. G., & Shah, S. P. (2007). Rheology of extruded cement-based materials. *ACI Materials Journal, 104*(3), 283–290.

Mazhoud, B., Perrot, A., Picandet, V., Rangeard, D., Courteille, E. (2019). Underwater 3D printing of cement-based mortar. *Construction and Building Materials, 214, 458–467.* https://doi.org/10.1016/j.conbuildmat.2019.04.134

Okamura, H., & Ouchi, M. (2003). Self-compacting concrete. *Journal of Advanced Concrete Technology, 1*, 5–15.

Palacios, M., & Flatt, R. J. (2016). Working mechanism of viscosity-modifying admixtures. In P.-C. Aïtcin & R. J. Flatt (Eds.), *Science and technology of concrete admixtures* (pp. 415–432). Woodhead Publishing/Elsevier. https://doi.org/10.1016/B978-0-08-100693-1.00020-5

Perrot, A., & Rangeard, D. (2019). 3D Printing in concrete: Techniques for extrusion/casting. In A. Perrot (Ed.), *3D Printing of concrete*. ISTE/Wiley. https://doi.org/10.1002/9781119610755.ch2

Perrot, A., Rangeard, D., Nerella, V. N., & Mechtcherine, V. (2018). Extrusion of cement-based materials – An overview. *RILEM Technical Letters, 3*, 91–97. https://doi.org/10.21809/rilemtechlett.2018.75

Roussel, N. (2018). Rheological requirements for printable concretes. *Cement and Concrete Research, 112*, 76–85. https://doi.org/10.1016/j.cemconres.2018.04.005

Sanchez, F., & Sobolev, K. (2010). Nanotechnology in concrete – A review. *Construction and Building Materials, 24*(11), 2060–2071. https://doi.org/10.1016/j.conbuildmat.2010.03.014

Varela, H., Barluenga, G., & Palomar, I. (2020). Influence of nanoclays on flowability and rheology of SCC pastes. *Construction and Building Materials, 243*, 118285. https://doi.org/10.1016/j.conbuildmat.2020.118285

Varela, H., Barluenga, G., Palomar, I., & Sepulcre, A. (2021). Synergies on rheology and structural build-up of fresh cement pastes with nanoclays, nanosilica and viscosity modifying admixtures. *Construction and Building Materials, 308*, 125097. https://doi.org/10.1016/j.conbuildmat.2021.125097

Wangler, T., Lloret, E., Reiter, L., Hack, N., Gramazio, F., Kohler, M., Bernhard, M., Dillenburger, B., Buchli, J., Roussel, N., & Flatt, R. (2016). Digital concrete: Opportunities and challenges. *RILEM Technical Letter*, 67–75. https://doi.org/10.21809/rilemtechlett.2016.16

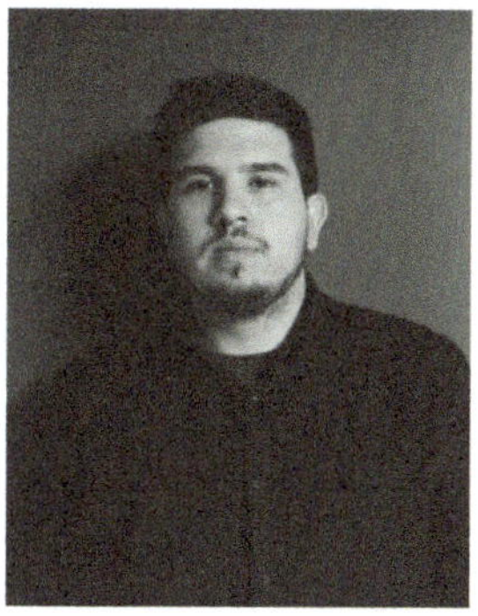

Hugo Varela is a researcher in Architecture at the University of Alcalá, Spain. He is working as Ph.D. Student in the research group of Professor Ph.D. Gonzalo Barluenga. The main research lines of the group are about rheology and fresh state of advanced construction materials cement-based, working on technologies such as self-compacting concrete and 3D Printing. He has been contributing in two national researches and one regional project of the group, publishing two journal papers (Q1), four book chapters and one patent. He has attended several international conferences with three oral communications associated and has carried out two international short stays granted by the University of Alcalá, under the supervision of Prof. Ph.D. A. Perrot (France, 2021) and Prof. Dr. M. Sonebi (UK, 2022).

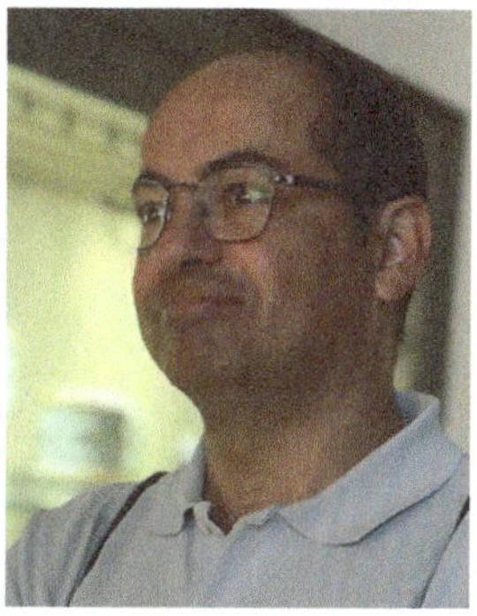

Gonzalo Barluenga holds a Ph.D. in Architecture and is Full Professor of Construction and Building Materials at the School of Architecture of the University of Alcalá (Madrid, Spain). His lines of research involve Sustainability and Resilience of Innovative Building Materials and Architectural Technologies. He has led multidisciplinary research projects on innovative, fire resistance, sustainable, energy efficient and resilient cement-based materials and construction technologies. He leads a research team that collaborates with several national and international groups on bio-based and bio-inspired smart building materials and digital fabrication for architecture. He has published more than 100 articles and book chapters and he holds several patents.

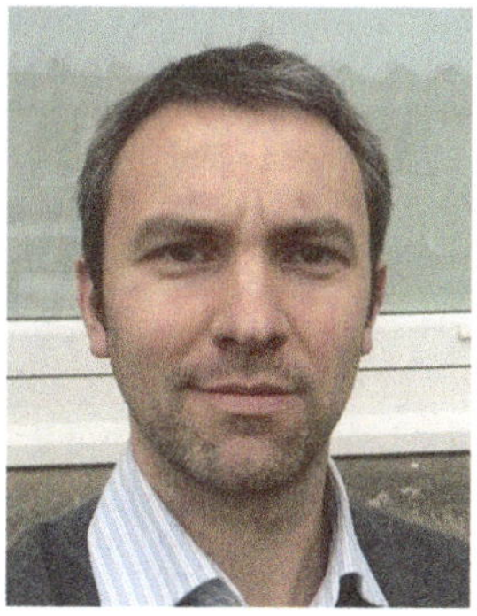

Arnaud Perrot holds a Ph.D. in Civil Engineering and is Full Professor at the University of Southern Brittany (Lorient, France). His research lines lead with rheology of cement-based and raw earth material on digital fabrication and underwater 3D printing. He is leading a research team that collaborates with many researchers and institutions worldwide involved in digital fabrication technologies with concrete. He has published numerous highly cited contributions (beyond 500 citations per year in 2021 and 2022) and has more than 130 co-authors in his background. In addition, he is a member of the editorial board of Materials and Structures journal (Springer) and has been in charge of a special issue about digital fabrication on the journal Materials (MDPI).

Bacteria Encapsulation Method for Achieving Post-Fire Self-Healing in Concrete

Ajitanshu Vedrtnam and Gonzalo Barluenga

Abstract

Fire is a noteworthy threat to human life and civil infrastructures. Civil engineers are craving for safer sustainable replacements of concrete that fulfill the fire safety requirements without compromising their intended design. Firstly, the article reports development of cement-based composites consisting on processed plastic (PET) bottle residuals with improved residual compressive performance. The PET bottles are processed using novel constraint oxygen heating and residual condensation process. The Scanning Electron Microscopy (SEM) micrographs were used to establish structure-property relations, and the reasons of improved residual compressive performance were recognized.

It is well known that the fire exposure deteriorates and selected bacteria can self-heal concrete. The concrete rehydration allows pore structures recovery to an extent but with minor improvement in the mechanical performance. The present work also reports the post-fire self-healing of concrete for the first time in literature. Novel encapsulation strategy is suggested for protecting bacteria incorporated into concrete samples during fire and later facilitating bacteria activation for self-healing. The pre- and post-fire residual compressive strength of samples was described. Concrete deterioration with fire duration and intensity along with recovery in mechanical performance due to self-healing for 1 month was assessed. Finally, a new research direction to the professionals working for fire protection of concrete structures was proposed.

Keywords

Bacteria · Concrete · Fire · PET · Self-healing · Waste

Introduction

The requirement for fire safe structural materials is evident due to consistent unfortunate fire disasters (www.nfpa.org). Although concrete structures are safer than other construction materials, fire exposure deteriorates concrete properties (Shah & Sharma, 2017; Haddad et al., 2008). Concrete performance in fire is widely studied as occupants and structural safety depend on concrete behavior in fire (Abid et al., 2017; Anand & Godwin, 2016; Li et al., 2004; Ma et al., 2015). Concrete deteriorates due to cement paste interface and aggregate debonding, aggregate deformation, thermal incompatibility of constituents, cement paste chemical transformation, CSH gel disruption, and internal pressure due to entrapped steam (Chu et al., 2016). The concrete performance deterioration depends also on concrete grade. The lower concrete grade deteriorates lesser in comparison to higher concrete grades (Shah & Sharma, 2017). The heating rate and duration also determine concrete performance (Thanaraj et al., 2019). Fire-resistant concrete is commercially available these days. Post-fire curing is successful to some extent. Besides, active fire control methods such as sprinklers also combat fire in buildings.

The plastic production is peaking all the time and solid plastic waste, posing a challenge for research community. For achieving circular economy and sustainability, it is essential to develop effective methods for converting waste plastics into useful products as presently more than 90% of solid plastic waste is either landfilled or incinerated. The efficient utilization of waste plastic in construction products is rewarding but challenging due to compatibility issues. In the present work, constrained oxygen heating and residual con-

A. Vedrtnam (✉)
Department of Mechanical Engineering, Invertis University, Bareilly, UP, India

Department of Architecture, University of Alcala, Alcalá de Henares, Spain
e-mail: ajitanshu.m@invertis.org; ajitanshu.vedrtnam@uah.es

G. Barluenga
Department of Architecture, University of Alcala, Alcalá de Henares, Spain
e-mail: gonzalo.barluenga@uah.es

M. R. Cervera Sarda et al. (eds.), *Architecture Inspired by Nature*, https://doi.org/10.1007/978-3-031-33144-2_19

densation (COH-RC) method (www.nationalgeographic.com) is utilized for producing the waste polyethylene terephthalate (PET) residual. The PET residual was utilized as reinforcement to produce cement-based composites (CBC) with an aim to achieve fire retardancy.

The application of self-healing concrete is growing commercially. Healing agents that consist of dormant bacterial spores are commercially available. The water penetrates in cracks of concrete structure and the dormant bacteria are activated. The bacterial activity results in calcium carbonate formation that self-heals the concrete up to specific dimensions (Vedrtnam et al., 2020). Since post-fire self-healing are yet not tried in cited literature, the use of bacteria for post-fire healing is considered as the objective of the present work. This is to be noted that the protection of bacteria during the fire and further controlled activation of bacteria for self-healing is challenging. Novel encapsulation method is attempted in the present work to achieve this in CBC with bacteria. The present work reports fire-resistant construction products with post-fire self-healing features. These construction products will be a promising alternative with superior safety features to the existing concrete.

Material and Methods

Cubical samples of reference concrete, CBC-PET (10% PET by volume) and CBC-PET samples having 10% (by volume) proportions of bacteria + calcium lactate + encapsulation were prepared for fire and compression testing, by casting them in wooden molds (150 mm^3) with the assistance of a vibrator. The samples were removed from the wooden molds after 24 hours and cured in water up to 28 days at 26–32 °C (ambient conditions). The standard ASTM test procedures are used for evaluating the physical properties of concrete.

Cement, sand, aggregate (>19 mm), and water were 345 kg/m^3, 881 kg/m^3, 1081 kg/m^3, and 228 kg/m^3, respectively, in the concrete samples. Water to cement ratio was 0.6. Cement, natural sand, and aggregate conformed to ASTM C150 and ASTM C33-16, respectively. The specific gravity and loose density (kg/m^3) of natural sand and aggregate were 2.39 and 2.61, 2002.4, and 1818.1, respectively. CBC samples included PET plastic bottle residual as a replacement for natural sand in 10% by volume. The particle size of PET residuals was less than 1.18 mm, as measured by the sieve analysis.

The PET residual was produced using constraint oxygen heating and residual condensation (COH-RC) method. Figure 1 shows a schematic diagram and products of plastic recycling used in COH-RC method. COH-RC method includes heating plastics at different temperatures into a closed container with air but no air supply during the process and directing the exhaust gases into another chamber filled with water that allows the condensing exhaust gases. The temperature range was 500 ° C to 600 ° C for 30 minutes. The porous plastic residual includes carbon-rich content that is not converted into exhaust gases during constraint oxygen heating. The porous plastic residual was left in the container where plastic was heated. The gases are condensed into another closed container filled with water. This condensation process results into three separate layers inside the container of grease, oil, and water. The porous plastic residual from PET is shown in Fig. 2. The residual oil and grease are used for biodiesel/foam production and as lubricant, respectively. However, they are yet not characterized fully to identify their further applications. The residual oil was having flammability and may have many other possible applications. An intensive experimentation is required to achieve the stated objectives in the next section.

The PET residual was converted into a powder before being used as reinforcement in CBCs.

All the samples were heated for 1 hour in fire (Fig. 3a–c) and also in furnace following ISO 834. Figure 3d shows the thermal loading scheme. Fig. 4 shows comparison of time-temperature variation in furnace and recommended by ISO834.

The surface temperature of the samples was measured using an infrared and K-type thermocouples. The nonuniform surface temperature was post-processed to determine average temperature at the mid of each sample face. In Fig. 3d:

- T1, average temperature at bottom (exposed) face
- T2, T3, T4, and T5, four-side cube face temperatures
- T6, denoting the top face temperature

A standard compression testing machine (loading rate = 3 KN/min) was used for determining the residual compressive strength of samples. SEM micrographs of the fractured samples were used to establish structure-property relationship.

The CBC-PET samples with gelatin capsules for encapsulating the bacteria as a replacement for the natural sand and coarse aggregates based on filler sizes were casted, the 400 mg, "#1" size of gelatin capsules immobilized bacteria. The gelatin capsules were filled with 70% bacteria and 30% calcium lactate mixture in the first encapsulation method. These capsules were coated with cement paste as shown in Fig. 5 before their incorporation into the concrete samples.

Results and Discussion

The average natural, submerged, and saturated weights of concrete, CBC-PET, and CBC-PET with bacteria encapsulation were 2230, 1278, 2276 g; 2180, 1270, 2210 g; and 2076, 1242, 2164 g. Figure 6 shows the average time-temperature

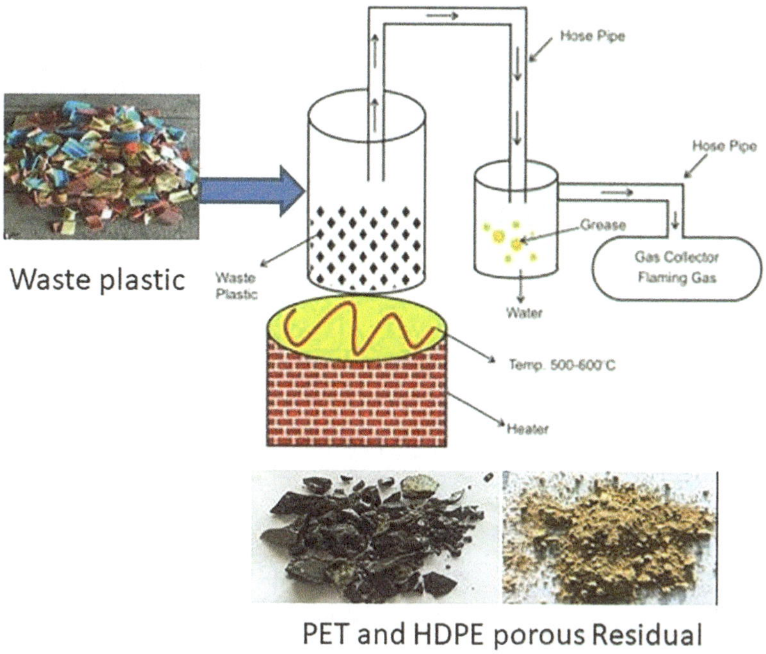

Fig. 1 Schematic diagram of COH-RC method and photograph of residual products

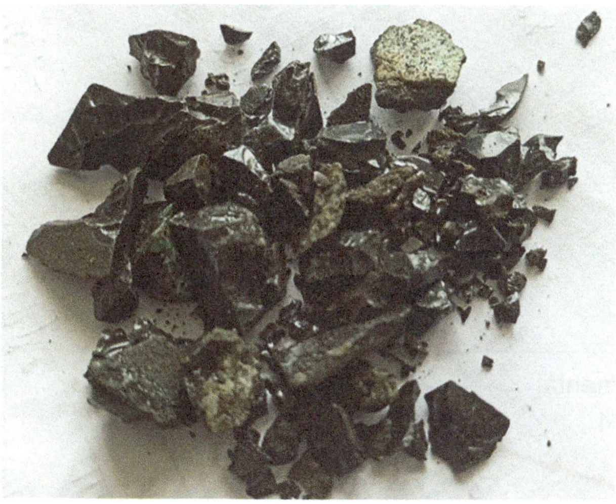

Fig. 2 PET bottle residual

variation of all faces of the samples when subjected to fire exposure for an hour.

Figure 7 shows two infrared thermometer images of the samples during fire exposure. Along with K-type thermocouple infrared, thermocouple images were also used for observing the temperature across the dimensions of the samples.

Figure 8 shows the images of sample after fire exposure. It is evident from the images that the bottom face of the samples was severely deteriorated due to fire exposure.

Table 1 shows the average compressive strength of the samples (minimum three of each type) pre- and post-fire exposure and also after keeping sample for post-fire self-healing for 28 days in ambient conditions. Concrete samples were having highest compressive strength followed by CBC-PET and CBC-PET with bacteria-encapsulated samples. The post-fire average residual compressive strength was the maximum in CBC-PET samples followed by CBC-PET samples with encapsulated bacteria. For both furnace heating and fire exposure, the concrete samples were degraded most.

The possible reason for better post-fire performance of CBC samples with PET residual could be the limited deterioration due to presence of carbon-rich burnt porous residual that allows passage of entrapped steam and observed limited deterioration during fire. The samples were kept in ambient condition for 28 days post-fire exposure. The most significant recovery is observed in the samples with bacteria.

The post-fire recovery in CC-PET bacteria samples was greater in fire-exposed samples than in furnace-heated samples. The possible reason of survival of encapsulated bacteria could be the protection by the cement coating on gelatin capsules. During fire, the water in cement paste might have lost that causes cracks in the coating. The gelatin might have melted during fire. Later, the cracks in cement coating might have allowed penetration of moisture and activation of bacteria for self-healing. However, additional investigation is required to confirm the same.

Further, the weight of samples post-fire and post-fire recovery were measured. Up to 2% weight gain is observed in CBC-PET with bacteria self-healed samples in comparison to weights just after fire exposure. Insignificant change in weights post fire and after keeping the samples 28 days for recovery was observed in concrete and CBC-PET sample.

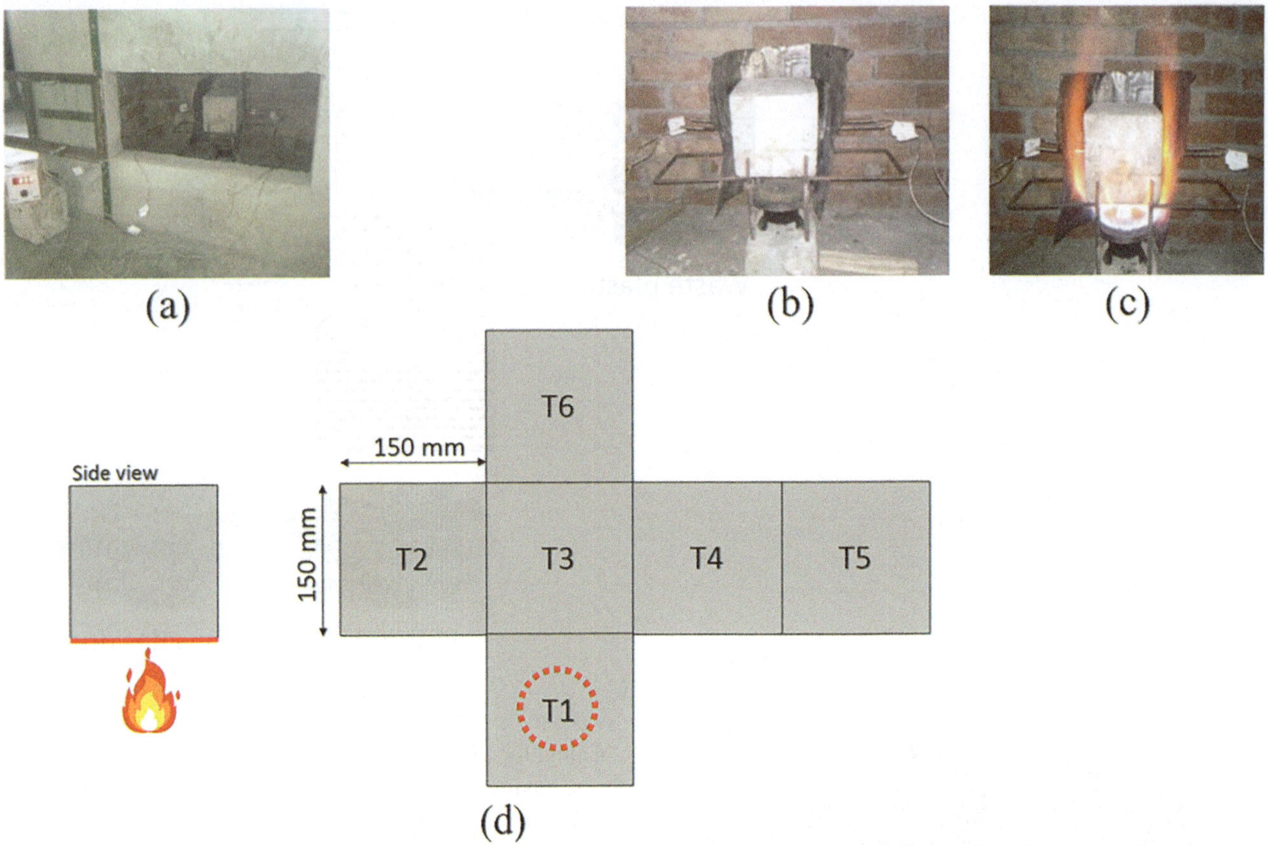

Fig. 3 Selected photographs of fire testing for the CBC specimens: (**a**) furnace view, (**b**) setup, and (**c**) heating stage, with (**d**) thermal loading scheme (Vedrtnam et al., 2020)

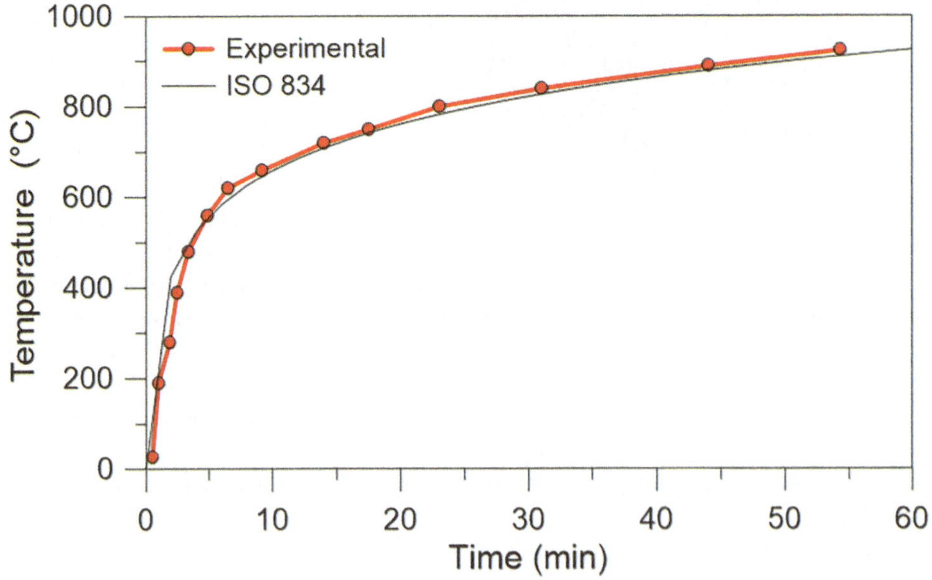

Fig. 4 Time-temperature in experiment vs. standard ISO 834 fire curve (Vedrtnam et al., 2020)

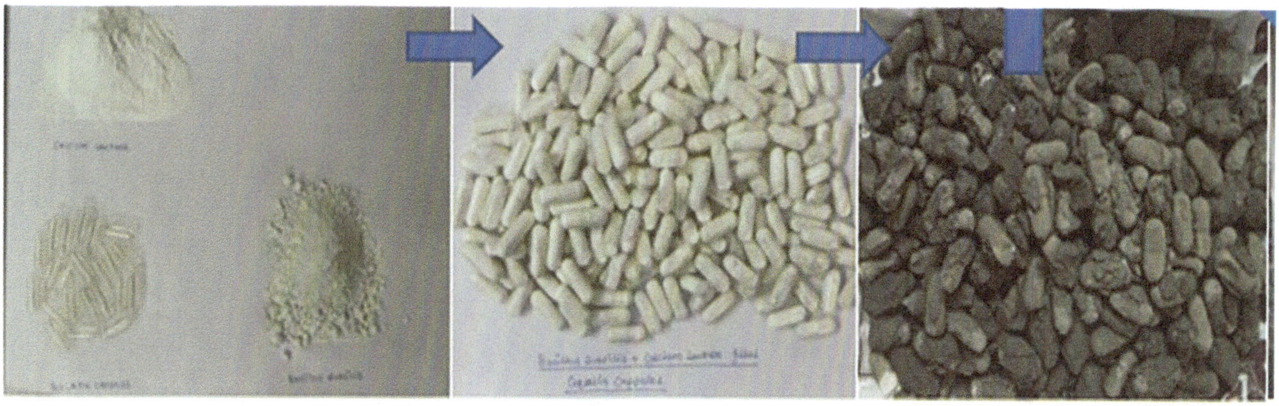

Fig. 5 Cement paste-coated gelatin capsules encapsulating bacteria

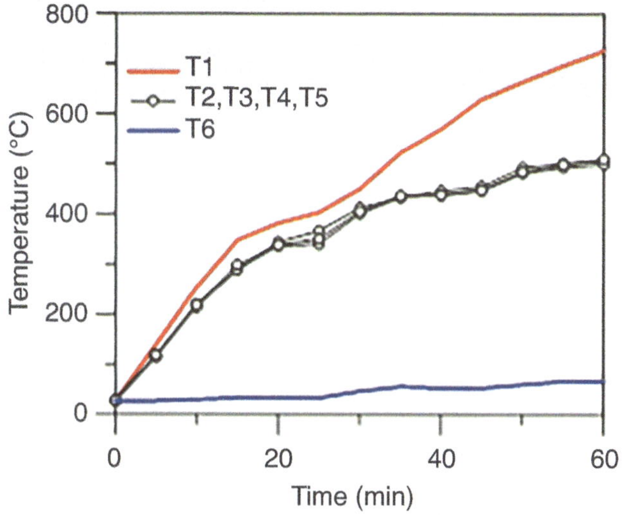

Fig. 6 Average time-temperature variation at all edges of samples subjected to fire

The increase in weight of self-healed CBC-PET with bacteria samples gives an indication of bacteria activation and formation of calcium carbonate in these samples that accounts for better residual compressive strength.

Figure 9 shows the SEM images of concrete, CBC-PET, and CBC-PET with encapsulated bacteria samples. The SEM image of CBC-PET sample shows porous phases in comparison to concrete sample SEM image. This support the logic that entrapped steam during fire might be escaped from the CBC-PET samples that resulted in lesser undue internal stresses and lesser degradation of these samples. Figure 9c shows the presence of coated gelatin capsules. The irregular surface texture is further exaggerated due to the cement paste coating on the gelatin capsule in CBC-PET bacteria encapsulated samples. The post-fire recovery achieved in current investigation encourages for a detailed investigation. A further detailed study is needed to establish mechanism of post-fire self-healing of encapsulated CBC-PET samples.

Conclusions

The present work reports utilizing PET residual produced by COH-RC method for producing fire-retardant cement-based composites and possibility of post-fire self-healing by innovative encapsulation strategies. The performance of concrete, CBC-PET, and CBC-PET with encapsulated bacteria was reported during ISO 834 furnace heating and nonuniform direct fire exposure for 1 h. The following are the main conclusions:

- Concrete samples have the highest compressive strength followed by CBC-PET samples and CBC-PET with bacteria-encapsulated samples.
- CBC-PET bottle residual have highest residual compressive strength followed by CBC-PET with encapsulated bacteria and concrete samples.
- The concrete and CBC-PET samples have shown more surface cracks in nonuniform fire exposure compared to furnace heating for same fire duration.
- CBC-PET samples with encapsulated bacteria gained up to 2% of weight 28 days after curing in ambient conditions from post-fire weight.
- The CBC-PET samples with encapsulatedbacteria regained 7.76% compressive strength in furnace-heated samples and 7.96% in fire-exposed samples.
- The SEM micrographs have shown pores in the microstructure of CBC-PET sample that has resulted in lesser compressive strength but superior residual compressive strength in CBC-PET samples compared to concrete samples.

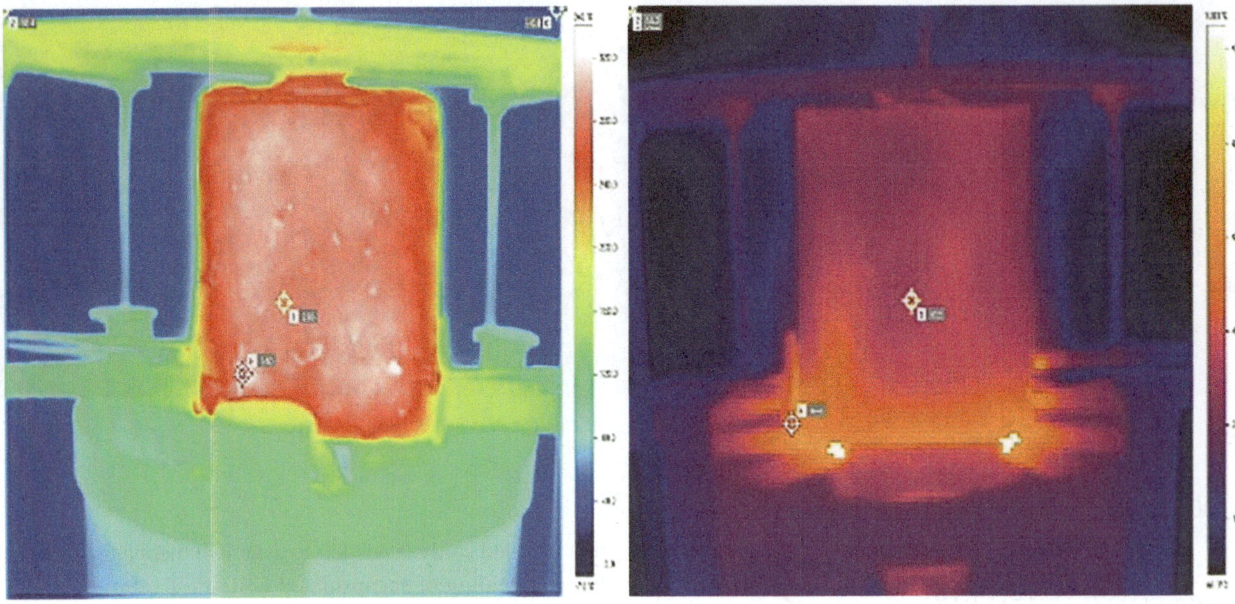

Fig. 7 Infrared thermometer images during fire exposure

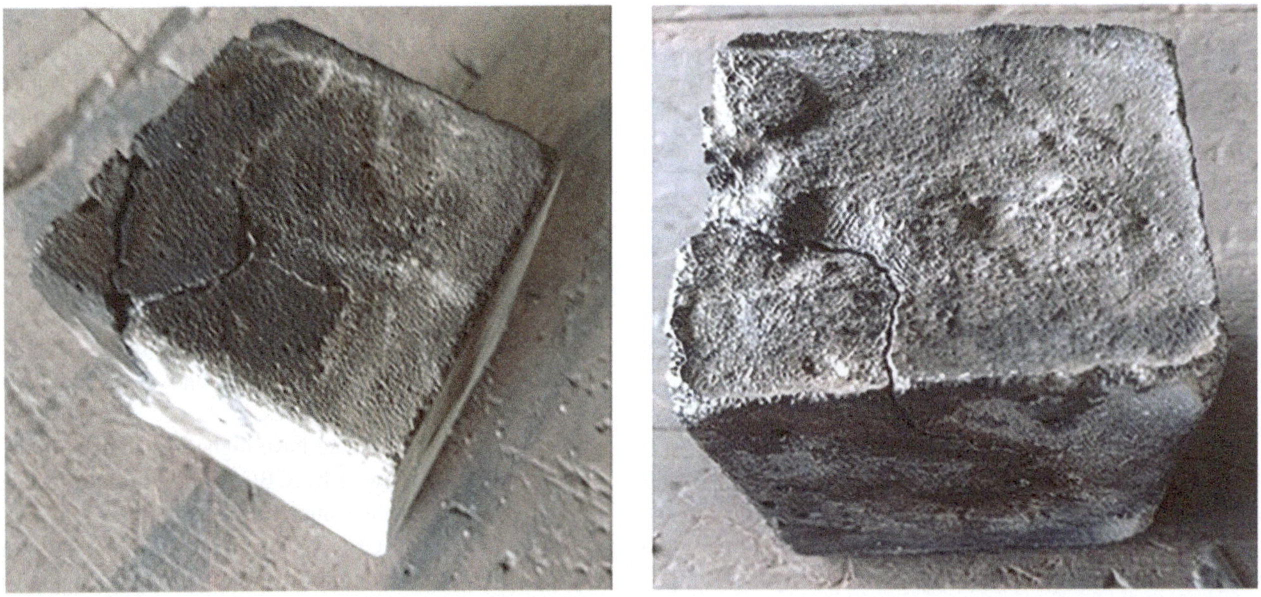

Fig. 8 Sample images post-fire exposure

Table 1 Compressive strength of samples pre- and post-fire exposure and post self-healing

	Average compressive strength (MPa)		
	Concrete	CBC	
		PET	PET-bacteria
Temperature (°C)/proportion (%)	–	10	10
27 (ambient)	19.21	18.6	18.4
1 h heating in furnace	6.11	7.9	7.4
1 h heating in fire	5.77	7.1	6.9
28 days post fire (furnace samples)	6.12	7.96	8.78
28 days post fire (fire exposed samples)	5.81	7.17	8.96

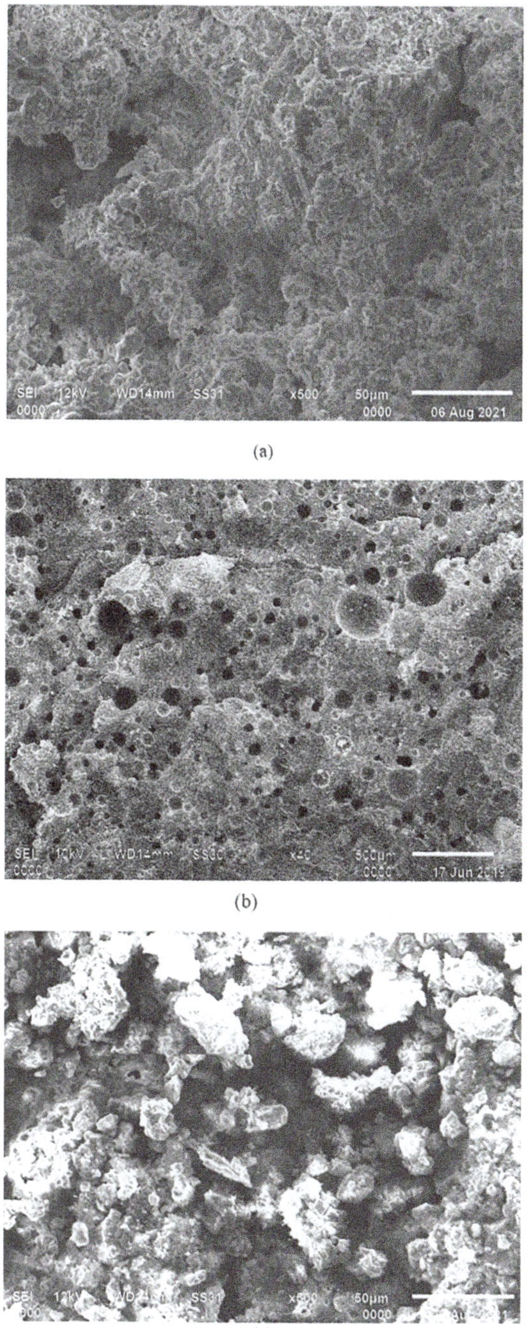

(a)

(b)

(c)

Fig. 9 SEM images of (**a**) concrete, (**b**) CBC-PET, and (**c**) CBC-PET with encapsulated bacteria

Acknowledgments This project has received funding from the European Union's Horizon 2020 research and innovation program under the Marie Skłodowska-Curie Grant Agreement No. 754382, GOT ENERGY TALENT. The content of this article does not reflect the official opinion of the European Union. Responsibility for the information and views expressed herein lies entirely with the author(s). The waste PET recycling section of the study is supported by the Department of Science and Technology, India.

References

Abid, M., Hou, X., Zheng, W., et al. (2017). High temperature and residual properties of reactive powder concrete – A review. *Construction and Building Materials, 147*, 339–351.

Anand, N., & Godwin, A. (2016). Influence of mineral admixtures on mechanical properties of self-compacting concrete under elevated temperature. *Fire and Materials, 7*(40), 940–958.

Chu, H. Y., Jiang, J. Y., Sun, W., et al. (2016). Mechanical and physicochemical properties of ferro-siliceous concrete subjected to elevated temperatures. *Construction and Building Materials, 122*, 743–752.

Haddad, R. H., Al-Saleh, R. J., & Al-Akhras, N. M. (2008). Effect of elevated temperature on bond between steel reinforcement and fiber reinforced concrete. *Fire Safety Journal, 43*, 334–343.

Li, M., Qian, C., & Sun, W. (2004). Mechanical properties of high-strength concrete after fire. *Cement and Concrete Research, 34*, 1001–1005.

Ma, Q., Guo, R., Zhao, Z., et al. (2015). Mechanical properties of concrete at high temperature—A review. *Construction and Building Materials, 93*, 371–383.

Shah, A. H., & Sharma, U. K. (2017). Fire resistance and spalling performance of confined concrete columns. *Construction and Building Materials, 156*, 161–174.

Thanaraj, D. P., Anand, N., Prince Arulraj, G., & Zalok, E. (2019). Post-fire damage assessment and capacity based modeling of concrete exposed to elevated temperature. *International Journal of Damage Mechanics*, 105678951988148. https://doi.org/10.1177/1056789519881484

Vedrtnam, A., Bedon, C., & Barluenga, G. (2020). Study on the compressive behaviour of sustainable cement-based composites under one-hour of direct flame exposure. *Sustainability, 12*(24), 10548.

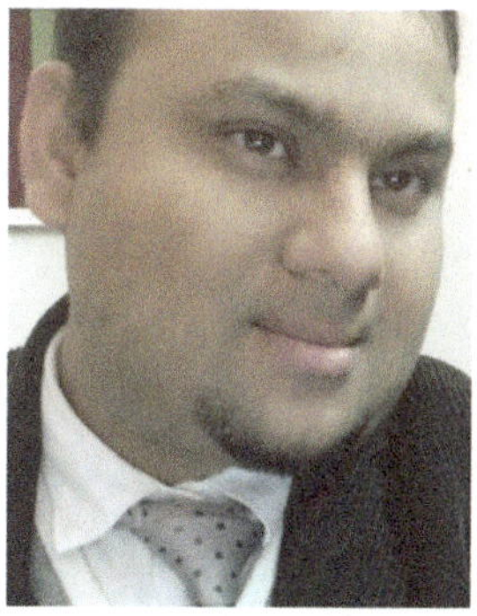

Ajitanshu Vedrtnam has more than 18 years of research and teaching experience. He has completed B.E. (Mechanical Engineering), MBA (Marketing), M.E. (Manufacturing)-Gold Medalist, GATE (M.E) and Ph.D. (Applied Mechanics). Vedrtnam has published more than 100 articles including 50 articles in high impact journals, holds 2 patents and authored 2 books. Vedrtnam has been actively involved in research for building state-of-the-art technological systems to handle key challenges related to structural materials, energy supply and environmental issues. Vedrtnam is the recipient of prestigious fellowships including Marie-Curie grant, post-doctoral fellowships by University of California, USA, Technion Israel institute of Technology, Israel, Institute of advanced Materials, Sweden, Ministry of HRD, Govt. of India, etc. He serves as a scientific evaluator, member of experts in several panels and committees to various entities; member of Ph.D. committees, and has been invited for 100+ plenary, keynote and invited lectures in international conferences. He has guided a good number of students on Cement-Based Composites, Energy storage materials, Bio-fuels, Structural glass, Additive Manufacturing, Composite Materials, and Aerodynamics. He is serving as editor-in-chief, editor, editorial board member and reviewer for many peer-reviewed international journals. Vedrtnam was PI/Co-PI of multiple research projects sponsored by the Department of Science and Technology, Govt. of India, IC IMPACT, Canada, Ministry of HRD, India, NAVA, Poland, etc.

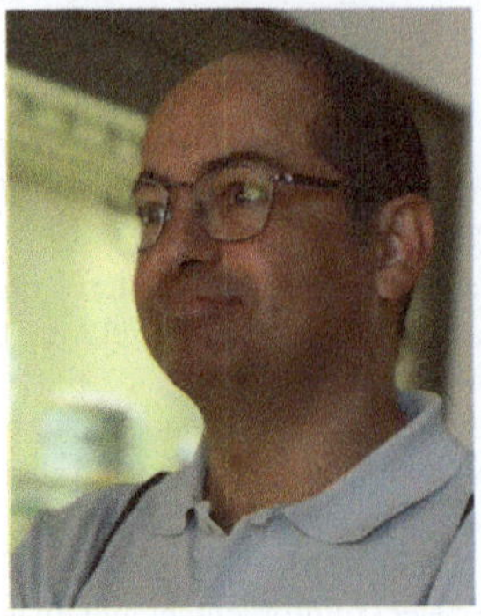

Gonzalo Barluenga holds a Ph.D. in Architecture and is Full Professor of Construction and Building Materials at the School of Architecture of the University of Alcalá (Madrid, Spain). His lines of research involve Sustainability and Resilience of Innovative Building Materials and Architectural Technologies. He has led multidisciplinary research projects on innovative, fire resistance, sustainable, energy efficient and resilient cement-based materials and construction technologies. He leads a research team that collaborates with several national and international groups on bio-based and bio-inspired smart building materials and digital fabrication for architecture. He has published more than 100 articles and book chapters and he holds several patents.

Adaptative and Bioinspired Materials: Cement-Based Materials with Self-Modulating and Self-Sensing Properties

Javier Puentes, Irene Palomar, Gonzalo Barluenga, and Cynthia Guardia Martin

Abstract

The approach to sustainability in architecture can be considered as the reduction of the use of resources during the whole life of buildings. For cement-based materials, the strategy is focused on material optimization related to raw materials extraction, manufacturing process, life span, and finally the possibility of reuse or recycling. This chapter shows the possibility of achieving the objective through novel and alternative ways where the environment-material interaction is the basis of this development for new cement-based materials. This adaptability results in materials with self-responsive properties: self-sensitive, self-healing, self-modulating, and self-cleaning. In the present paper, two self-responsive materials are shown. In the first case, material design is based on reducing energy consumption through energy efficiency, by accumulating heat in a dynamic environment. The second material investigates the self-sensitivity capacity that can diagnose and identify changes of mechanical properties that could compromise durability, saving maintenance resources.

In both studies, the search for new properties is achieved using different additions that incorporate new capabilities to the cement-based materials. Phase change materials (PCM) were used as self-modulating materials in cement-lime mortar for thermal regulation, while micro and nano carbon-based fibers (CNF) were used in self-compacting concrete (SCC) to vary the electrical properties, turning the material into a real-time sensor. These nano- and microscale modified materials are presented as real alternatives in the search for dynamic materials applied to sustainable construction.

J. Puentes (✉) · I. Palomar · G. Barluenga · C. G. Martin
Universidad de Alcalá, Alcalá de Henares, Spain
e-mail: javier.puentes@uah.es; irene.palomar@uah.es; gonzalo.barluenga@uah.es; cynthia.guardia@edu.uah.es

Keywords
Cement-based material · Energy efficiency · Phase change material · Self-sensing materials · Carbon-based fibers

Introduction

The development of materials with self-responsive properties is an alternative within the strategy for the reduction of energy consumption in construction materials (Bekzhanova et al., 2021). To reach this objective, the life cycle of the material must be considered, from its design, its placement on site, the extension of its life span and reduction of maintenance, and finally its reuse or recycling, reducing the overall environmental footprint.

Self-responsive materials are useful once the material has been placed in the element/object, collaborating to the reduction of energy consumption.

Accordingly, self-responsive materials are the example of the search for sustainability and durability in construction through these three aspects:

1. Expanded life span
2. Reduction of energy consumption with durable materials and constructions
3. Application throughout its life span of the benefits that are part of a system to reduce the energy footprint without sacrificing the comfort of the user

These initiatives, either directly or indirectly, are the alternatives proposed together with the human actions of the maintenance or use of the architectural element that pursue energy efficiency.

Self-responsive materials can be classified based on their main purpose, mode of action, and form of action. This categorization comprises the different types of materials, as

M. R. Cervera Sarda et al. (eds.), *Architecture Inspired by Nature*, https://doi.org/10.1007/978-3-031-33144-2_20

well as the latest technological developments. The self-responsive materials presented in this article are categorized by the type of response action: self-modulating materials and self-sensing materials.

The paper describes the development of two types of cement-based materials. The first one is a cement-lime mortar for multilayer wall coatings and its thermal performance by taking advantage of the heat storage capacity of the material (Cabeza et al., 2011). The second material is a self-compacting concrete (SSC) with self-sensing capabilities. The material incorporates carbon-based material additions, carbon nanofibers and carbon microfibers, to provide electrical properties and transform SCC into a real-time sensor of the architectural element.

These studies present two research lines for different building material applications seeking energy efficiency through different routes. Nevertheless, they present a single objective: energy efficiency and savings approached from totally different visions in the development of new materials research.

Self-Modulating Cement-Lime Mortar

To achieve self-modulating capacity, the reference mortar incorporates a phase change material (PCM) in the form of microencapsulated paraffin wax. This mortar performs as a self-modulating material that provides energy efficiency by reducing energy consumption, offering thermal comfort. This is the role of self-modulating materials, where the objective is to provide the material with an adaptability that allows responding to temperature and humidity changes without requiring energy consumption. Paraffin wax-based PCM main characteristics are thermal and chemical stability, noncorrosive, recyclable, and low cost in the commercial market. The ease of mixing microencapsulated materials with other components allows the cement-lime mortar for rendering applications to absorb and accumulate energy due to external temperature variation, balancing the heat variations produced by day/night cycles as shown in Fig. 1.

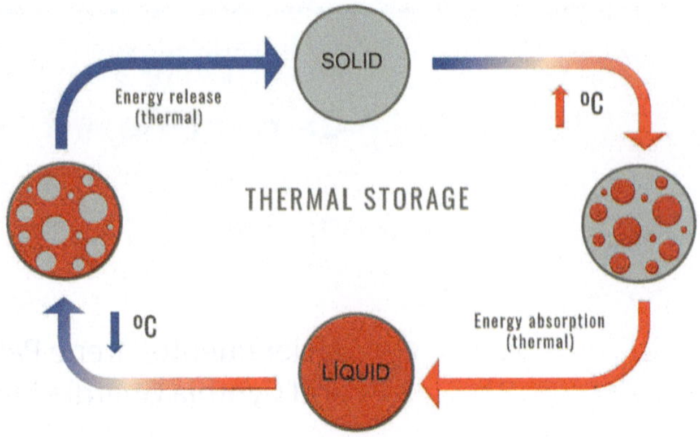

Fig. 1 Phase change process and operation of a PCM (Guardia et al., 2019)

Materials

Micronal DS 5040X microencapsulated paraffin waxes supplied by BASF were used in this study, with microcapsule sizes ranging 50–300 μm. The melting point for the phase change of this commercially available wax is 23 °C, which is near the human comfort temperature (Fig. 2). The main properties of the microcapsules are presented in Fig. 3.

A 10% and 20% PCM was added to a reference cement-lime (type BLII/B-L 32.5 N and CL90-S, respectively) rendering mortar, with a silica sand 0–4. Cellulose fibers (F) and perlite (LWA) were also incorporated to the mortar in different proportions to improve the thermal insulation of the mixture. A total of 12 mortars were studied.

Methodology and Testing

The testing techniques used in this study focused on the evaluation of the energy storage capacity developed by the PCM mortars when the melting temperature was reached during heating and cooling cycles. The experimental work comprised three phases. The first phase involved the characterization of the physical, mechanical, and thermal properties of the mortars (Palomar et al., 2019; Guardia et al., 2019). In the second phase, an analysis of the thermal behavior of the mortar under different environmental conditions was carried out. Finally, the mortar was evaluated incorporated in a multilayered building enclosure subjected to different environmental conditions (Guardia et al., 2020).

The fresh and hardened properties of the material, physical properties of density, open porosity, capillary absorption, and water vapor permeability were characterized. Also, the mechanical strength in compression, flexural strength, and elastic constants of the material were evaluated. Finally, the thermal properties of conductivity and enthalpy were also assessed (Guardia et al., 2019). Within the series of tests, an evaluation of the heat flow through the multilayer wall, simulating a multilayered enclosure, was carried out using a climate chamber as shown in Fig. 4.

Twelve different mortar mixtures were used in combination with a multilayer enclosure building system composed by a hollow brick masonry layer, PCM mortar, and extruded polystyrene insulation (XPS), where the mortar was the variable of the system in order to compare the performance of the different mortar compositions under different indoor/outdoor environmental conditions as cycles of heating and cooling. Each cycle lasted 1400 min.

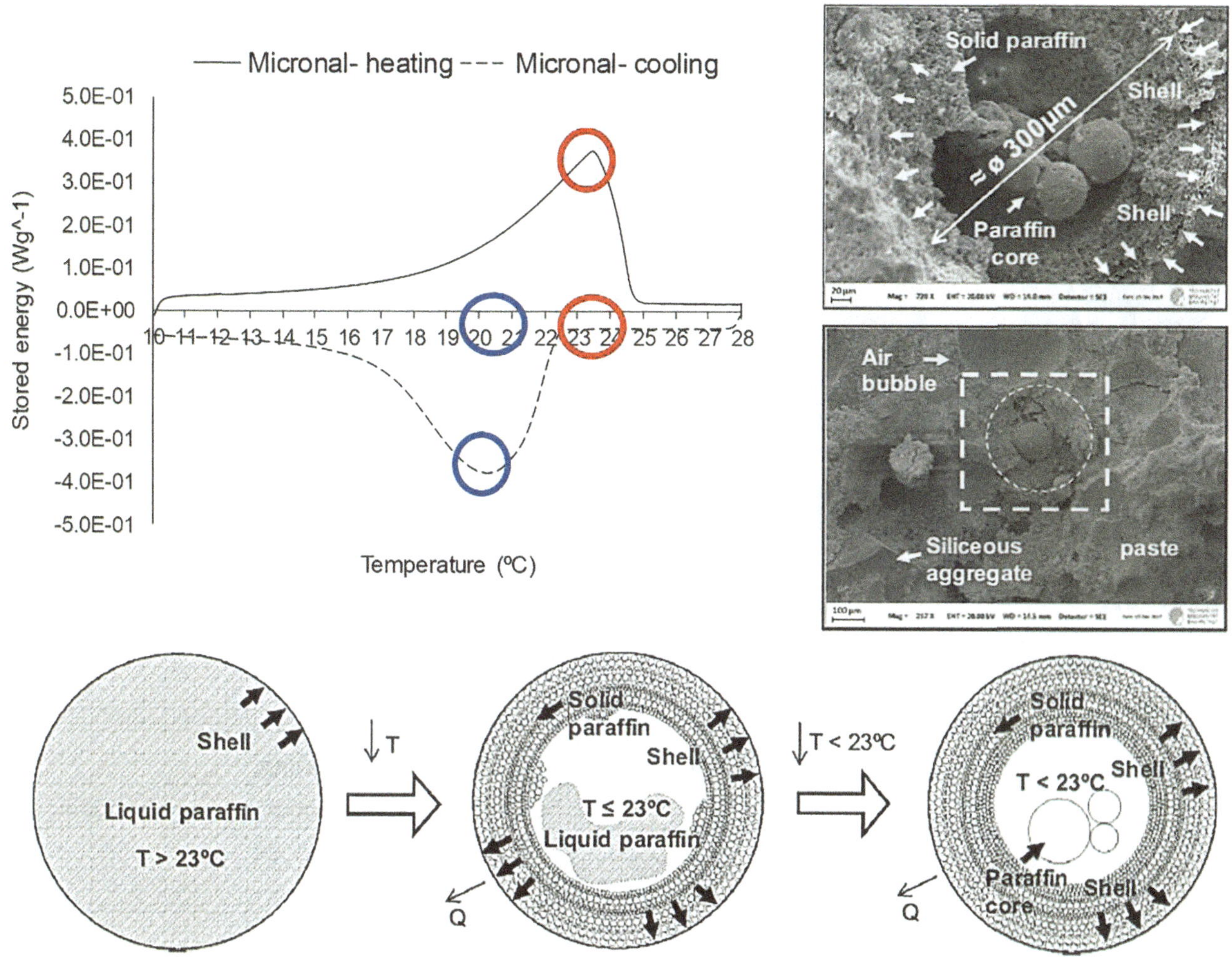

Fig. 2 PCM performance characteristics (Guardia et al., 2019)

Paraffin wax (MICRONAL DS 5040X)	
Appearance	Powder
Bulk density	300-400 kg/m3
Melting point	23ªc
Enthalpy	96 KJ/kg
Heat capacity (10ºC-30ºC)	136KJ/kg
Size	50-300µm

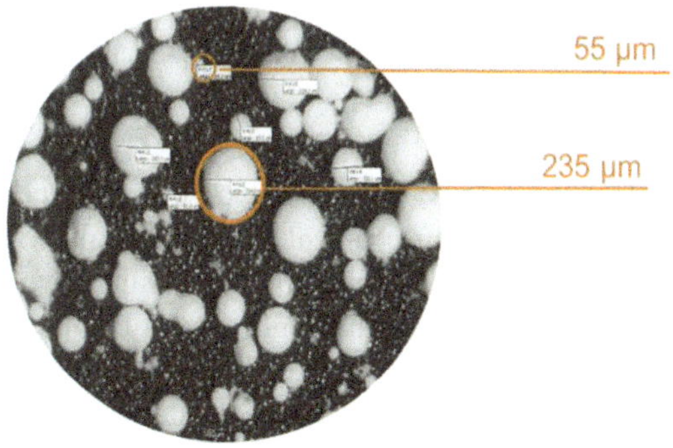

Fig. 3 Basic properties of paraffin wax microcapsules

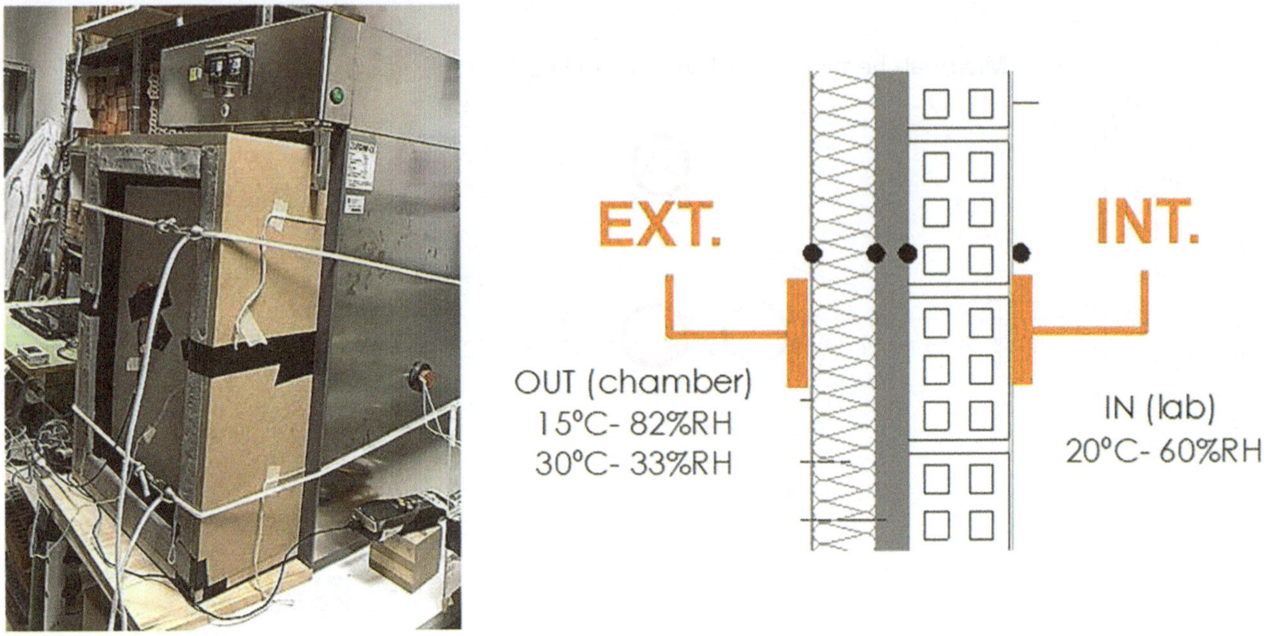

Fig. 4 Experimental setup of a constructive solution at different temperature conditions (Guardia et al., 2020)

Results

Table 1 summarizes the main experimental results for temperature and heat flux (HF) for enclosure specimens with an intermediate mortar layer with 20% of PCM, LWA, and cellulose fibers under heating and cooling cycles, compared to the reference mortar without PCM. The multilayer wall with cement-lime mortar with 20% PCM and LWA (CL20) showed the best thermal performance, storing more heat, taking advantage of the material storage capacity as shown in Table 1. The increase of heat storage and heat wave delay regarding the reference mortar is plotted in Fig. 5.

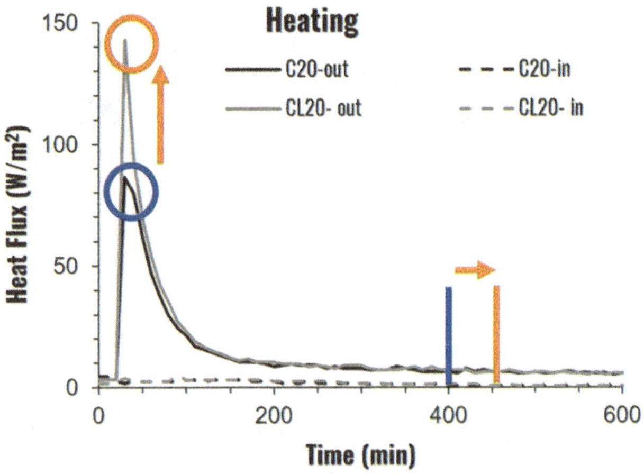

Fig. 5 Heat flux of multilayer cladding with lime mortar C20 and lime mortar with lightweight aggregates CL20. (Guardia et al., 2020)

Conclusions

The addition of PCM to a reference cement-lime mortar modifies physical, mechanical, and thermal properties, reducing density and strength while increasing the heat storage capacity. The use of lightweight aggregates and cellulose fibers improved the PCM mortar properties. Multilayered enclosures can benefit from the thermal behavior of PCM cement-lime mortar, improving inner comfort during heating/cooling cycles without additional energy consumption, increasing building energy efficiency.

Table 1 Evaluation results of heating-cooling processes for multilayer enclosure with different types of PCM mortars in climatic chamber cycles

	C	C20	CF20	CL20	CLF20
Heating (15–30 °C)					
HF max. ext. (W/m^2)	105	90	95	140	110
Inflection point int. (min)	370	400	580	440	540
Cooling (30–15 °C)					
HF max. ext. (W/m^2)	100	90	110	95	105
Inflection point int. (min)	320	450	270	540	270

Self-Compacting Concrete with Self-Sensing Properties

Concrete can develop self-sensitivity through the incorporation of additions that can modify its electrical properties (Han et al., 2015). The objective is to transform a material into a real-time sensor that allows data collection and analysis to detect damages and to locate and quantify them in a predictive and preventive way. This property would allow to predict and to increase the life span of the structure with efficient performance levels. This type of application could determine changes in the material by normal use and accidental changes by external aggressive agents. For this type of material, a number of factors must be taken into account, as (a) parameter to be measured, (b) aggressiveness of the environment, (c) time and sequence of monitoring, and (d) structural typology.

Real-time monitoring is not a new approach of diagnostics (Taheri, 2019). The sensors used in concrete can be embedded or bonded to the surface with the limitations of use they may have (Fig. 6). However, the development of sensors where the material itself is the sensor is the result shown in the present work (Azhari & Banthia, 2012).

This material is known as intrinsic self-sensing concrete (ISSC) and refers to a material that uses its microstructure to self-monitoring without any connection or remote-integrated sensors (Han et al., 2015). Piezoresistivity (PZR) relates the electrical properties of concrete to its mechanical properties, linking the action-reaction effect to a tensile stress.

This property is incorporated in cement-based materials with conductive additions that contribute to the concrete/mortar the piezoelectric capacity (PZE). This incorporation is made to the concrete paste and is limited to the percentage of paste in the total content of the mixture. That is why the

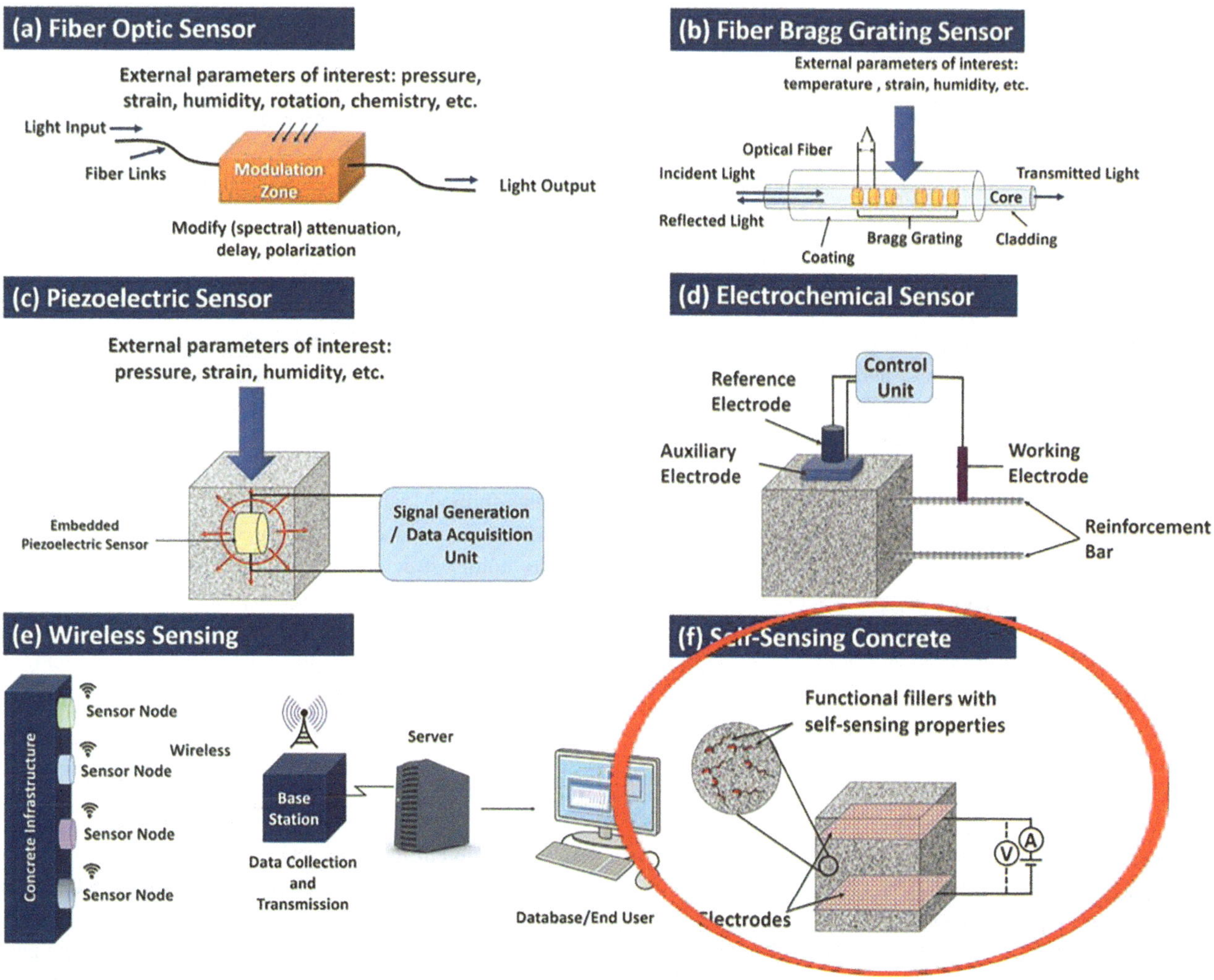

Fig. 6 Different types of control and monitoring systems for structures (Taheri, 2019)

use of SCC, with higher paste content than conventional concrete, allows a greater volume of addition with regard to the other components of the concrete mix. The use of carbon-based nanocomponents improves the electrical conductivity of the material based on its size, improving the percolation path from the distribution in the matrix and electron hopping between the different CBM added (Bekzhanova et al., 2021).

To measure the material electrical and PZE properties, electrical resistance and resistivity (current-voltage) tests were carried out to measure the variation of the direct electrical resistivity and PZR (current-voltage and charge) based on Ohm's law (1):

R concrete (Ω) : R = V/I

$$R\text{concrete}(\Omega.\text{m}) = \text{R}(\text{S}/\text{L}) \quad (1)$$

This parameter is used to define the variation of the electrical properties of the material, which are directly related to the piezoresistivity properties to be achieved by the modified concrete.

Materials

To achieve this objective, several research groups have used carbon-based materials (CBM) as particles or fibers and in nanometric or micrometric sizes. The addition can be a combination of different types, producing a hybrid solution. In addition to the different types of fibers, CBM as particles such as carbon black and graphene oxide are also used (Azhari & Banthia, 2012).

The experimental studies were carried out with carbon nanofibers, carbon nanotubes, and carbon microfibers in sizes of 6 and 12 mm (Table 2). The self-compacting concrete was made with CEM 1 42.5 and complies with the fluidity requirements of an SCC.

Tests were performed on cylindrical specimens of 150 × 300 mm and prismatic specimens of 40 × 40 × 160 mm as show in Fig. 7. Electrical resistance measurements can be performed on both specimen geometries. For prismatic specimens, stainless-steel mesh-embedded terminals are used for the connection terminals as shown in Fig. 9. For cylindrical specimens, wire terminals bonded with conductive paint on the specimen surface were used (Fig. 7, right).

Table 2 Main properties of nano and micro carbon-based fibers (CNF)

Type	CMF	CNF
Length	6/12 mm	30 μm
Diameter	7 μm	20–80 nm
Carbon content	95%	
Tensile strength	4.0 GPa	
Resistivity (Ω.m)	1.5×10^{-4}	1×10^{-3}
Density (g/cm^3)	1.80	1.97

Methodology and Testing

The self-sensing test was performed following the direct method test, Standard UNE 83988-1, used as shown in Fig. 8, in order to determine the variation of the electrical resistivity according to the addition.

The test can quantify the variation of the electrical resistivity and the improvement versus the amount of CBM in percentage with regard to cement weight. The amount of addition ranged from 0% to 3%, limited by the workability and the compromise of the consistency and the self-compactibility of the sample (Fig. 8).

Prismatic compression specimens with loading and unloading cycles were used for the PZR test. The loading speed was 0.1 kN/s and the loading interval for each cycle varied between 2 kN and 10 kN. Current (2 V) was supplied between the outer terminals as shown in Fig. 9 (left). The variation of resistance and resistivity was measured during the loading and unloading cycles according to the previously published equations (Alonso & Puentes, 2020).

Results

The results presented in Fig. 8 (right) show the reduction in resistivity with the combined use of nanofibers and microfibers in the added proportions of up to 3%. A trend line is observed where the resistivity is reduced indicating an improvement in the electrical conductivity of the material in around an order of magnitude.

It was observed that the use of microfibers is limited to percentages around 2%. For higher percentages, the self-compactibility of the mixture was compromised.

Regarding PZR, a pattern of variation of the resistivity of the material as a function of the applied load can be observed in Fig. 10. A correlation between the electrical conductivity and the applied load can be observed. The microstructural modification produced by the force produced the alignment of the atomic structure, allowing an easier transmission of electrons as an effect of the PZR property. Figure 10 shows the relationship between the load-deformation-resistivity variables and their evident proportionality with respect to the magnitude of the applied load.

Conclusions

An improvement in the electrical conductivity properties of concrete has been obtained with the addition of CBM. Accordingly, it was possible to set the property of PZR in a self-compacting concrete by introducing the property of self-sensitivity. However, the incorporation of larger

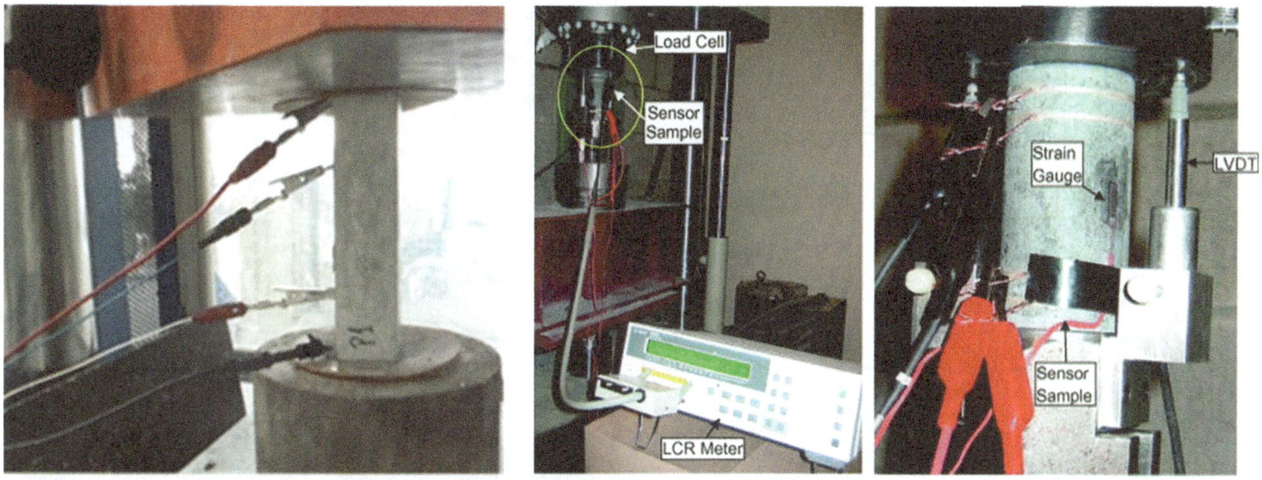

Fig. 7 Left: specimen of 40 × 40 × 160 mm (Alonso et al., 2021). Right: cylindrical specimen of 150 × 300 mm (Azhari & Banthia, 2012)

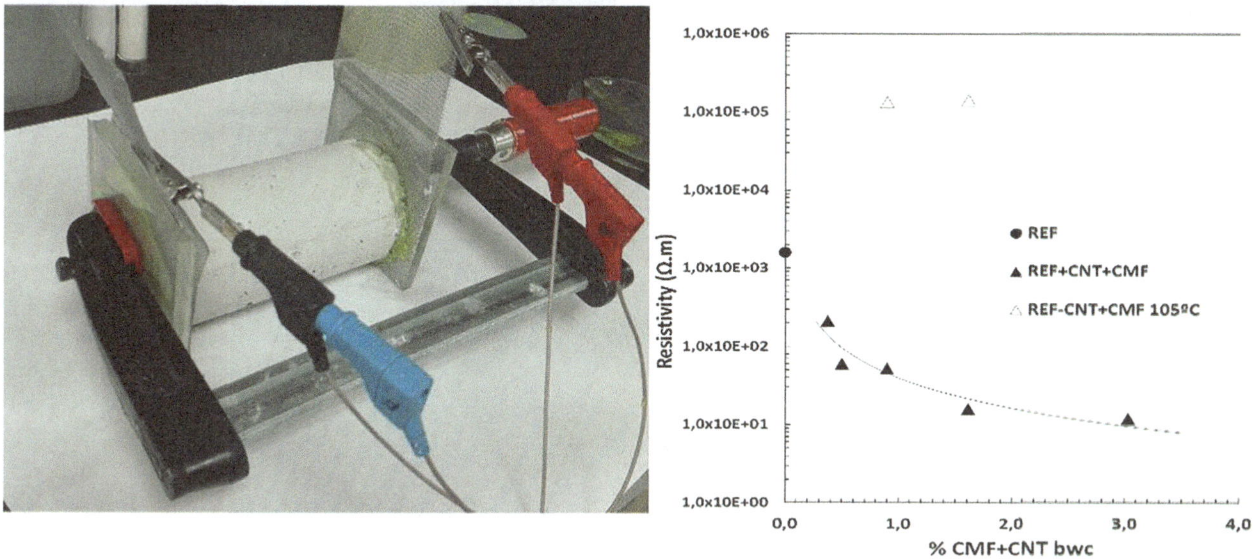

Fig. 8 Electrical resistance measurement, direct methods (Alonso & Puentes, 2020)

amounts of carbon-based components could compromise the concrete self-compactibility.

Final Conclusions

Advances in cement-based materials show the ability to approach the objectives of the search for adaptive materials as an alternative to modern needs. These materials are the example of an answer to the search for energy efficiency and the saving of material resources from two totally different points of view.

Phase change materials (PCM) were used as self-modulating materials in cement-lime mortar for thermal regulation, seeking energy efficiency to reach a state of equilibrium in energy costs, through the reduction of consumption. A promising initial result have been obtained with these types of mortar and PCM additions.

Self-sensitive materials include a large number of technological possibilities, using the material as a sensor. They make it possible to extend the service life of materials by introducing efficient maintenancesystems for infrastructures, rationalizing resources. The objective is to improve the performance of materials for a more eco-sustainable future. The future of materials is focused on improving their performance from the ability to interact with the environment and respond to external inputs.

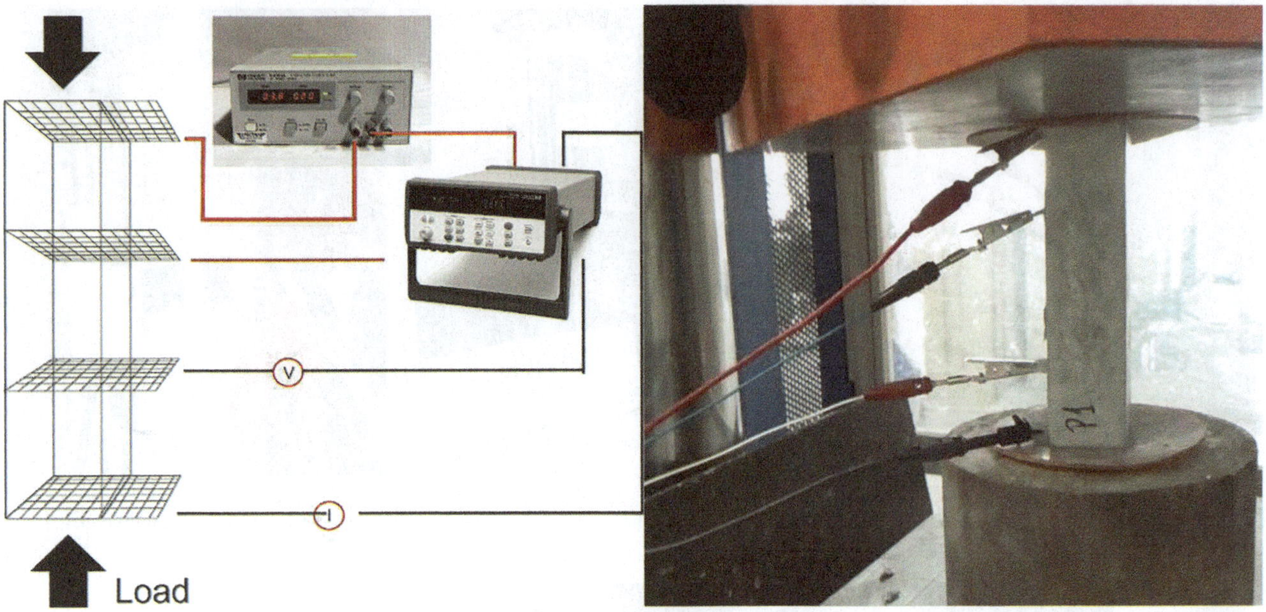

Fig. 9 PZR test setup measurement of variation of resistance under load stress (Alonso & Puentes, 2020)

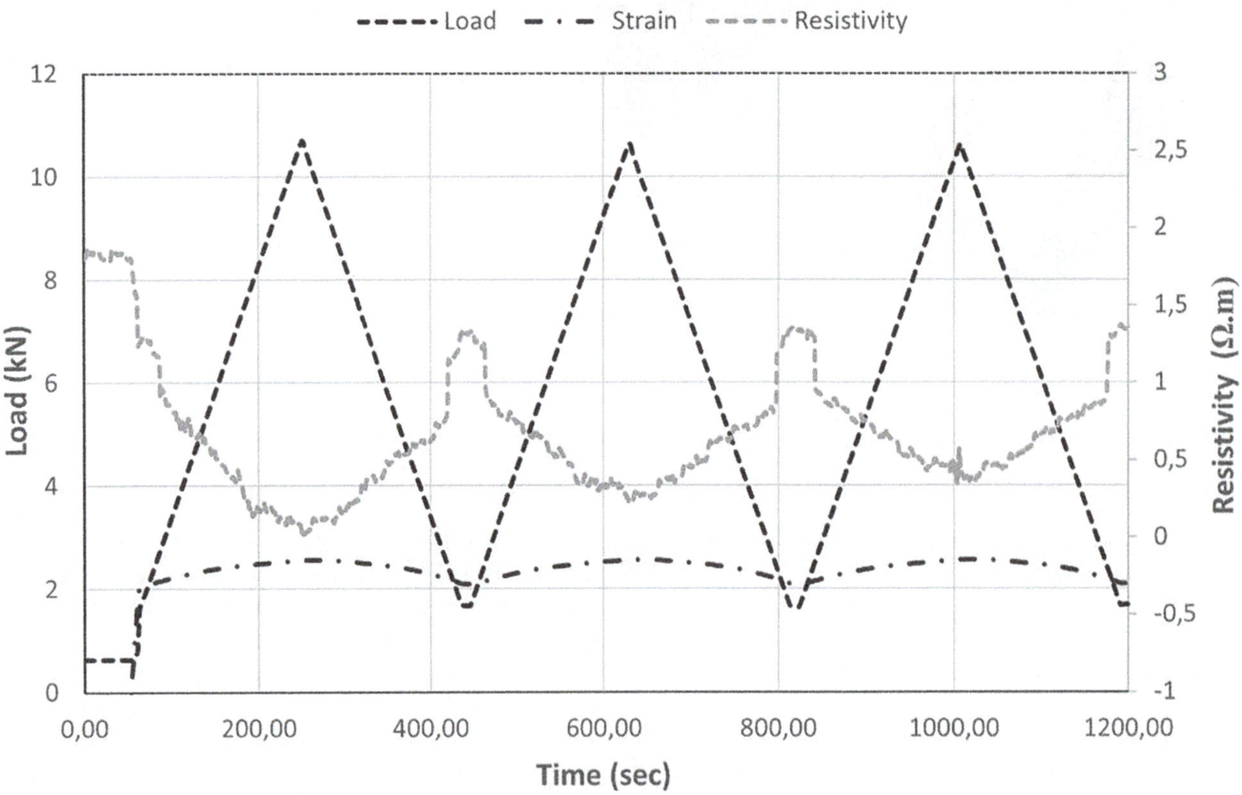

Fig. 10 Compressive load cycles and electric conductivity proportional variations

Acknowledgments The authors acknowledge the funds provided by the Research Program for the Promotion of Young Researchers, co-funded by Comunidad de Madrid and the University of Alcala (Spain), as part of the project IndoorComfort (CM/JIN/2019-46).

Also, this study has been partially funded by the European Union by the action HORIZON-TMA-MSCA-SE "BEST-Bio-based Energy-efficient materials and Structures for Tomorrow" (grant number 101086440).

References

Alonso, M. C., & Puentes, J. (2020). Self-compacted concrete with self-protection and self-sensing functionality for energy infrastructures. *Materials, 13*(5), 1106.

Alonso, M. C., Puentes, J., Gómez, A., Chozas, V., & Vera, J. (2021). Health monitoring of failure detection of functionalized cementitious composites with self-diagnosis additives of CNT+CMF. In *Proceedings of the international conference on SHMII-10* (Porto, Portugal), 251–258.

Azhari, F., & Banthia, N. (2012). Cement-based sensors with carbon fibers and carbon nanotube for piezo resistivity sensing. *Cement and Concrete Composites, 34*, 866–873.

Bekzhanova, Z., Memon, S. A., & Kim, J. R. (2021). Self-sensing cementitious composites: Review and perspective. *Nanomaterials (Basel), 11*(9), 2355.

Cabeza, L. F., Castell, A., Cabeza, L. F., De Gracia, A., & Fernandez, A. I. (2011). Materials used as PCM in thermal energy storage in buildings: A review. *Renewable and Sustainable Energy Reviews, 15*, 1675–1695.

Guardia, C., Barluenga, G., Palomar, I., & Diarce, G. (2019). Thermal enhanced cement- lime mortars with phase change materials (PCM), lightweight aggregate and cellulose fibers. *Construction and Building Materials, 221*, 586–594.

Guardia, C., Barluenga, G., & Palomar, I. (2020). PCM cement-lime mortars for enhanced energy efficiency of multilayered building enclosures under different climatic conditions. *Materials, 13*, 4043.

Han, B., Ding, S., & Yu, X. (2015). Intrinsic self-sensing concrete and structures: A review. *Measurement, 59*, 110–128.

Palomar, I., Barluenga, G., Ball, R. J., & Lawrence, M. (2019). Laboratory characterization of brick walls rendered with a pervious lime-cement mortar. *Journal of Building Engineering, 23*, 241–249.

Taheri, A. (2019). A review on five key sensors for monitoring of concrete structures. *Construction and Building Materials, 204*, 492–509.

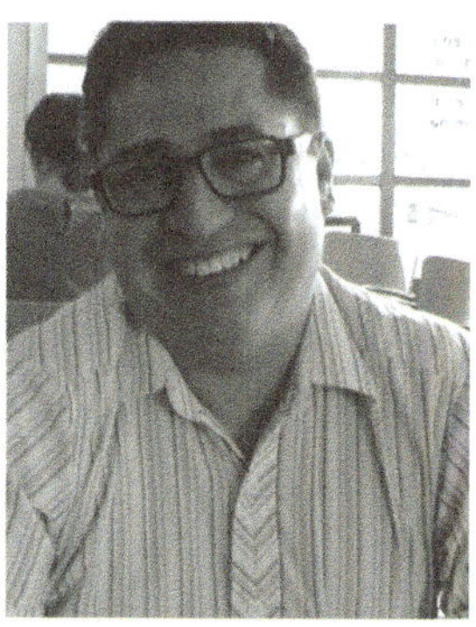

Javier Puentes is Ph.D. Arch. His research experience focuses on the topic of construction materials. At the predoctoral stage (before 2015), the research focused on the study of the use of nano additions to control and modify the hydration action of concrete. The processes that affect the durability of concrete, the extension of service life through efficient and controlled hydration processes for new cements and additions. Currently his area of research is complemented by the design of concretes with new properties through the incorporation of carbon-based nanomaterials, nanofibers and supplementary mineral additions. He was selected for a postdoctoral research position at the CSIC-IETcc in a European project LORCENIS, N° 685445. Long-lasting Reinforced Concrete for Energy Infrastructure under Severe Operating Conditions.

Currently, he is working as a Lecturer at the University of Alcalá, Spain, at the Architectural School. He participates as a teacher in the Master of Architecture Program, in different Universities in Spain and Latin America. The research work was complemented with the professional activity in a I+D+I laboratory for construction materials in the private sector. It allowed Javier Puentes a specific specialization and a direct relation with the constructive processes.

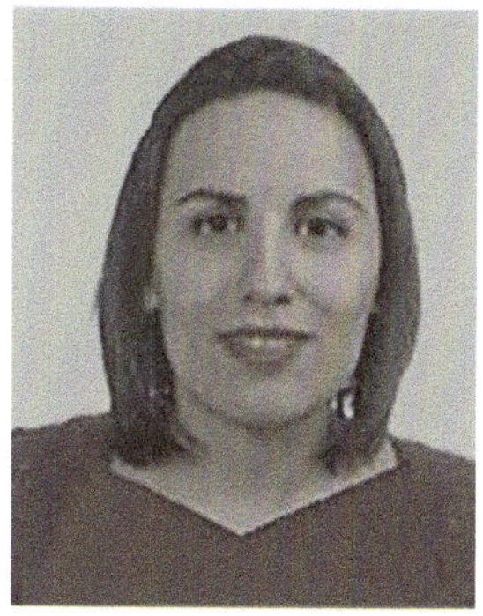

Irene Palomar is an Assistant Professor in the Department of Architecture at the University of Alcalá, Spain, since 2019. Her research interests include advanced building materials, physical-mechanical characterization of cement-based building materials, sustainable materials and building technologies for architectural applications and refurbishment. She worked in eight projects as a part of research group leaded by Prof. Gonzalo Barluenga. She has participated in international and national conferences and co-authored 18 papers in indexed journals.

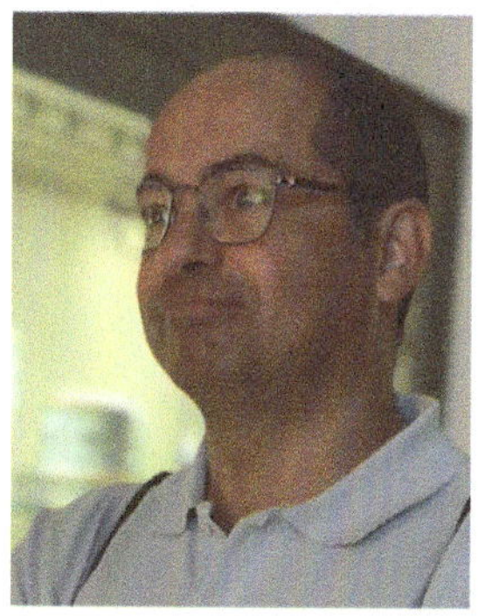

Gonzalo Barluenga holds a Ph.D. in Architecture and is Full Professor of Construction and Building Materials at the School of Architecture of Alcalá (Madrid, Spain). His lines of research involve Sustainability and Resilience of Innovative Building Materials and Architectural Technologies. He has led multidisciplinary research projects on innovative, fire resistance, sustainable, energy efficient and resilient cement-based materials and construction technologies. He leads a research team that collaborates with several national and international groups on bio-based and bio-inspired smart building materials and digital fabrication for architecture. He has published more than 100 articles and book chapters and he holds several patents.

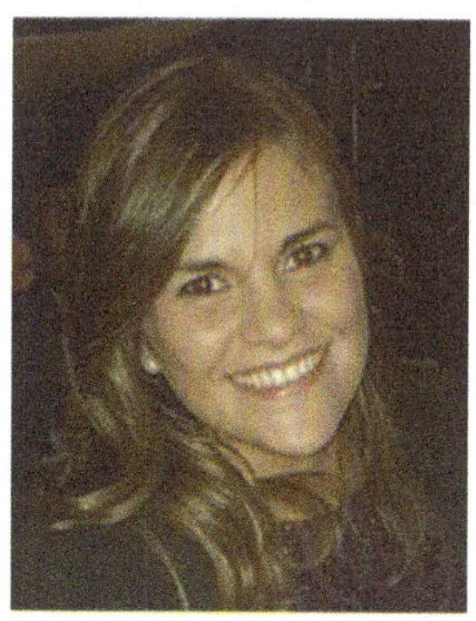

Cynthia Guardia Martin obtained the Ph.D. in Architecture with "cum laude" and international mention in Architecture at the University of Alcalá (UAH, Spain) in 2021. She has a Master's Degree in Pathology and Sustainable Refurbishment of Heritage from Universidad Europea (Madrid) and is an architect from IE University.

After her doctoral thesis, she obtained a research contract in competitive call of Margarita Salas Grants for young researchers with Next Generation European funds.

Her research lines focus on the study of innovative, energy-efficient and sustainable eco-building materials, systems and technologies for architectural applications. She is also interested in the NDT techniques.

As PhD student, she studied in the Polytechnic University of Madrid and Technical University of Darmstadt at Institute of Construction and Building Materials.

She was visiting professor at Universidad Francisco Marroquin from Guatemala. She has also directed Master's Final Projects at UEM.

Since 2015 she works as part of research group lead by Prof. Gonzalo Barluenga at the University of Alcalá.

Experimental Bionics Workshops

Mark P. Sarkisian, María Rosa Cervera Sardá,
Elena-Codina Duşoiu, Ana Mohonea, and Tana Nicoleta Lascu

Introduction

Mark P. Sarkisian and R. Cervera

Experimentation and exploration are essential to creative architectural design. Research leads to new ideas. Investigations should range from broad urban plans to specific technologies embedded in small architectural details. Students with limited influence from traditional methodologies are key to the development of inventive ideas through the influence of natural organization. With the proper guidance of instruction, results can be remarkable. Tangible techniques such as physical modeling and freehand drawing allow students to image form, use, and ultimately construction. Careful considerations for architectural components and structural and mechanical systems are key to a more sustainable future. Buildings, that one day could become self-sustaining or even regenerative, is the goal for design. These buildings could be both expressive of structure and building services systems that exhibit a natural response to the environment. Interconnected development in districts or cities could heal individual building deficiencies of energy, water, or managing waste through locally conceived infrastructure networks.

The practical work of the workshops opens new doors to the approach to architectural, structural, and constructive design, always starting from a view of nature. Learning from nature does not imply copying its forms. The wings that serve a bird to fly would not serve an airplane, just as a tree is not a skyscraper. To learn from nature is to discern the ability of living species to save energy and use matter efficiently. To learn from nature is to look for new ways to use natural resources and to innovate in recycling and reusing processes. Nature is an open book, placed at our disposal, which contains all the answers and is the ideal framework where scientists, philosophers, and artists can, through observation, ideas, and thoughts, write together the new pages of eco-philosophy and biotechnology.

The workshops' assignments are fast and intuitive and use simple and direct resources and materials that lead the young student, designer, or architect to shed their clichés in their approach to the project. Fresh and stimulating solutions for artistic production, with a new range of aesthetic solutions and optimization of resources, are explored in each workshop with the nodal themes that articulate them. The aim is a production that is in harmony with the human being and his home: the planet Earth.

The original version of this chapter "Experimental Bionics Workshops" has been revised and we have added all missing figures to it. The correction to this chapter can be found at https://doi.org/10.1007/978-3-031-33144-2_22

M. P. Sarkisian (✉)
Structural Engineering Partner, Skidmore, Owings & Merrill, San Francisco, CA, USA
e-mail: mark.sarkisian@som.com

M. R. Cervera Sardá
Universidad de Alcalá, Alcalá de Henares, Spain
e-mail: rosa.cervera@uah.es

E.-C. Duşoiu · A. Mohonea · T. N. Lascu
University of Architecture and Urbanism "Ion Mincu", Bucharest, Romania

M. R. Cervera Sarda et al. (eds.), *Architecture Inspired by Nature*, https://doi.org/10.1007/978-3-031-33144-2_21

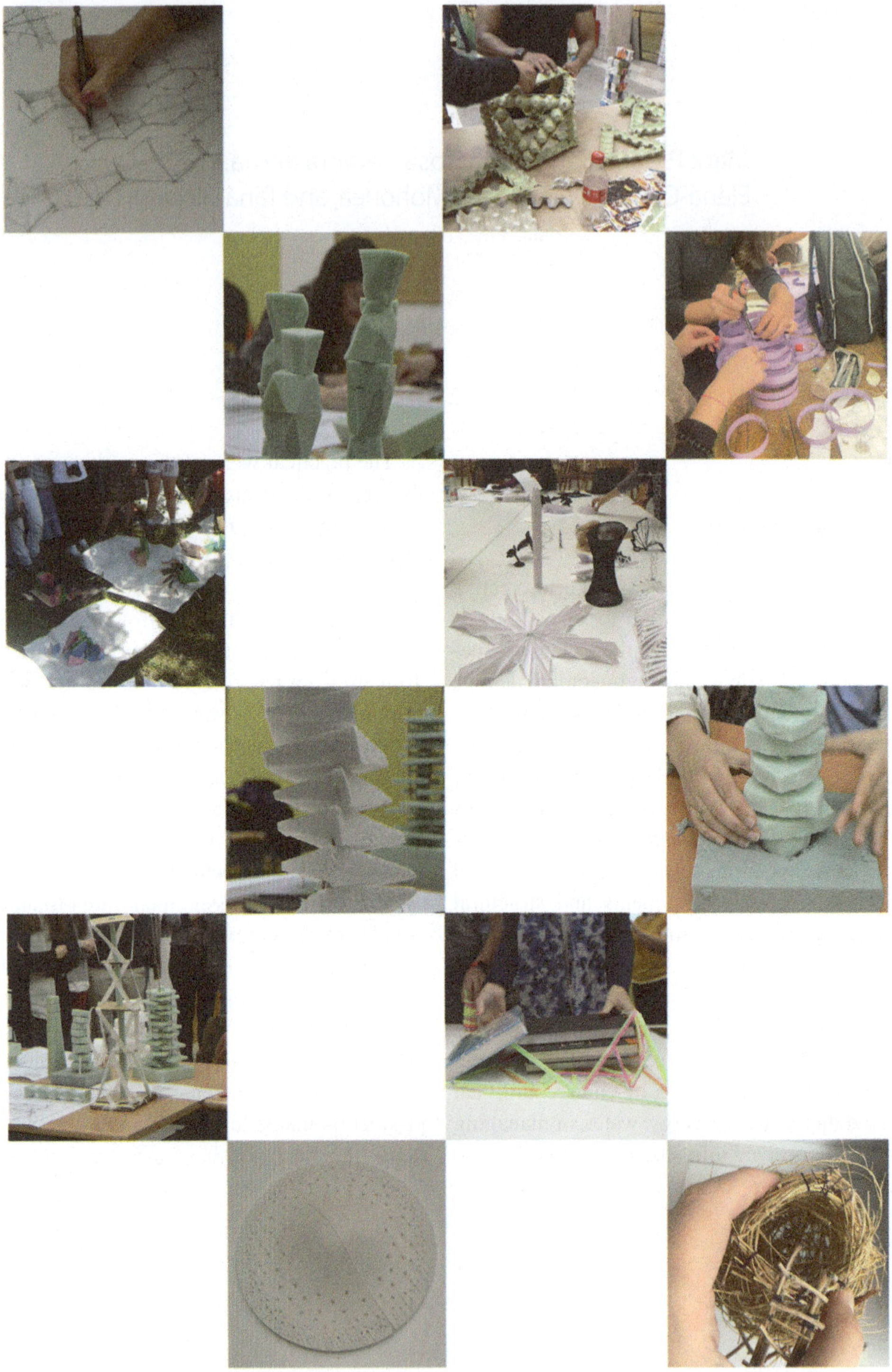

Vertical Structures

María Rosa Cervera Sardá

The human aspiration to build vertically dates back to prehistoric times. Nowadays, with the exponential urban growth that is predatory of nature, it is necessary to find alternatives that allow the vertical development of the city in environmentally sustainable and human-friendly models. Nature offers us inspiring models of vertical construction of surprising efficiency. Trees, which can reach up to 100 m in height, are able to withstand winds and earthquakes and live for hundreds of years. Their internal structure allows them to transport water and nutrients vertically and horizontally without the need for any external energy. And their base material, wood, is of very low density and little mass yet great strength.

In the search for alternatives for a new vision of high-rise construction, different families of solutions are proposed. Paper and foam make it possible to work with cumulative modules in an innovative approach to verticality. Wooden rods or wires build three-dimensional frameworks of great lightness and stability that explore heights and shapes. The examples are reinterpreted in the light of architecture, generating proposals that, inspired by the principles of nature, achieve new expressiveness.

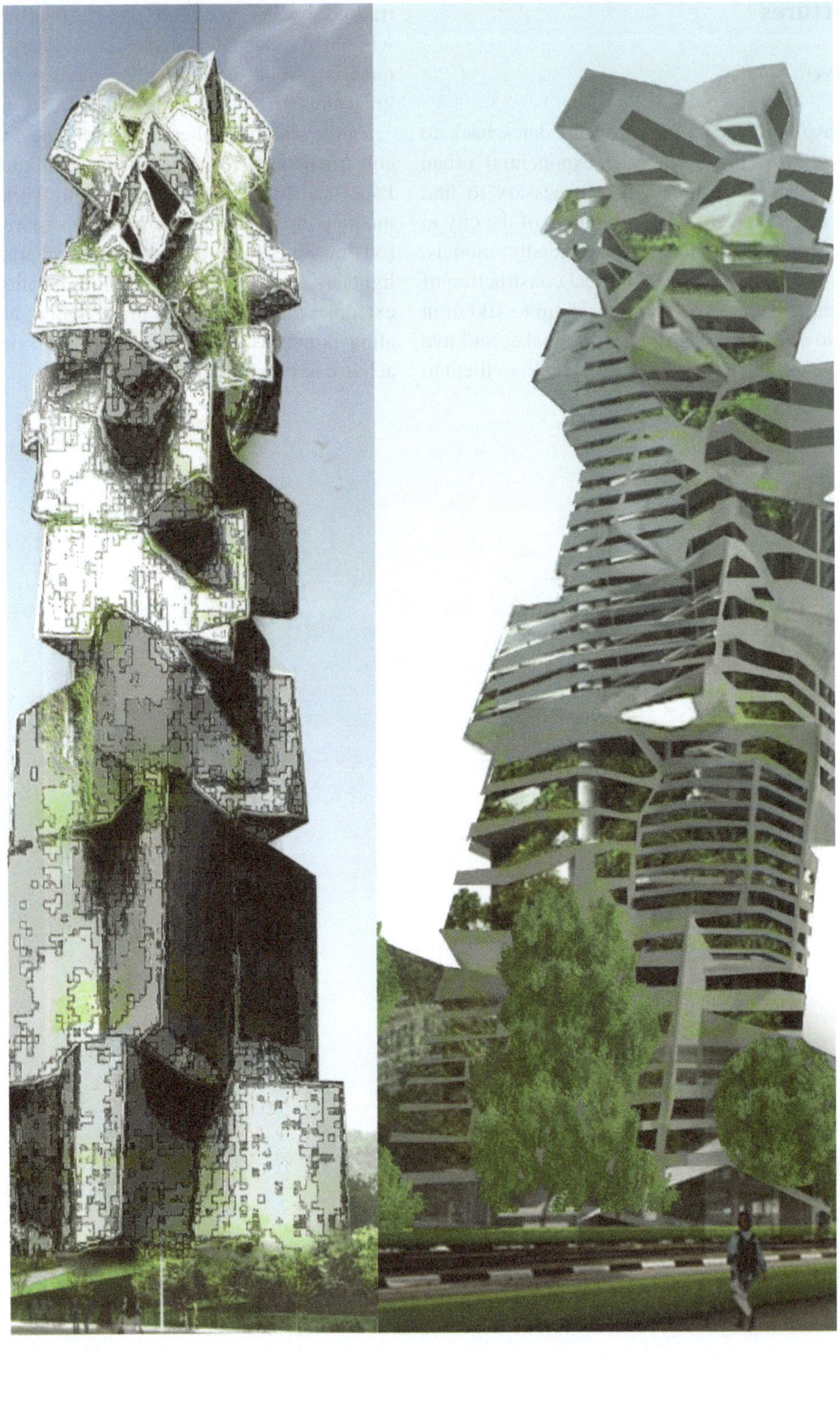

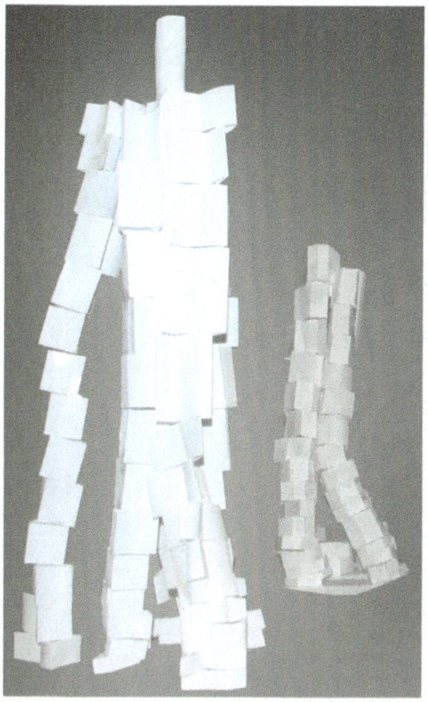

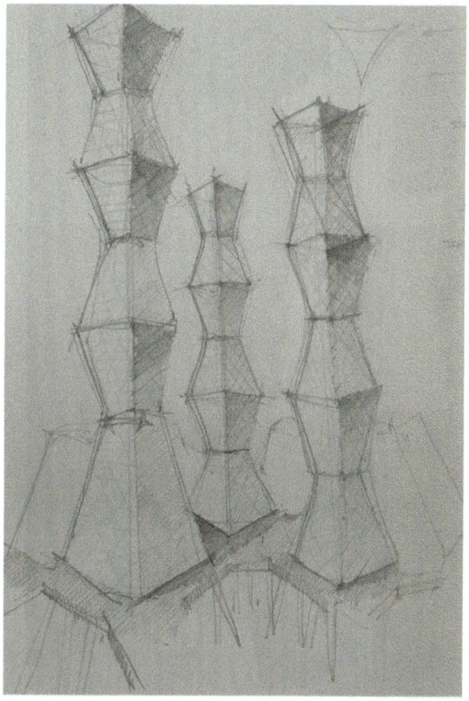

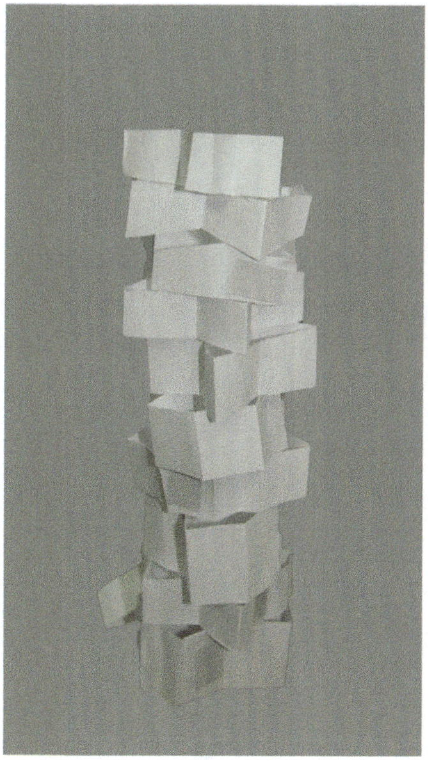

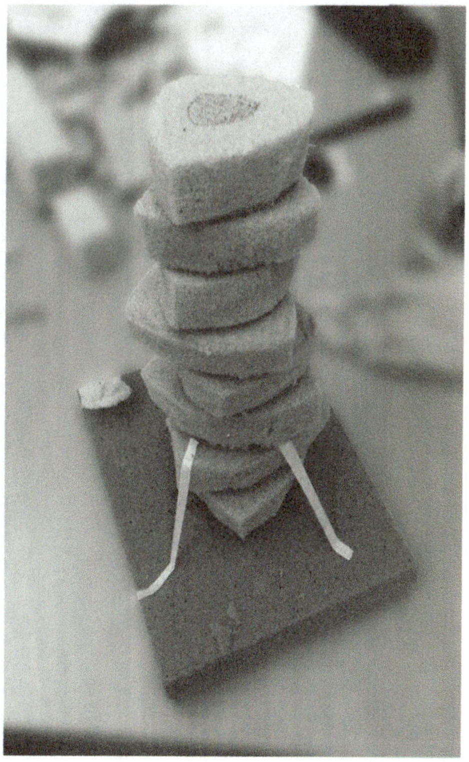

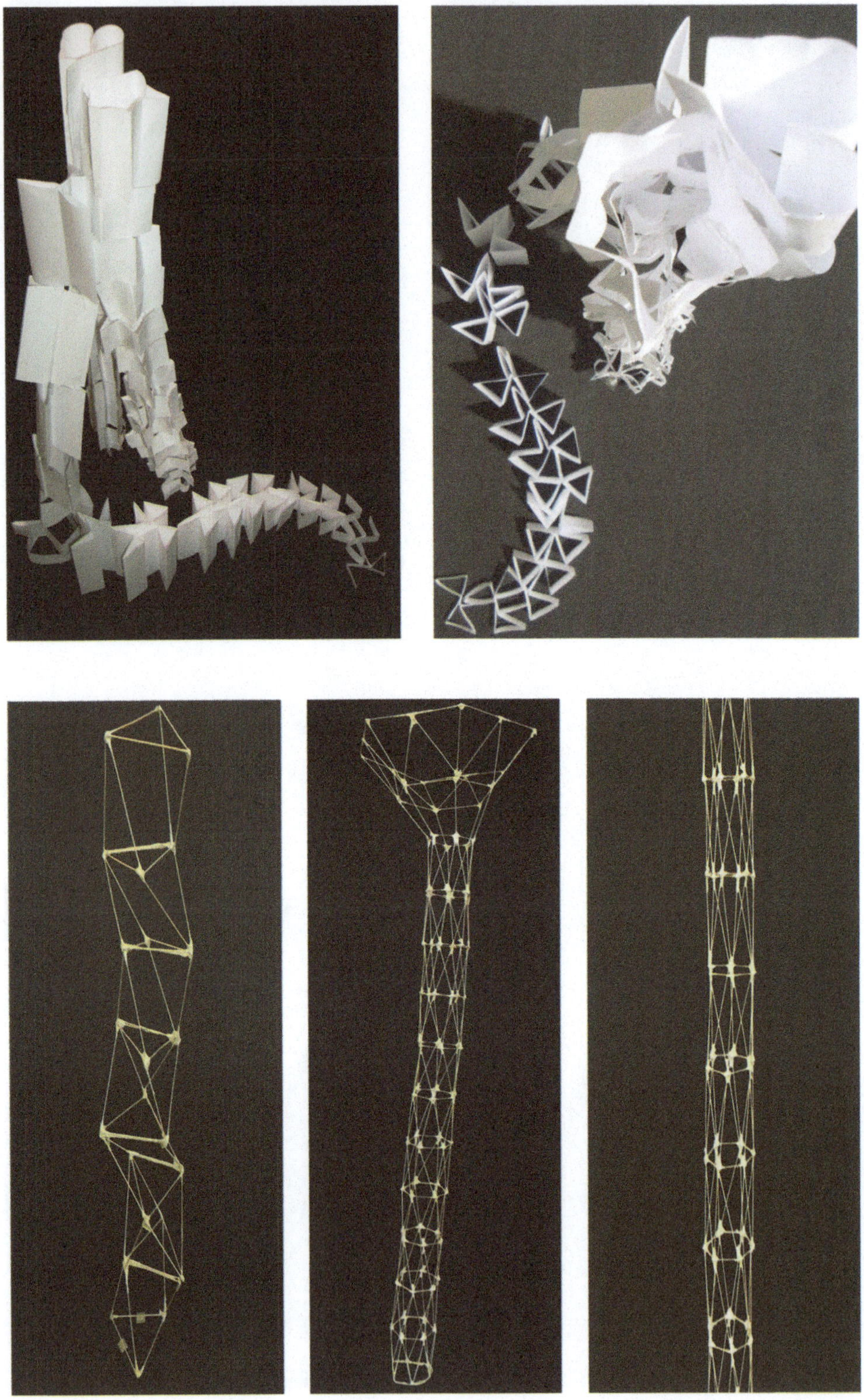

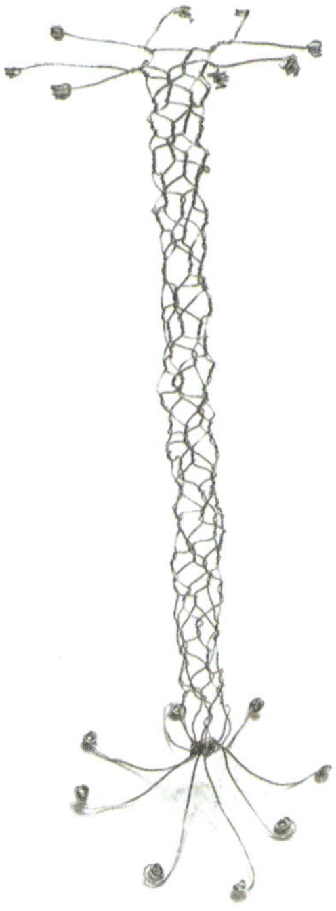

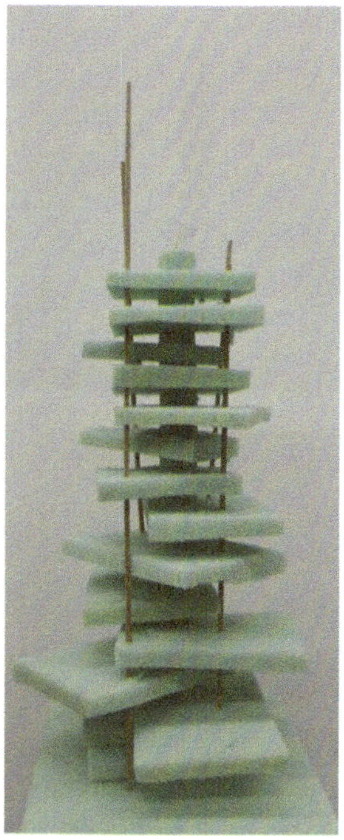

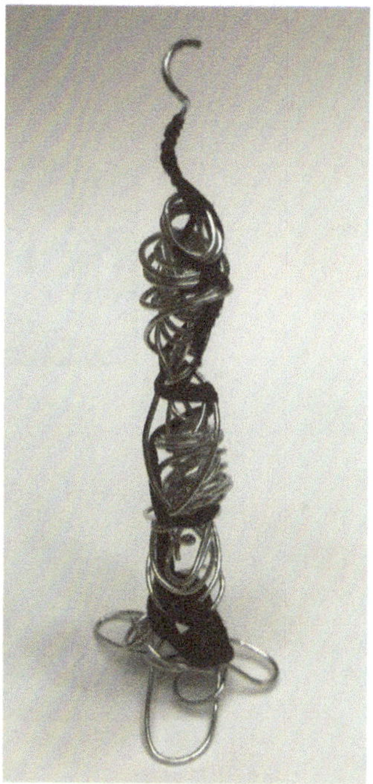

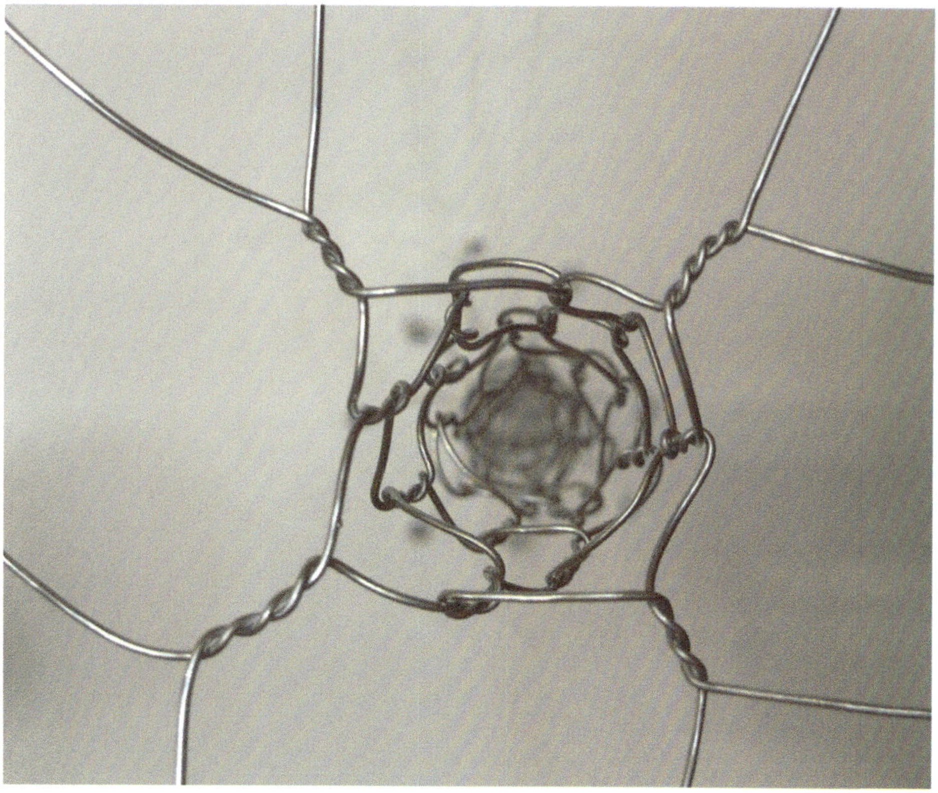

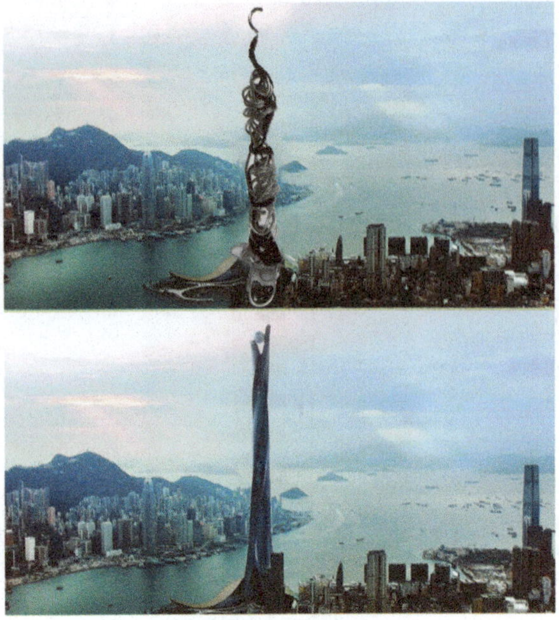

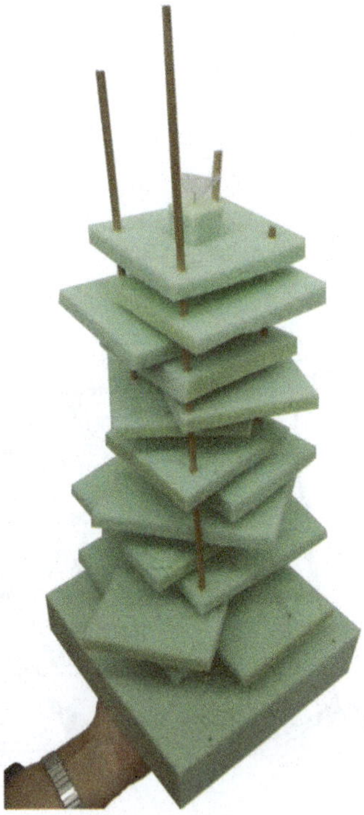

Bioinspired Forms

Elena-Codina Duşoiu and Ana Mohonea

The studies aim to research functioning mechanisms from nature realized on the basis of spatial structures that are both rational and aesthetic, for example, an inflorescence, a fruit, an insect wing, crustaceans, a leaf, the skeleton of a fish or scales of a fish, etc. The transformation of a natural organism into a structure with architectural application (support, roofing, partition, wall frame, etc.) can be realized through essentialism and abstraction. The details of the chosen object (joints of the limbs, pincers, ramifications) can also be studied and will be adapted as joints for the proposed structure.

Further on, the research can be applied in a more detailed, practical and focused way. The documentation of a natural structure can be continued, focusing on the details of some specific organisms (e.g., the claws of a lobster, an insect's legs, the role of bones in a skeleton, the compliance of woody tissue of a tree, etc.). Details can be later taken into architectural context, considering a variety of scales (from landscape and urban planning to architectural structures, interior design, furniture, etc.)

The objective of this approach is to gather documentation on the structures and growing laws of nature. Catching details of the elements they form can be used as possible devices in product design or considered for details in architectural structures. The documentation related to self-supporting structural forms is based on extracting structural principles related to the used materials, natural geometry and structures, mechanisms, and physiological laws.

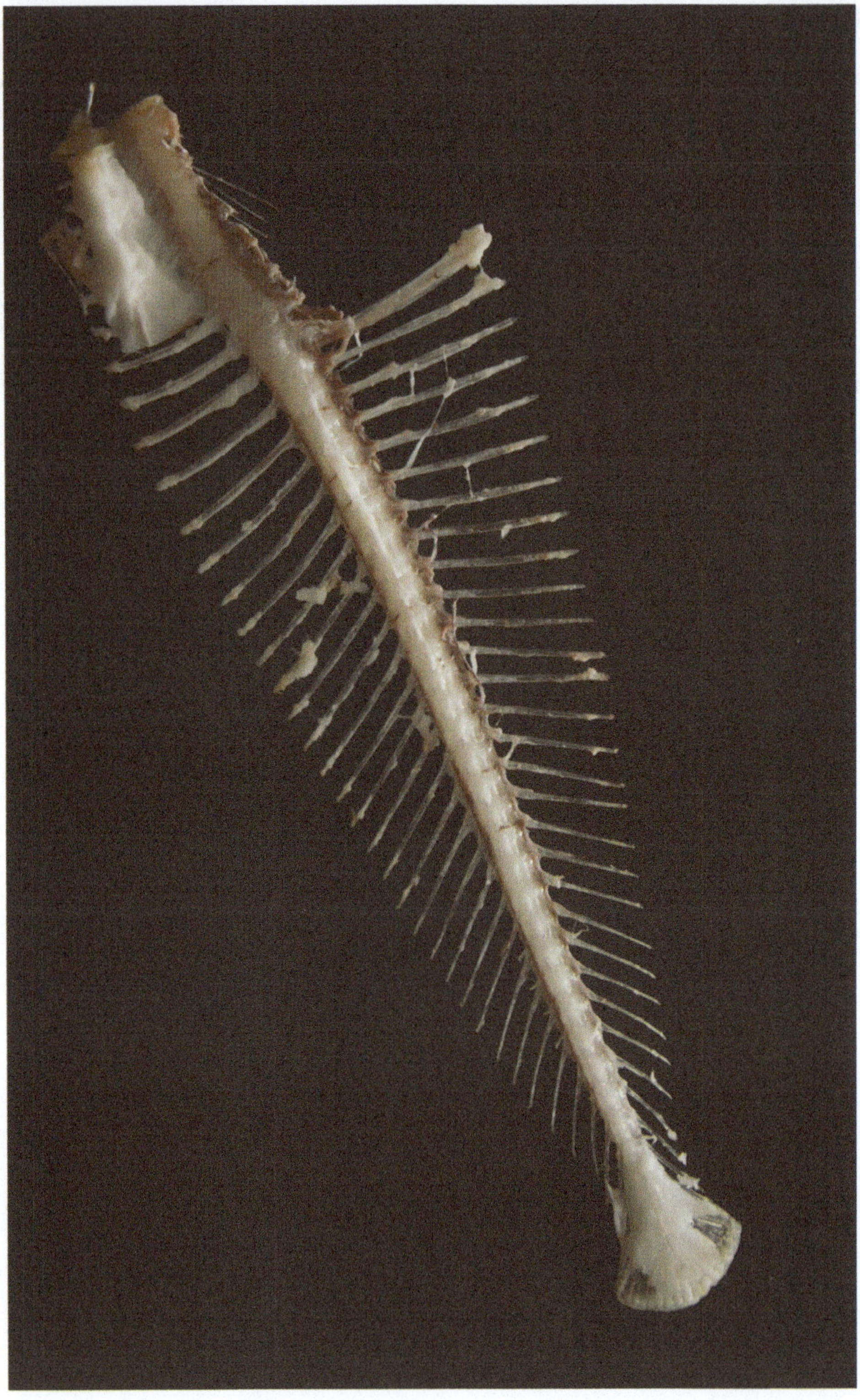

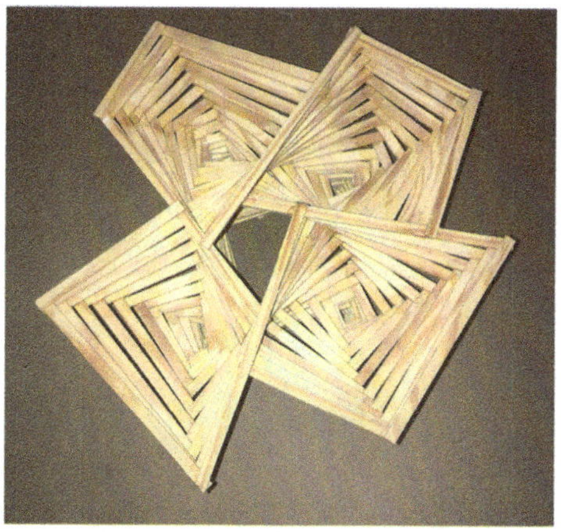

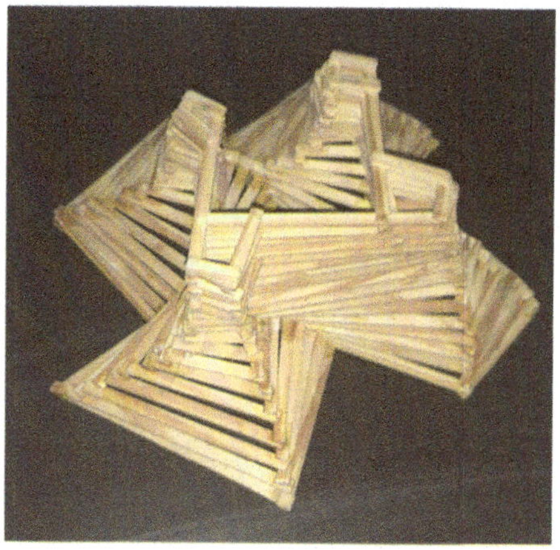

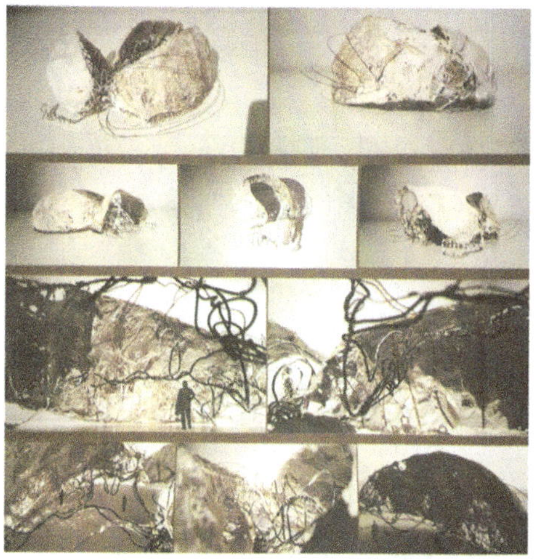

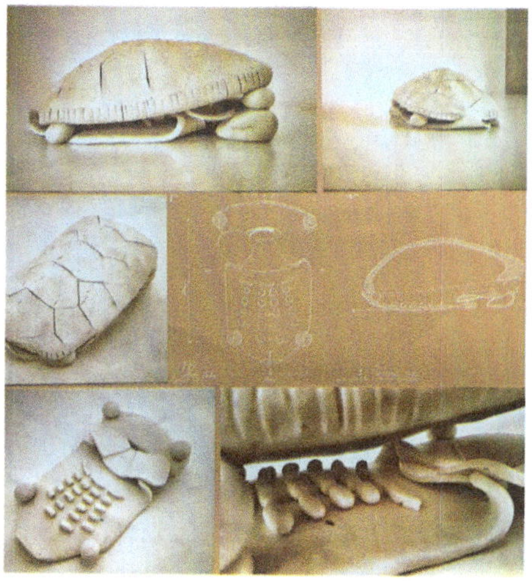

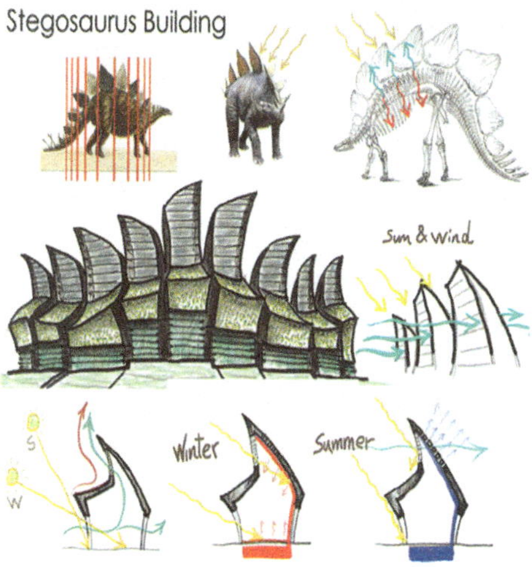
Stegosaurus Building
Sun & Wind
S
W
Winter
Summer

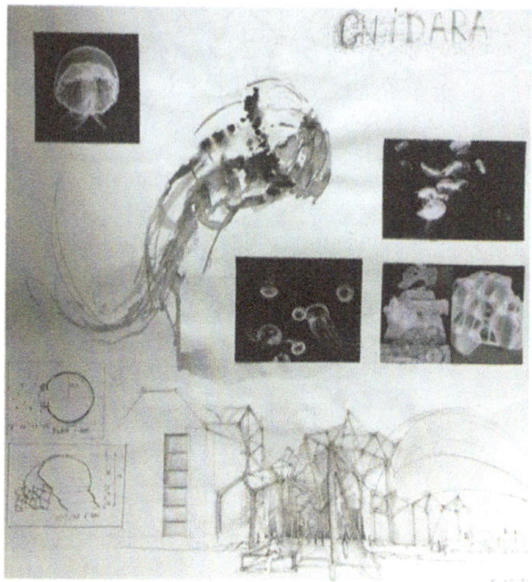

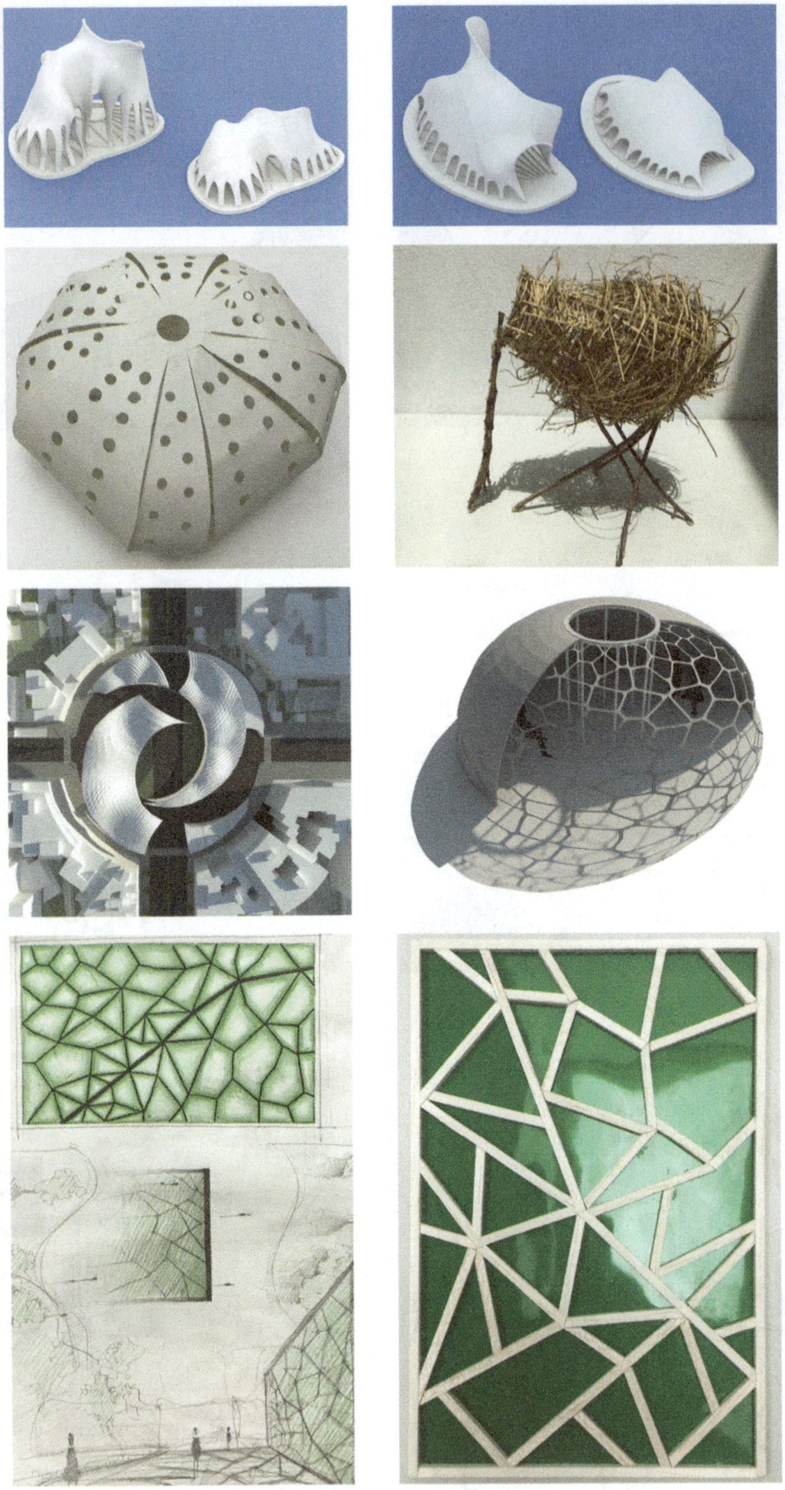

Fluid Topological Compositions

María Rosa Cervera Sardá

Topology is the mathematics of position (situs geometry) and distortion, which deals with the fundamental geometrical properties of objects in which relationships remain constant, regardless of any distortion to which they are subjected. To put it in mathematical terms, topology deals with the transformation of invariant relations. In topology, two points on a line maintain the same continuity relationship to each other, whether the line is straight or curved; that is why, topologically, a doughnut, or torus, is the same as a sewing needle, or a circle is the same as an ellipse. It is a geometry that does not depend on shape. The shape can vary, stretch, shrink, and deform, and as long as it maintains its connecting relationships, it remains identical to itself. And architecture is nothing more than the continuity and discontinuity of spaces, which is why it allows the exploration of forms from a topological approach.

Folds, stretching, or shrinking of a surface or a volume does not modify the relationship of continuity of the points in them, even if the distance is altered, and does not modify the form from a topological point of view, while cuts imply discontinuities and, therefore, formal alterations. Probing the effects of simple actions on volumes and sufaces allows us to explore compositions beyond Cartesian Geometry. If architectural form is no longer conceived from its fixed coordinates and positions, it can be adapted, with fluid formal variations, to the changing conditions of the environment, just as nature does.

Exercises with folded paper, with papyro-flexible distortions, with elastic meshes, and with pieces adaptable to variable and changing forms introduce us to resilient composition.

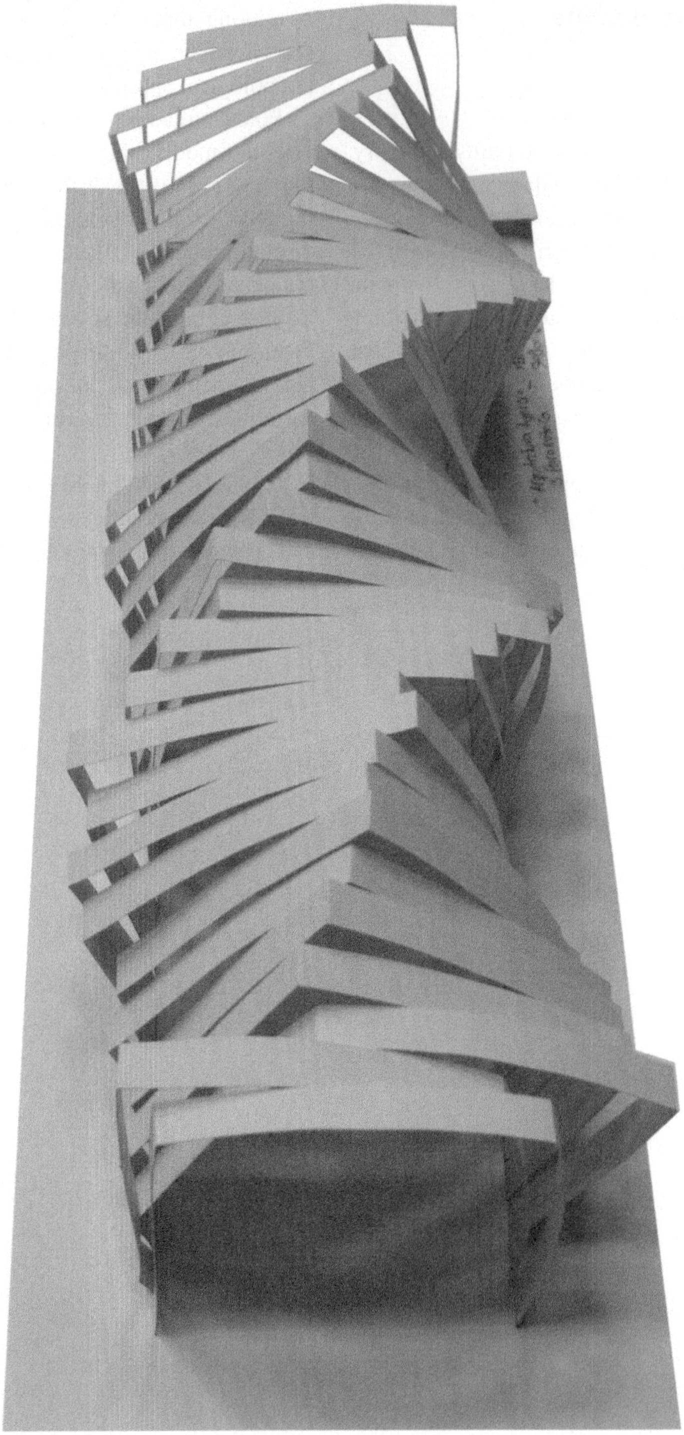

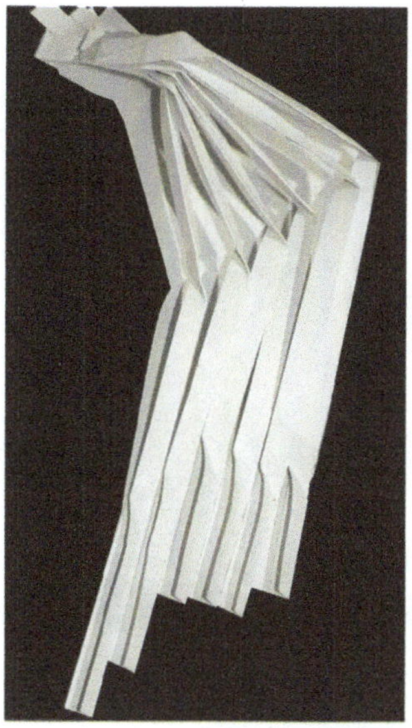
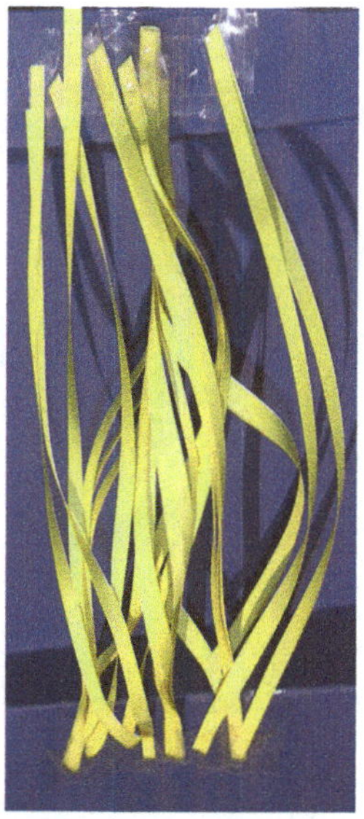
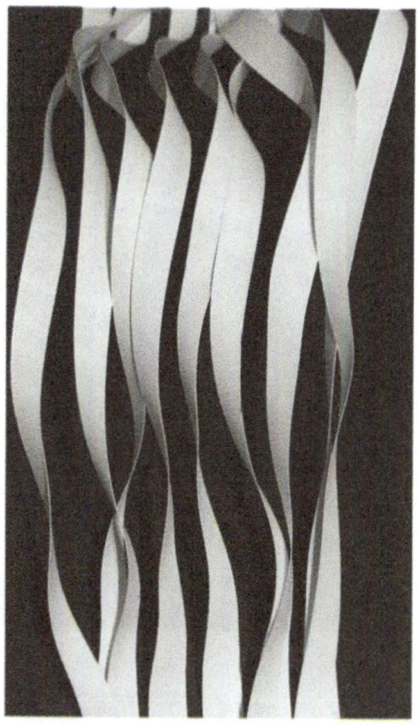

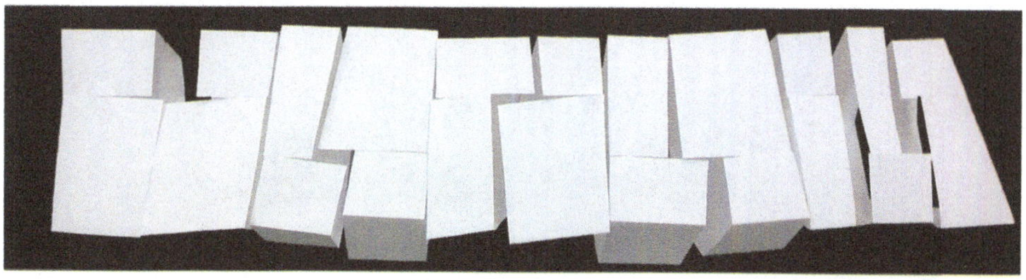

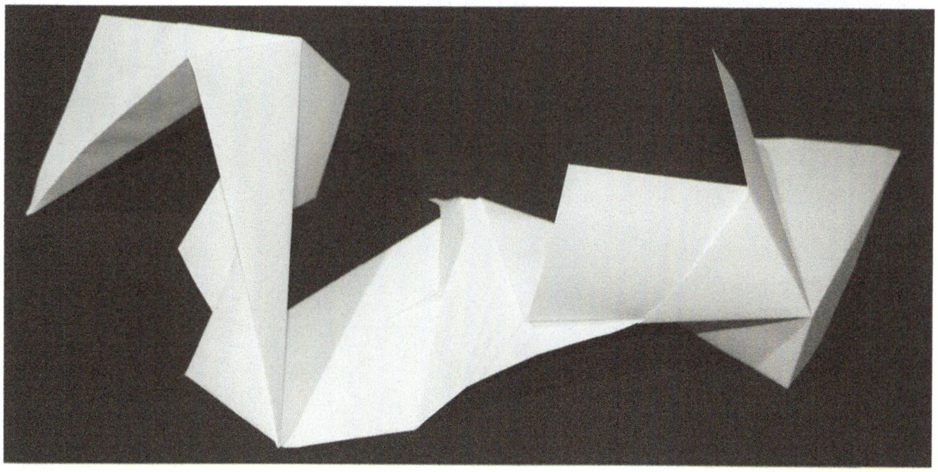

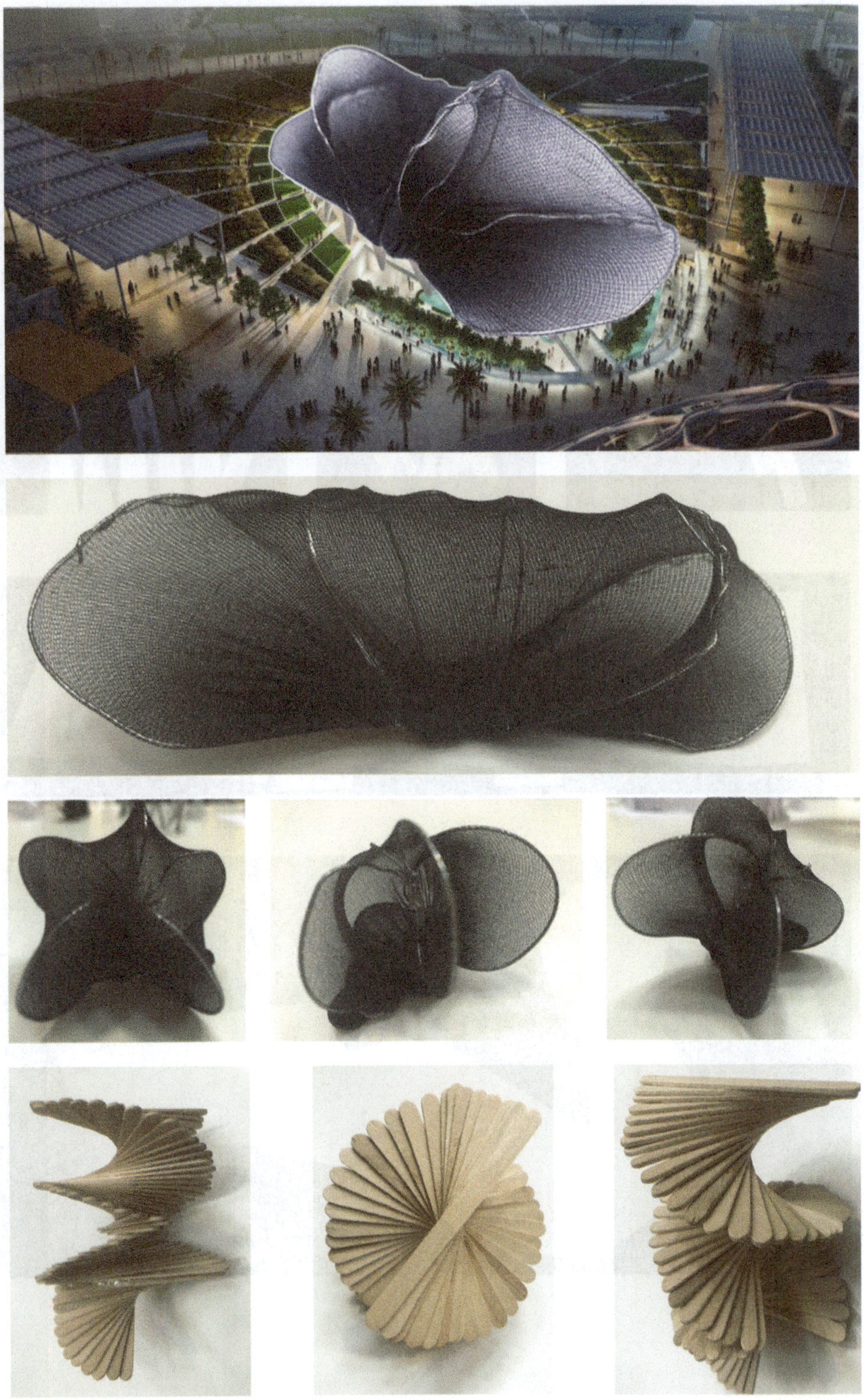

Designing with Fractal Geometry

María Rosa Cervera Sardá

Fractal geometry interprets the formal complexity of nature. Its scaling and iterative character, with continuous repetition of a form at different scales, replicates natural processes of self-organization. The forms in nature are always the result of a process of evolution and growth, which demands geometry of a dynamic nature capable of responding to variation. Nature functions as a complex system in which predictions are not possible and in which determinism has great limitations. It must therefore be built with systems that admit change as intrinsic to its reality. That is why fractal geometry allows us a new approach to the understanding of natural forms. Its flexible nature, admitting change as something intrinsic, connects with the dynamic being of the universe. Simple codes and laws, which can be represented by simple mathematical functions, transmit the information to each unit, however small, and to the whole. The result is a coherent form that allows for variation and adaptability.

Simple exercises based on laws of formal self-similarity with dimensional variation make it possible to achieve a great display of forms of expressive richness. The formal exuberance, however, is based on the repetition of elements, which guarantees the minimization of efforts and, therefore, savings and economy. The choice of the module or base piece and its geometric configuration is the initial step, and the establishment of the law of scalar variation is the second action. From there, we proceed to the self-organized generation of the form, which, after a certain number of iterations, ends up reaching its maximum complexity and opening the door to different modes of architectural composition.

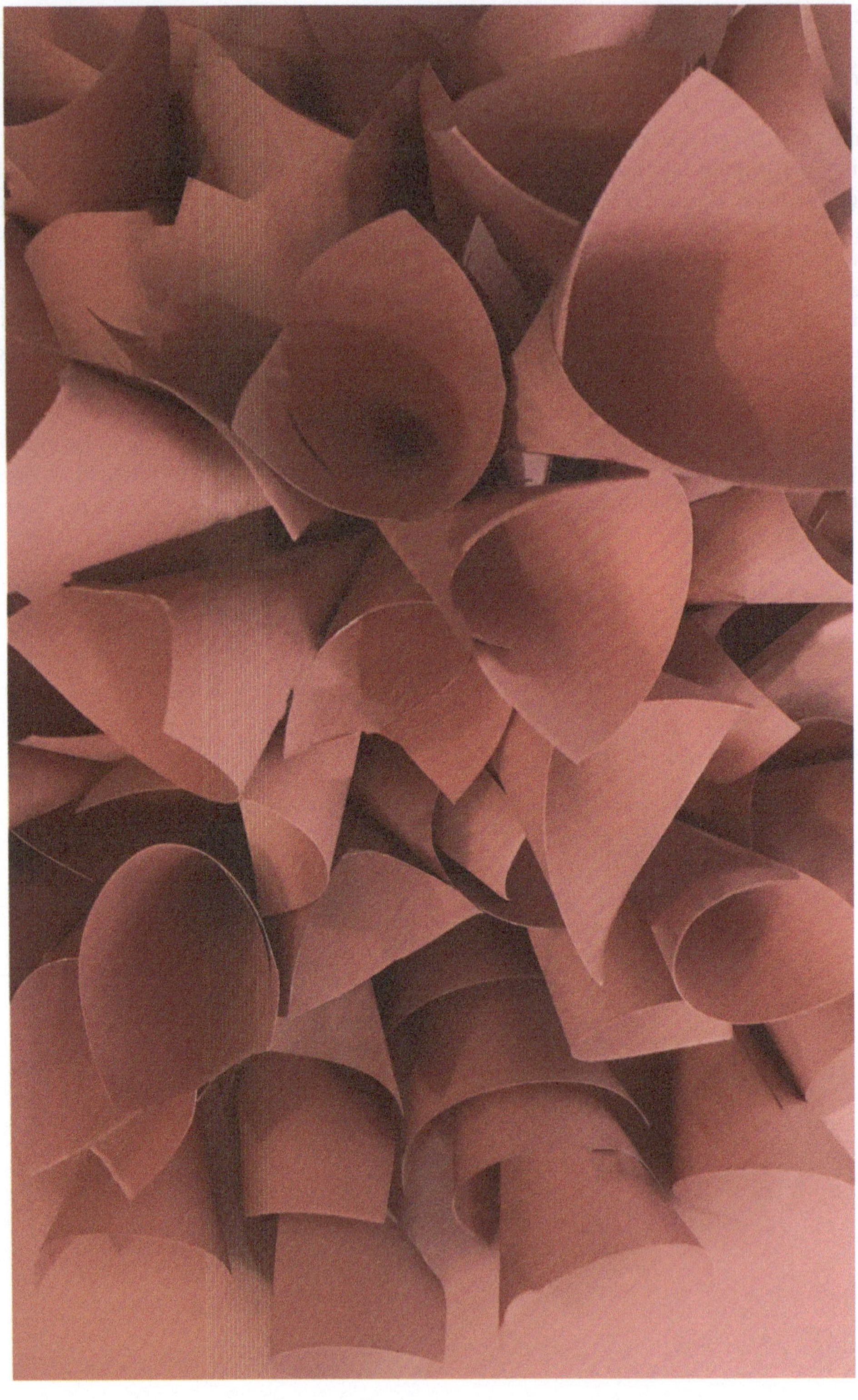

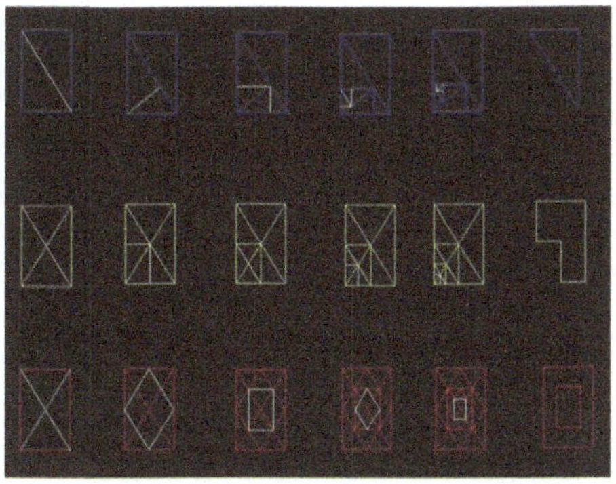
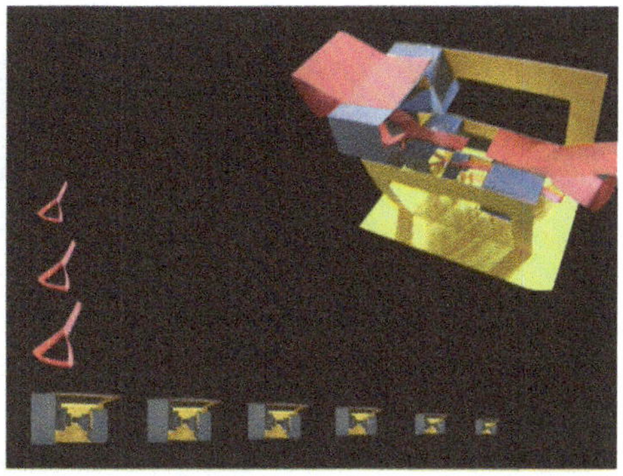
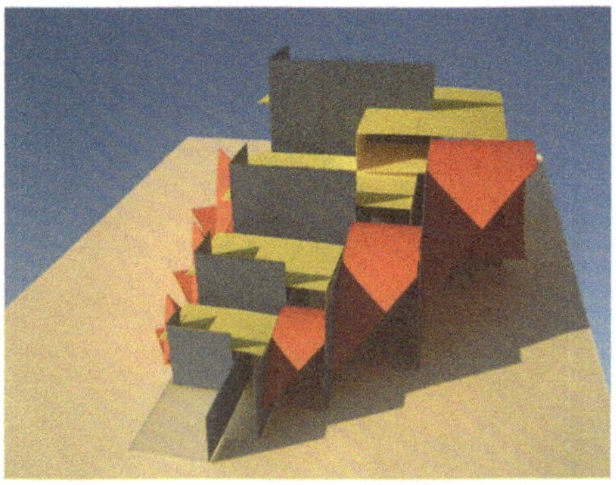
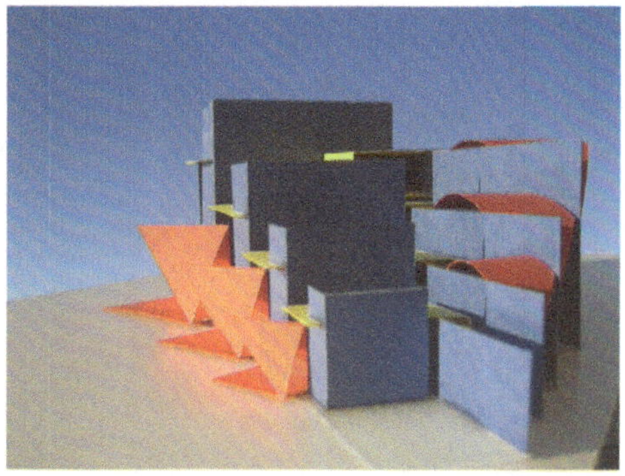
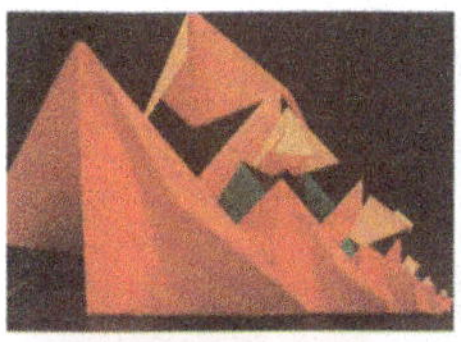
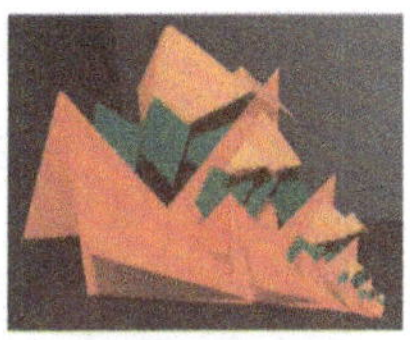
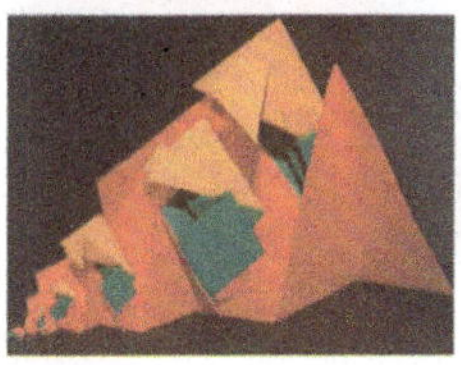
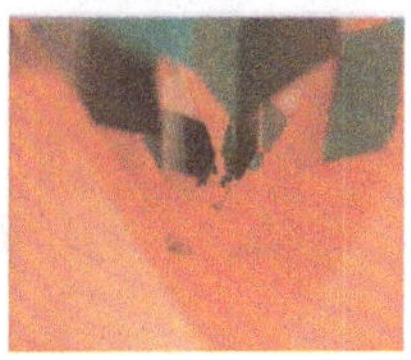
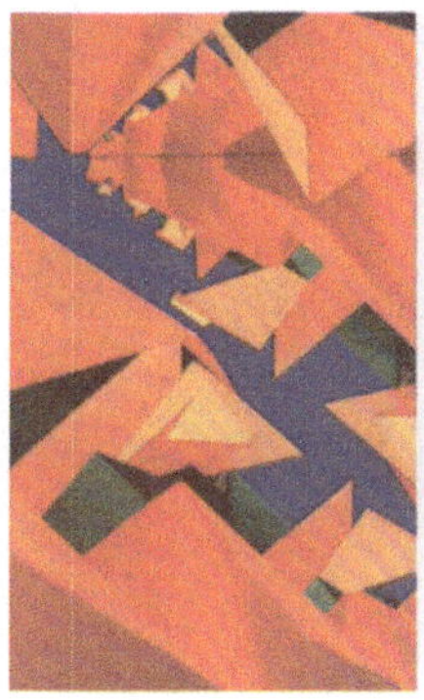
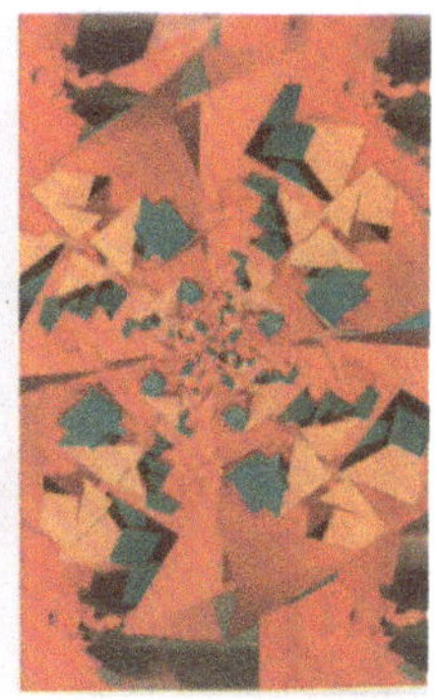
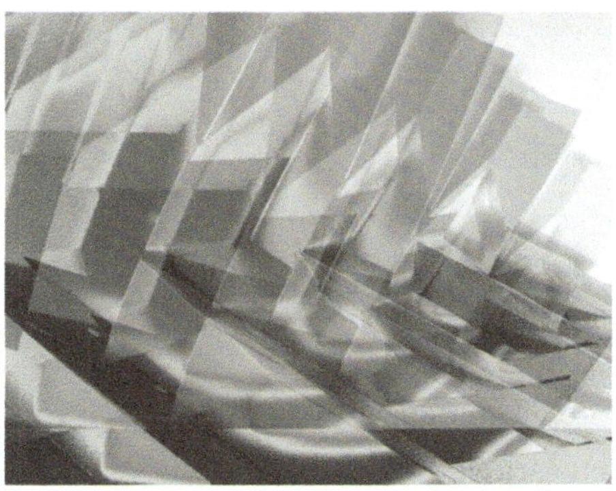
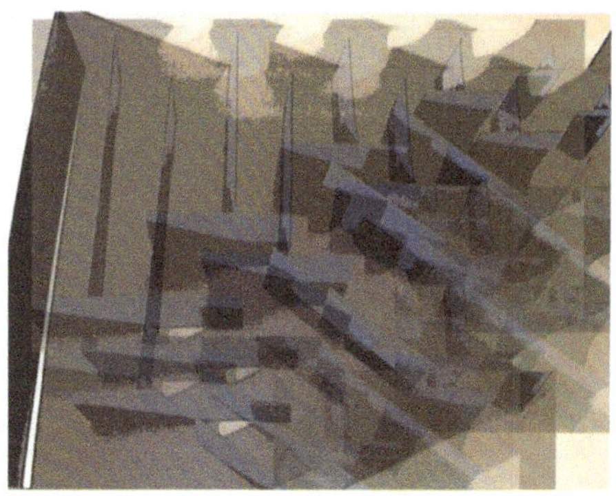

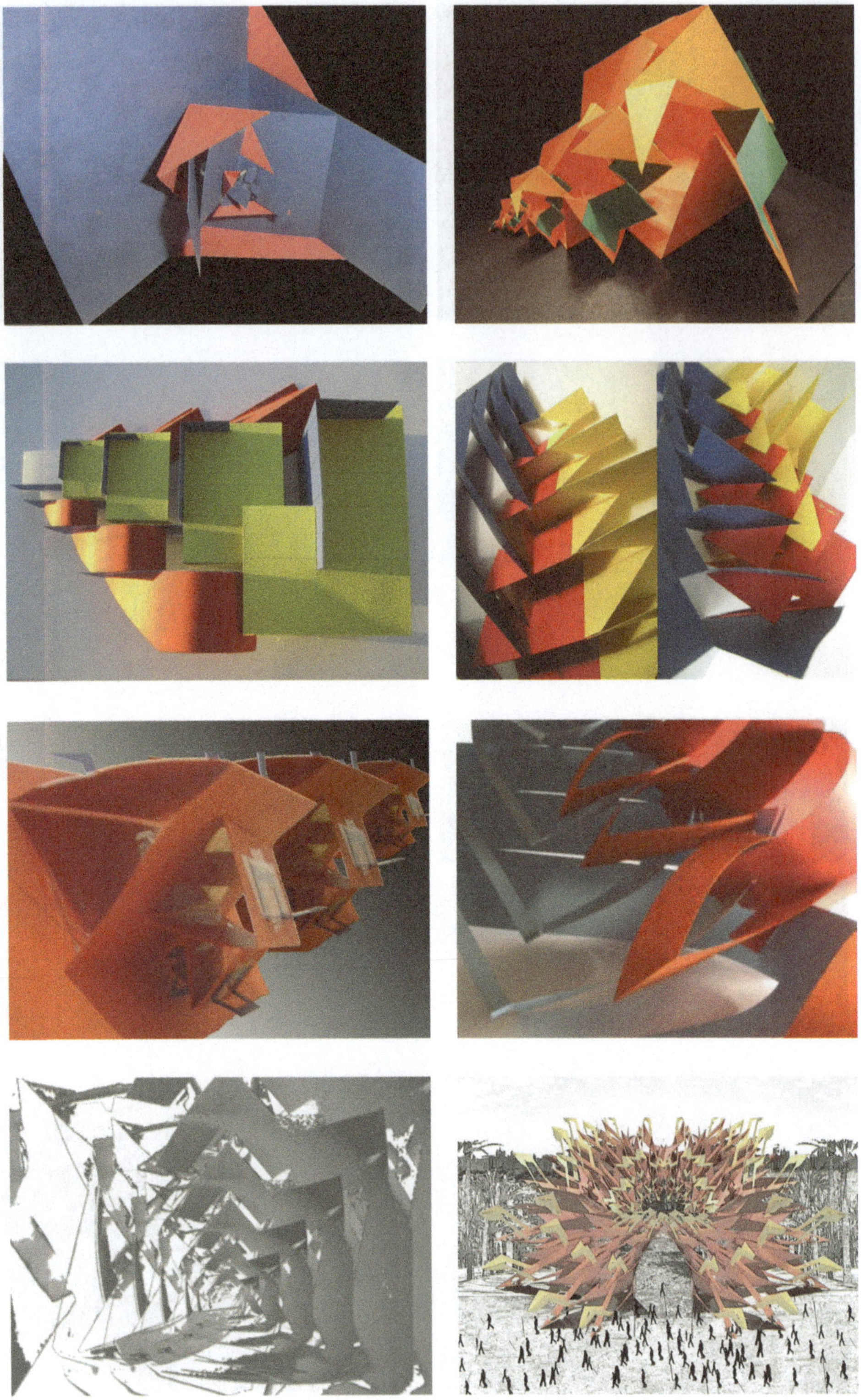

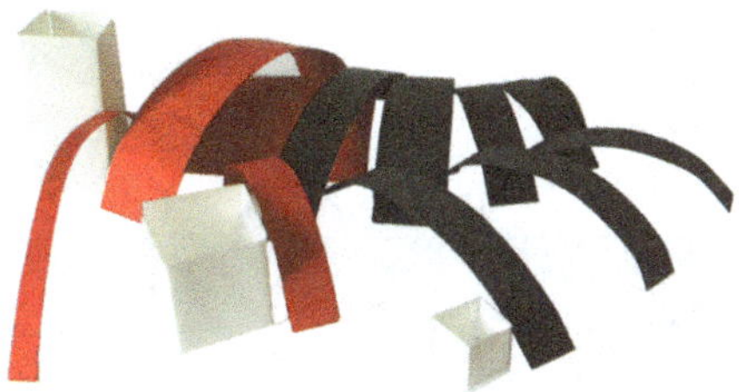

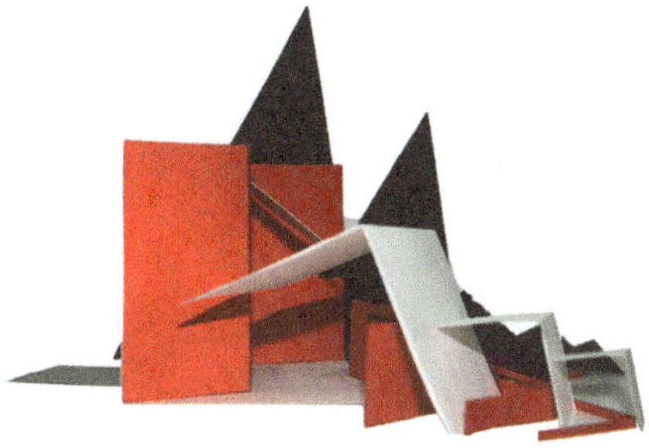

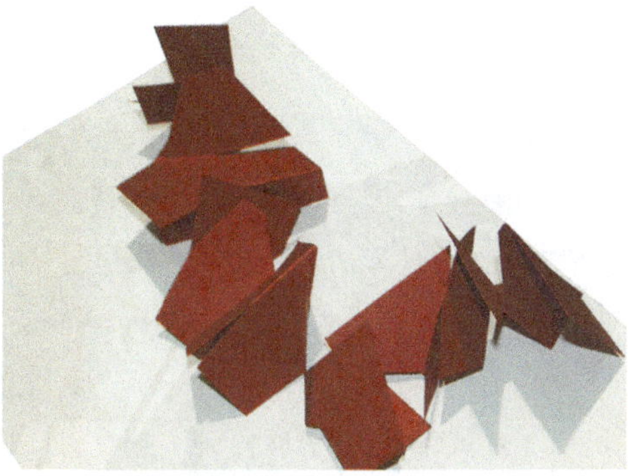

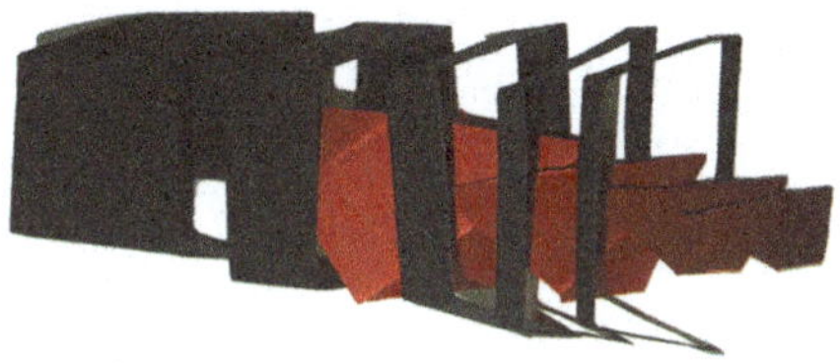

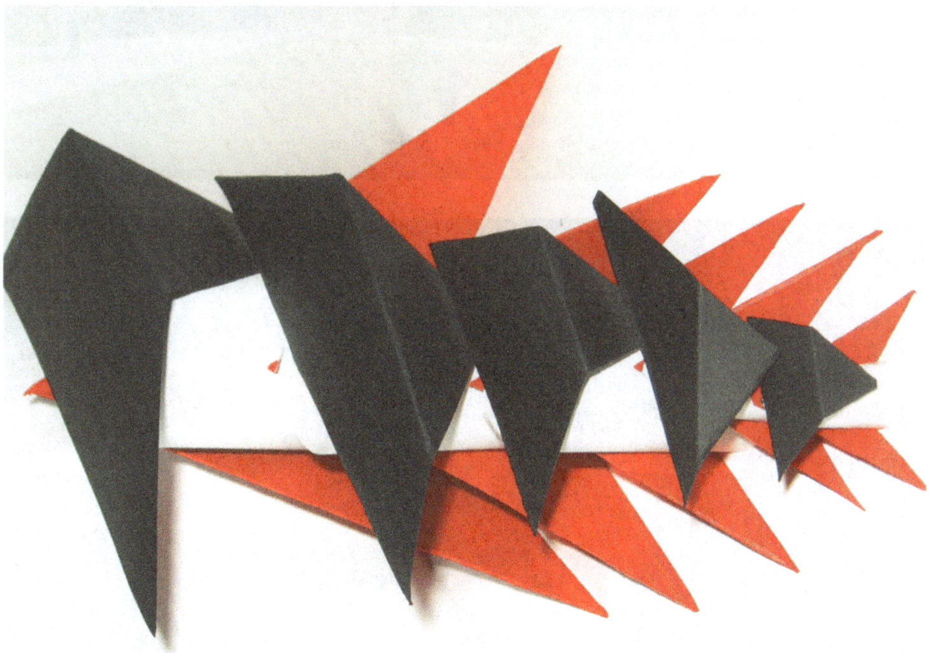

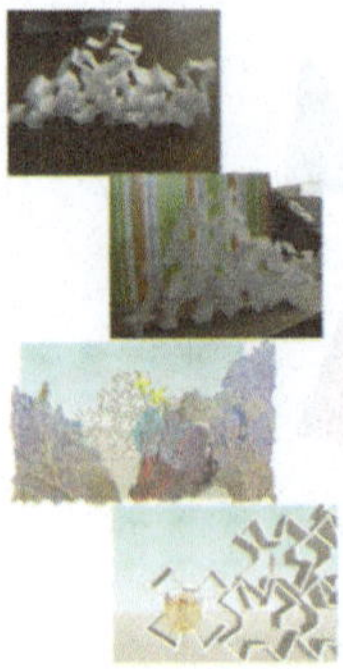

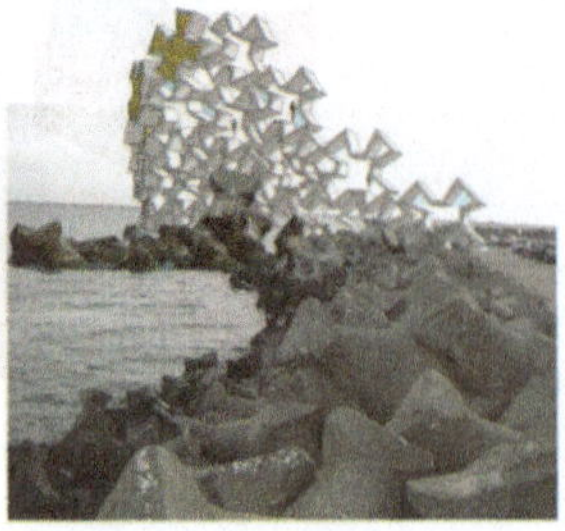

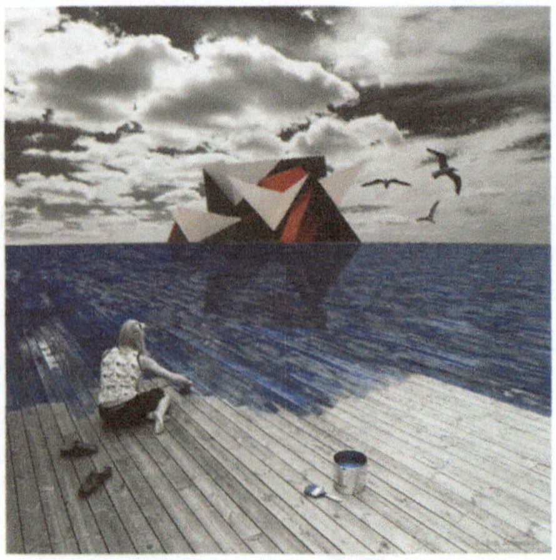

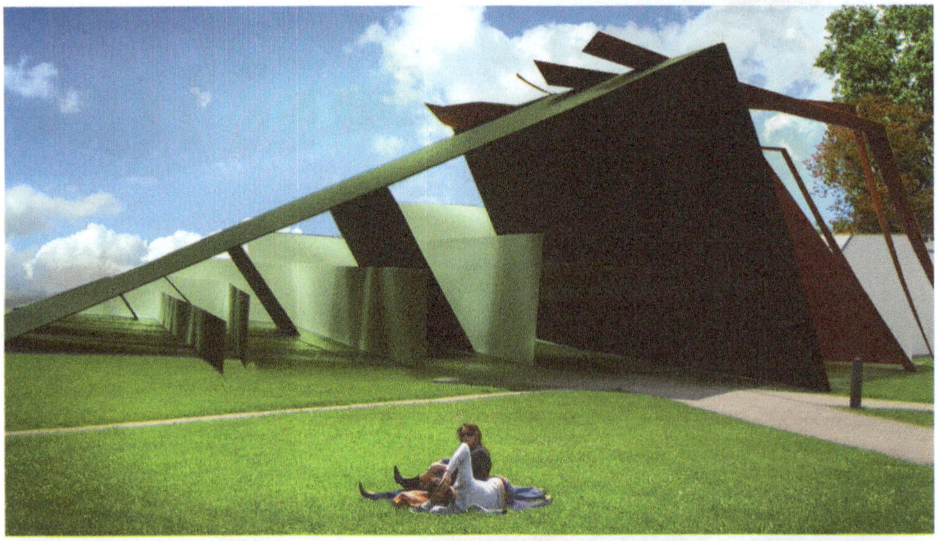

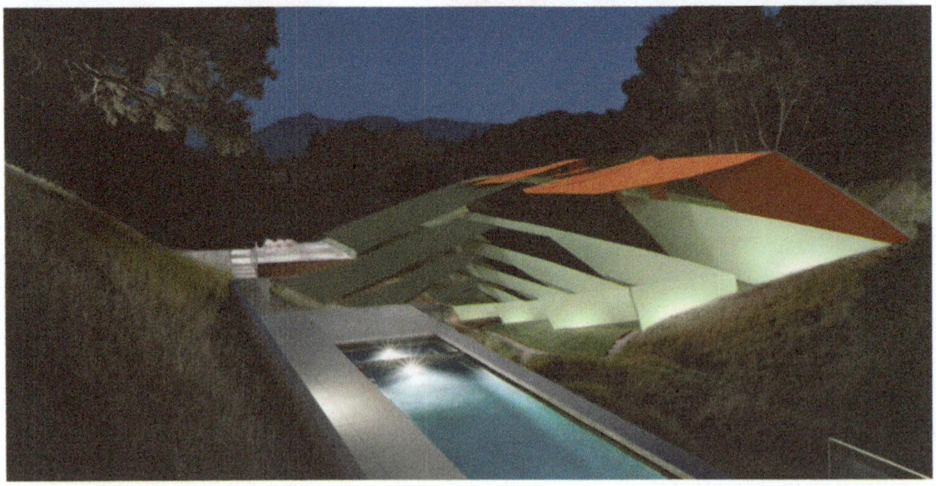

Building with Modules

Tana Lascu

If at first glance nature appears to us as a marvelous but incomprehensible and irregular world of forms, the reality is that they all come from a codified system that is capable of organizing and ordering everything that exists on the basis of simple laws. Thus, in the process of morphogenesis, nature develops strategies that we could call "genetics of the minimum," leading to simplifying and economizing through order. Concepts that we usually associate with art, and therefore with human production, such as symmetry, rhythm, proportion, etc., are nothing more than refined mechanisms of efficiency. For example, symmetry is a system of savings, which means that once an optimum form has been defined, it is repeated as many times as necessary. Rhythm, like symmetry, is another efficiency mechanism, now developed in the temporal field instead of the spatial field. Rhythmic regularity not only organizes but also facilitates efforts. Thus, the cosmos tends toward a process of cadentiality, just as the everyday life of the human earthly world is governed, in both physiological and voluntary acts, by temporal symmetry. The module or piece to be repeated is a formal shaping strategy of nature. Based on the efficient and economical concept of the module, maximum expressiveness can be achieved.

Compositions made from simple modules and their repetition make it possible to generate fabrics, self-supporting surfaces, and volumes of surprising aesthetic richness as inspiration for new architecture.

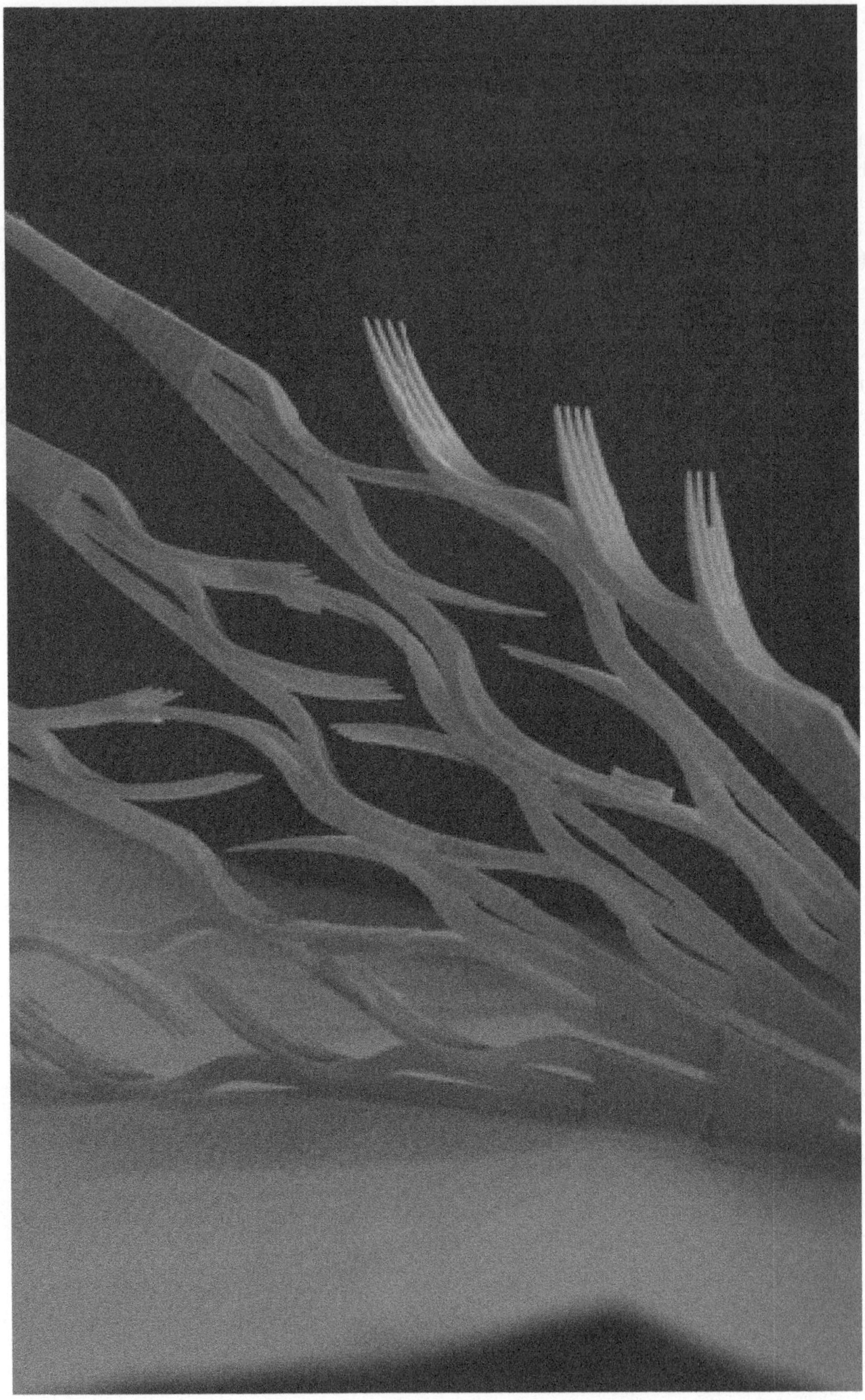

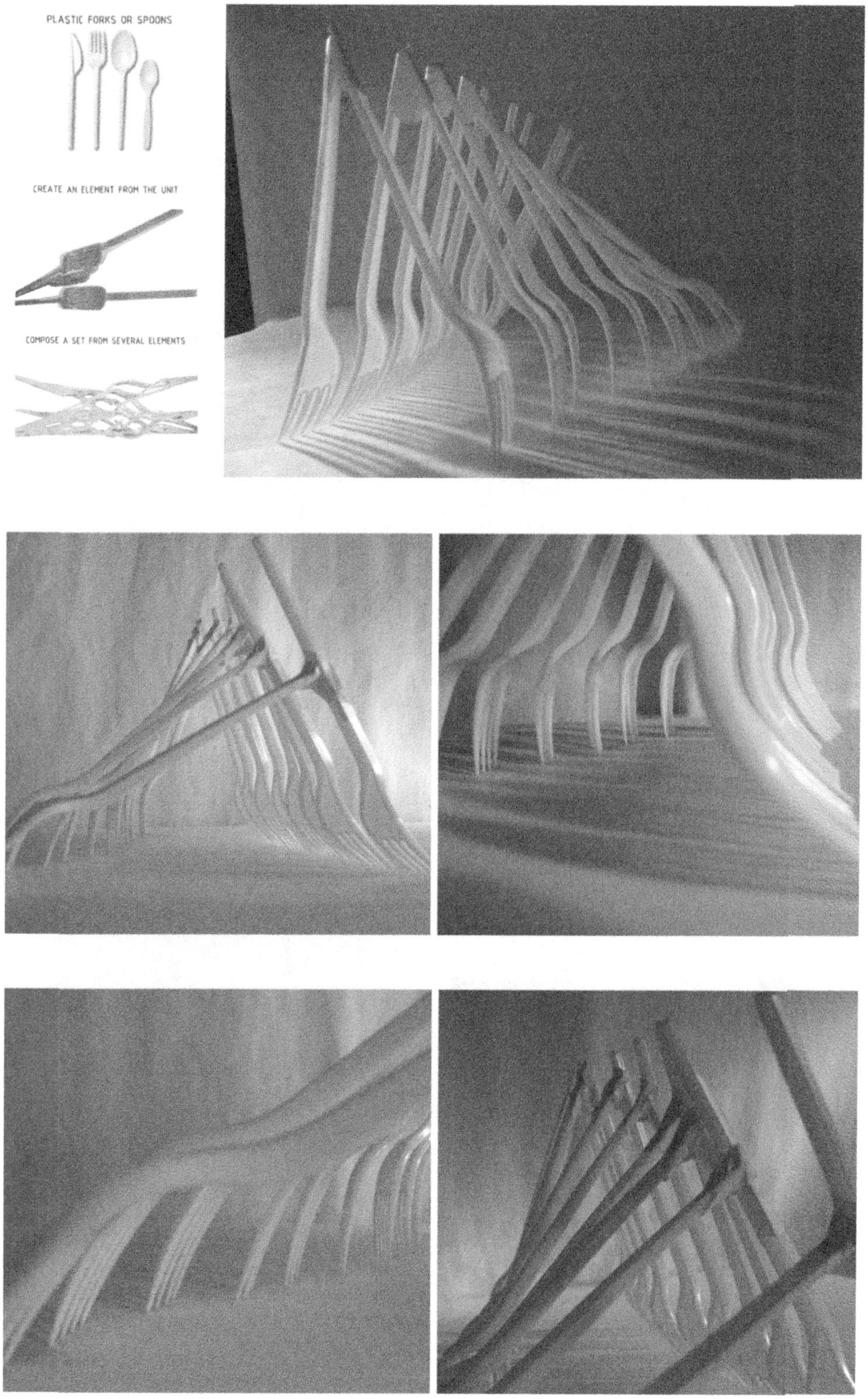
PLASTIC FORKS OR SPOONS
CREATE AN ELEMENT FROM THE UNIT
COMPOSE A SET FROM SEVERAL ELEMENTS

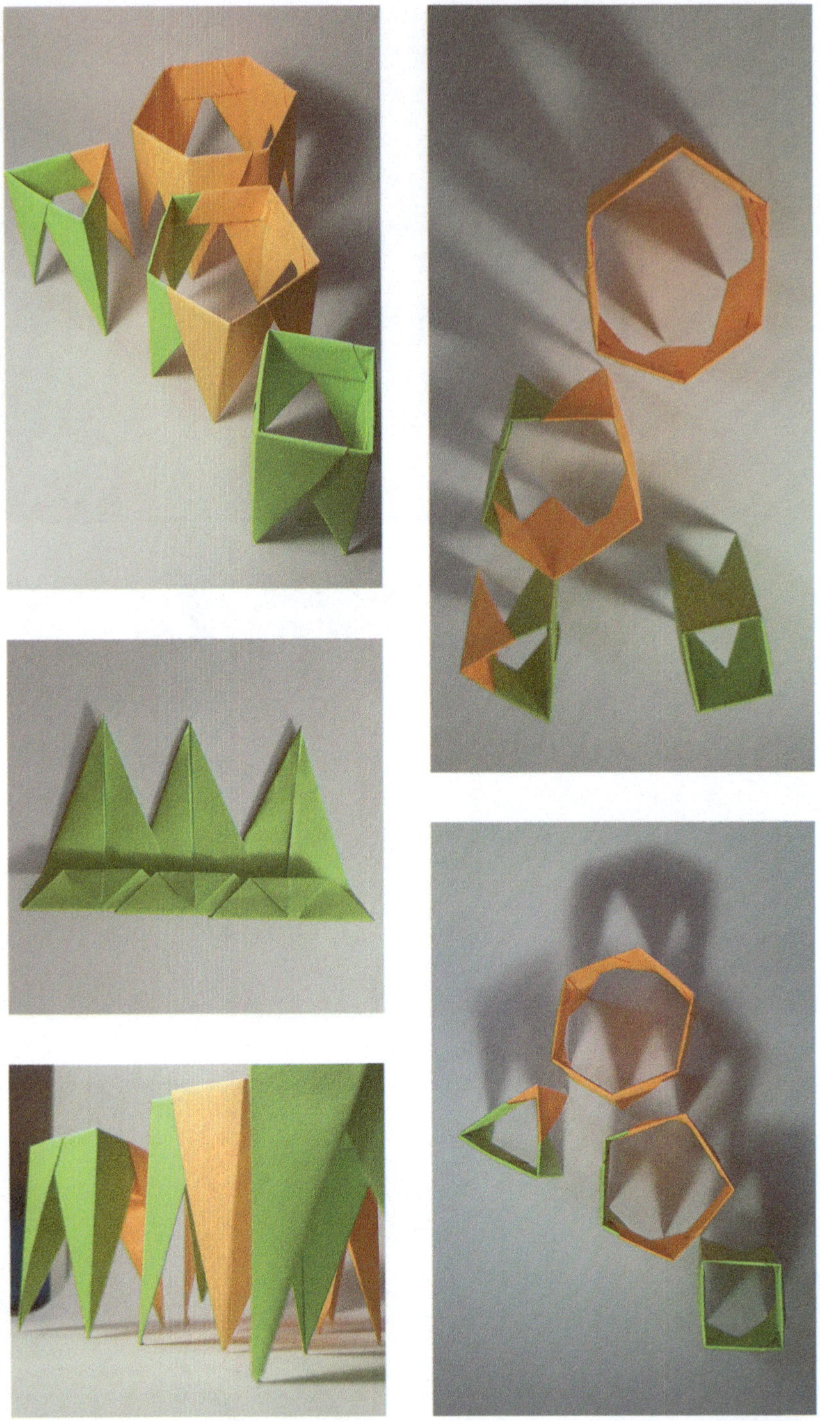

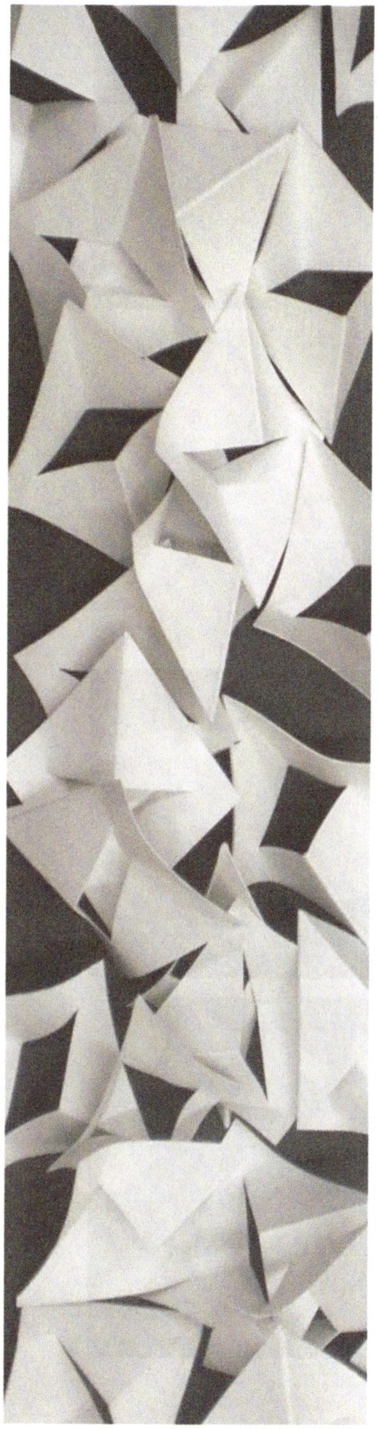

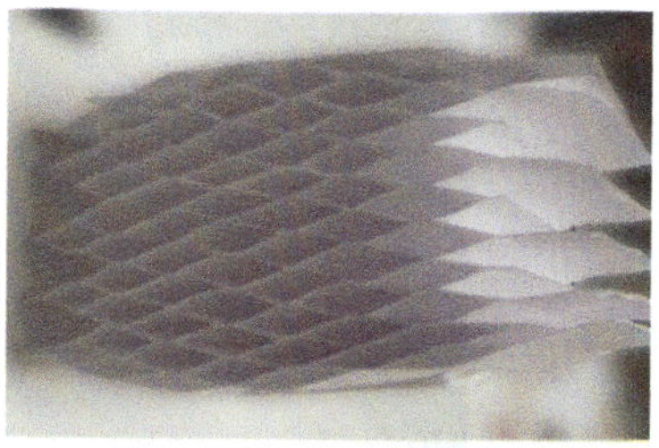

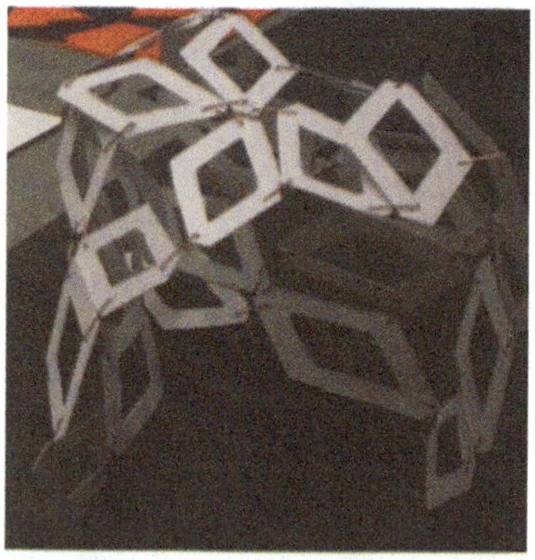

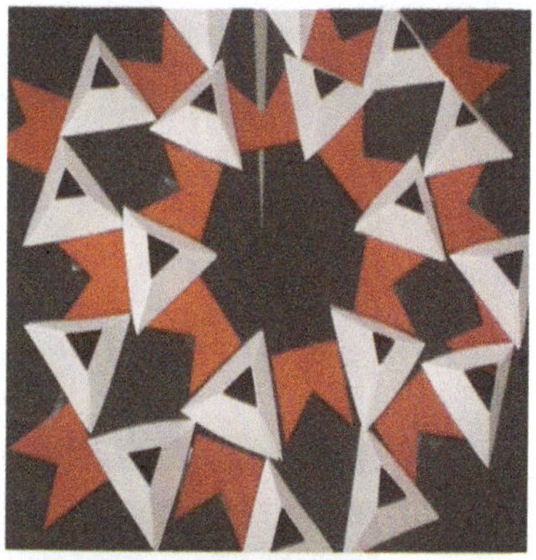

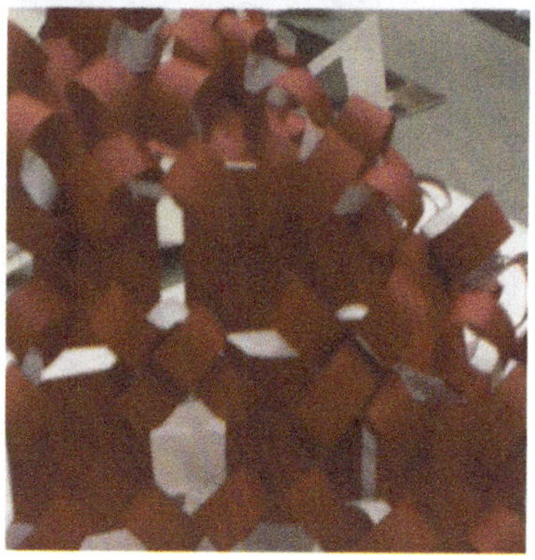

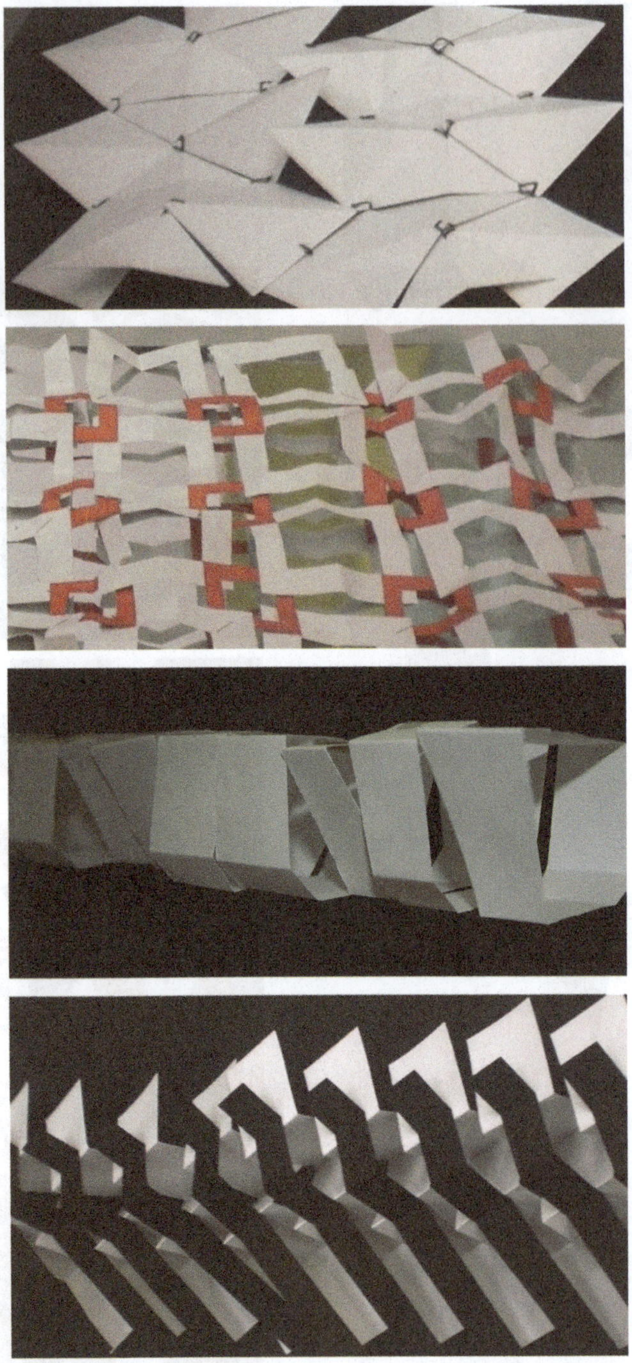

Recycling: A Tool of Nature Adapted to Design

Elena-Codina Duşoiu

Sustainability, ecology, and sustainable development are just notions trying to express a reality that we are obliged to live with: ensuring the continuity of life on our planet (and its quality). From this perspective, urbanism, architecture, and design are requested to bring solutions for life in harmony with the environment, and art rarely sounds the alarm regarding the ecological disasters that menace our planet, for example, by trying to reuse raw material that we are unfortunately used to call "garbage" (e.g., wrappers, boxes, residues, and other objects that may be used again). Such gestures aim to educate people in the spirit of natural and cultural values of the planet, which may constitute the true definition of ecology.

We can tell the history of recycling in architecture and construction from antiquity until now. The historical reuse of building materials and construction elements, known in architectural history with the Italian term "Reimpiego dei materiali," is nothing else but large-scale recycling, transforming architecture into a huge mosaique of pieces and techniques.

Modern and contemporary architects and designers produced original styles starting from recycling, such as A. Gaudi, who used small ceramic and glass pieces taken literally from garbage, or Shigeru Ban, who recently received the Pritzker Prize for imposing an architectural conception based on recycling and reuse of containers, boxes, modules, and other materials.

Recycling can be adapted to a variety of scales, from urban conception to the architectural scale (e.g., walls including bottles, rubber wheels, pets, etc.) and design. There are several famous contemporary prototypes of furniture conceived and made solely with recycled materials (e.g., Frank Gehry, Shigeru Ban, etc.). Urban furniture has embraced the idea of recycling as well, promoting this type of conception in public space and turning the city into a setting for education. Visual artists have also produced many installations and sculptures created by assembling recycled products and containers. With the use of common waste materials that belong to our everyday life, such as cans, cartons, plastics, disposable cups etc., artistic and architectural composition can be stimulated while, at the same time, the importance of recycling is promoted.

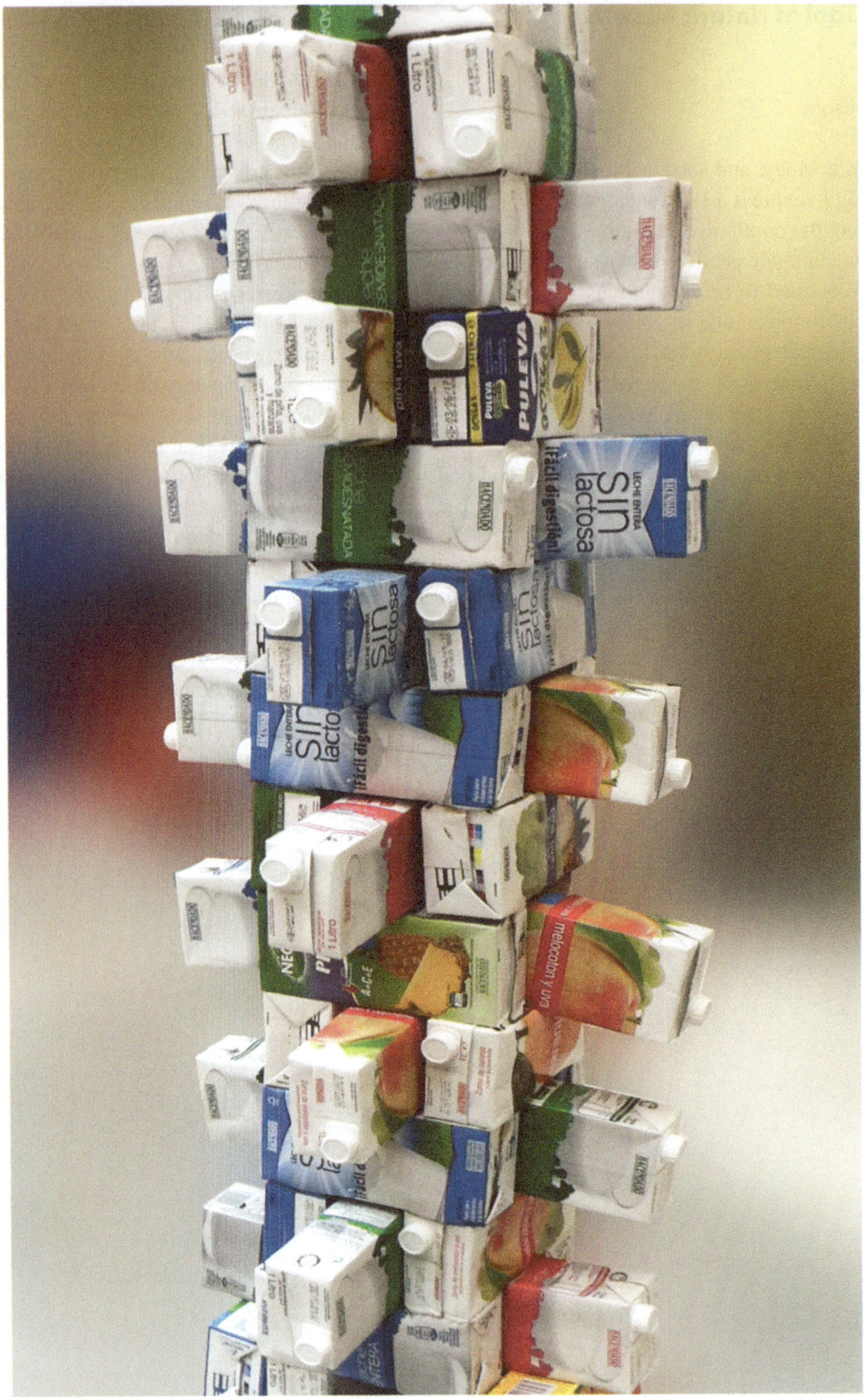
1 Litro
PULEVA
LECHE ENTERA
sin
lactosa
fácil digestión
melocotón y uva

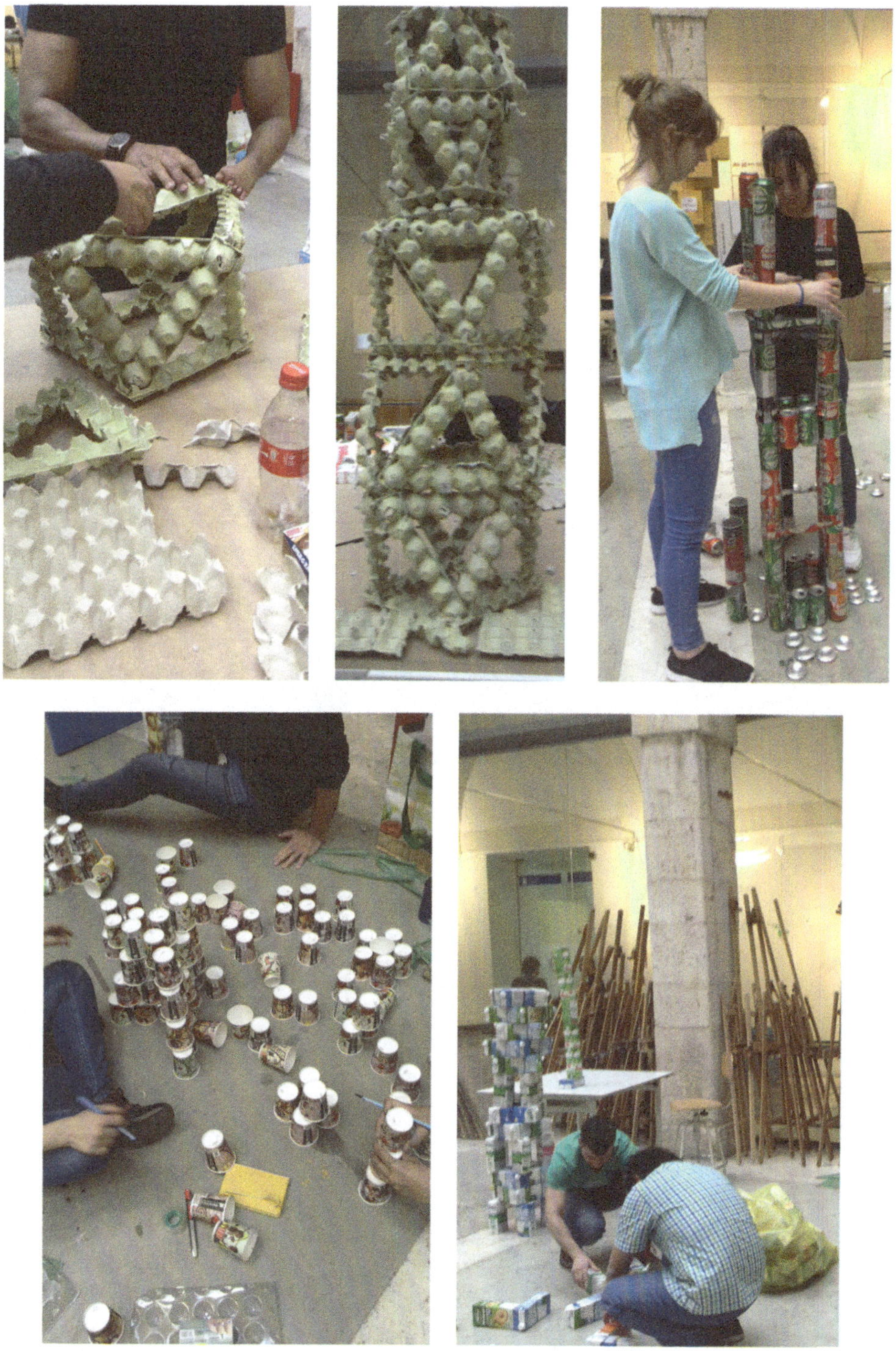

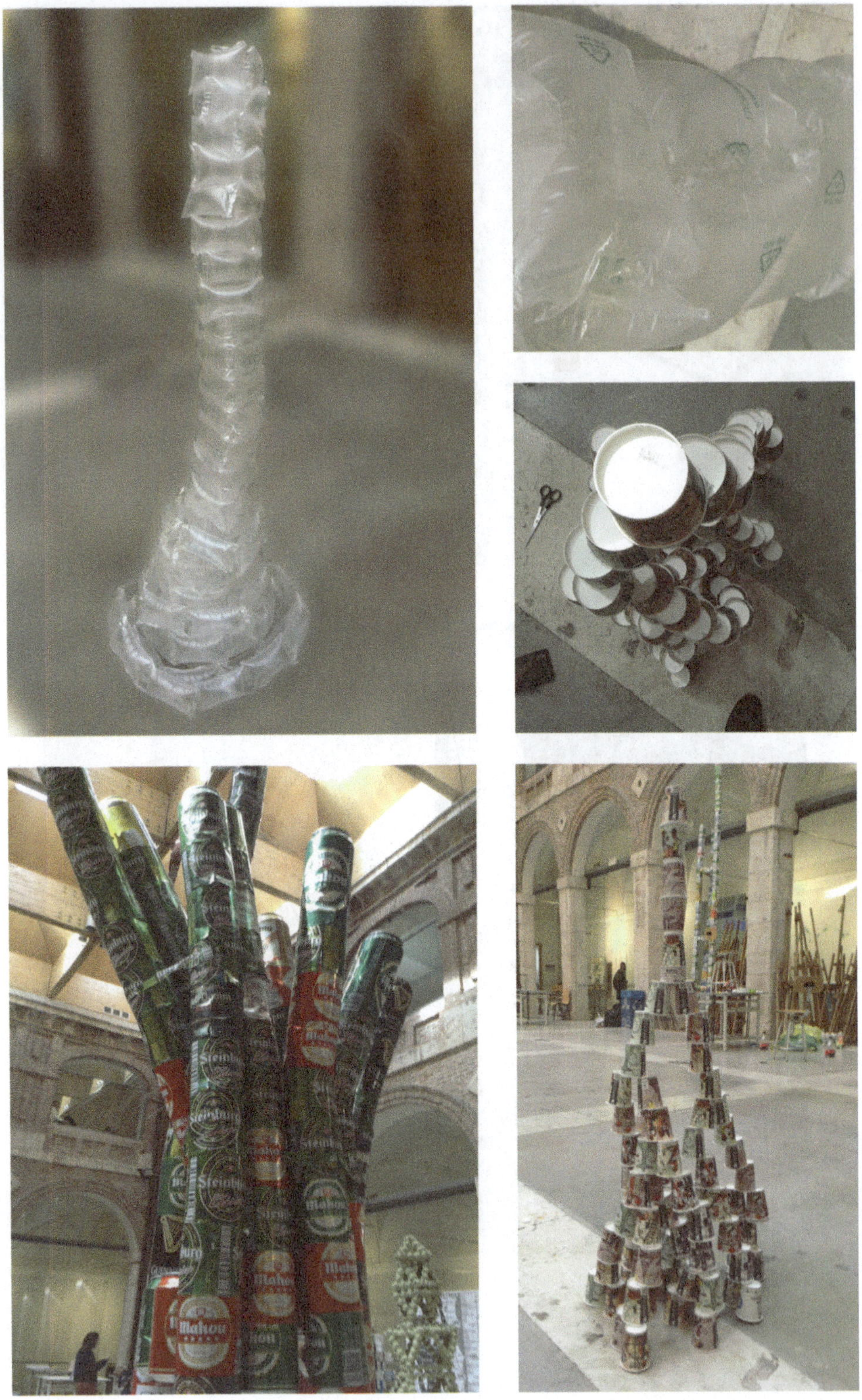
Mahou

Bionic Perception of Light: From Nature to Design

Elena-Codina Duşoiu

The presence of light is fundamental in the perception of architecture and public space. Le Corbusier included light as an essential element in the definition of architecture: "Architecture is the masterly, correct and magnificent play of masses brought together in light".

We can also question ourselves how deep our perception of light is and how it influences our representation of space.

The human sight is based on the detection of different wavelengths of light using three receptors from our eyes (cone cells): shortwave (blue light), mediumwave (green light), and longwave (red light). And the specific stimulation of these types of receptors provokes the colored ambience we perceive. This has been the basis of the digital, computer-controlled image (RGB spectrum). However, there are species of animals endowed with much more accuracy. Mammifers as the mandrills, birds as the peacock, or insects as some species of butterfly have four types of cone cells instead of three, which provides them with a much more accurate perception of the colors. Other species are provided with a complete system of sight different to the camera-type eye of humans (e.g., the composed eye of the insects, which are made out of a number of tiny lenses, or the multitude of ocelli, which spread over the body of certain species of insects and mollusks). Some "eyes" of reptiles also have thermoregulating functions (as is the case of the tuatara). It is a challenge to translate all these patterns to architectural space, studying it in terms of color spectrum, multiplication of image, focalization, polarization, geometric 180 or 360 degrees perception, etc.

Six different approaches of working with light in architectural space are presented: control of different wavelengths; light conceived as a flexible membrane; the relation between sun and shadow; and light reflected in the sky, for example, the pattern of clouds; the relation of the skin with light; and light in the dark ("The myth of the cavern" by Plato).

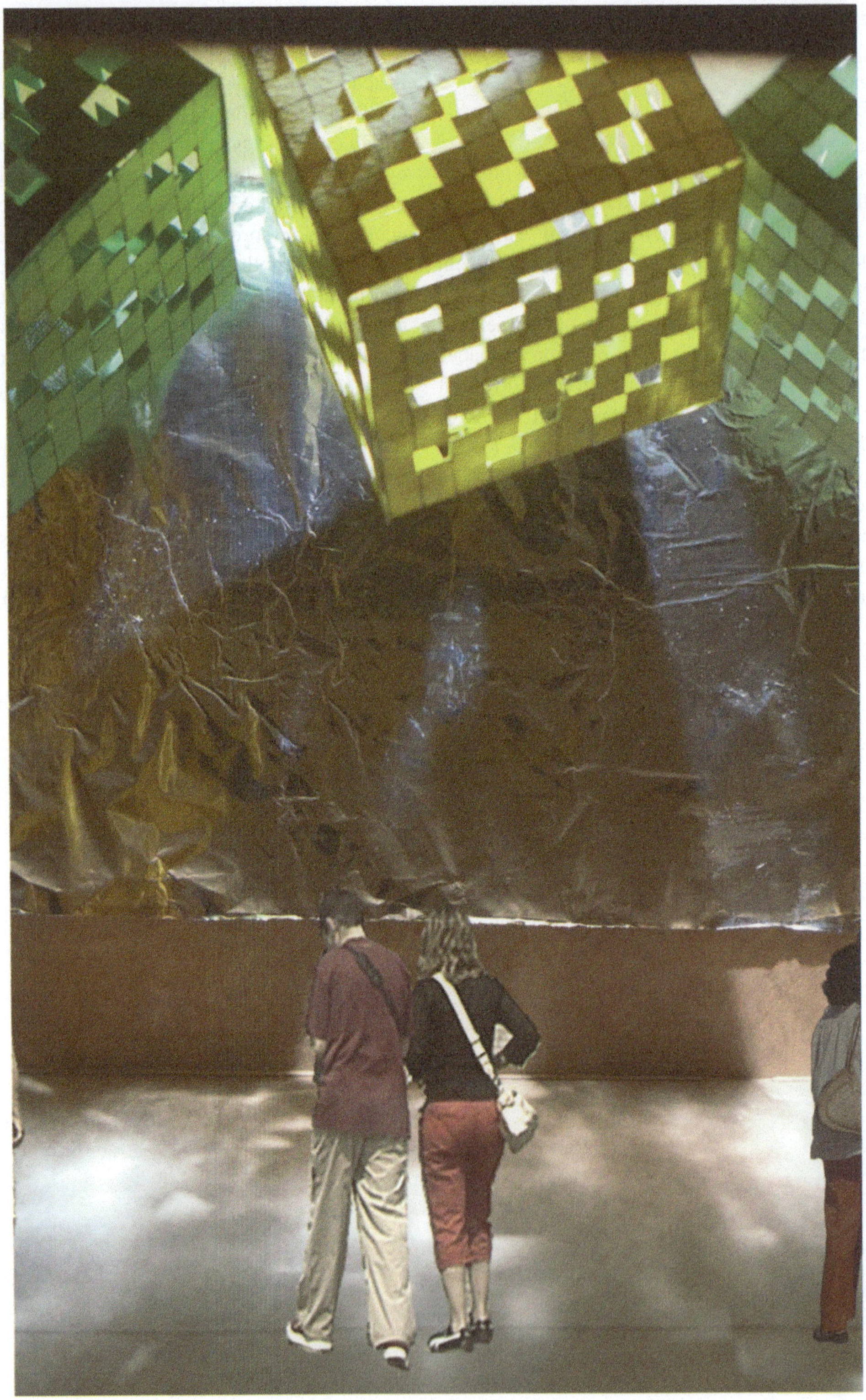

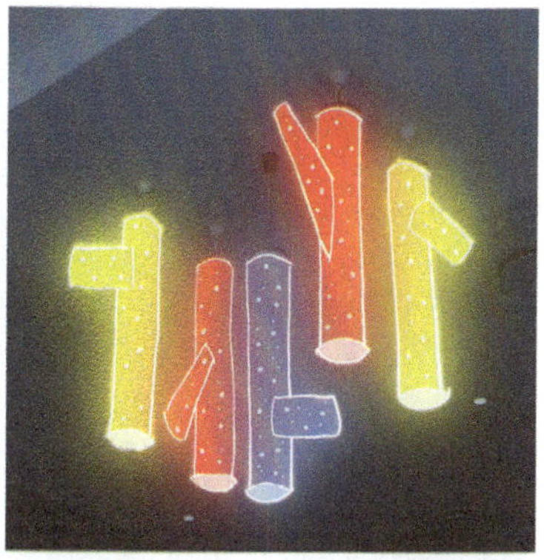
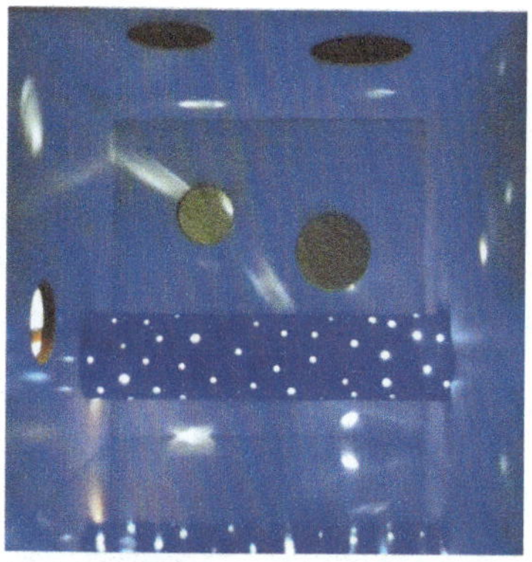
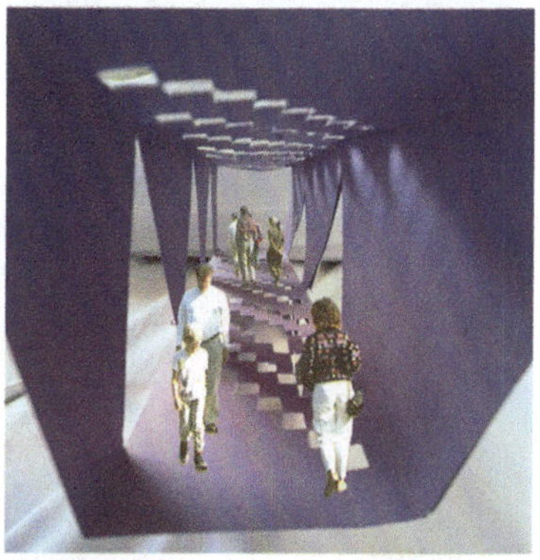
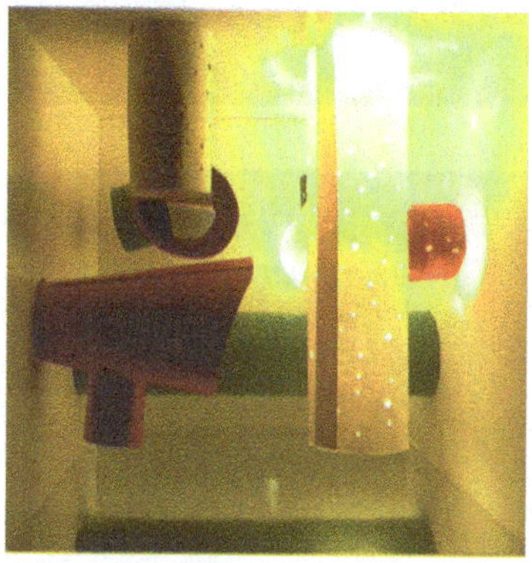
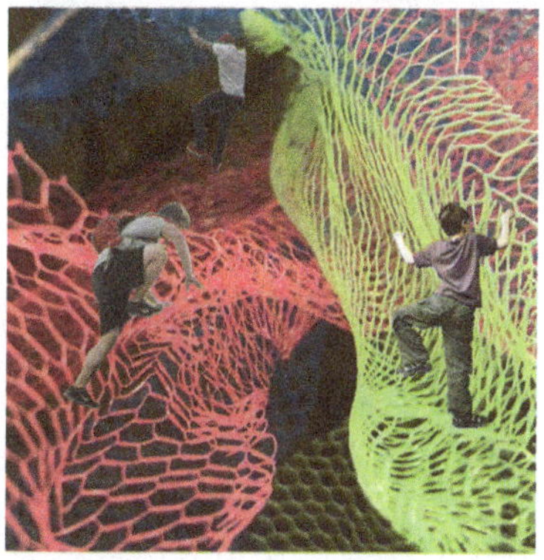
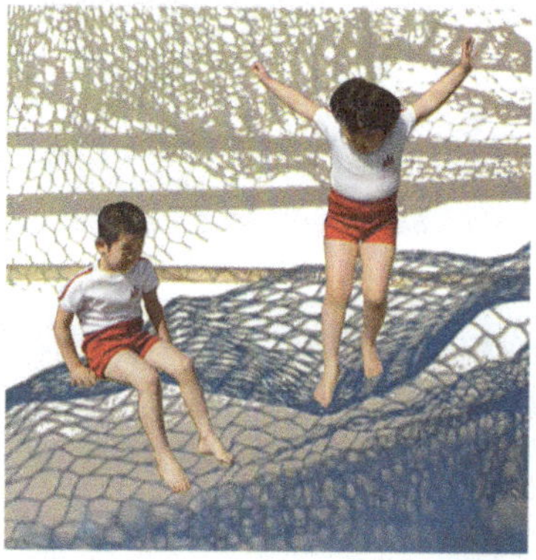
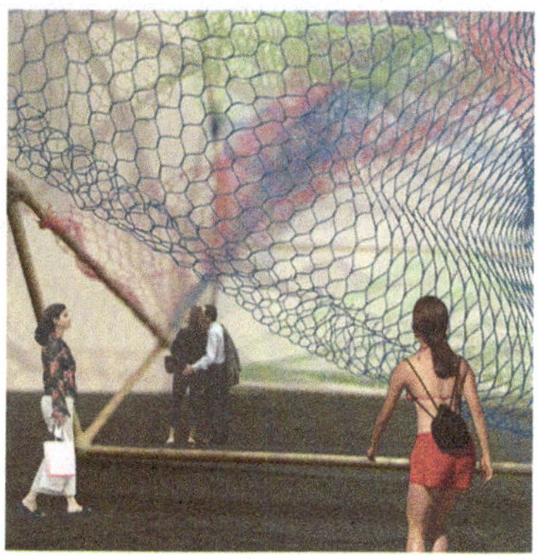
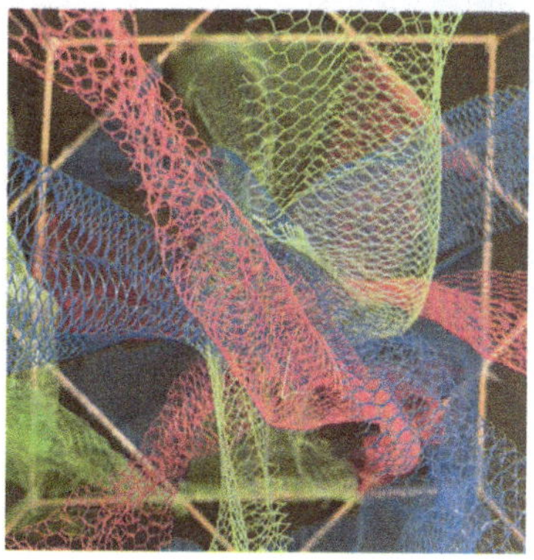

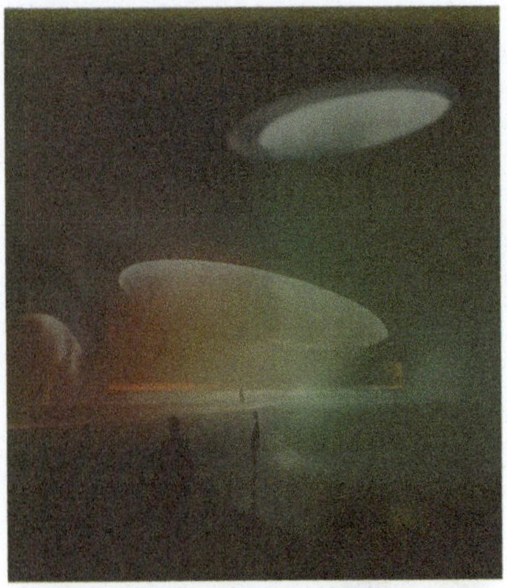

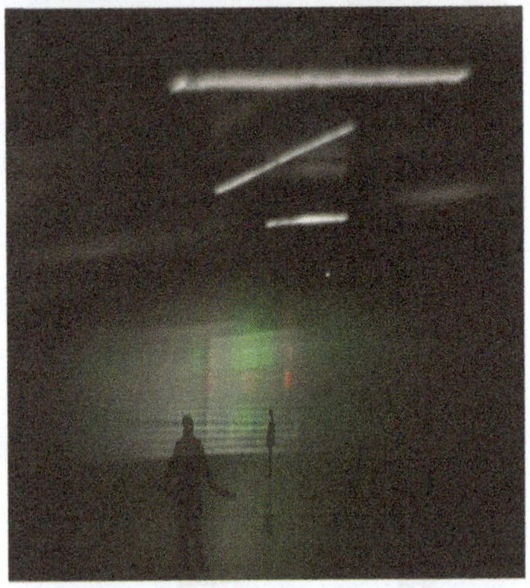
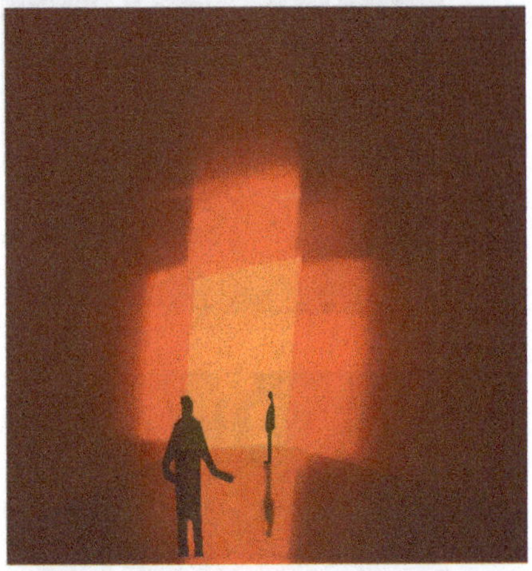
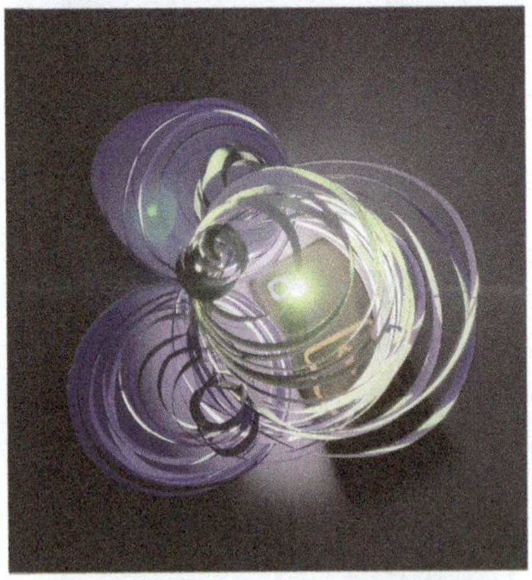
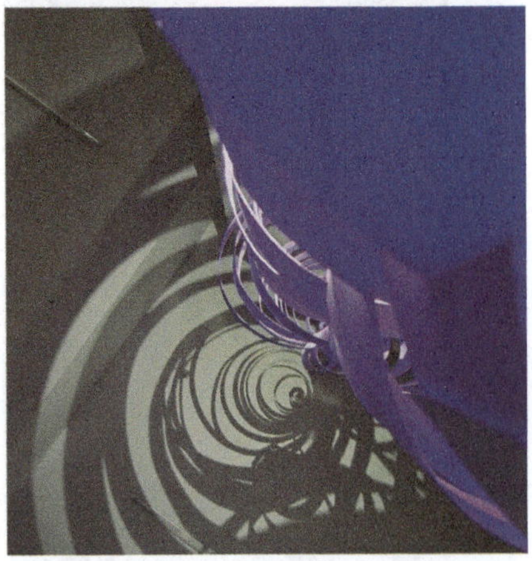

Lightweight Structures

María Rosa Cervera Sardá

Rarely do we find massive and heavy structural forms in nature. Generating matter consumes a great deal of energy and nature tends to minimize stress. For this reason, their forms are spongy, using air as a cooperating element of the structure. And for this reason, their shapes are made up of multiple light and cooperating elements. The principle is very simple, not to start from any excessive gesture and to achieve the maximum performance with the minimum. If we apply this theory to the structural field, we will realize the existence of a system, which we can call bio-structure, in which the concept of "all-resistant" cooperation prevails, generating an interweaving of multiple resistant elements in such a way that, as the solicitations increase, this framework implements the number of its structural pieces and reduces its size, becoming a structure of great resistance and, at the same time, of great lightness. The structures of bones, plant tissues, and wood fibers are all made up of fine elements of fractal arrangement, where the void is as important as the matter.

Resistance can be obtained with extremely light materials, such as a sheet of paper or a soda straw. Simply by the geometric organization or by the collaboration of the elements, unexpected weights and loads can be supported, which opens doors to new understandings of structural systems.

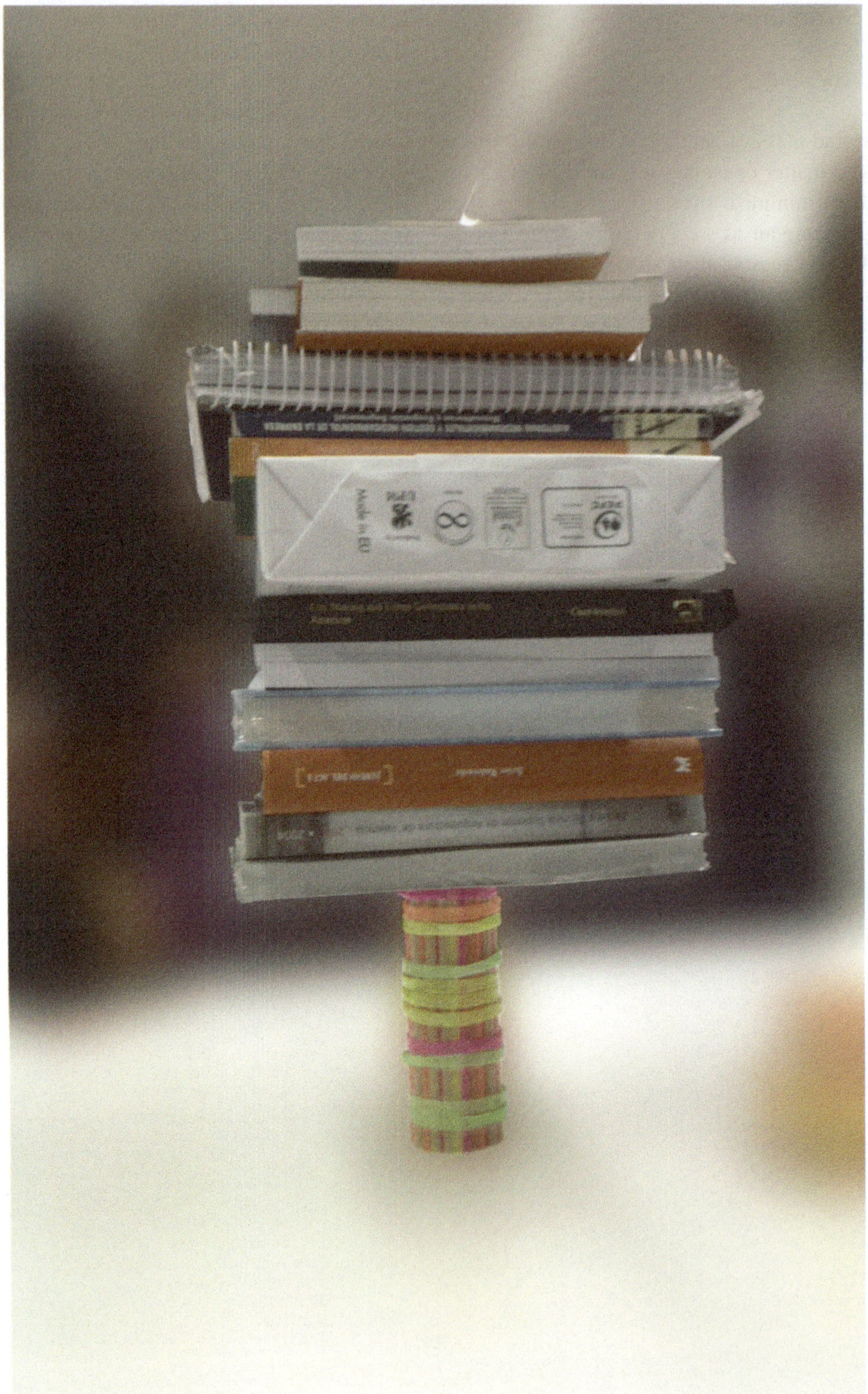

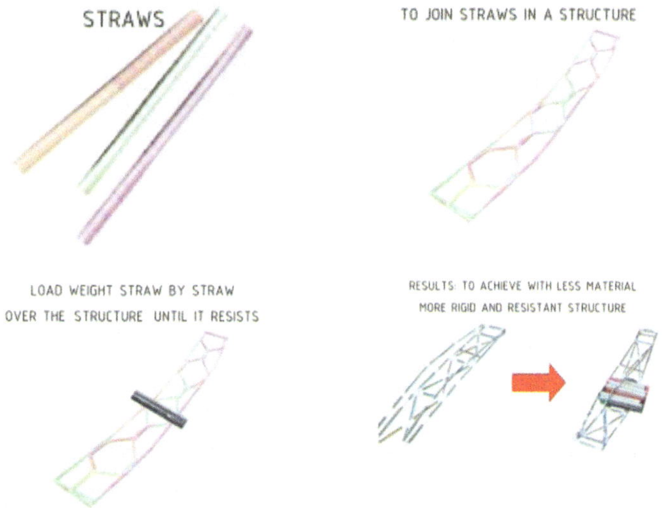
STRAWS
TO JOIN STRAWS IN A STRUCTURE
LOAD WEIGHT STRAW BY STRAW
OVER THE STRUCTURE UNTIL IT RESISTS
RESULTS: TO ACHIEVE WITH LESS MATERIAL
MORE RIGID AND RESISTANT STRUCTURE

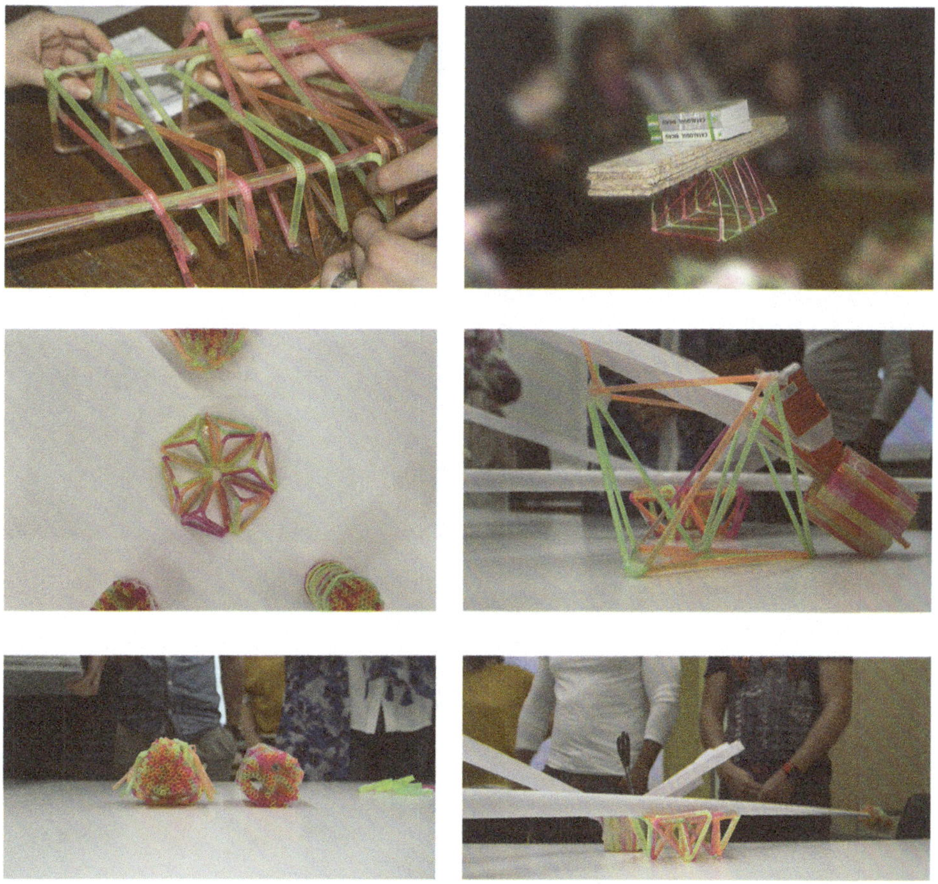

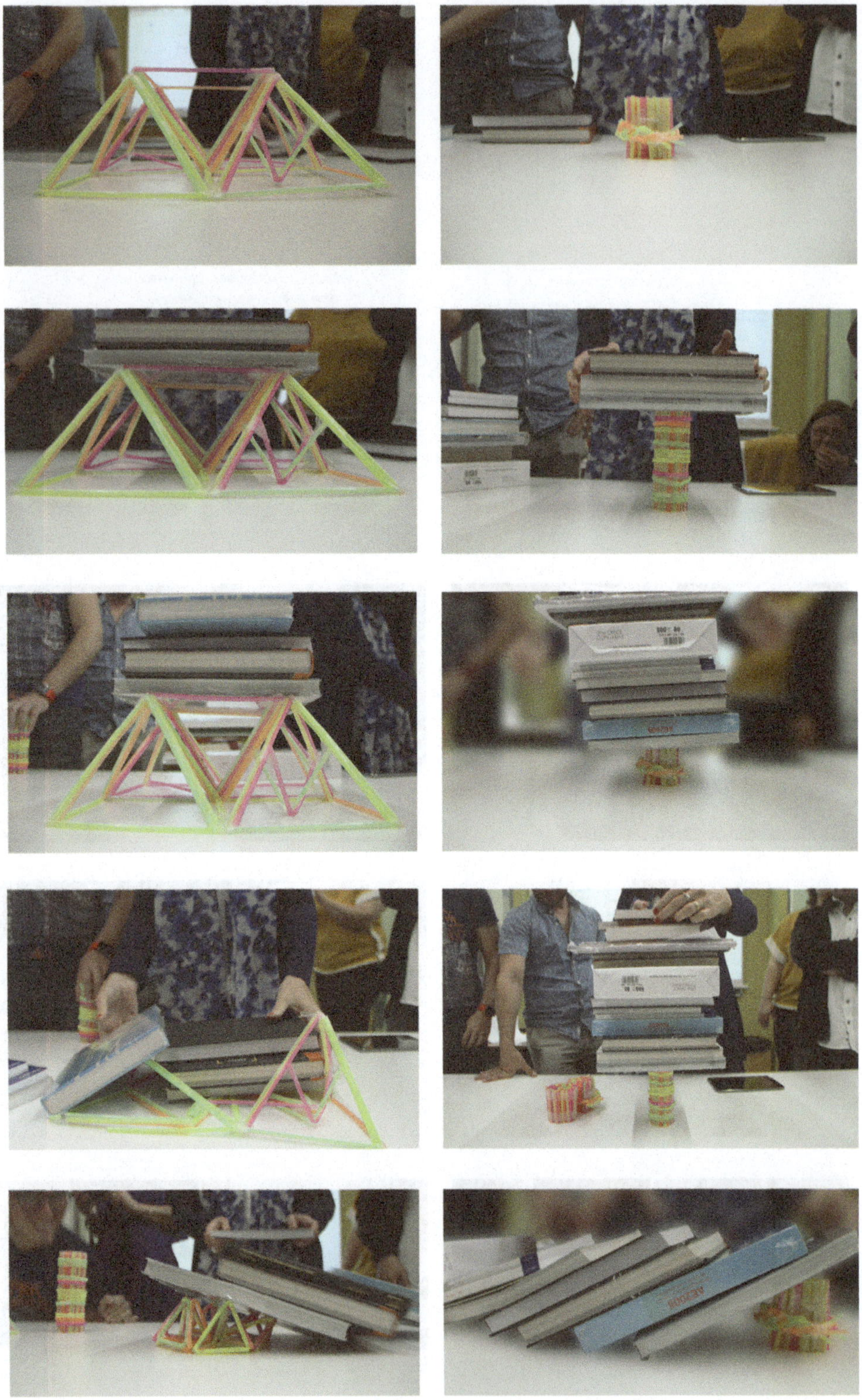

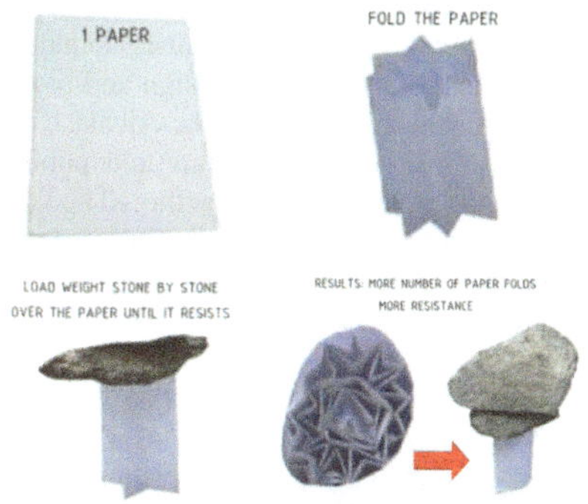
1 PAPER
FOLD THE PAPER
LOAD WEIGHT STONE BY STONE
OVER THE PAPER UNTIL IT RESISTS
RESULTS: MORE NUMBER OF PAPER FOLDS
MORE RESISTANCE

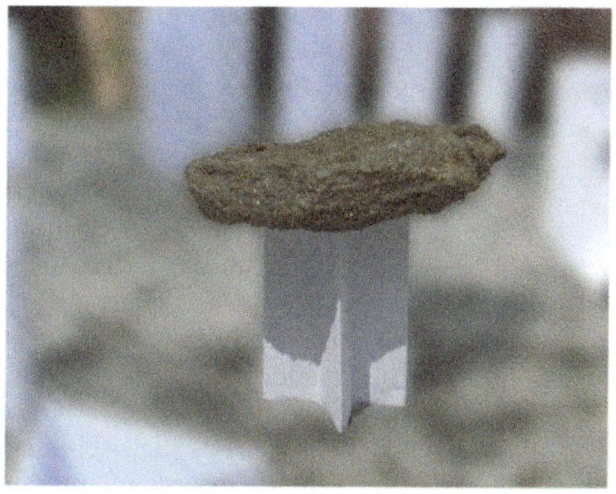

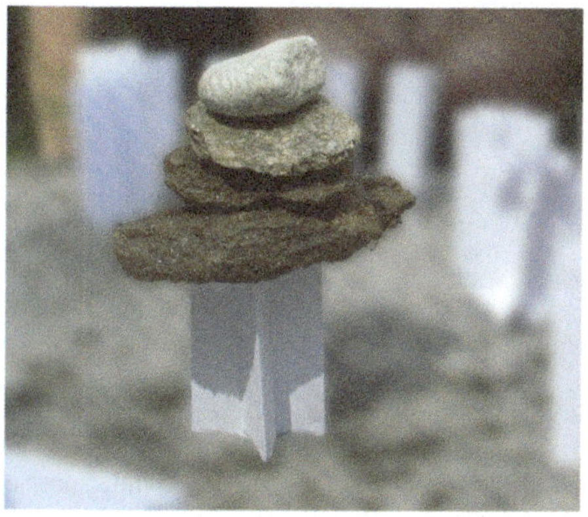

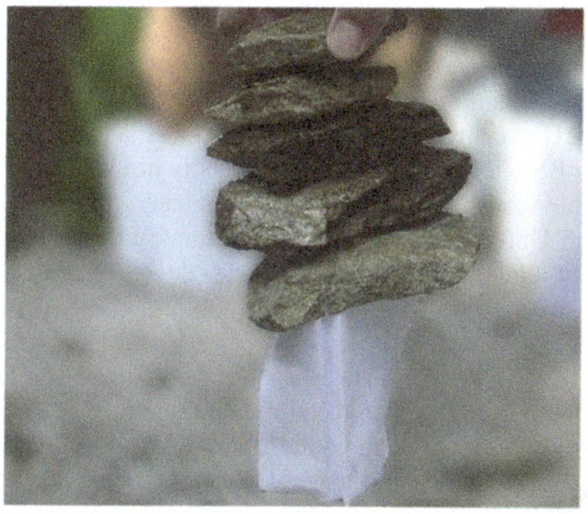

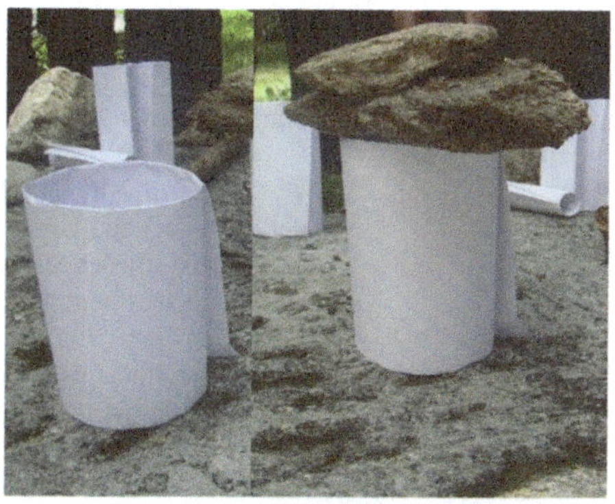

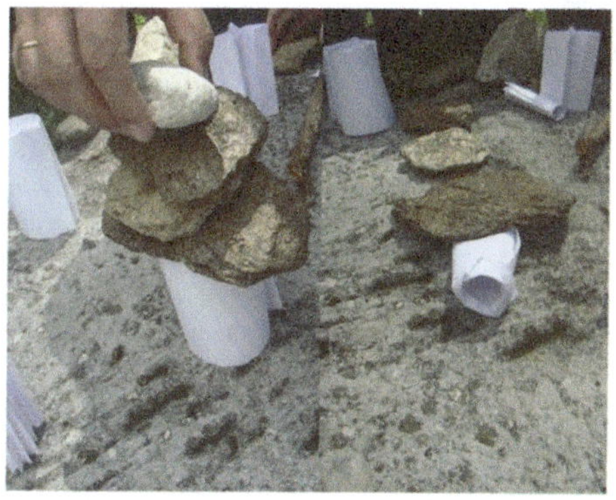

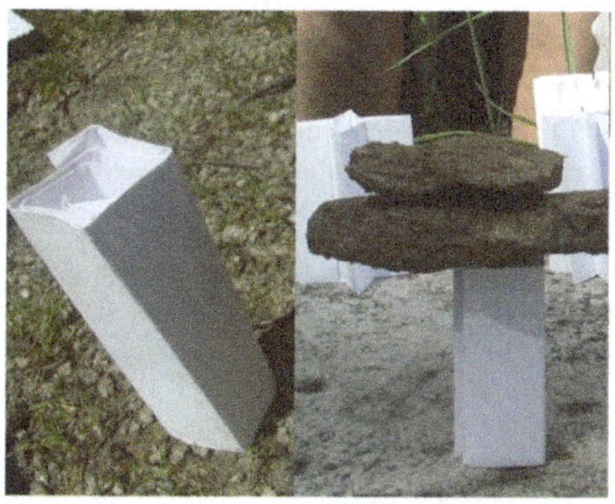

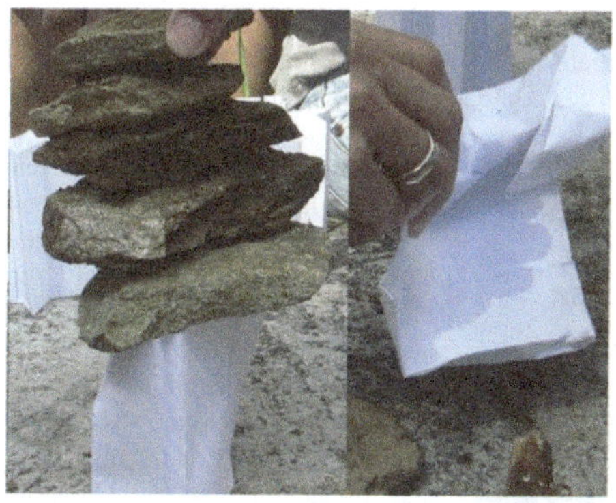

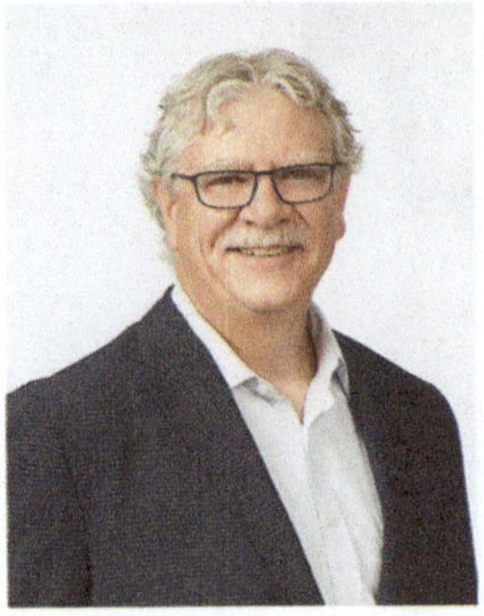

Mark P. Sarkisian PE, SE, NAE, LEED BD+C, Partner of Seismic and Structural Engineering in the San Francisco office of Skidmore, Owings & Merrill LLP, has developed innovative engineering solutions for over 100 major building projects around the world including some of the tallest and most complex. Mark holds ten US Patents and five International Patents for high-performance seismic structural mechanisms designed to protect buildings in areas of high seismicity and for seismic and environmentally responsible structural systems. In 2021, Mark was elected to the prestigious United States' National Academy of Engineering. He is the author of *Designing Tall Buildings – Structure as Architecture*, and teaches at UC Berkeley, California College of the Arts, Stanford University, Cal Poly, Northeastern University, NC State University, the Pratt Institute and Alcala University in Madrid. He has a BS-CE Degree from the University of Connecticut and is a Fellow of the Academy of Distinguished Engineers, an MS-SE Degree from Lehigh University, an Honorary Doctorate Degree from Clarkson University and an Honorary Master's degree from the Politecnico di Milano.

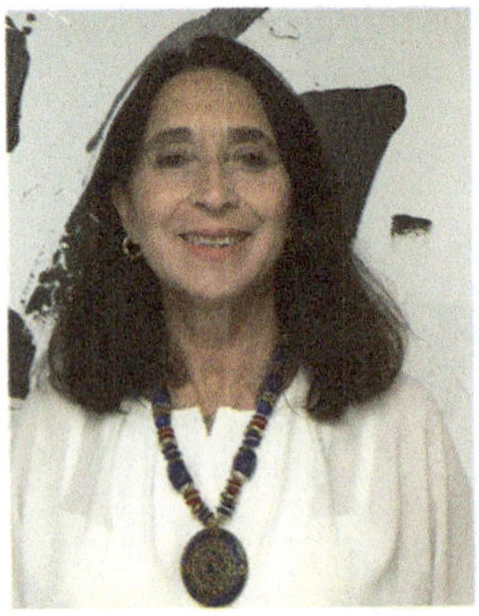

María Rosa Cervera Sardá Ph.D. Architect, Professor and former Dean of the School of Architecture of the University of Alcalá. Current Director of the Master's degree in Advanced Architecture and City Projects. She is the author of writings and books on architecture, among which we highlight: *Bionics, Biomimetics and Architecture* (2019); *Space and Time in Architectural Composition* (2018); *Recycling Mumbai. Re-envisioning the Slum* (2012); *Madrid, Recycled City* (2011); *Iron in 19th Century Madrid Architecture* (2006). A regular speaker in professional and academic circles, she has given lectures in Spain, China, India, USA, Bolivia, Peru, El Salvador, Venezuela, Romania, Italy, etc.

As an architect, Rosa Cervera has received several awards – the "Antonio Maura" award, the COAM award, the "Transfer of Knowledge" award – and has won several international architecture competitions.

Rosa Cervera is a pioneer in the research of Bionics, Biomimetics and the application of biological structures to innovative and efficient architecture and urban design. A direct result of these studies is the Self-Sustainable Vertical Garden City, Bionic Tower.

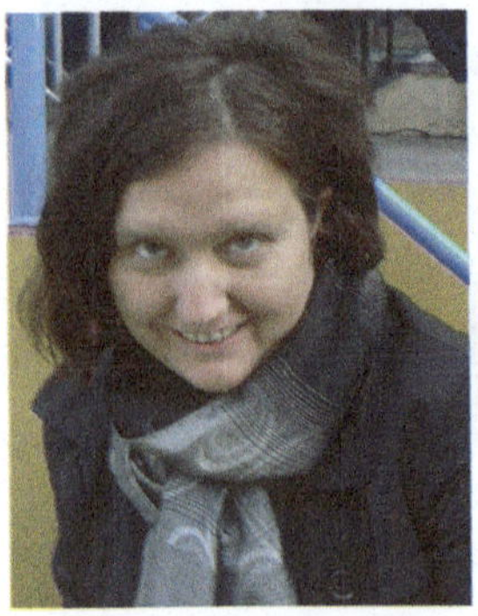

Elena-Codina Dușoiu is an Architect (graduated in 2000, IMUAU) and Professor within the "Ion Mincu" University of Architecture and Urban Planning from Bucharest, Romania. She also accomplished study periods within the Polytechnic University of Catalonia, Barcelona (Master in Restoration of Monuments, 2001) and Venice International University (*Dottorato di Eccelenza. Storia della Citta, dell'Architettura e del Restauro,* 2004). Her main fields of research are: rehabilitation and conversion of buildings (published books: *The Dynamics of the Sacred Space. The Influence of Function*, *3 Breweries and Their Destiny*, Ion Mincu Publishing House, 2009), vernacular architecture and ecology (organizer of 12 editions of Spanish-Romanian workshops on vernacular architecture 2006-2018), design and bionics (research studies realized within the IMUAU and the Alcalá University, 2021–2022). She received various research grants and published about 100 scientific articles and 8 authored and co-authored books in the mentioned research fields. Outstanding activity as visiting professor in Spain, Italy, Greece, Czech Republic, Liechtenstein, Argentina etc. Several national and international prizes and nominations in architectural research and architectural design, owner of a personal architecture studio since 2003.

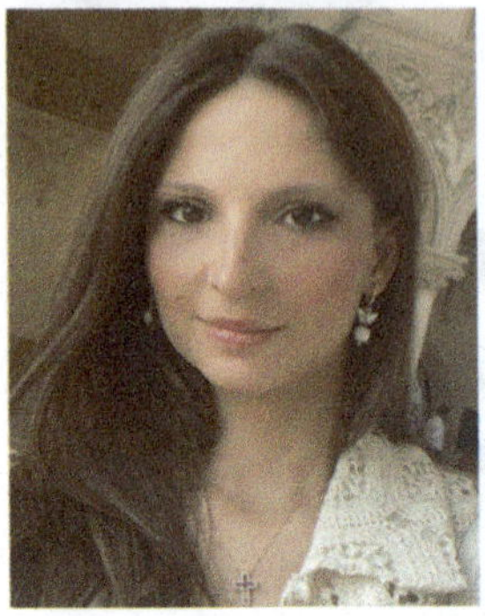

Ana Mohonea is an Interior Architect, graduated from the Faculty of Interior Architecture of the "Ion Mincu" University of Architecture and Urban Planning in Bucharest, Romania.

She studied her undergraduate, master and doctorate at the "Ion Mincu" University, where she is currently working as an Assistant Prof. Ph.D. Arch. She was awarded the "Bene Merenti" distinction for the high grades obtained in her studies and the Medal of the "Ion Mincu" University for her diploma project.

Ana Mohonea was a pioneer of research in bionic architecture in Romania. Her doctorate thesis is titled: "Bionic Architecture. Premises for Sustainable Building", IMUAU.

She is a member of the Interior Architecture Society from Romania, collaborator with ROGBC (Romania Green Building Council) and former BREEAM international assessor. She participated in many competitions, projects, summer schools, architecture biennales and international workshops and received many awards, among which was the first Prize at National Level of the 8th edition of the Design competition for avant-garde furniture for her *Guitar Chair*.

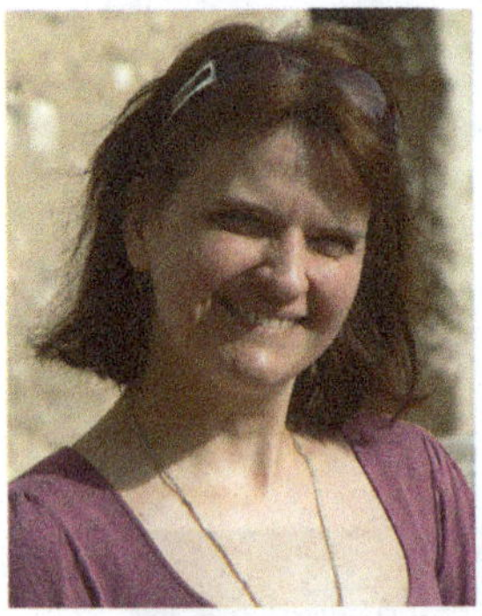

Tana Nicoleta Lascu Ph.D. Architect and Urban Planner, Lecturer at the Basics of Architectural Design Department, Faculty of Architecture of the "Ion Mincu" University of Architecture and Urban Planning from Bucharest.

Graduated at IMUAU in 1992 and having a postgraduate specialization in restoration and conservation of monuments and historic sites, she has been involved in several restoration projects in Romania, France, and Italy, thereafter developing more than 100 projects during the 7-year activity at Cornelis de Jong Architektenburo bna, Middenbeemster, Noord-Holland, the most significant in Urk, De Rijp, Marken and Beemster, elaborating the project for the listing of Beemster polder on the UNESCO World Heritage List.

She was a Visiting Professor at the University of Liège (2014), the University of Architecture in Venice – IUAV, and the University of Alcalá (2021).

Since 2012, she has represented IMUAU within The Network of Universities for Studies and Education according to the European Landscape Convention – UNISCAPE.

Her doctoral thesis "Landscape as an Integrated Concept in the Sustainable Development", finalized in 2011, and her studies published in four books and over 20 international conference Proceedings, as well as the three research grants and more than 50 international curated exhibitions, seminars and workshops emerged as a result of her constant interest in connecting architecture and landscape within a transdisciplinary integrated approach.

Publisher Correction to: Experimental Bionics Workshops

Mark P. Sarkisian, María Rosa Cervera Sardá, Elena-Codina Duşoiu, Ana Mohonea,
and Tana Nicoleta Lascu

Publisher Correction to:
Chapter 21 in: M. R. Cervera Sarda et al. (eds.), *Architecture Inspired by Nature*,
https://doi.org/10.1007/978-3-031-33144-2_21

The original version of the chapter "Experimental Bionics Workshops" was inadvertently published without figures. The chapter has now been updated with all figures.

The updated original version for this chapter can be found at
https://doi.org/10.1007/978-3-031-33144-2_21

M. R. Cervera Sarda et al. (eds.), *Architecture Inspired by Nature*, https://doi.org/10.1007/978-3-031-33144-2_22

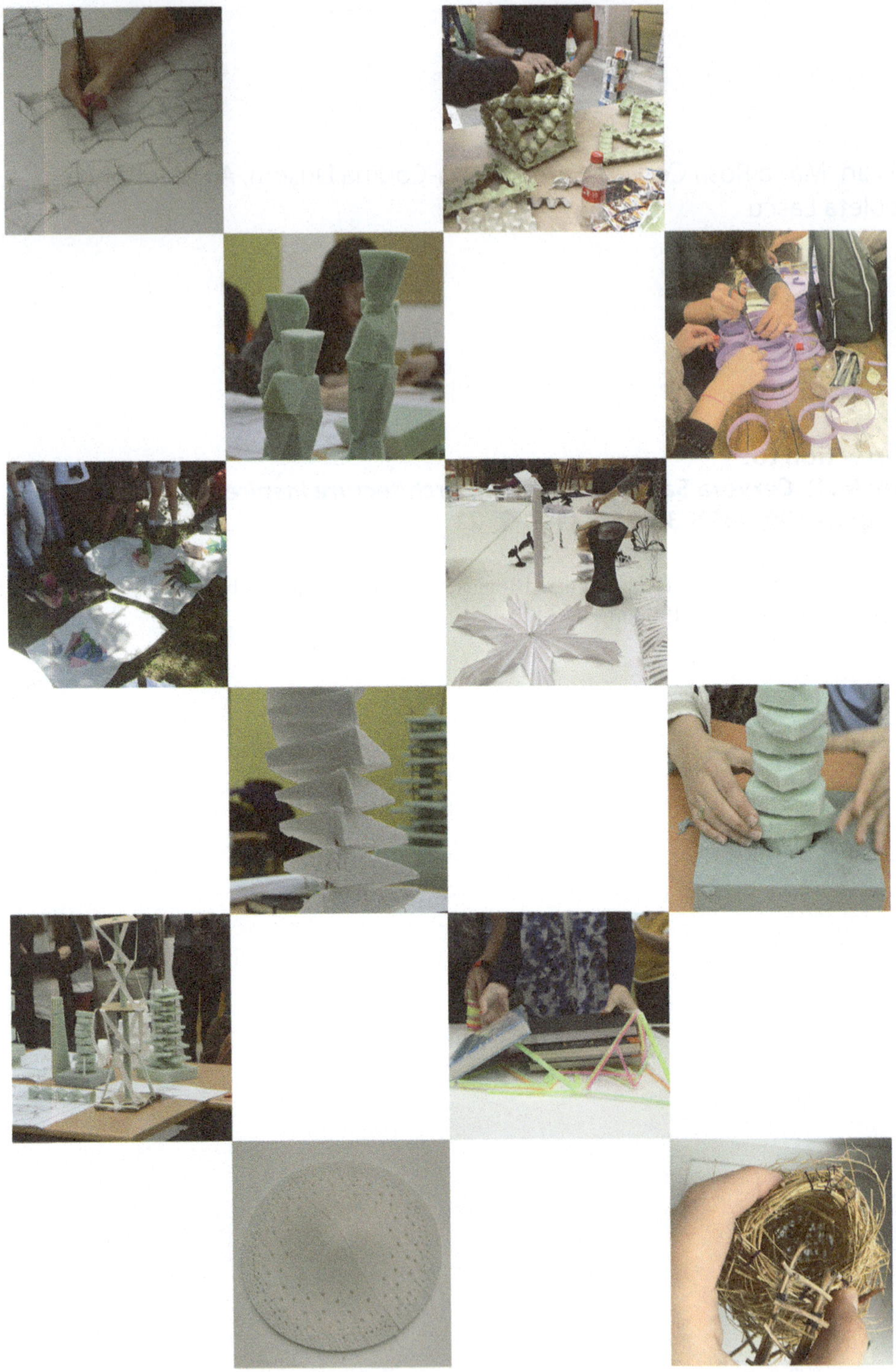

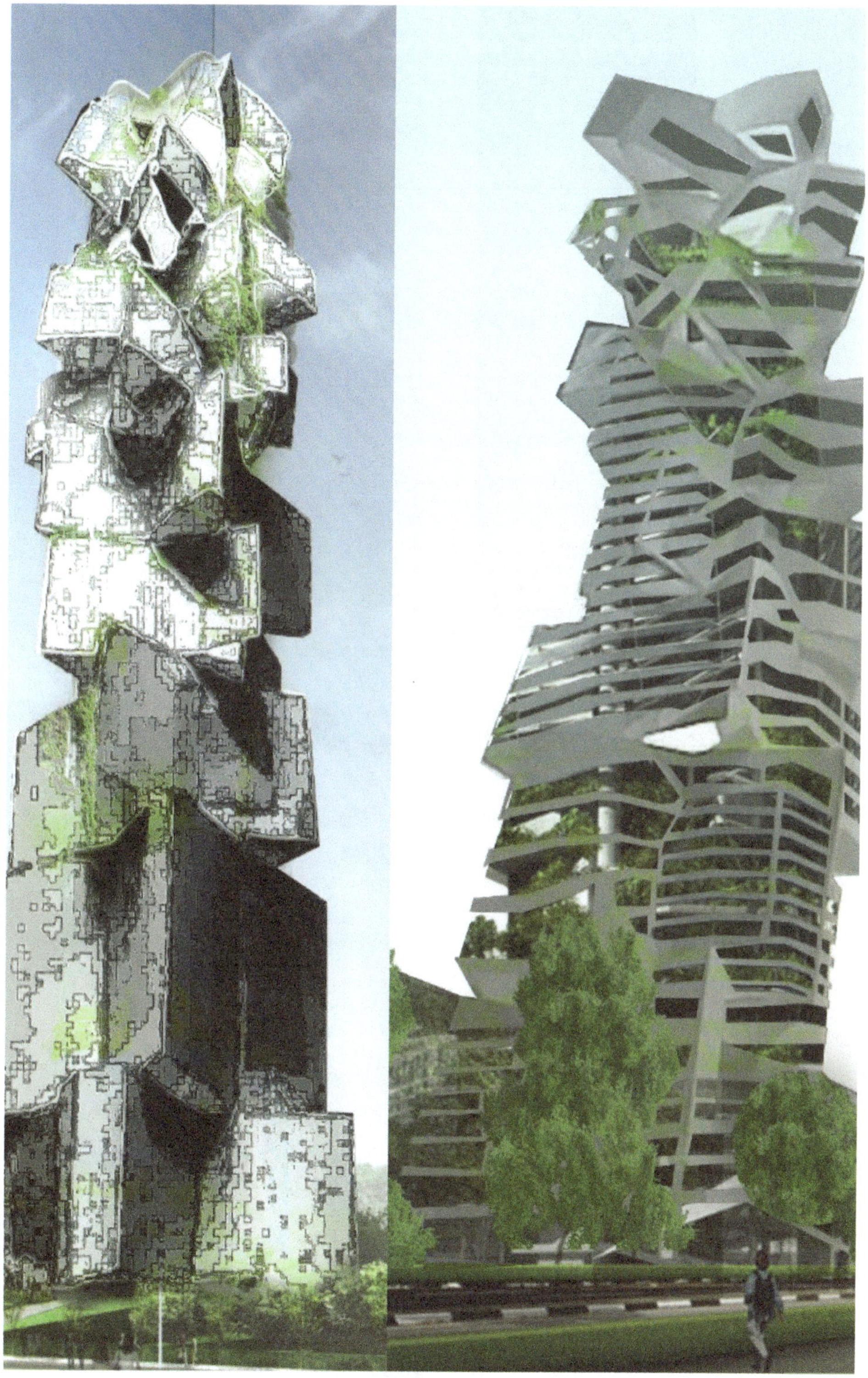

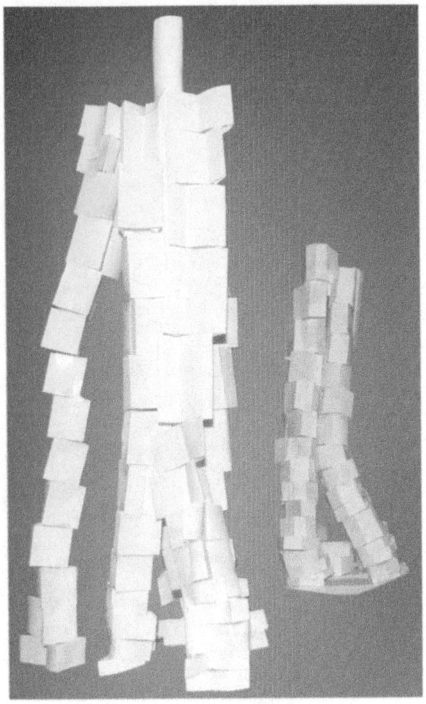

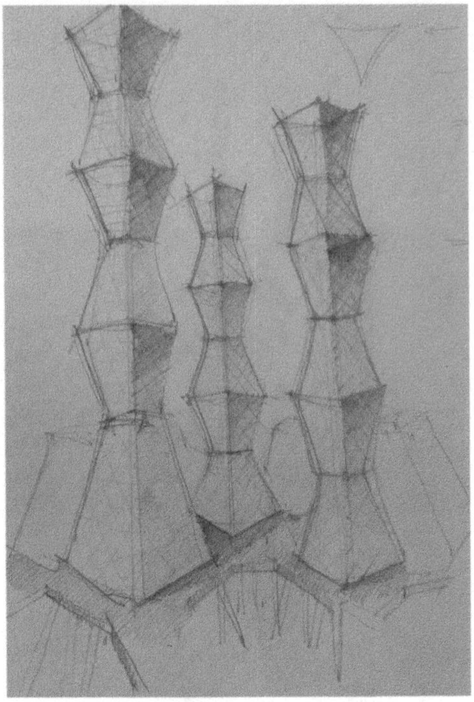

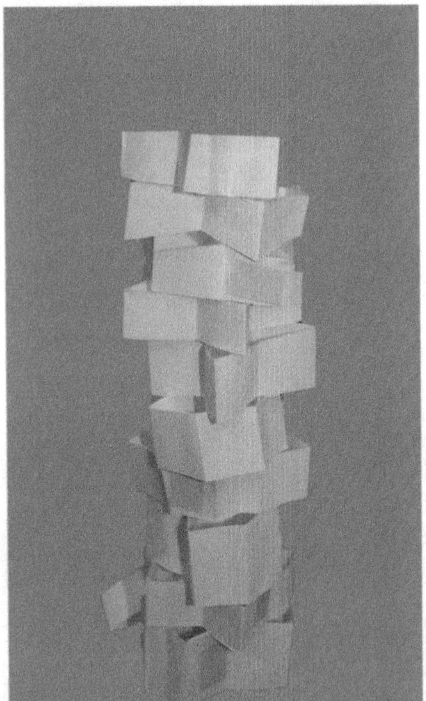

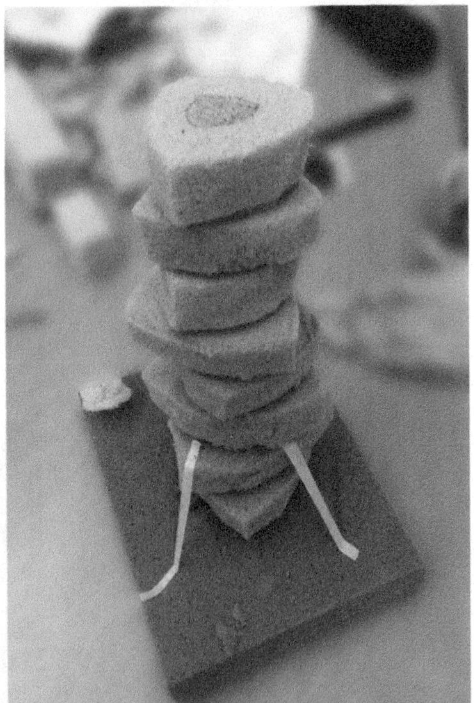

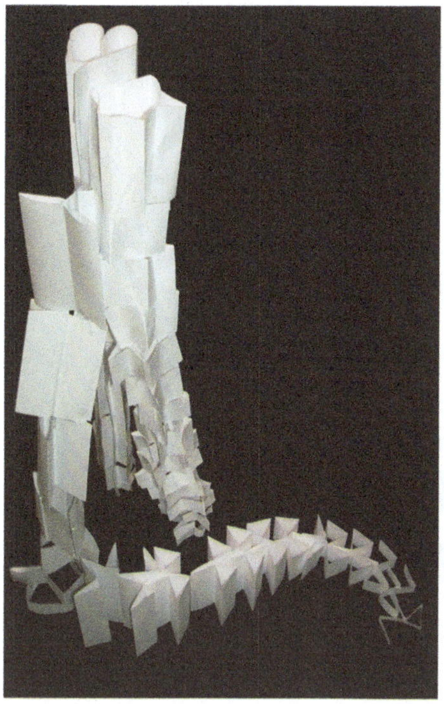
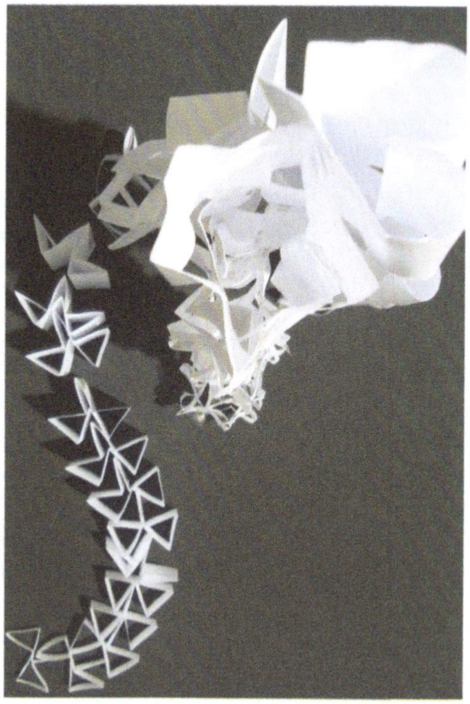
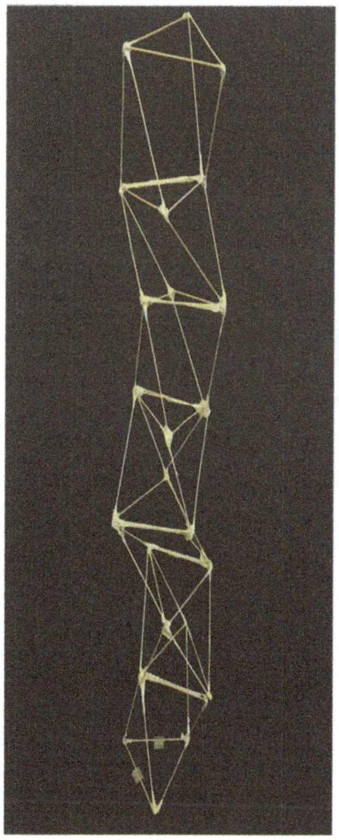
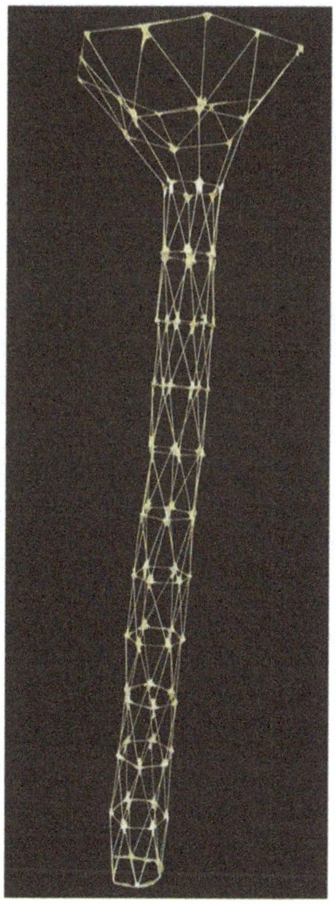
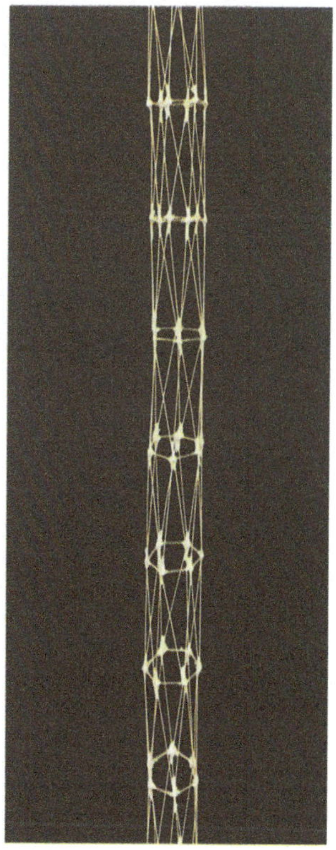

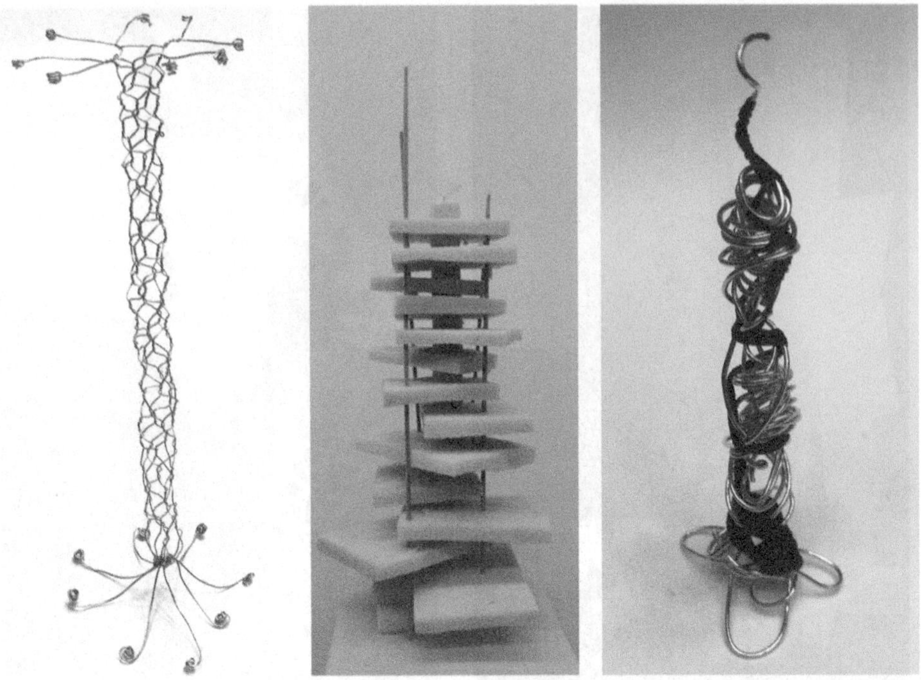

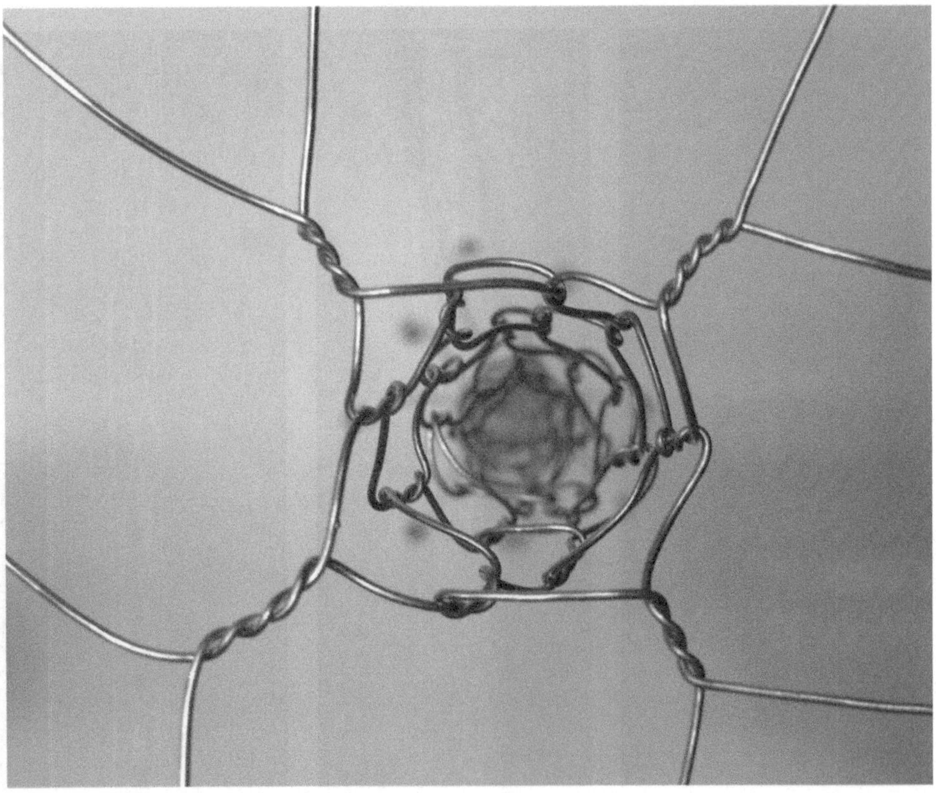

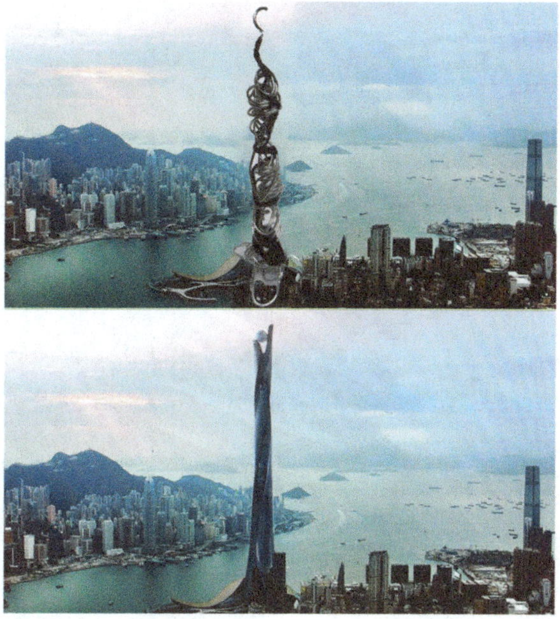

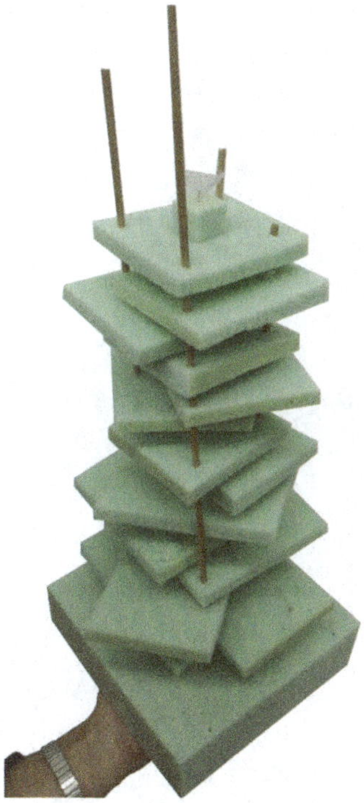

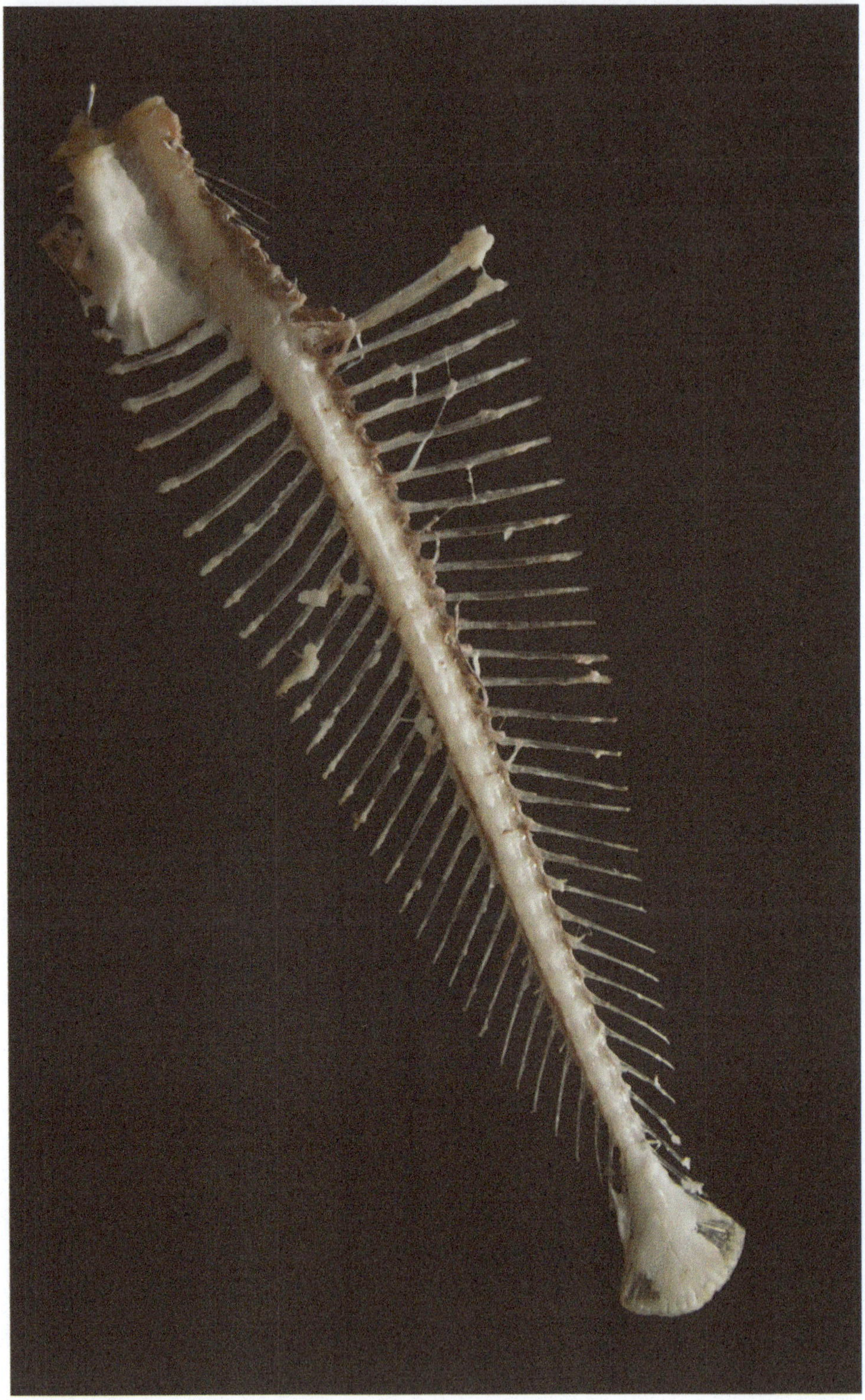

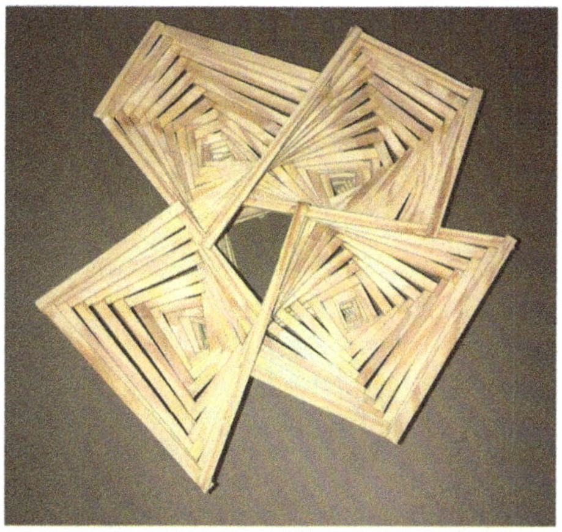

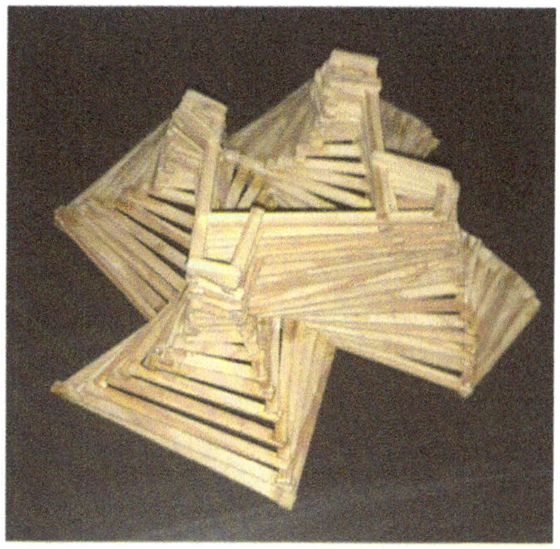

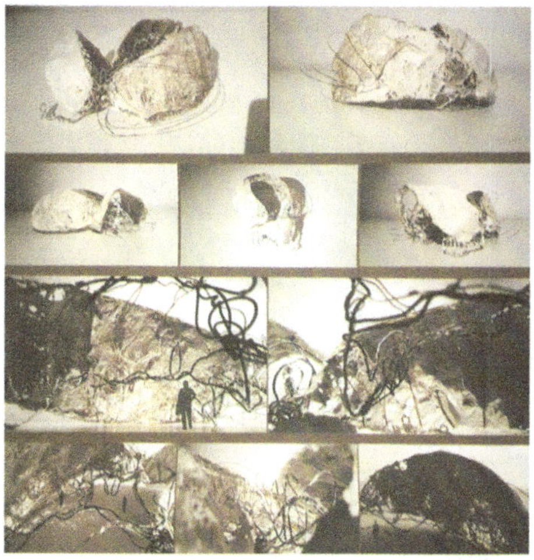

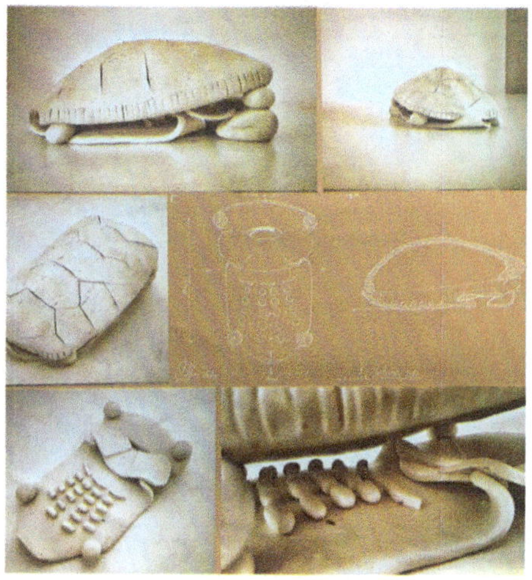

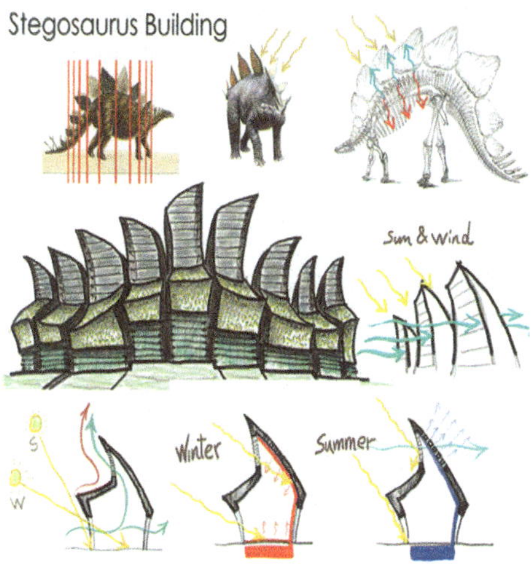
Stegosaurus Building
Sun & Wind
S
W
Winter
Summer

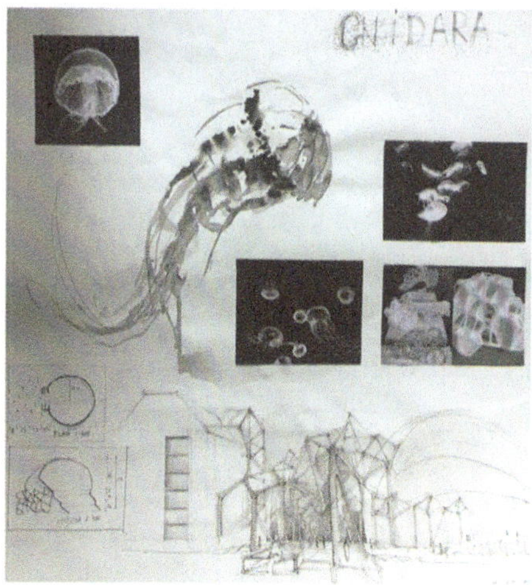

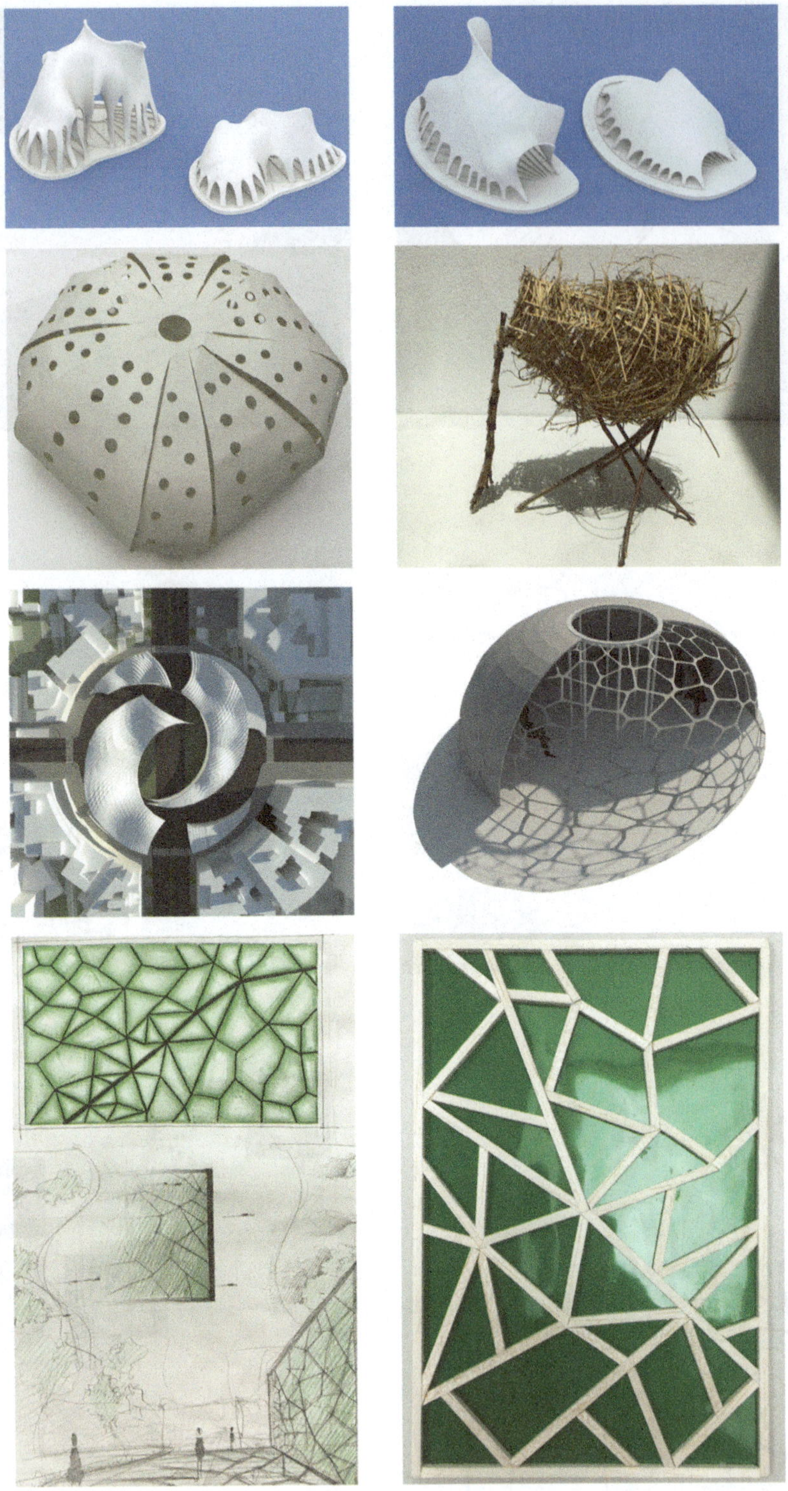

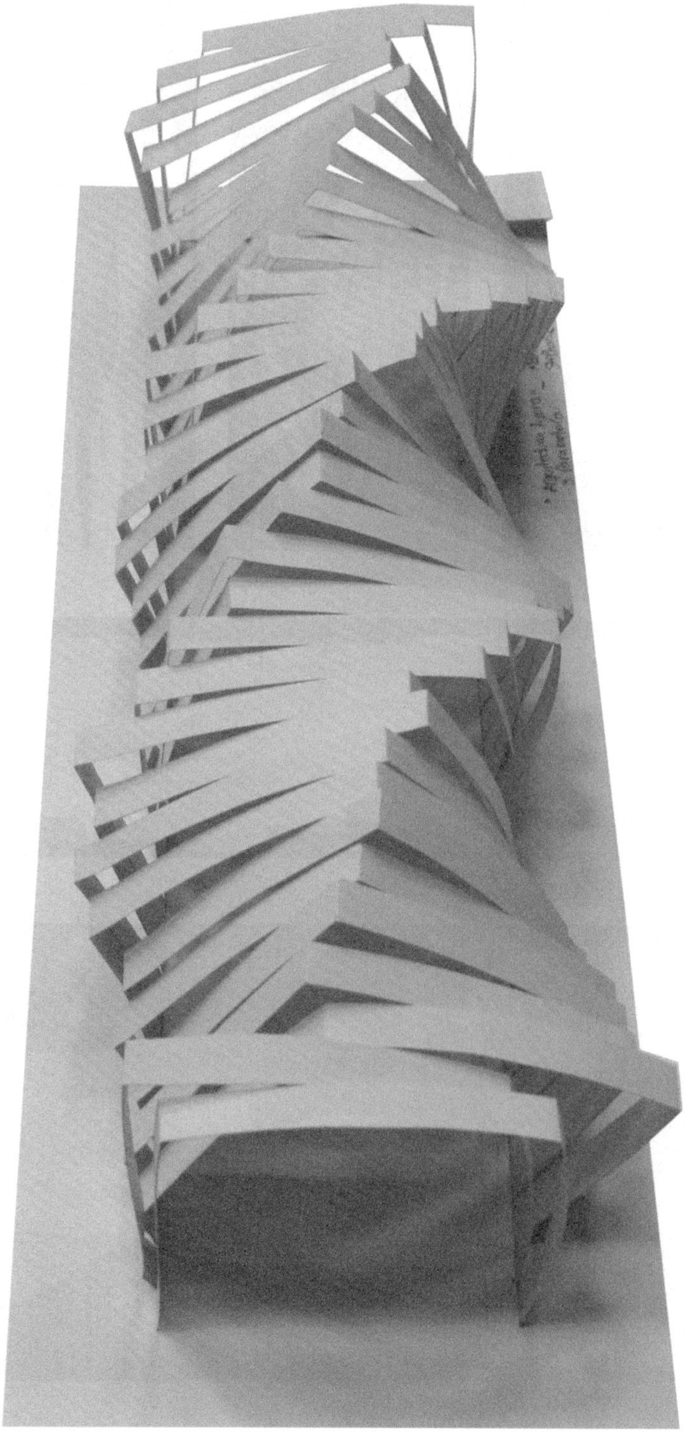

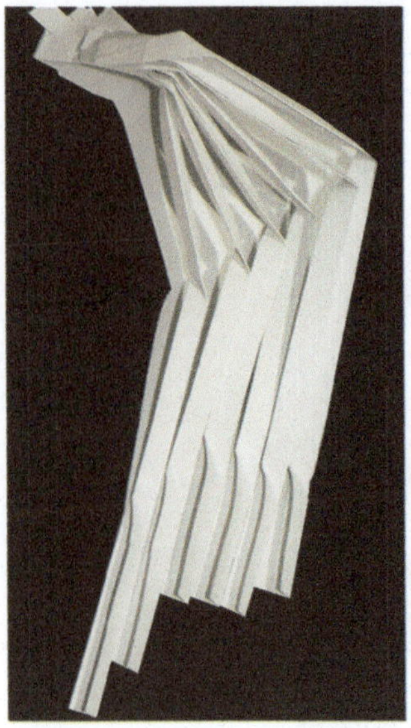

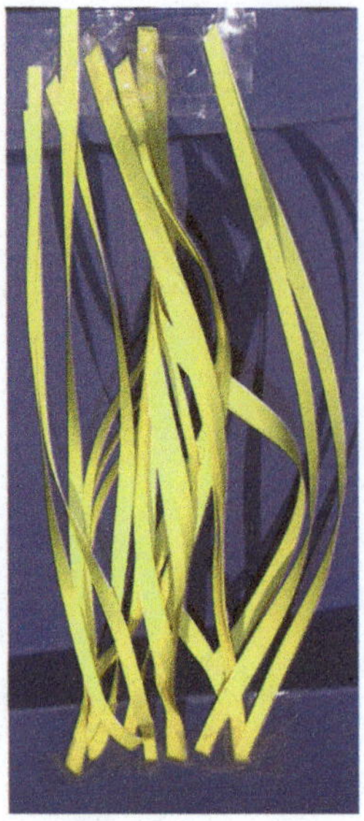

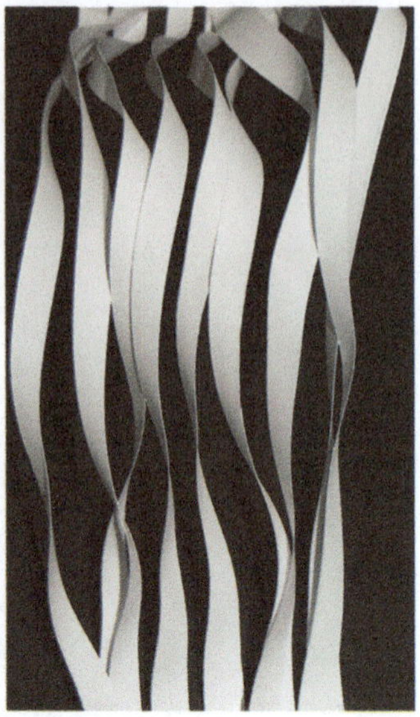

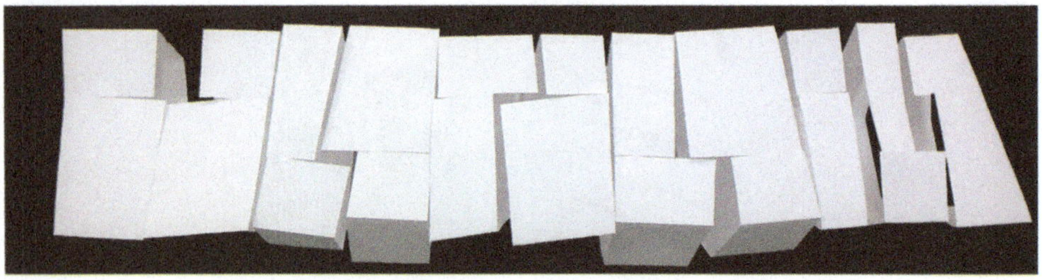

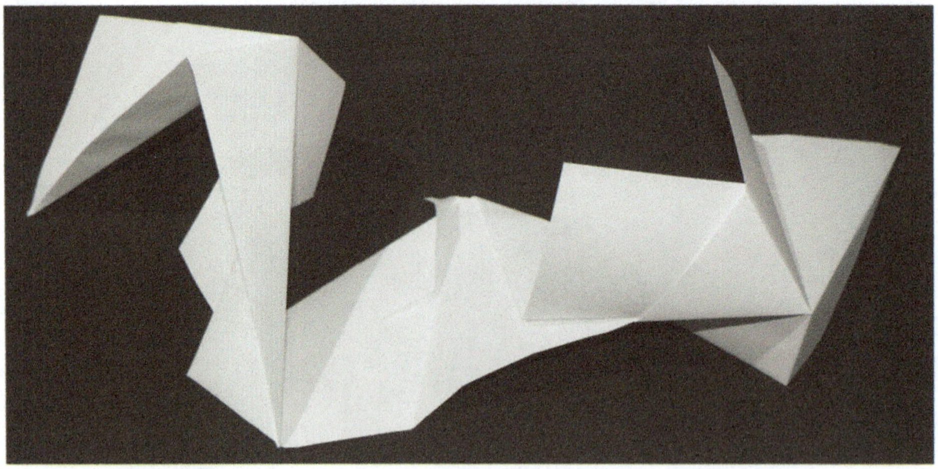

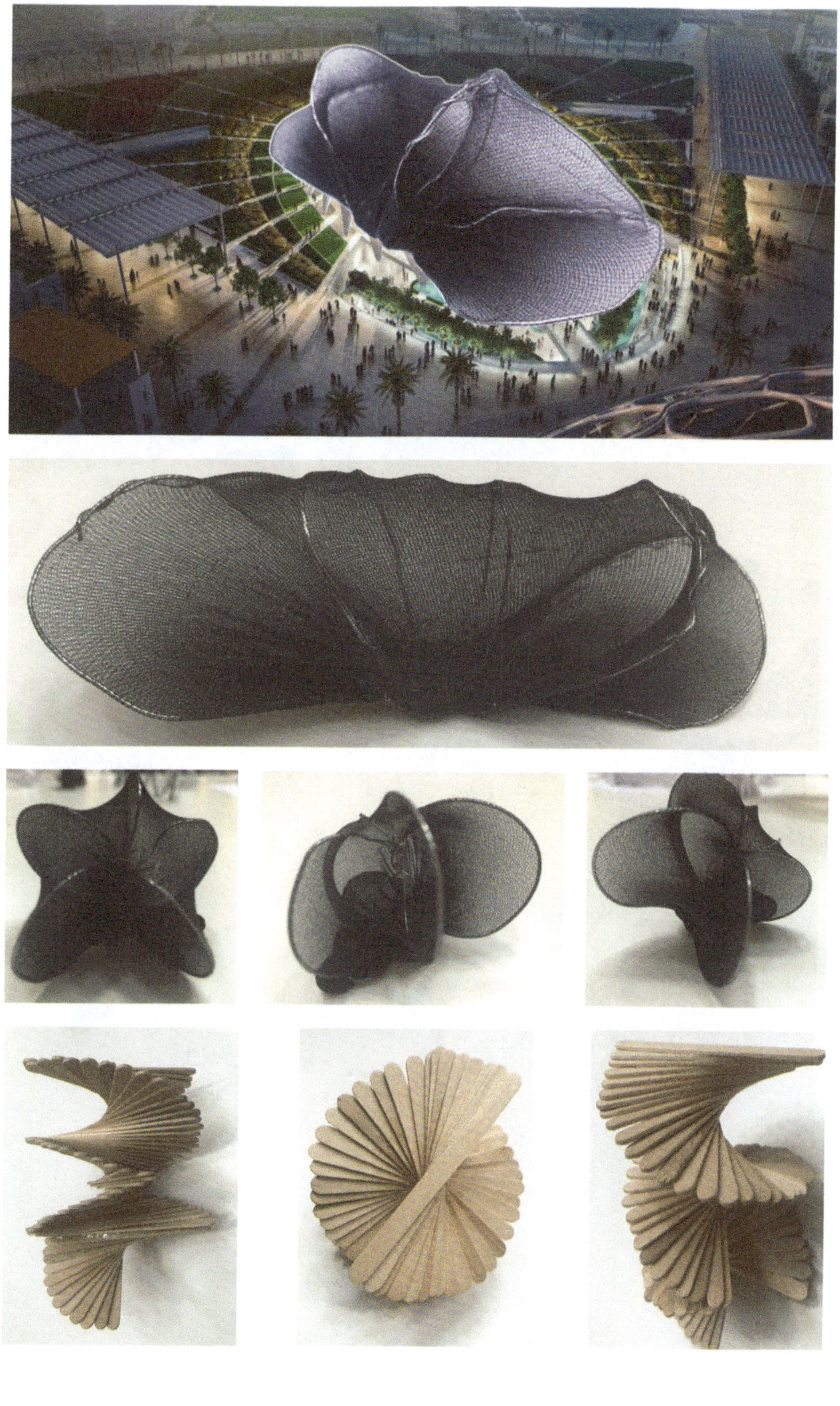

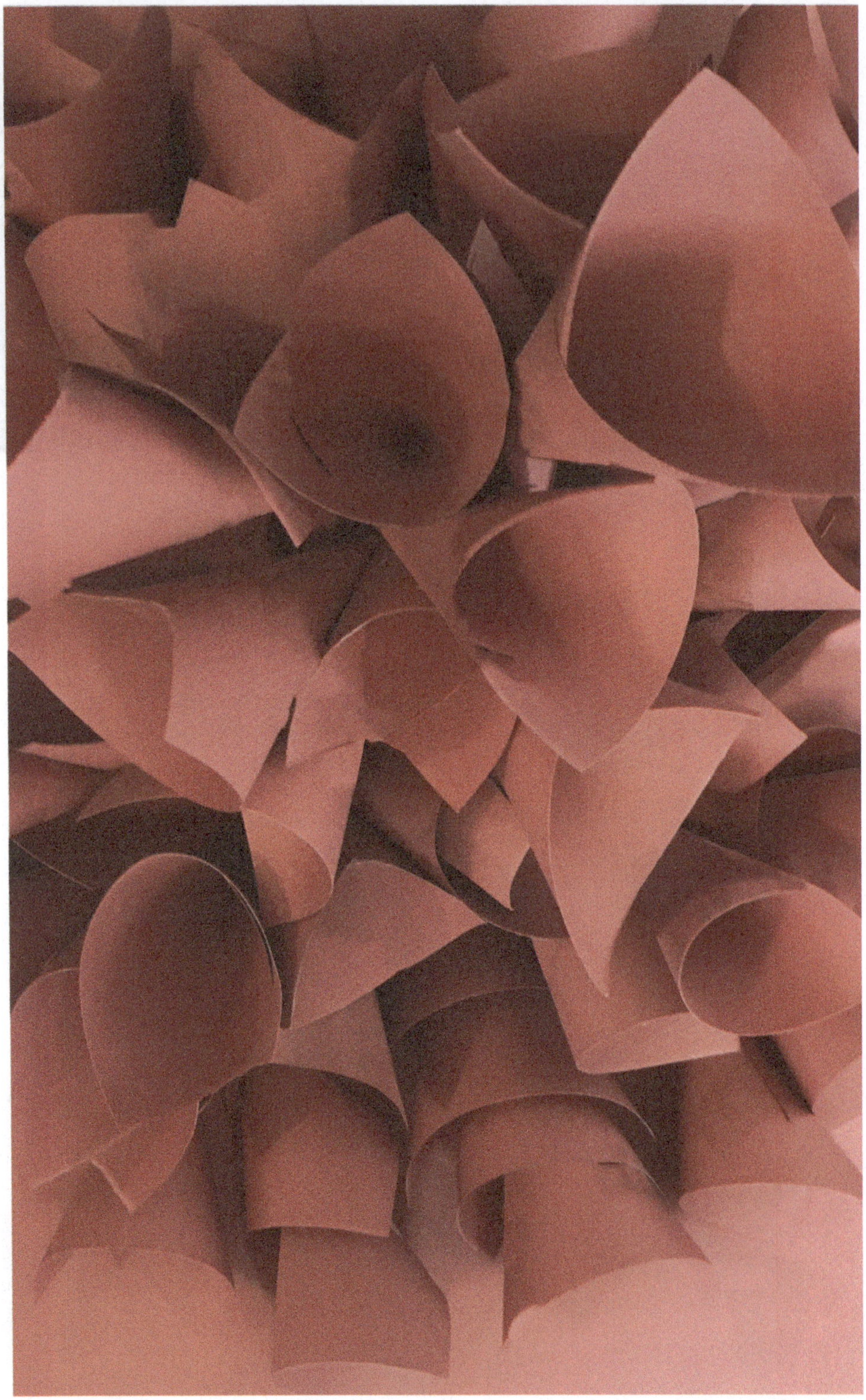

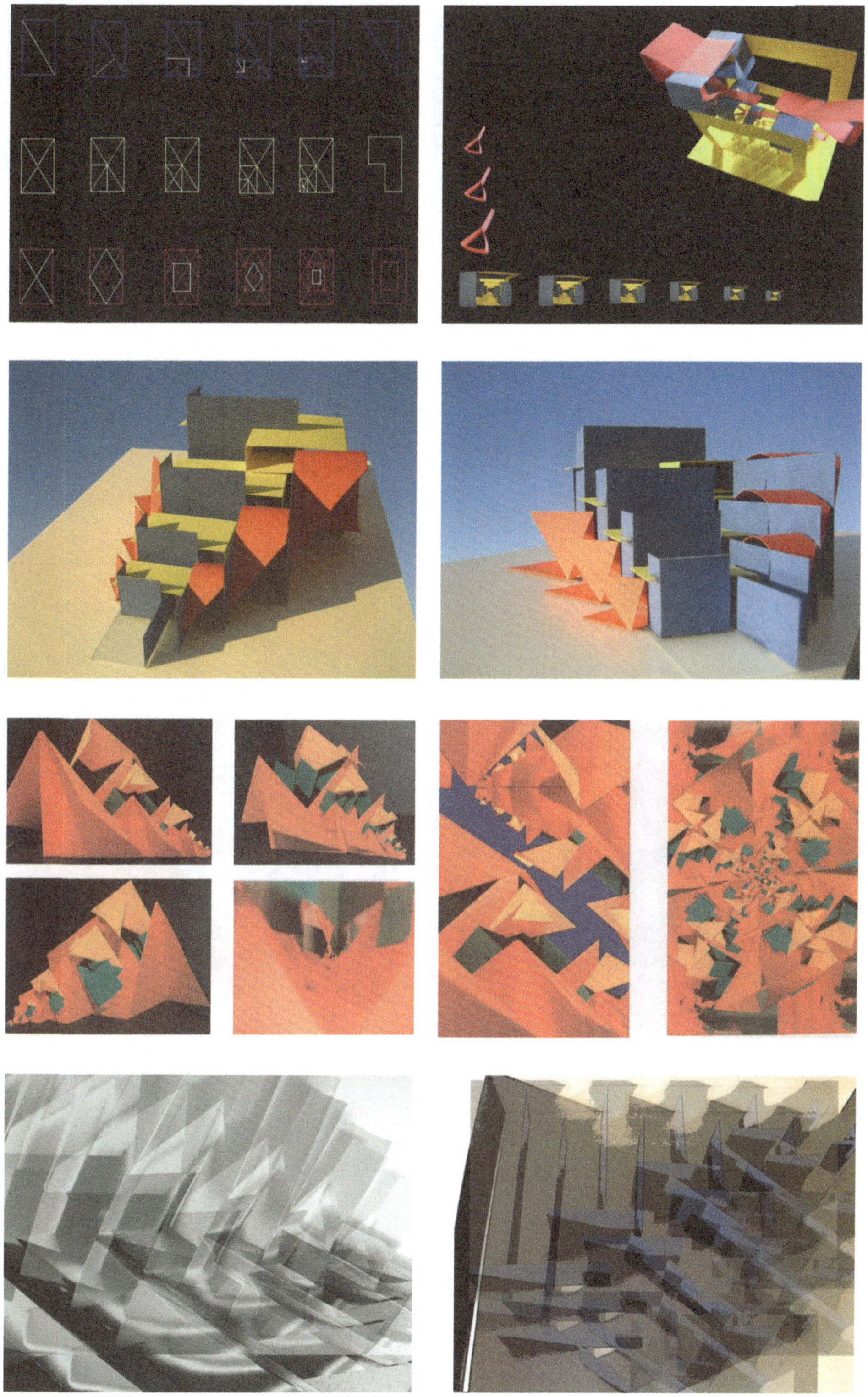

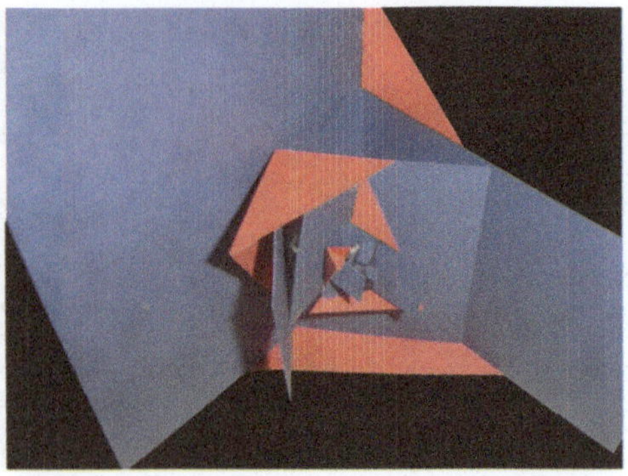
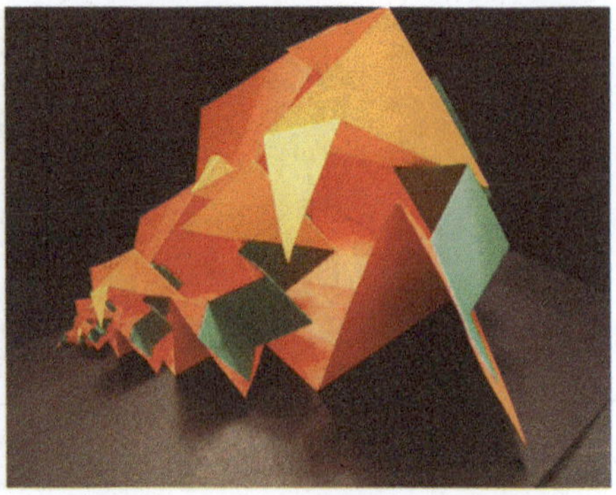
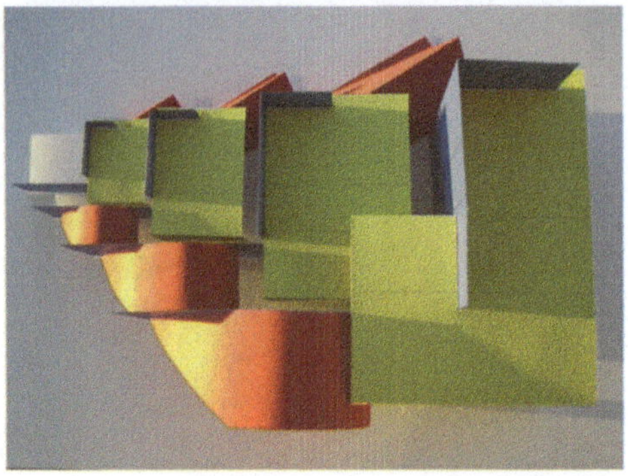
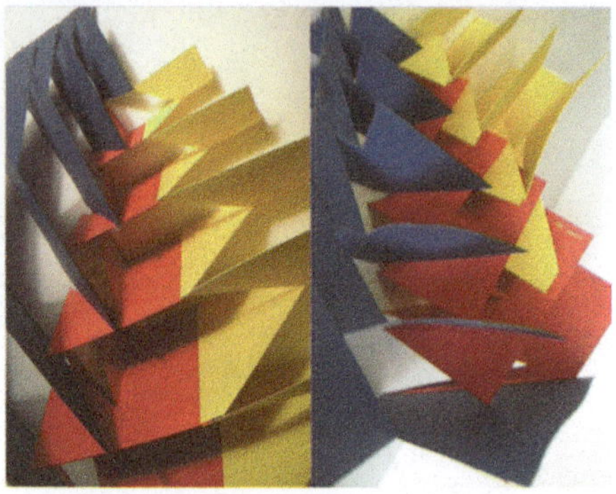
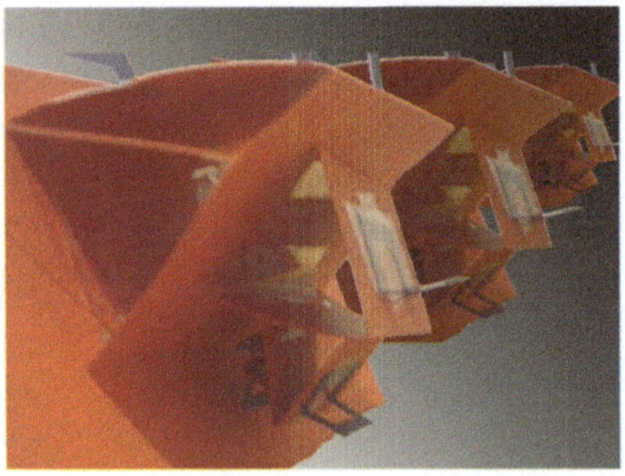
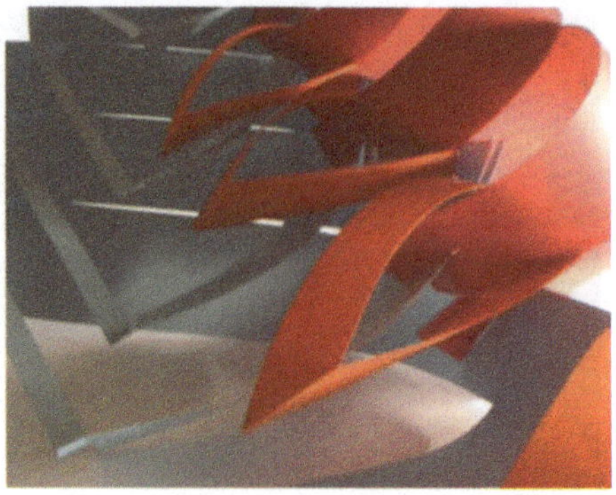
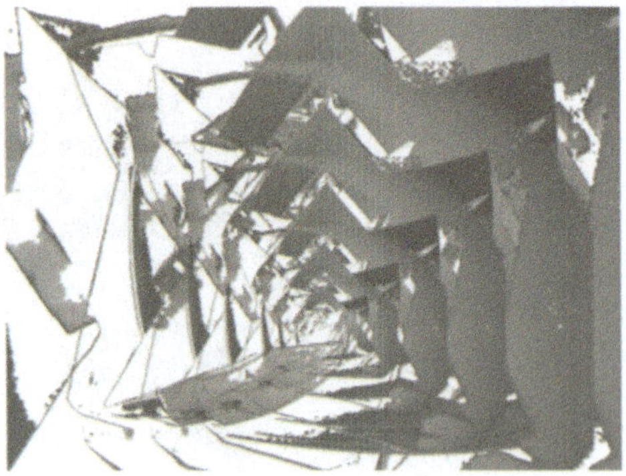
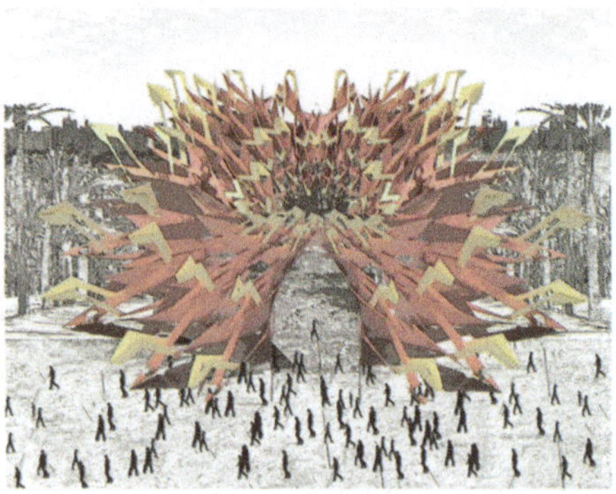

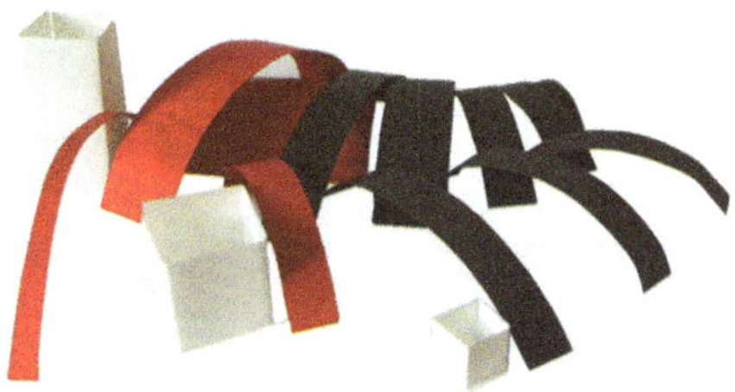

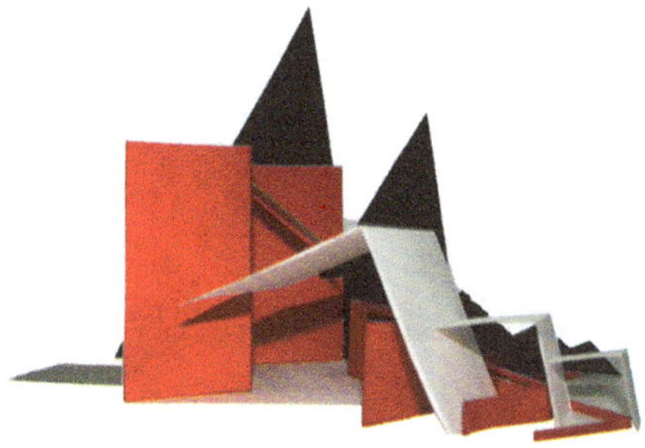

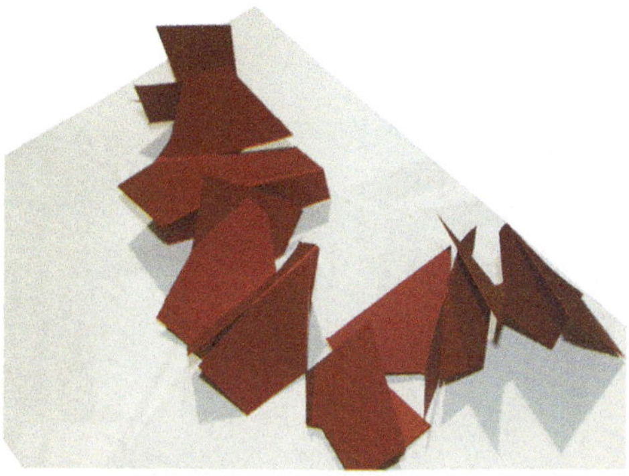

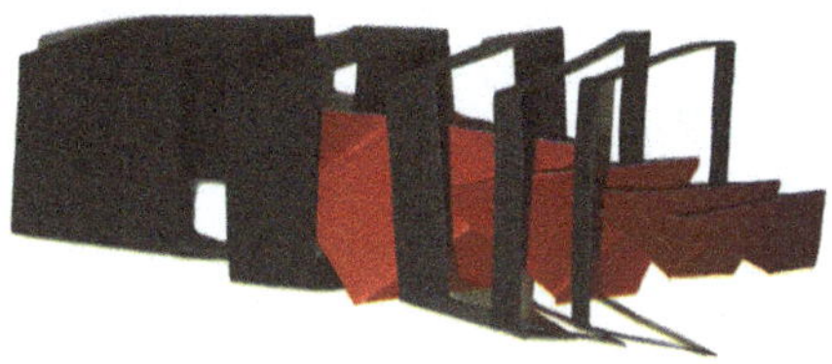

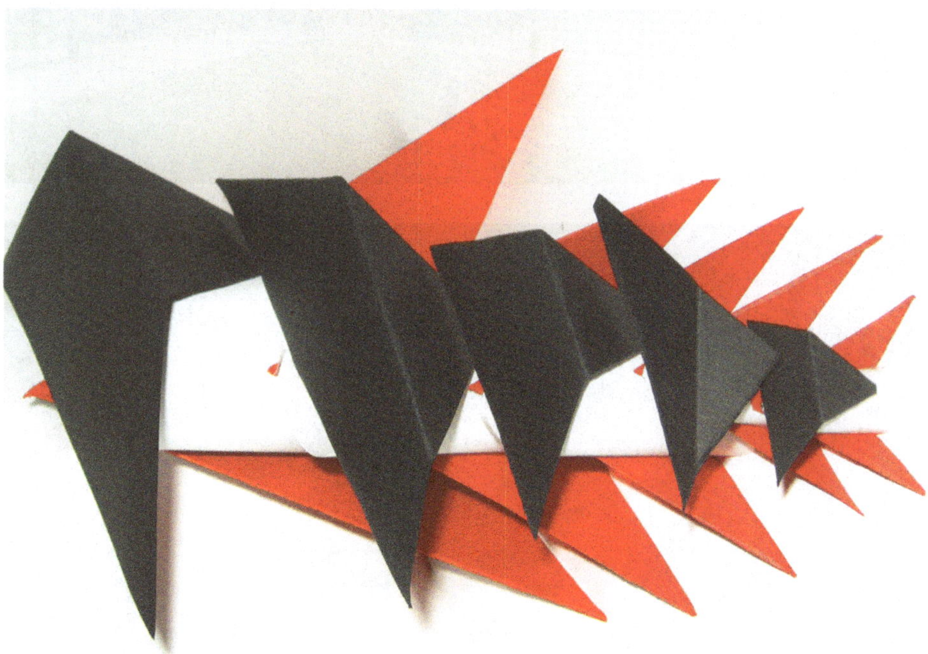

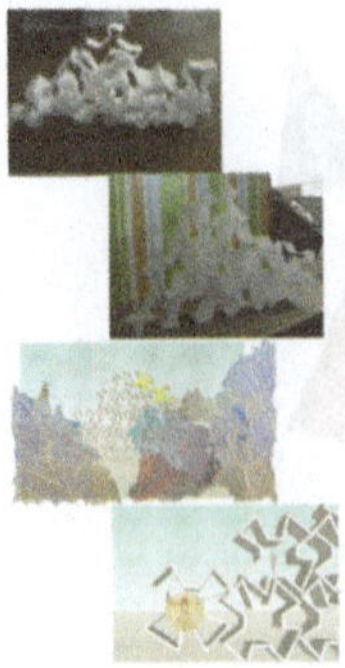

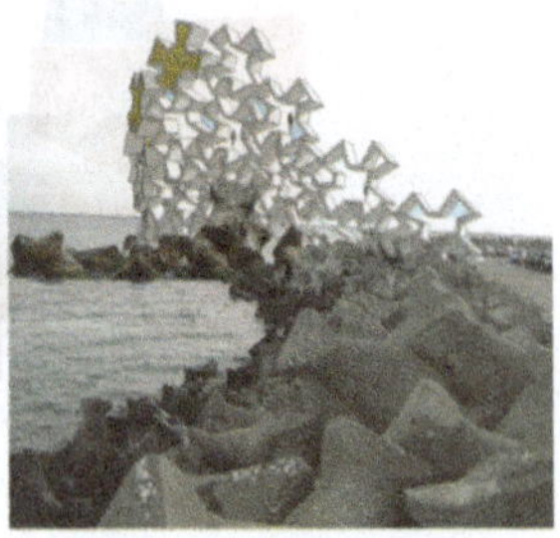

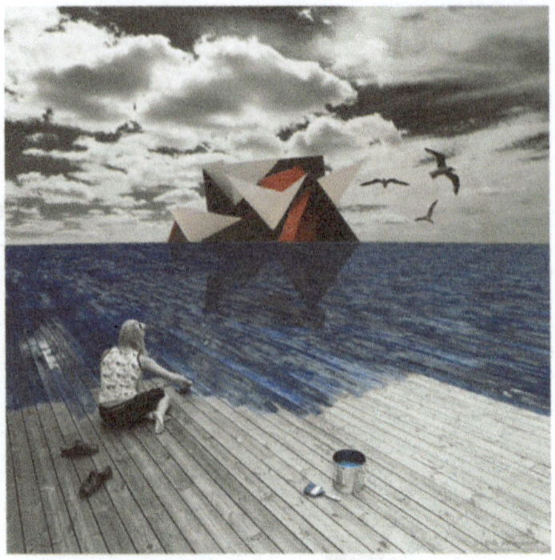

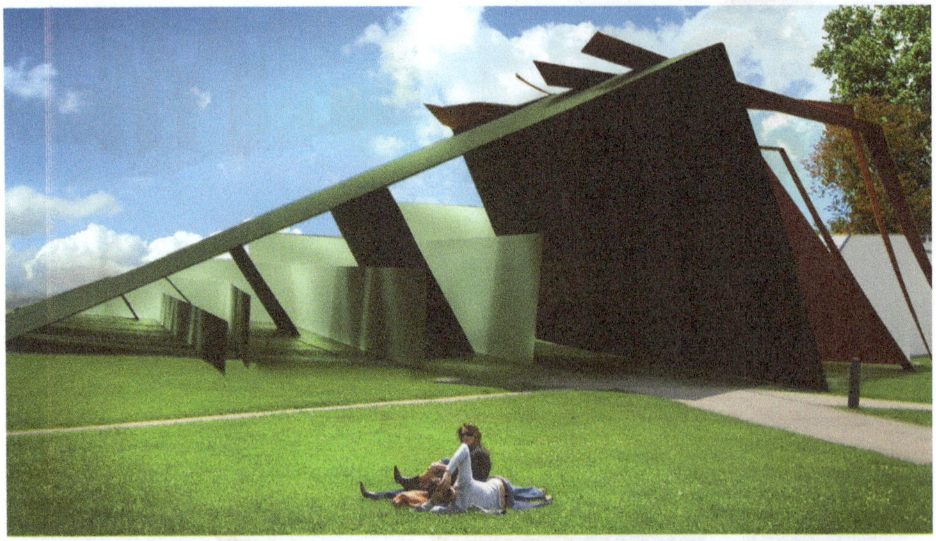

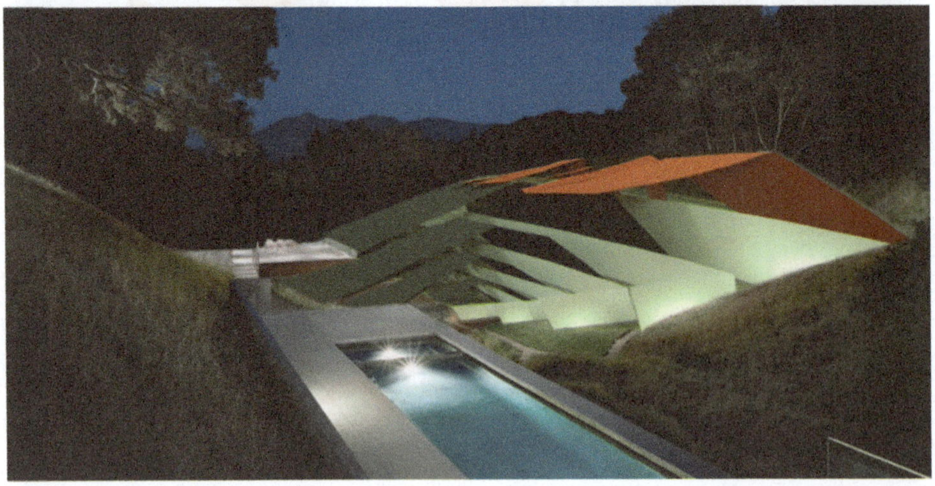

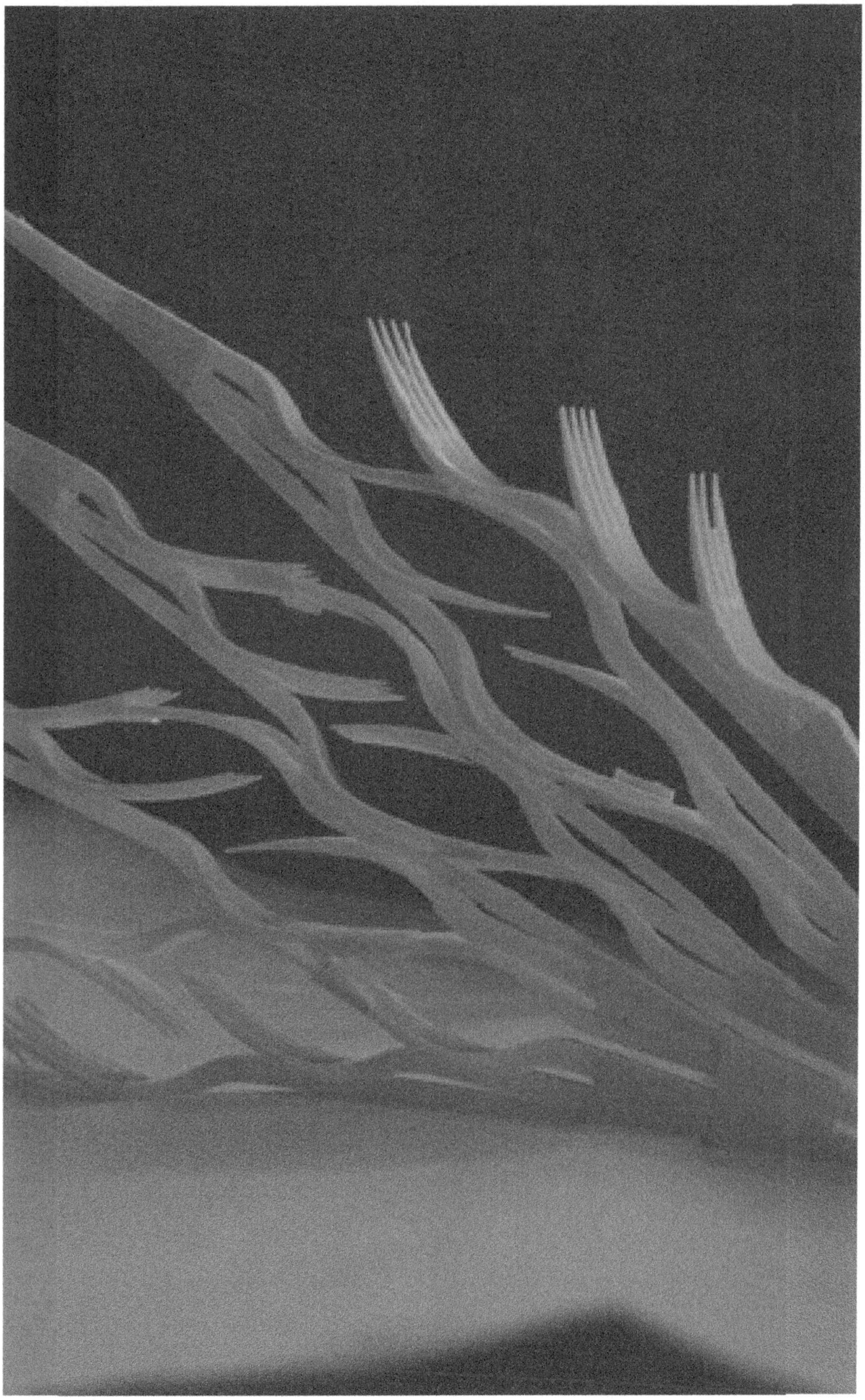

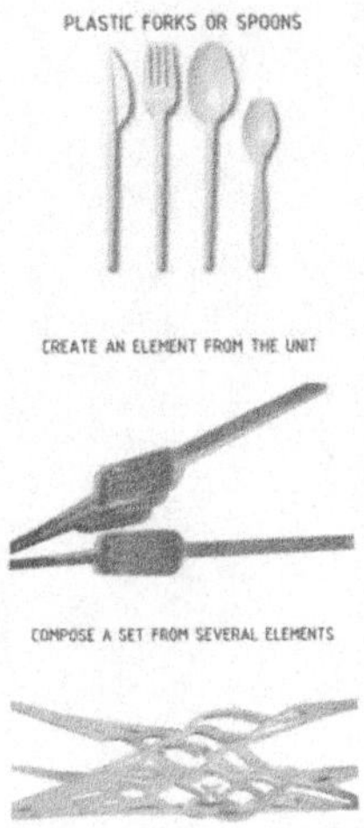
PLASTIC FORKS OR SPOONS
CREATE AN ELEMENT FROM THE UNIT
COMPOSE A SET FROM SEVERAL ELEMENTS

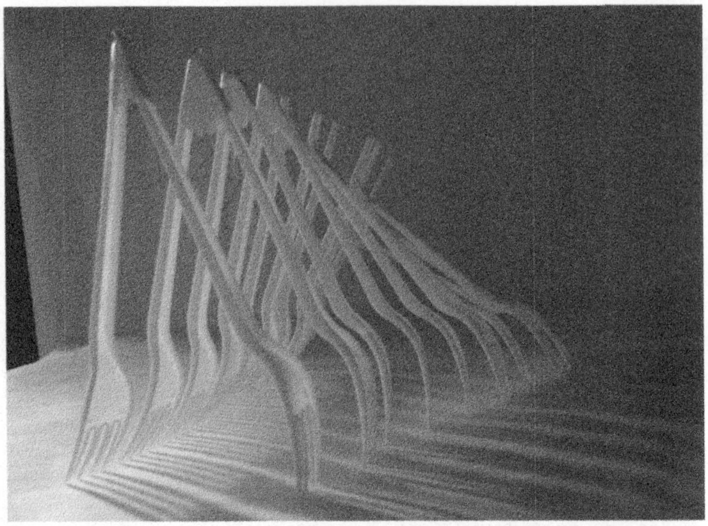

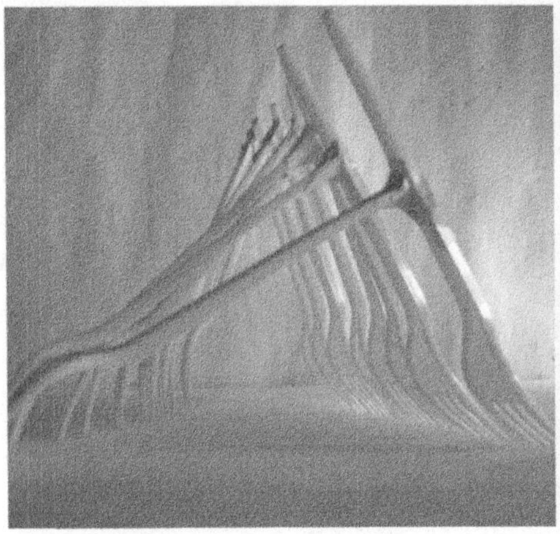

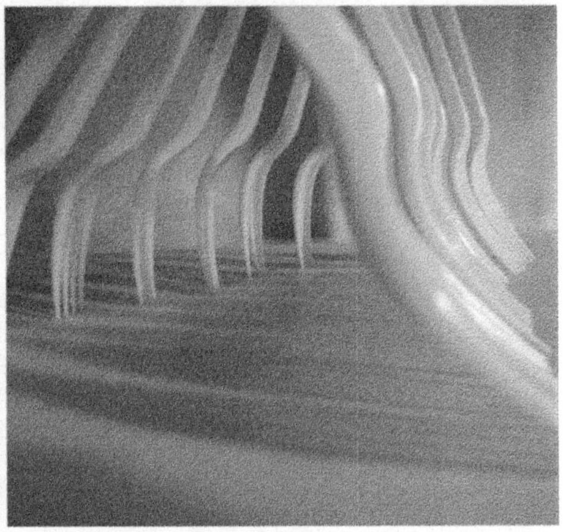

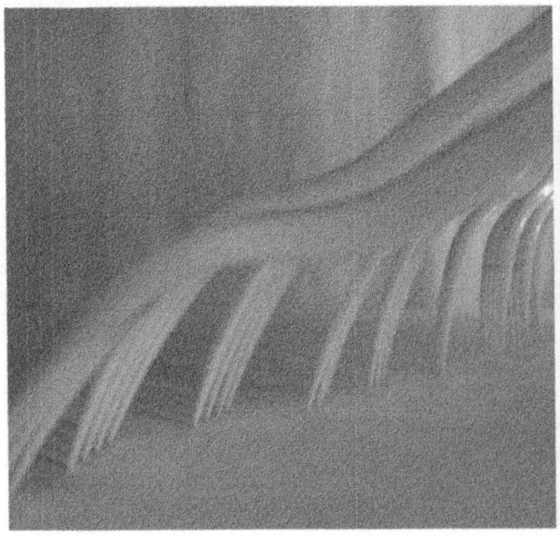

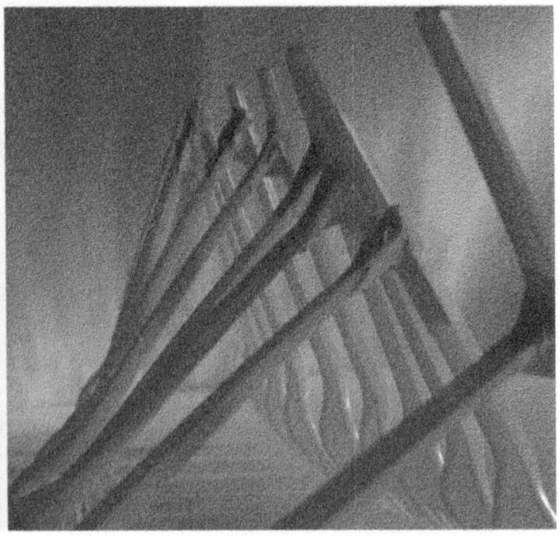

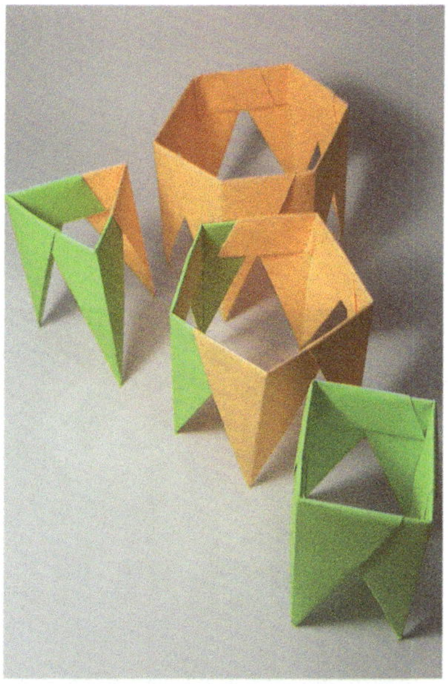

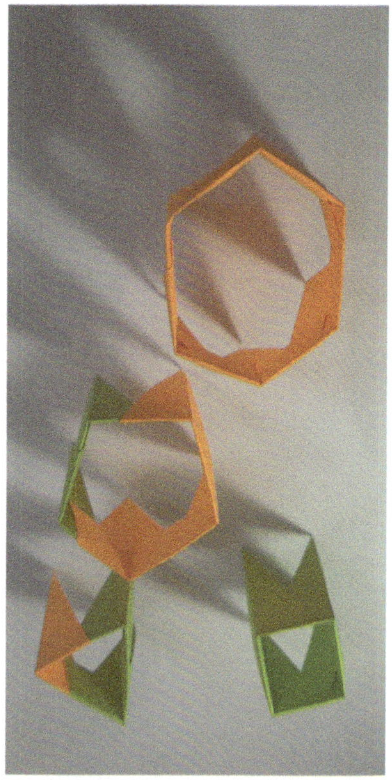

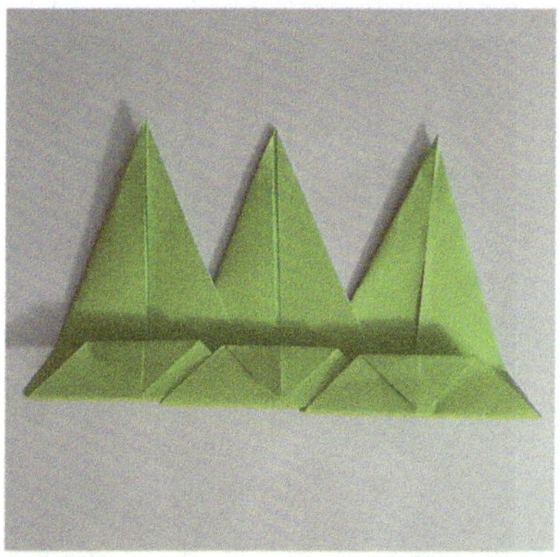

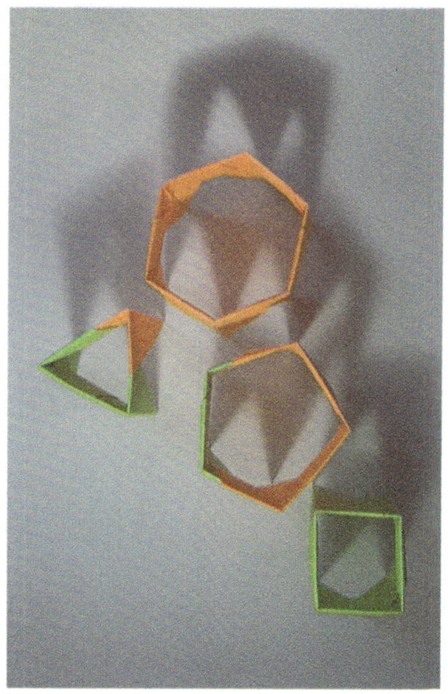

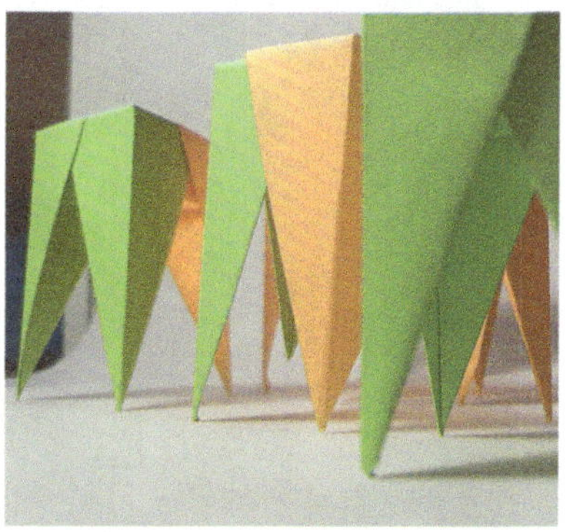

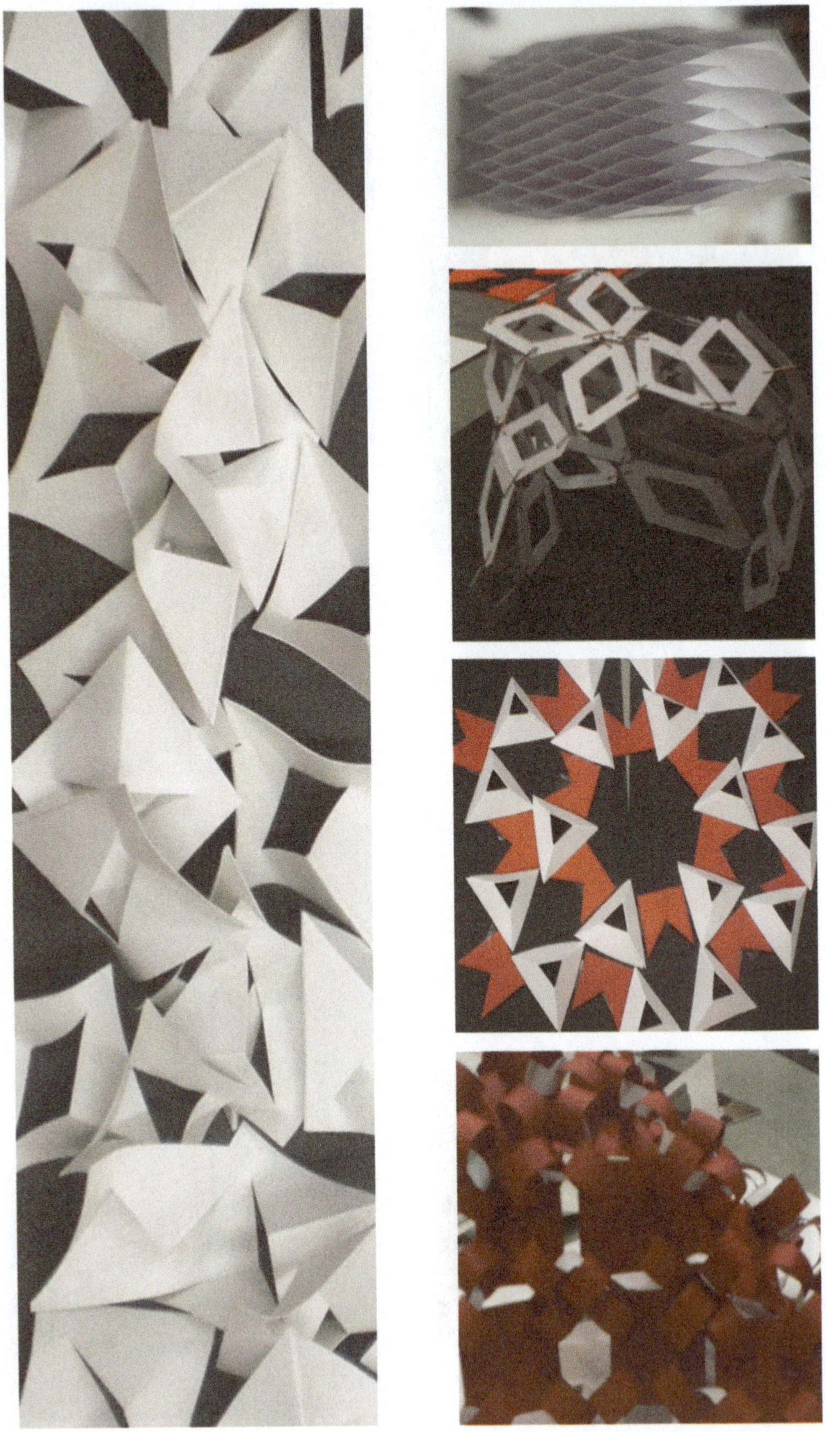

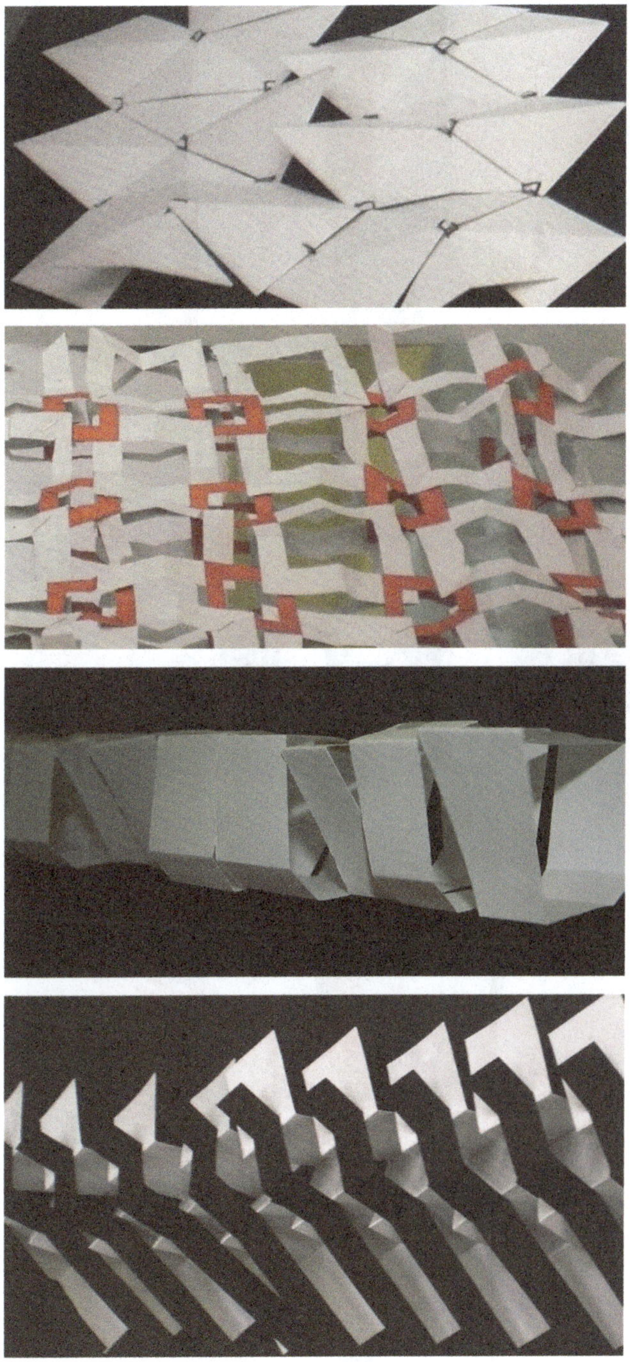

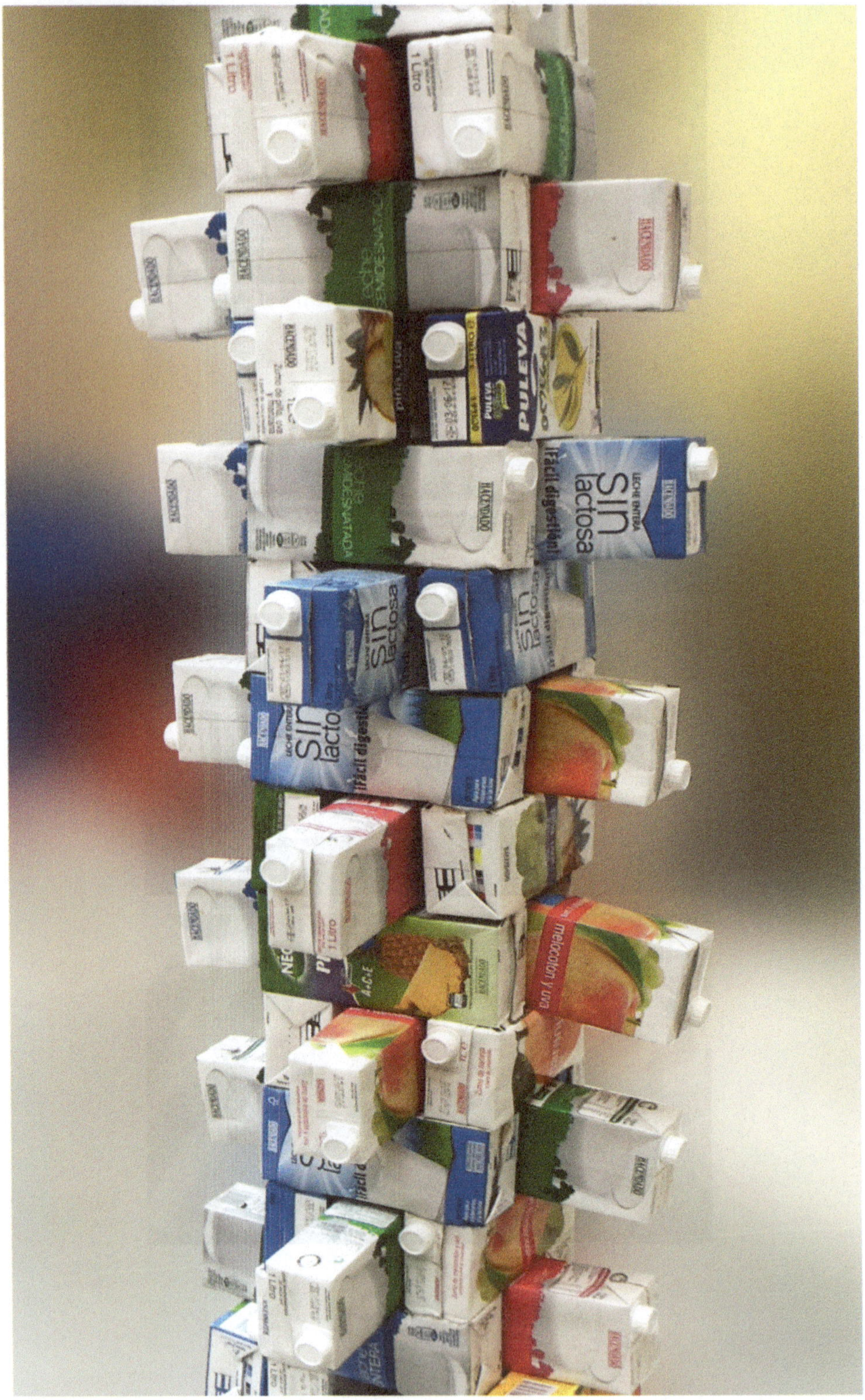
1 Litro
PULEVA
LECHE ENTERA
sin
lactosa
fácil digestión
melocotón y uva

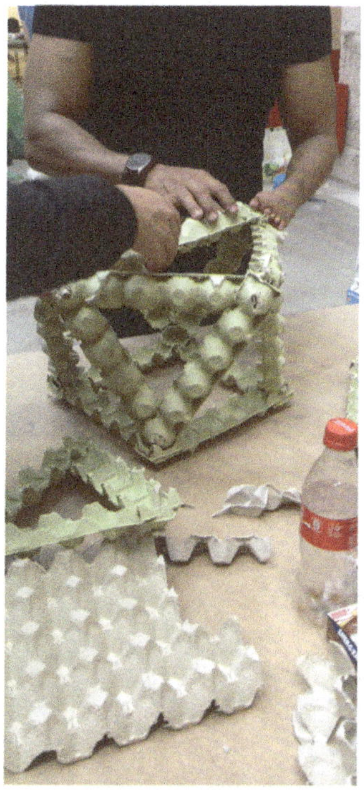
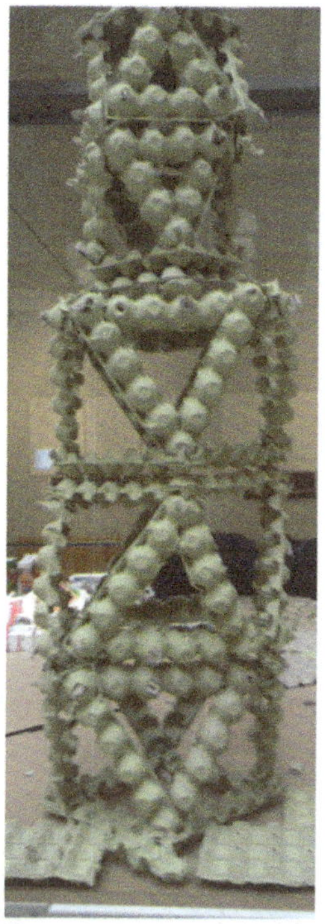
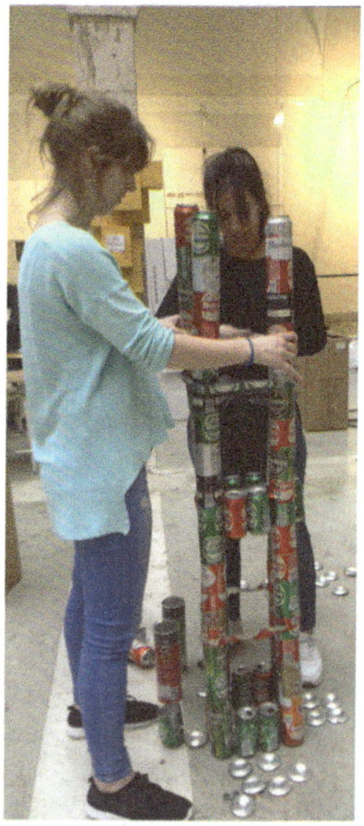
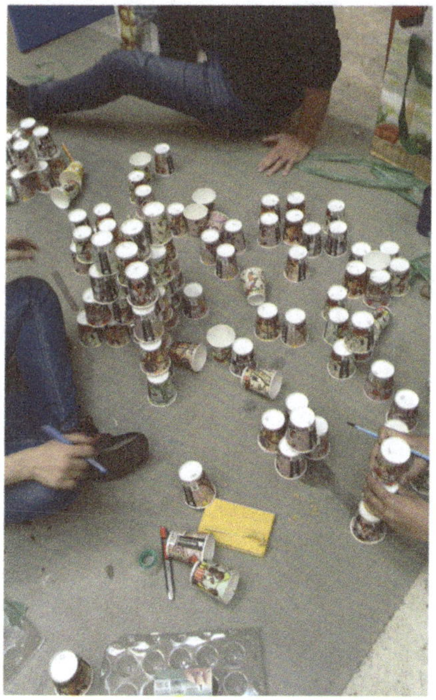
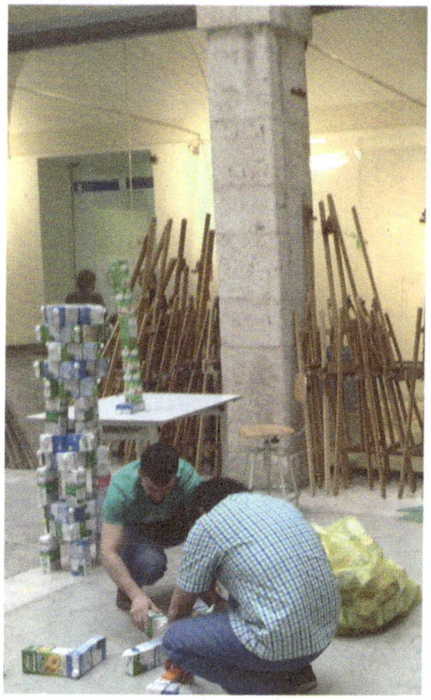

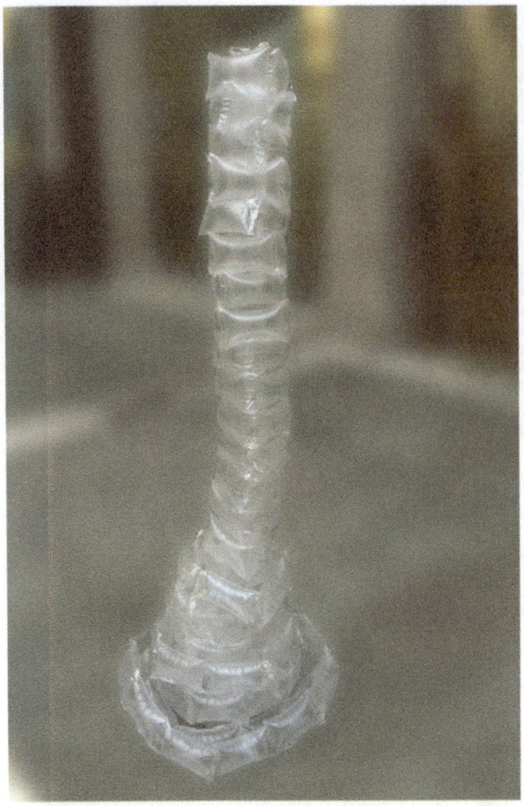

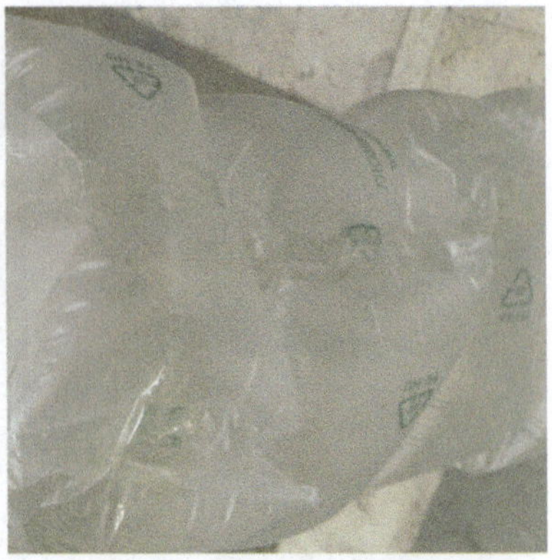

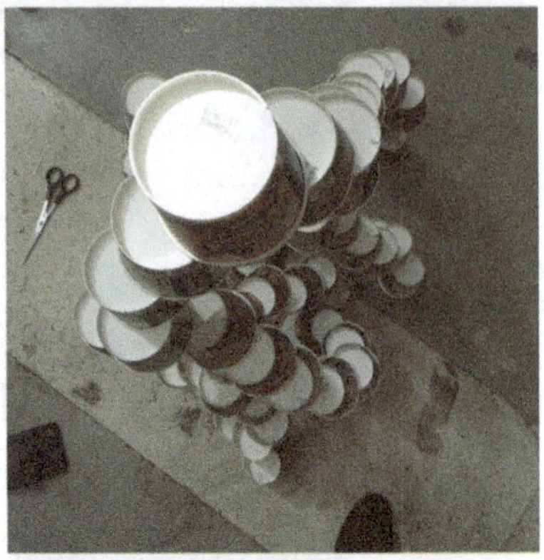

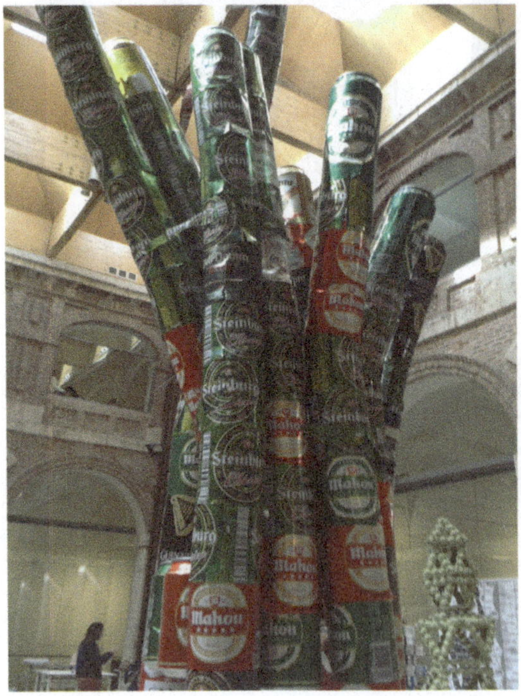
Mahou

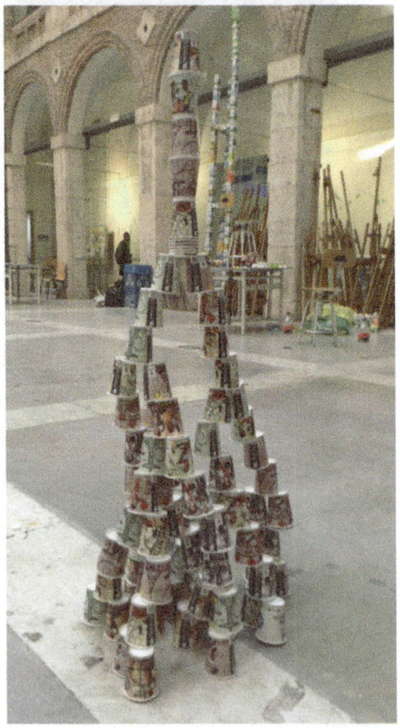

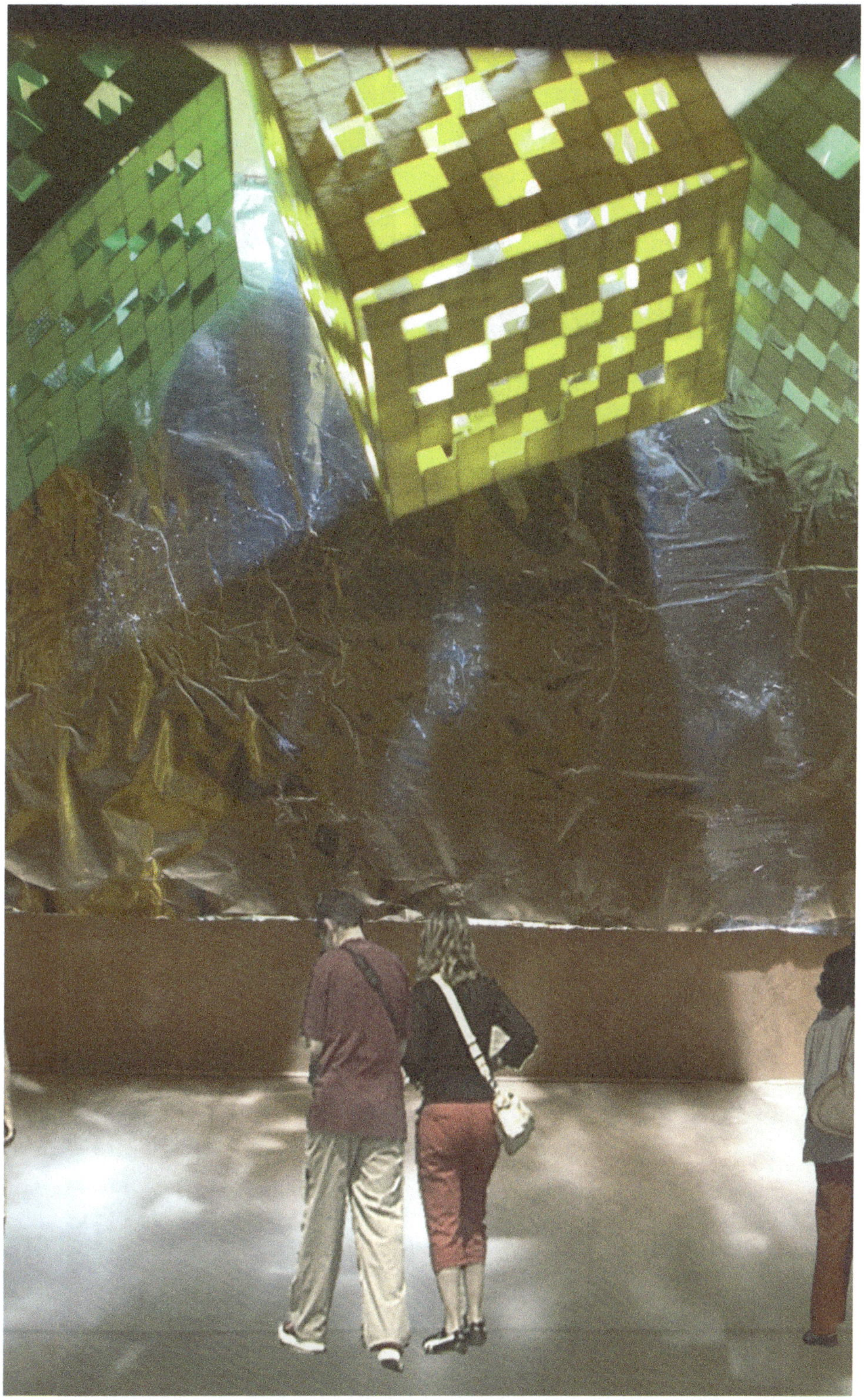

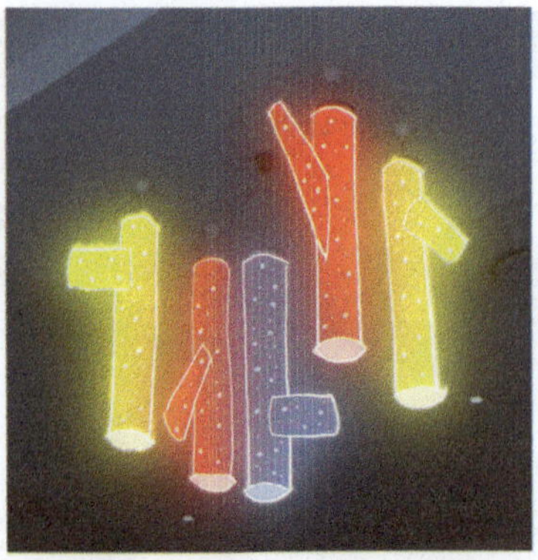
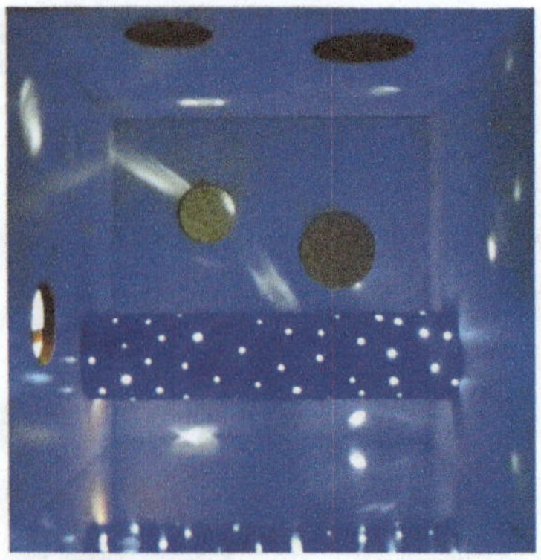
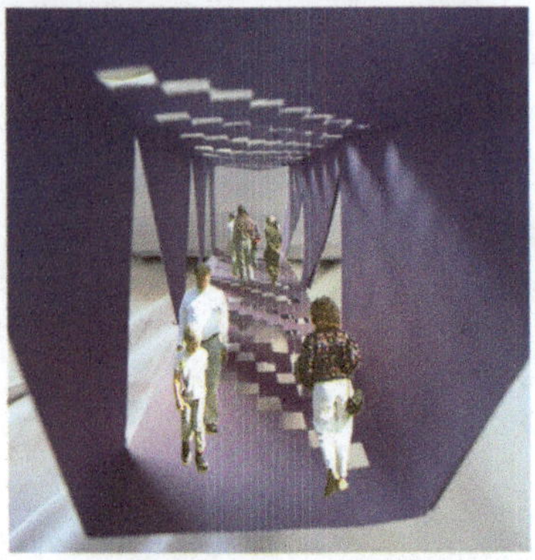
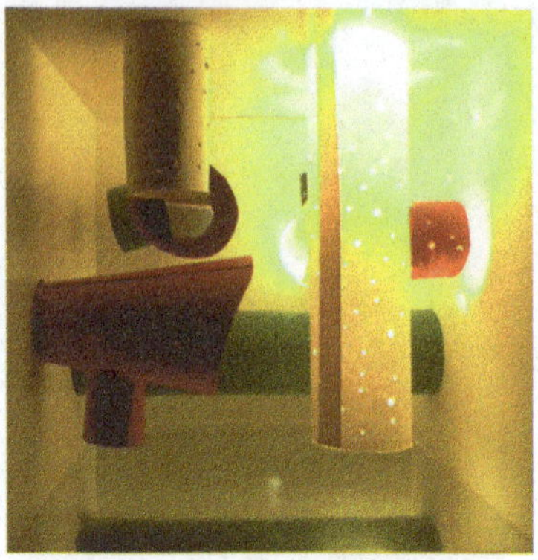
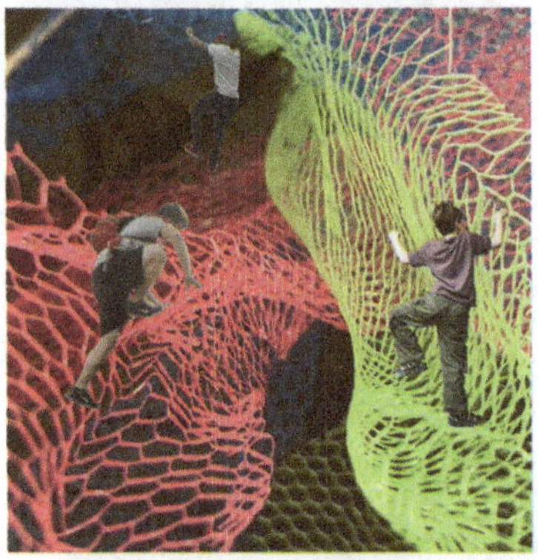
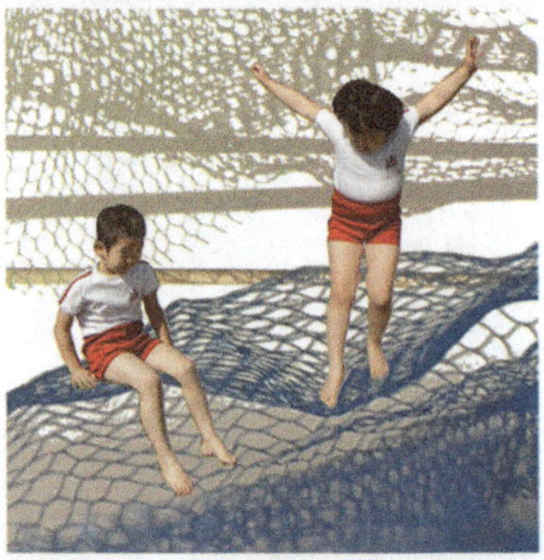
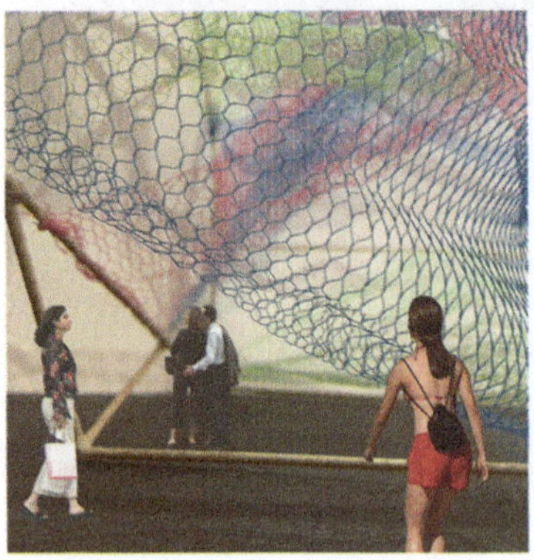
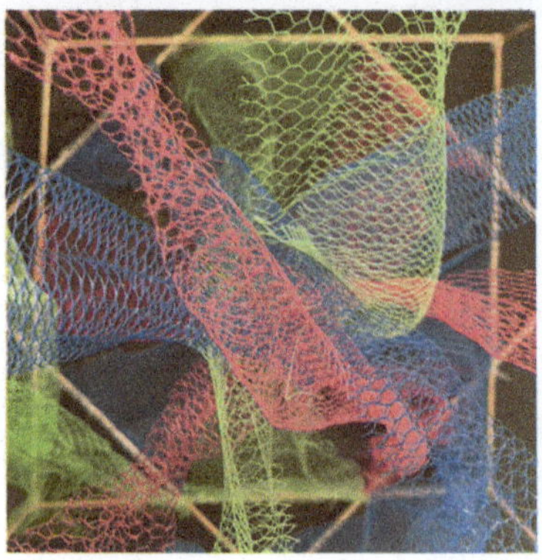

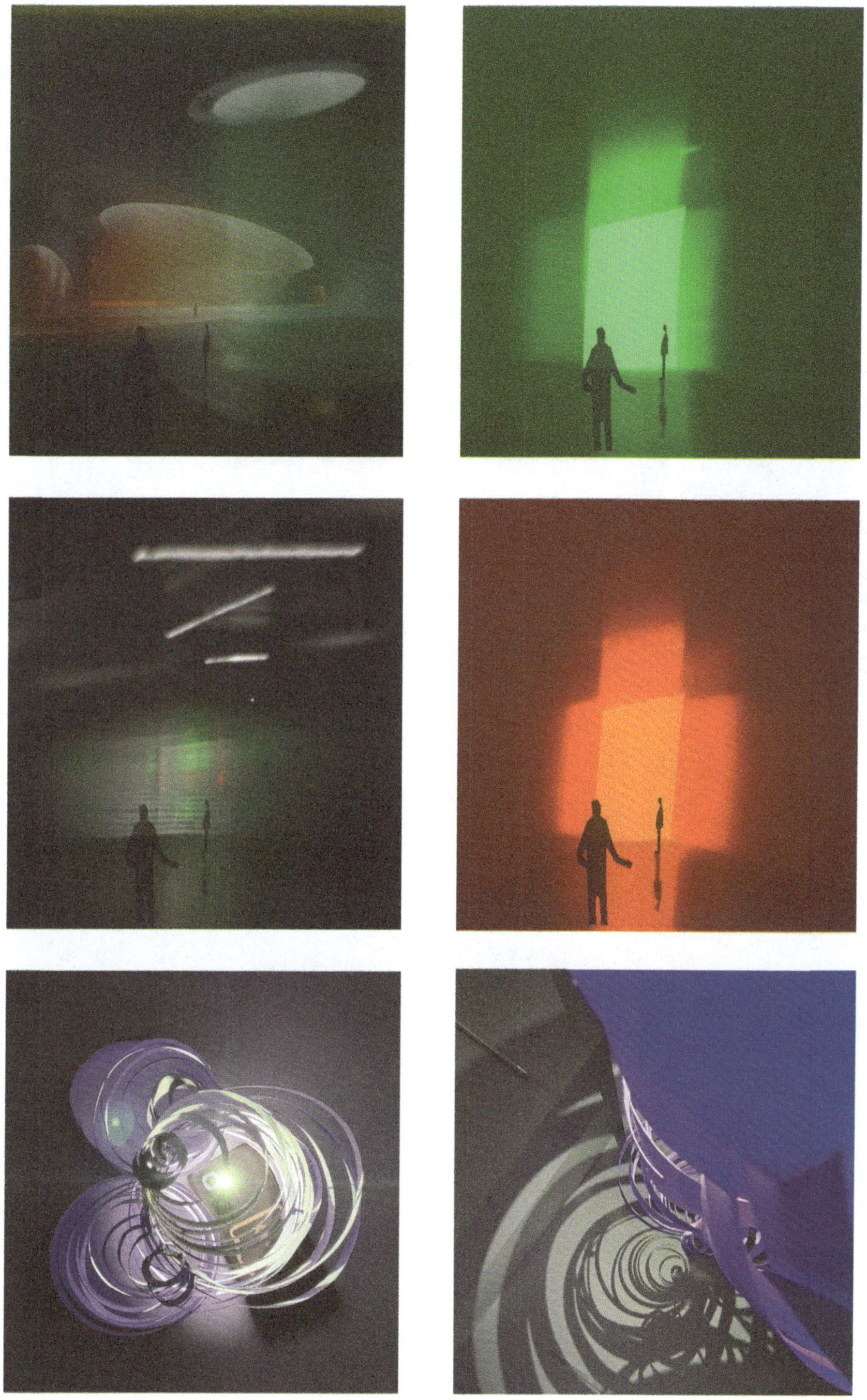

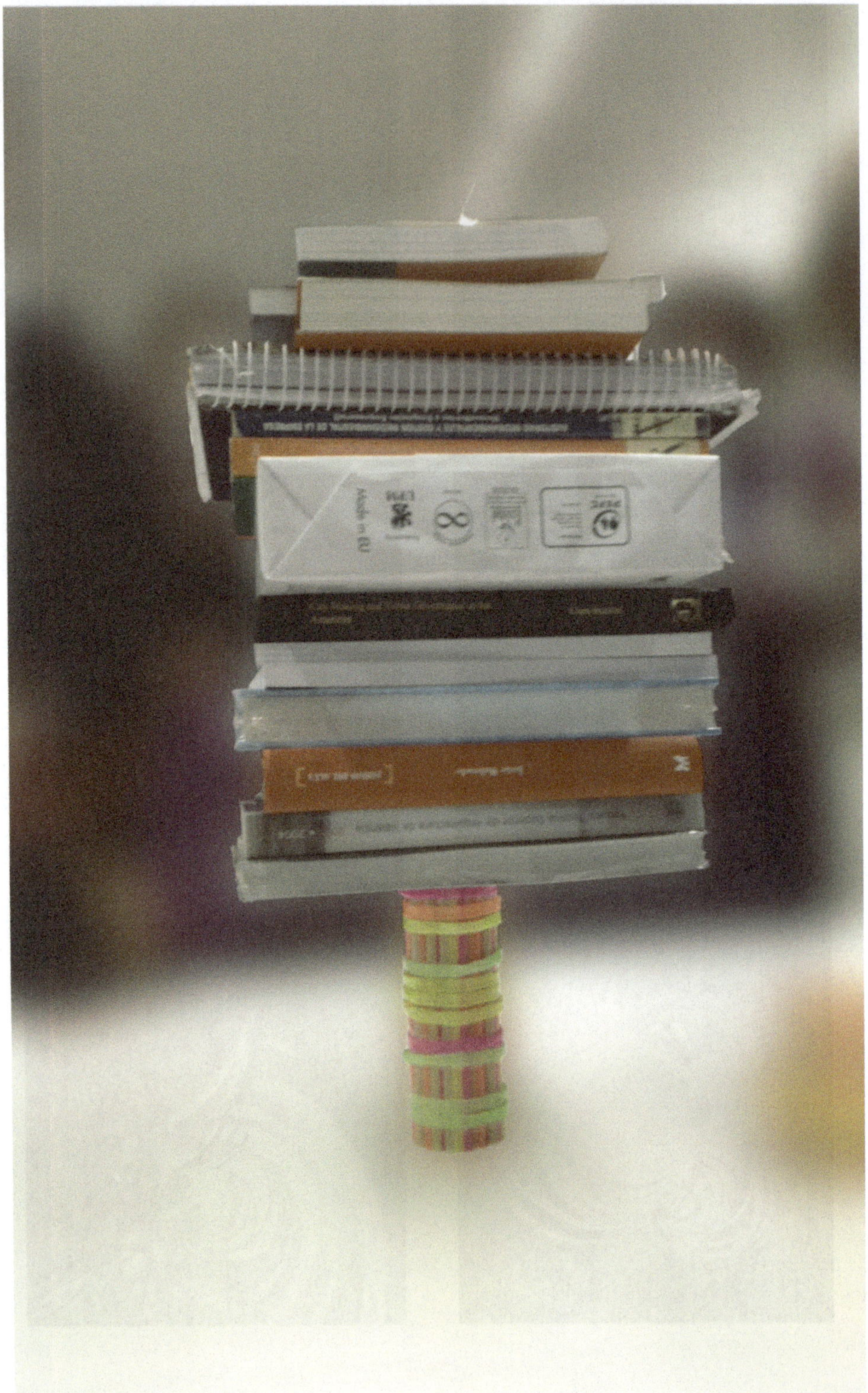

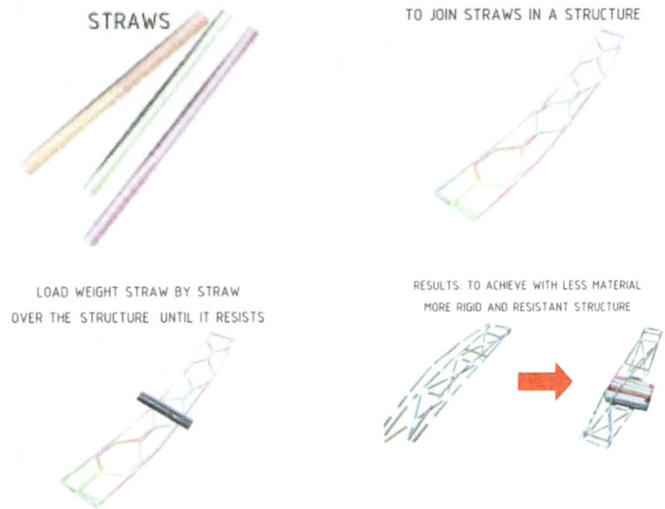
STRAWS
TO JOIN STRAWS IN A STRUCTURE
LOAD WEIGHT STRAW BY STRAW
OVER THE STRUCTURE UNTIL IT RESISTS
RESULTS: TO ACHIEVE WITH LESS MATERIAL
MORE RIGID AND RESISTANT STRUCTURE

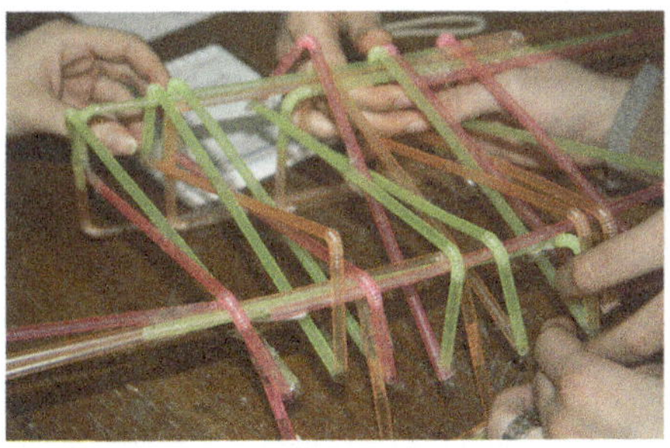

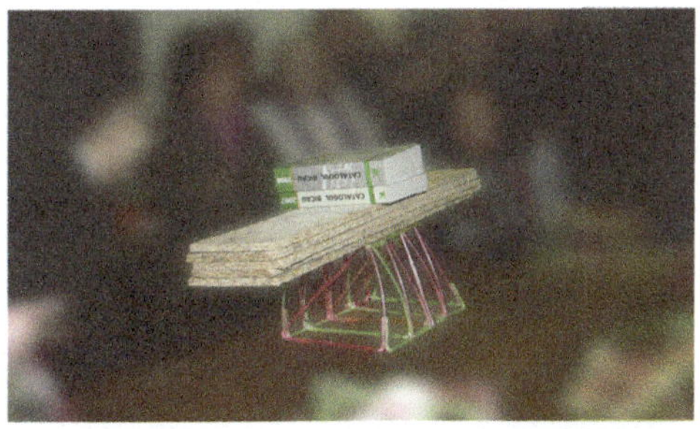

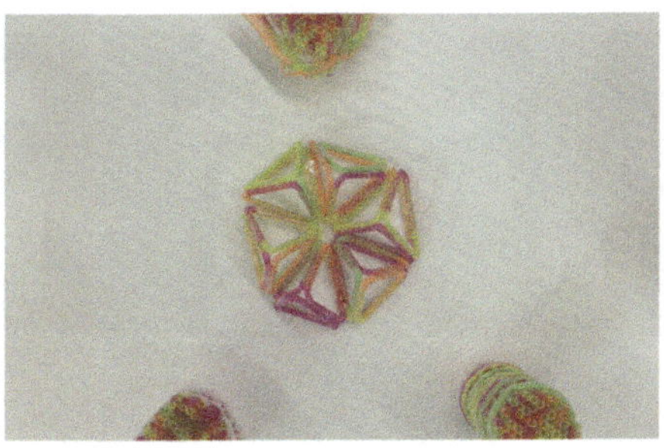

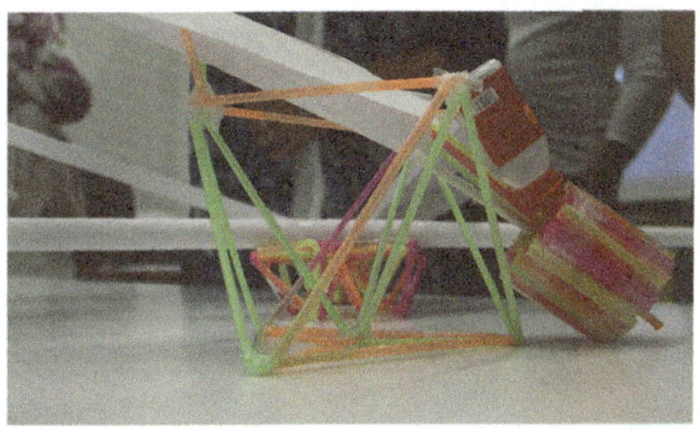

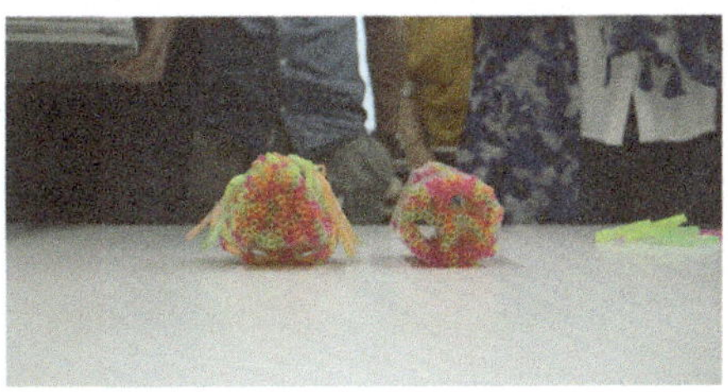

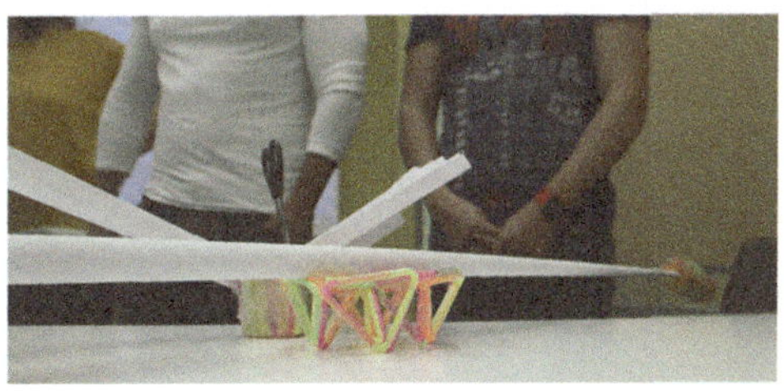

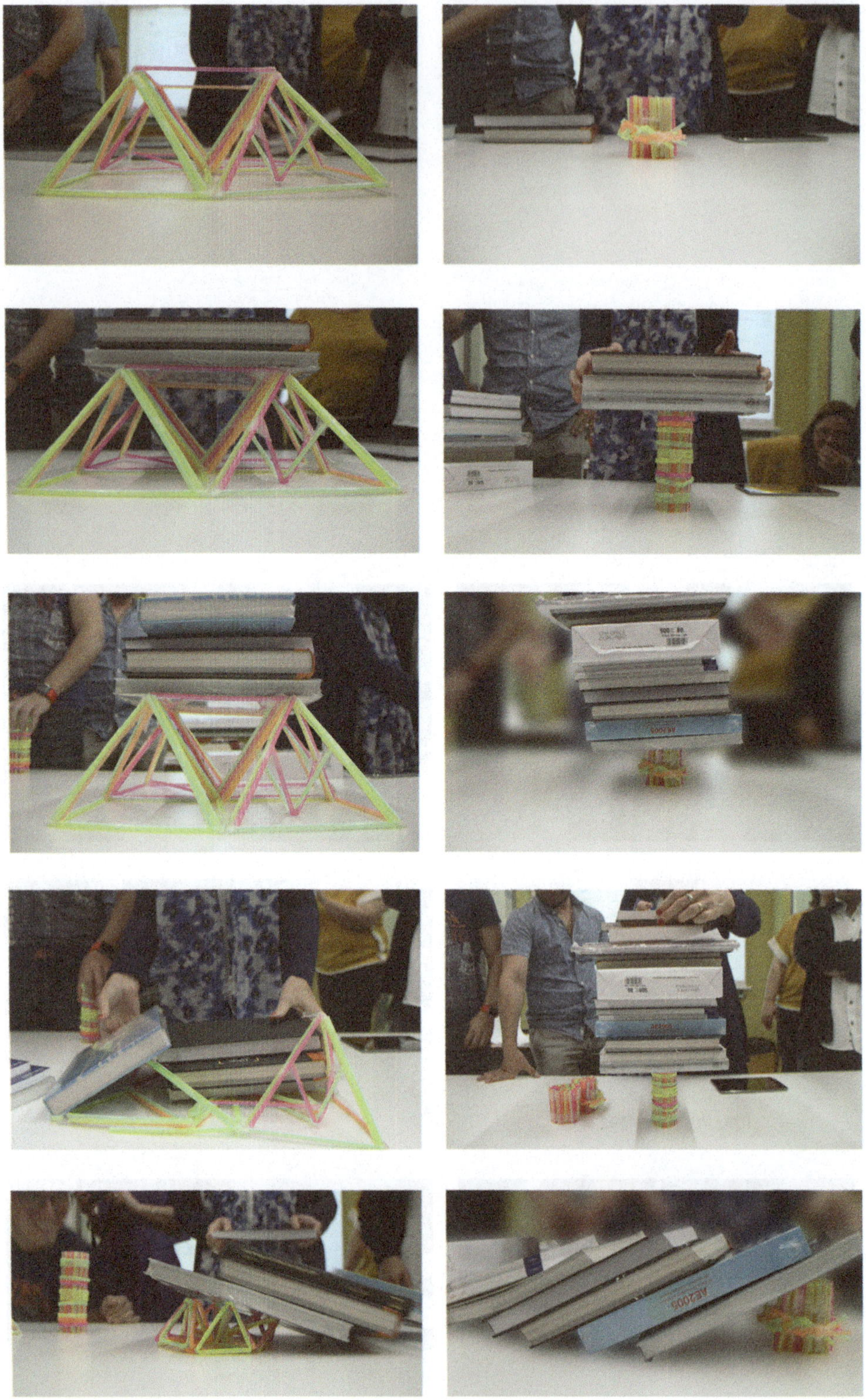

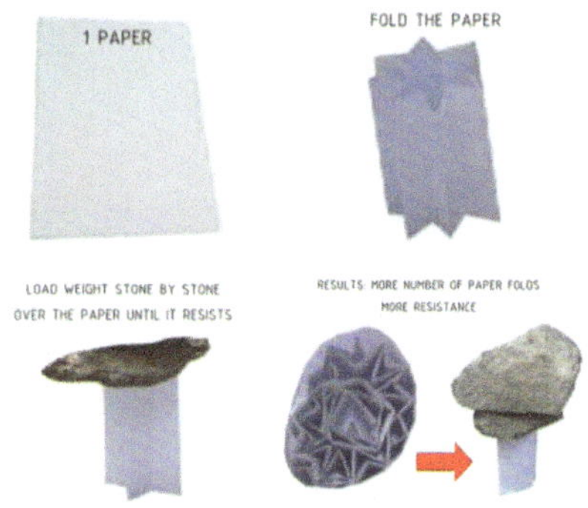
1 PAPER
FOLD THE PAPER
LOAD WEIGHT STONE BY STONE OVER THE PAPER UNTIL IT RESISTS
RESULTS: MORE NUMBER OF PAPER FOLDS MORE RESISTANCE

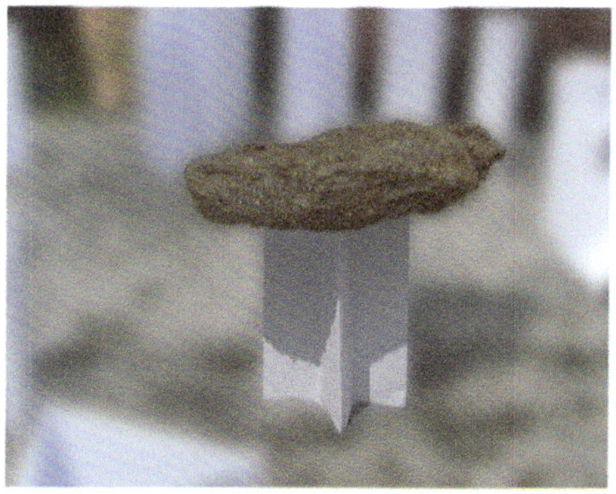

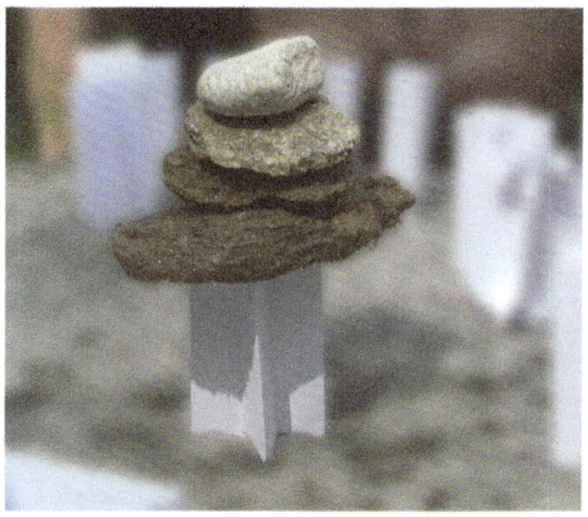

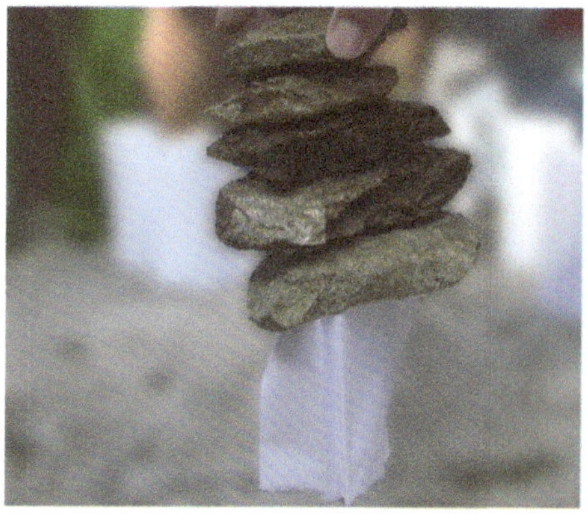

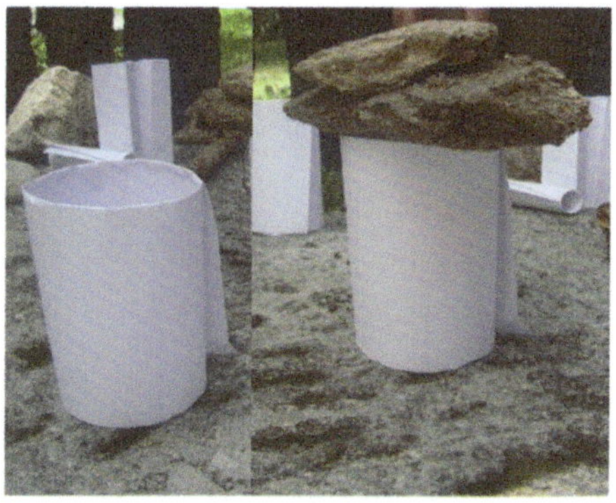

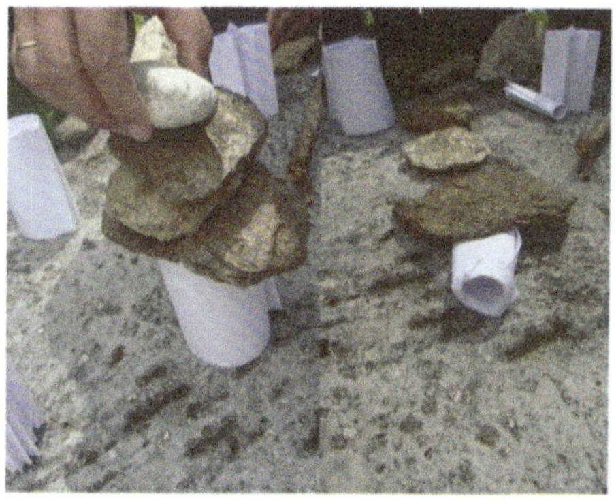

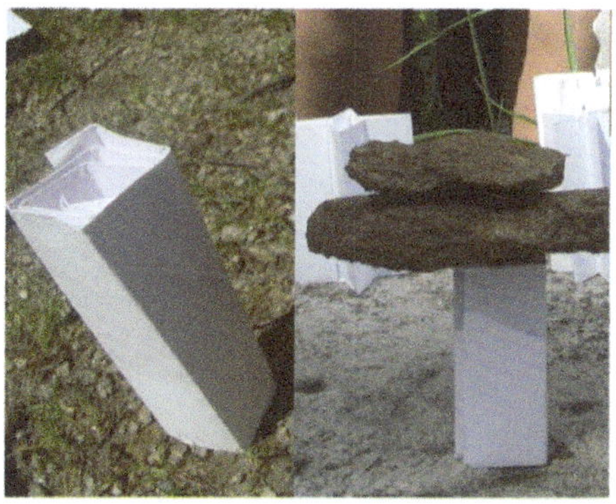

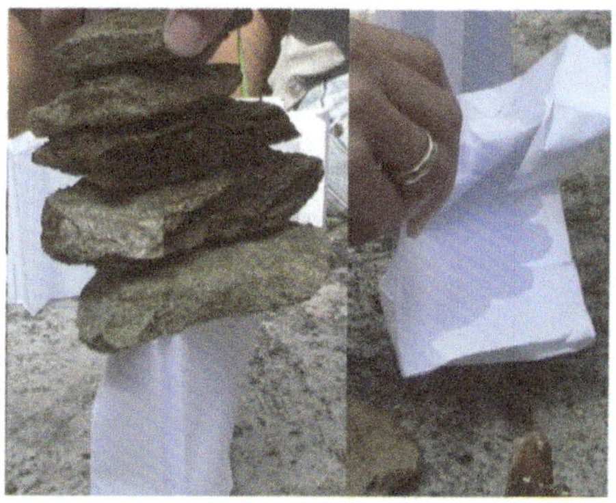

Index

M. R. Cervera Sarda et al. (eds.), *Architecture Inspired by Nature*, https://doi.org/10.1007/978-3-031-33144-2

SPRINGER NATURE

GPSR Compliance

The European Union's (EU) General Product Safety Regulation (GPSR) is a set of rules that requires consumer products to be safe and our obligations to ensure this.

If you have any concerns about our products, you can contact us on ProductSafety@springernature.com

In case Publisher is established outside the EU, the EU authorized representative is:

Springer Nature Customer Service Center GmbH
Europaplatz 3
69115 Heidelberg, Germany

The manufacturer's authorised representative in the EU is Springer Nature Customer Service Centre GmbH, Europaplatz 3, 69115 Heidelberg, Germany. If you have any concerns regarding our products, please contact ProductSafety@springernature.com

Printed and bound by CPI Group (UK) Ltd, Croydon, CR0 4YY
30/09/2026
02225477-0001